The Ancient Tradition of Geometric Problems

WILBUR RICHARD KNORR

DOVER PUBLICATIONS, INC.
New York

Copyright © 1986 by Birkhäuser Boston.
All rights reserved under Pan American and International Copyright Conventions.

Published in Canada by General Publishing Company, Ltd., 30 Lesmill Road, Don Mills, Toronto, Ontario.
Published in the United Kingdom by Constable and Company, Ltd., 3 The Lanchesters, 162–164 Fulham Palace Road, London W6 9ER.

This Dover edition, first published in 1993, is an unabridged, corrected republication of the work first published by Birkhäuser, Boston, in 1986.

Manufactured in the United States of America
Dover Publications, Inc., 31 East 2nd Street, Mineola, N.Y. 11501

Library of Congress Cataloging-in-Publication Data

Knorr, Wilbur Richard, 1945–
    The ancient tradition of geometric problems / Wilbur Richard Knorr.
        p.   cm.
    Originally published: Boston : Birkhäuser, 1986.
    Includes bibliographical references and indexes.
    ISBN 0-486-67532-7
    1. Geometry—History. 2. Geometry—Problems, Famous. I. Title.
QA443.5.K58   1993
516.2'04—dc20   92-41718
                                                                               CIP

Table of Contents

Preface   v

1   Sifting History from Legend   1

2   Beginnings and Early Efforts   15
    (i)   The Duplication of the Cube   17
    (ii)   The Quadrature of the Circle   25
    (iii)   Problems and Methods   39

3   The Geometers in Plato's Academy   49
    (i)   Solutions of the Cube-Duplication   50
    (ii)   Geometric Methods in the Analysis of Problems   66
    (iii)   Efforts toward the Quadrature of the Circle   76
    (iv)   Geometry and Philosophy in the 4th Century   86

4   The Generation of Euclid   101
    (i)   A Locus-Problem in the Aristotelian Corpus   102
    (ii)   Euclid's Analytic Works   108
    (iii)   The Analysis of Conic Problems: Some Reconstructions   120
    (iv)   An Angle-Trisection via "Surface-Locus"   128
    (v)   Euclid's Contribution to the Study of Problems   137

5   Archimedes—The Perfect Eudoxean Geometer   151
    (i)   Circle-Quadrature and Spirals   153
    (ii)   Problem-Solving via Conic Sections   170
    (iii)   Problem-Solving via *Neuses*   178
    (iv)   An Anonymous Cube-Duplication   188
    (v)   The Impact of Archimedes' Work   194

6   The Successors of Archimedes in the 3rd Century   209
    (i)   Eratosthenes   210
    (ii)   Nicomedes   219
    (iii)   Diocles   233
    (iv)   On the Curve called "Cissoid"   246
    (v)   Dionysodorus, Perseus and Zenodorus   263
    (vi)   In the Shadow of Archimedes   274

7 Apollonius—Culmination of the Tradition 293

- (i) Apollonius, Archimedes and Heraclides 294
- (ii) Apollonius and Nicomedes 303
- (iii) Apollonius and Euclid 313
- (iv) Apollonius and Aristaeus 321
- (v) Origins and Motives of the Apollonian Geometry 328

8 Appraisal of the Analytic Field in Antiquity 339

- (i) The Ancient Classifications of Problems 341
- (ii) Problems, Theorems and the Method of Analysis 348
- (iii) "... and many and the greatest sought, but did not find." 361
- (iv) Epilogue 367

Bibliography 382

Indices 392

Preface

Within the ancient geometry, a geometric "problem" seeks the construction of a figure corresponding to a specific description. The solution to any problem requires for its completion an appeal to the constructions in other problems already solved, and in turn will be applied to the solutions of yet others. In effect, then, the corpus of solved problems forms an ordered sequence in which each problem can be reduced to those preceding. The implications of this simple conception struck me, as I was completing a paper on Apollonius' construction of the hyperbola (1980; published in *Centaurus*, 1982), for it served to unify a diverse range of geometric materials I had then been collecting for some five years.

That the ancient problem-solving effort took on such a structure is much what one would expect, given the prominence of the role of "analysis" for discovering and proving solutions; for this method seeks in each case to reduce the stated problem to others already solved. We possess one ancient work, Euclid's *Data*, which organizes the materials of elementary geometry in this manner; but there survives none which attempts the same for the more advanced field. The *Conics* of Apollonius, for instance, is by its own account ancillary to this effort, providing the essential introduction to the theory of conics through which the so-called "solid" problems might be solved; but it undertakes the actual solution of such problems only to a very limited extent. This salient omission from the extant record thus defines the project of the present work: to exploit the materials extant from Archimedes, Apollonius, Pappus and others in order to retrieve a sense of the nature and development of this ancient tradition of analysis.

The present effort is conceived as an exploratory essay, intended to reveal the opportunities which the evidence available to us provides for an interpretation of the ancient field. A definitive survey overreaches its scope, however, and may ultimately be unattainable in view of the gaps in our documentation. Similarly, an exhaustive survey of the immense secondary literature on the history of geometric constructions could not be attempted here. The sheer bulk of this literature has made omissions inevitable, but I have endeavored to include references to those contributions which seemed to me historically and technically stimulating, as well as directly pertinent to my special objectives. If through inadvertence I have omitted discussions of worth, I apologize for that and welcome information which will prevent similar lapses in the future.

Since the field of ancient geometry, in particular, the work of Archimedes, has been an area of special interest to me for several years, the present study occasionally touches on issues discussed in previous publications of my own. Thus, I sometimes have reason to summarize positions elaborated in detail elsewhere, or to allude in passing to treatments of related issues. In such cases I have striven to avoid mere repetition of previous efforts, and the present work, but for these occasional overlaps, consists of new material.

In my interpretive efforts here I will emphasize the mathematical and historical aspects of the ancient writings, while the final chapter will include aspects of philosophical interest as well. Another aspect, that dealing with textual issues of documentation, while often overlooked in studies of this type, sometimes becomes indispensable for understanding the ancient work. I thus include commentary along these lines, when it bears on specific matters taken up here. The broader inquiry, of surveying the ancient documents and working out an account of their production and transmission, is not attempted here, but forms the project of a separate volume currently in preparation. I intend to examine in that second work a set of Greek and Arabic texts on geometric problems, thereby to reveal the general character of the scholarship which has preserved testimony of the ancient geometric field.

Grants from the American Council of Learned Societies and the Institute for Advanced Study enabled me to undertake these researches at the Institute in 1978-79, and a grant from the National Science Foundation, administered by the American Academy of Arts and Sciences, provided support for continuation of this work in 1979-80.

I have profited considerably from the criticisms and suggestions of students in the Department of Philosophy at Stanford University, in particular, Susan Hollander, Henry Mendell and David O'Connor. I wish also to thank my colleagues in the Departments of Philosophy, Classics and Mathematics for their comments, and, in particular, Sol Feferman, Halsey Royden, Hans Samelson and Patrick Suppes, for reading draft chapters, and Jody Maxmin for advice in the presentation of the plates. Ian Mueller (University of Chicago) provided extensive and valuable comments on the work in its later stages. I especially appreciate the respect and courtesy with which they all received my effort, even when my positions differed sharply from their own.

I am endebted to a number of museums and research libraries for permission to reproduce photographs of items in their collections. These are noted in the captions to the frontispiece and the plates. A grant from Stanford University, through the Office of its Dean of Graduate Study and Research, has contributed toward the costs of publishing these photographic materials.

Finally, I wish to acknowledge the support, patience and diligence of the president and editors of Birkhäuser Boston, Inc.

W.R.K.
Stanford, Calif.

CHAPTER 1

Sifting History from Legend

The problems of cube duplication, angle trisection, and circle quadrature have deservedly attracted major interest among scholars of ancient Greek geometry. The ancient literature on these problems constitutes one of the most richly documented fields of ancient mathematics, amounting to over two dozen different solutions, often in multiple versions and spanning the whole course of antiquity from the pre-Euclidean period through the Hellenistic, Graeco-Roman, and Byzantine periods even into the Arabic Middle Ages. Throughout, these problems were affiliated with researches in the more advanced fields of geometry.

A historical survey of these studies could thus be hoped to provide insight of value into the development of the techniques of geometry in antiquity: What were the sources of the interest in these problems? How did their study relate to prior efforts on the same or associated problems? How did the results of these studies contribute to the development of new geometric techniques for the solution of geometric problems? What were the precise specifications for the construction of solutions, and did these conditions themselves change with the introduction of new techniques and concepts? Ultimately, did the ancient geometers and philosophers view the quest for solutions as having succeeded?

Unfortunately, only a small part of this project now permits its presentation in the form of a straightforward narrative of how things happened. The extant technical literature has large gaps, so that we cannot claim to know all the solutions to these problems which the ancients worked out, nor even, in some cases, the geometers responsible for solutions which have survived. More difficult still is the assessment of motive, since technical documents rarely provide direct insight into the reasons which lead geometers to take up specific problems and treat them in specific ways. For this we are often compelled to resort to testimonies in nonmathematical writings. But we then face new difficulties, in that their authors may not fully comprehend the technical issues and, at any rate, will have their own literary or philosophical concerns. The latter will invariably discourage them from presenting the technical materials in the clear and detailed manner we would prefer.

One of the important witnesses to the early study of the cube duplication illustrates well the hazards of using such nonmathematical sources. It is a passage from the *Demon of Socrates*, a dramatization of the political conspiracies in the Greek city-state of Thebes in 379 B.C. by a writer best known for his philosophical and biographical works, Plutarch of Chaeronea (1st–2nd century A.D.).[1] The speaker here is Simmias, one of the Theban conspirators, who is reporting on a recent visit to Egypt he made in the company of Plato:

> As we traveled from Egypt, certain Delians came upon us near Caria and begged Plato, as a geometer, to solve for them an unusual oracle proposed by the god. The oracle was that there would be a respite from present ills for the Delians and the other Greeks after they had doubled the altar in Delos. But as they couldn't figure out the purpose and were faring ludicrously with the construction of the altar (for on account of their unfamiliarity with the proportion which provides the double in length, they failed to notice that when each of the four [!] sides was doubled, they were effecting an eightfold increase in the solid space), they summoned Plato for assistance in the puzzle. And he, recalling the Egyptian, said that the god was making sport with the Greeks for their neglect of education, as it were taunting us for our ignorance and demanding that we engage in geometry, and not just as a pastime. For surely it is a task not for an intellect which is inferior and perceives dully, but rather for one trained to the limit in [the constructions of] lines to take two means in proportion, by which alone the figure of a cubic body is doubled by being increased in the same way from every dimension. This [he said] Eudoxus of Cnidus or Helicon of Cyzicus would work out for them; but they shouldn't think that the god wished this, but rather that he enjoined all the Greeks to abandon war and its ills and to consort with the Muses, and by soothing the passions through reasoning and mathematics to live together profitably and without harm.[2]

Plutarch's chief interest here is to make the point that theoretical disciplines like mathematics and philosophy can be an effective diversion from conflict, both through their moderating influence on human aggressiveness and through the power that reasoning has for reconciling differences. This very point had been made earlier in a passage alluded to here by the phrase "recalling the Egyptian," in which an Egyptian priest had informed certain Greeks that the document they brought to him for deciphering contained an appeal to cultivate the Muses instead of war.

But what does Plutarch mean to say about the origins of this mathematical inquiry itself? At first glance, we might suppose he has in mind simply that the geometers in Plato's circle began to study the cube duplication in answer to the puzzle posed by the oracle. In effect, Plutarch would be subscribing to a form of the "externalist" view under which technical activities find their motivation in factors outside the technical field, for instance, from political or economic causes, or, as here, from philosophical and religious ones. Indeed, passages like this have been cited by those who would maintain that ritual practices formed an important stimulus for early mathematical studies or that philosophical discourse provided the first model of abstract reasoning for the Greek geometers.[3] But a closer reading reveals that Plutarch does not actually claim this, for his Plato already knows certain facts about this problem, specifically, that it is equivalent to the problem of finding the two mean proportionals. If the side of the given cube is A, and we have found two lines X, Y such that $A : X = X : Y = Y : 2A$, then by compounding ratios, we have $(A : X)^3 = (A : X)(X : Y)(Y : 2A)$, whence $A^3 : X^3 = A : 2A$ or $X^3 = 2A^3$.[4] From other sources, in particular Eratosthenes of Cyrene (3rd century B.C.), we learn that this reduction of the cube duplication to the finding of two mean proportionals was the discovery of Hippocrates of Chios, that is, around the turn of the 4th century B.C. and thus predated by several decades the events on which Plutarch's narrative is based.[5] Plutarch himself appears to depend on Eratosthenes for his account of the Delian oracle, for a parallel passage on the cube duplication in the *Mathematical Exposition* by Theon of Smyrna (early 2nd century A.D.) cites Eratosthenes' *Platonicus* for the same incident of Plato and the Delians with the same basic point:

> And Plato told them that not for want of a double altar did the god pronounce this to the Delians, but rather to accuse and reproach the Greeks for their neglect of mathematics and their slighting of geometry.[6]

Aware of Plutarch's source, we thus do well to receive with caution the details of his account. It is clear, for instance, that the chronological association of this story with the uprising in Thebes derives entirely from Plutarch's adaptation of the older version.

Similar precautions are necessary in dealing with another passage from Plutarch, providing a sequel to the Delian story by portraying how Plato reacted to the methods of solution worked out by the geometers. This passage is inserted into Plutarch's description of Archimedes' mechanical achievements, critical for the defense of Syracuse against the Roman siege of 215–212 B.C.

> This admired and much talked of instrumental art was first set in motion by those around Eudoxus and Archytas, who embroidered geometry with subtlety, but who supported by means of perceptible and instrumental models problems not permitting rigorous geometric demonstration; for instance, in connection with the problem about the two means in proportion, an element necessary for many of the geometric propositions, both of them resorted to instrumental constructions by fitting out certain means-drawing apparatuses from curved lines and segments. But Plato took offense and contended with them that they were destroying and corrupting the good of geometry, so that it was slipping away from incorporeal and intelligible things toward perceptible ones and beyond this was using bodies requiring much wearisome manufacture. In this way mechanics fell out, separated from geometry, and

for a long time it was overlooked by philosophy and became one of the military arts.[7]

Plutarch thus alleges a schism between the mechanical arts and the theoretical disciplines, bridged only by Archimedes, presumably for putting into practice his insights into pure geometry. The roots of this division Plutarch quite reasonably assigns to the idealist recommendations of the Platonic philosophy. Indeed, in Plato's *Republic*, Socrates criticizes geometers for their insensitivity to the true nature of their inquiry:

> They speak in a very ludicrous and restricted manner. For they make talk about squaring and applying and adding and all, in a way as if they were engaged in practice and making all their words for the sake of practice. But surely the entire discipline is studied for the sake of knowledge...and knowledge of what always exists, but not of anything that can ever be made or destroyed (527 a-b). ...They use the visible forms and make words about them, but their thinking is not about these things, but about what these things resemble, when they make their words about the square itself and the diagonal itself, but not about the one which they draw, and the others in this way; these very things which they fashion and draw, and which have shadows and images of their own, they can use them in turn as images in their search to see those things which no one could see other than by the intellect (510 d-e).[8]

It is this Platonic sentiment, this conviction of the purity of the objects of true geometric study, which Plutarch captures in his anecdote on the cube duplication. Nevertheless, we can perceive that here too he draws from Eratosthenes as an intermediary source. For the Archimedes-commentator Eutocius of Askalon (6th century A.D.) assigns to Eratosthenes a long account of the history of this problem, including the following passage as sequel to Plato's meeting with the Delians:

> And when they [sc. the geometers in the Academy] applied themselves diligently and sought how to take the two means of two given lines, Archytas of Tarentum is said to have discovered it by means of half-cylinders, while Eudoxus [did so] by means of the so-called curved lines. But it happened that all of them effected this apodictically, but couldn't construct it physically or put it into service, save to some small degree Menaechmus, and that with difficulty.[9]

Here, as in Plutarch's version, the geometers busy themselves under the watch of Plato, Archytas coming up with one solution, Eudoxus with another. Yet there is a profound discrepancy: Eratosthenes deems their solutions to be too abstract and impracticable, while Plutarch portrays them as overly mechanical and too little abstract. This kind of discrepancy is one form of evidence adduced by some scholars to discredit Eutocius' assignment of this writing to Eratosthenes.[10] We will examine this argument later. For now, it is enough to note that the passages can be reconciled in a straightforward way. Two different writings by Eratosthenes are at issue here: the document on the cube duplication and the *Platonicus*; it is the latter which Plutarch exploits as his source. The *Platonicus* was presumably a fictional essay, so that its strongly antimechanical position here could have resulted from Eratosthenes' adaptation of the other, more plainly historical account, where the early solutions had no pronounced mechanical element.

On the relative value of the mechanical and the theoretical elements in geometry, it is the attitude of the geometers themselves we would most like to know. The difficulties in learning this from the passages just cited are evident, since the glosses by Plutarch, Eratosthenes, and the 4th-century Platonists must first be separated out before the underlying mathematical views can be discerned. Although Plato assigned a prominent role to mathematical studies within his general program of philosophy, that in itself does not qualify him or his disciples to speak for the technical researchers on the issues which concerned them. Indeed, passages like those cited from the *Republic* suggest that Plato and other philosophers might set themselves apart as *critics* of the technical field. But if we might try to infer from such critiques that the geometers insisted on a this-worldly orientation in their studies, we have the variant testimony of Eratosthenes (in the account of the cube duplication) that the early studies were cast in an abstract manner. The mathematical texts preserved by Eutocius on the methods used by Archytas and Menaechmus bear Eratosthenes out on this point.[11] Yet Eratosthenes' own solution is accompanied by the detailed account of an actual physical device built on its design, and this overtly mechanical element is found in the methods used by Nicomedes and Diocles and others around the same time (late 3rd century B.C.).[12] Evidently, views on the importance of mechanical approaches varied over the course of antiquity and might sometimes be a matter merely of personal preference. In this lack of unanimity on issues of methodological commitment, the ancient field of mathematics would conform to a pattern familiar in more recent periods in the history of mathematics.[13]

These reflections thus reveal three major recommendations bearing on the present study of the ancient problem-solving efforts. First, we must attempt to establish the simple chronological sequence of the various solutions; only thus can one grasp for each period the complex of problems, methods, and aims which set the directions of geometric research at that time. On this basis one may hope to discern changes over the longer term in the character of the field of research. Typically, treatments of these materials have subscribed to a division by specific problems or by solving methods. Precedents can already be cited from antiquity in the compendium of solutions to the cube duplication presented by Eutocius in his commentary on Archimedes' *Sphere and Cylinder*, in the sections on cube duplication and angle trisection in Books III and IV of Pappus' *Collection*, and presumably also in the survey of circle quadrature in Sporus' *Keria*, and in the surveys of special curves in the encyclopedias of Menelaus and Geminus.[14] Among modern discussions, one is most likely to consult the extended chapter on "Special Problems" in T. L. Heath's *History of Greek Mathematics*, the detailed articles on cube duplication, angle division, and *neusis* constructions by R. Böker in the *Pauly Wissowa Real-Encyclopädie*, the long paper on ancient studies of curved lines and surfaces by P. Tannery, and the massive compendium on the history and properties of special plane curves by G. Loria.[15] Without questioning the great value of these accounts, as well as many others too numerous to cite here, I nevertheless believe that their principle of organization implicitly commits them to the view that the ancients likewise conceived of the study of these particular curves or these particular problems as distinct subfields. That might hold for an encyclopedist like Geminus, or even for commentators like Pappus or Eutocius, but not, it seems to me, for the original investigators like Archimedes and Apollonius and their contemporaries. By contrast, if we adopt a chronological scheme, we can suspend commitment on the ancient conception of the status of these problems and the various solving techniques, so that this conception will be allowed to emerge.

The "simple" chronological questions are in fact often quite difficult to determine, as we shall find. The restriction to the three "classical" problems is of course artificial itself and will not be adhered to strictly. Its obvious advantage lies in the ample documentary coverage of these problems, touching, in some form, on each of the major periods of the ancient geometry. This raises the question of why the three problems continued to attract attention, even after many different solving methods had been worked out. An important key to the answer is found in the changes affecting the general field of research from one period to the next. As new techniques were introduced, their applicability to the older problems would naturally come up for consideration: for instance, a special curve generated in the context of a method for the angle trisection might be recognized to have properties relevant for an alternative solution of the cube duplication as well.[16] It is thus clear that one cannot adequately comprehend the approaches adopted in the cases of the individual problems without full regard to the more general environment of research. This presents not a difficulty, however, but an opportunity. For the documentation extant representing the more advanced researches at various times is often quite sparse. One may thus view the special problems as a cross section, suitable for filling out one's portrait of the wider field. This investigation will occupy the next six chapters of the present study (Chapters 2–7), moving from the pre-Euclidean period through the generations of Euclid and Archimedes to the time of Apollonius.

A second recommendation is to devote separate treatment to several important metamathematical issues: how the ancients divided the geometric field according to the types of problems and solving methods; what they viewed the special role of problems to be, especially in relation to that of theorems, and how they associated these with the important methods of analysis and synthesis; what conditions they imposed on the techniques admissible for the solution of problems, and whether they judged that satisfactory solutions for the three special problems had actually been found. Historical accounts of these materials almost invariably assume at the outset that the ancients had at a very early time distinguished compass-and-straightedge constructions from the others and were continually urged on in the quixotic search for constructions of this type for the three special problems.[17] As we now know, through results obtained in the 19th century, the ancients were in a position neither to produce such solutions nor even to show, save through persistent failure, that they could not be produced. For the latter is essentially an algebraic question not amenable to investigation under the geometric methods of the ancients.[18] I believe, however, that we cannot assume that the ancient views on these metamathematical issues were just the ones familiar within the modern field, or that even if certain geometers at certain times held views somewhat like those now familiar, all other geometers at all other times necessarily held the same views. Surely our goal must be to discover what the ancient views were by means of a consideration of the relevant historical evidence. This is the project undertaken in Chapter 8. However, as our look at Plutarch's passages on the cube duplication suggests, the search for answers will be far from straightforward. Most of the passages we may consult come from later authors whose primary interest is in the explanation of methodological observations made by philosophers, most notably Plato and Aristotle. For instance, the neo-Platonist element pervades the important Euclid commentary by Proclus (5th century A.D.), while much of our evidence must come in the form of interpretations of specific passages from Aristotle by Alexander of Aphrodisias, Themistius, Philoponus, Simplicius, and others. Recall that our passages from Plutarch derive from his discussions of Plato, and the same is true for his source, Eratosthenes. Even with writers who one might suppose are closer to the actual mathematical tradition, such as Pappus, one can readily detect the influence of certain philosophical commitments on their mathematical pronouncements.

This aspect of the commentators helps us to understand an emphasis commonly but, I believe, mistakenly made in discussions about the ancient geometry. One seems typically to assume that metamathematical concerns were the effective motivating force underlying the efforts of geometers, for instance, that the hallmark of their tradition was their organization of geometric findings into tight structures of deductive reasoning, as if their primary ambition was the production of treatises like the *Elements* of Euclid and the *Conics* of Apollonius. But this surely cannot be correct. The writing of textbooks is the *end* of mathematical research only in the sense that death is the end of life: it is the last term in a sequence, but not the purposive element which explains why one engages in the activity of progressing through the steps in this sequence.[19] As far as the ancient geometers are concerned, the goals underlying the activity of research are in most cases a matter for surmise. In general, I will find most convincing the "internalist" position that technical research is directed toward the solution of problems arising from previous and current research efforts. In effect, the formulation of problems keeps a field open by guiding new research. To be sure, the goal to rigorize the proofs of known results can stimulate one form of research; the most notable example of this among the ancients may be found in the mathematical insights of Eudoxus on which were built the formal theories of proportion and limits.[20] But with the exception of Eudoxus, the context of the interesting mathematical insights by the ancient geometers lies not in such formal concerns, but rather in the investigation of problems. In this respect, I believe that an intuition founded on the most recent period of mathematical history can be misleading, for metamathematical inquiries have crystallized into a recognized subfield, the study of foundations, commanding expertise of an entirely mathematical kind. But this did not happen in antiquity. If, for instance, the deductive project embodied in Euclid's *Elements* entails subtle questions of axiomatics, the ancients seem to have rest content with Euclid's formulation and with the Aristotelian theory to which it loosely relates, for both remain the focus of later discussions of foundational issues.[21]

The third recommendation is to take up the textual issues bearing on our use of the writings of authors from later antiquity. Their editorial efforts have preserved for us most of what now survives of the ancient geometric tradition. In some instances, they produced the extant editions of the treatises of the great geometers like Euclid, Archimedes, and Apollonius; in others, they compiled miscellanies of geometric results drawn from works now lost. It is thus extremely important for us to gain a sense of how these writers handled texts: Where did they obtain their texts? What changes did they make in them? What were their purposes in selecting and presenting these texts? In this regard, the texts on the three special problems, in particular, that of the cube duplication, provide an ideal instrument, since they survive in multiple versions from several writers. It is remarkable that so little along these lines has been attempted up until now. Even those prominent scholars, J. L. Heiberg and P. Tannery, for whom philological issues were a particular concern, did not much explore those questions which the comparisons of these texts might elucidate. It is as if the later writers were of no intrinsic interest to them, and that once one of several related texts could be accepted as the "best," the others might be disregarded.[22] In a sequel following the present volume an attempt will be made to display the interrelations among the surviving texts of these solutions. One can well anticipate that these writers are not highly original in their mathematical contributions. But the degree to which they depend on sources, not only for mathematical techniques and results, but even for the very wording of the texts they present, I believe, will be surprising even for those who have made a study of the ancient geometry. This dependence applies as well to Pappus as to the other commentators, despite the diversity and advanced level of the materials in his *Collection*. Further, the pattern of textual methods will be seen to carry over among Arabic writers of the 9th and 10th centuries A.D. The discussion of several texts on the angle trisection and cube duplication from Arabic authors, in comparison with selected items from the earlier Greek tradition, will occupy a part of the sequel and its Appendices. There I will attempt to show that many of these Arabic texts were produced as translations, or at least close paraphrases, of the Greek, even in certain cases where one might have supposed they were original Arabic writings or reworkings of Arabic prototypes. Thus, the derivative character both of this part of the Arabic mathematical tradition as well as of the Greek editorial tradition from the Graeco-Roman and Byzantine periods becomes clear and suggests that one might hope to exploit them in new ways for gaining insight into the earlier creative phases of Greek geometry. This last is of course a project extending well beyond the scope of the present study.

In this survey of the ancient problem-solving efforts, attention to matters of textual evidence will predominate in Chapter 8 and the sequel study. Elsewhere, the essential mathematical aspects of the different technical contributions will provide the principal instrument for revealing their historical relations. A difficulty here is that the ancient writers (especially those in the later editorial tradition) preferred the synthetic mode of exposition in their formal treatments of geometry. That is, one derives a claimed result (e.g., theorem or construction) as the conclusion of a deductive chain of reasoning which starts from the given terms of the theorem and otherwise utilizes only axioms and other first principles or previously proven theorems and constructions. To mathematicians of the 17th century, indeed, even to some of the ancients, this format was notorious for obscuring the essential line of thought, particularly in the cases of more complicated results.[23] But when the discovery of solutions for geometric problems was at issue, the ancients exploited an alternative method: that of *analysis*. In the case of a problem of construction, for instance, one here posits the desired figure as having already been effected and then deduces properties of this figure until an element of it emerges which is known from prior results to be constructible. The formal *synthesis* would then begin from these constructible terms and proceed through a deductive sequence in approximately the reverse order to that of the analysis until the desired construction has been produced. Since the analytic method can be applied fruitfully even where the solution of the problem is not yet known, it is especially well suited for the purposes of geometric research. If the ancients sometimes followed this analytic method in their presentations of known solutions as well, this would doubtless owe to the appropriateness of exposing learners to the use of this method in preparation for their own research efforts later. It also has advantages for exegesis, seeing that the analysis of a problem exposes in a natural and well motivated way the rationale behind each step of the construction. By contrast, when only the synthesis is given, these steps often appear arbitrary and without clear motivation until much later in the proof. Thus, in my accounts I will prefer analytic presentations to synthetic ones. In cases where only the synthesis is extant, I will fashion the analysis corresponding to it. Given the close parallelism between these two parts of the treatment of problems, one can recover such analyses with virtually absolute confidence. In doing this, moreover, one can sometimes discern reasons for otherwise puzzling features in the extant syntheses. Nevertheless, the reader should be aware that this is an interpretive element in the account which follows. My intent is not to paraphrase or otherwise reproduce what is already available in the extant primary literature, but rather to bring forward the essential, usually simple, geometric idea underlying each result and thus to provide an appropriate introduction for further investigation of that literature.

A few terms admitting a variety of meanings in common usage are so prominent throughout this discussion of ancient geometry that I do not dare employ them in any sense other than their formal mathematical one. This applies to "analysis" and "synthesis," for instance, which will here signify only those special geometric methods mentioned in the preceding paragraph. Similarly, the term "problem" will refer only to the geometric form of a problem, that is, to the type of proposition which seeks the construction of a figure in answer to certain specified conditions or the determination of certain properties of a specified figure. It will never be a mere synonym of "difficulty," for in fact a geometric problem need not be difficult at all. Its essence lies in its being a "project" (from *proballein*, "to cast forward"), that is, a demand to do something not yet done or to discover something not yet known. Another such term is "prove" along with its cognate noun "proof" and the synonyms "demonstrate" and "demonstration," respectively. Here these refer only to those deductive forms of reasoning employed in mathematics and logic. I will forego their use in metamathematical and historical contexts, where more flexible terms like "show," "reveal," "make evident," and so on, will be used in their stead. The serious student of ancient mathematics knows well that the speculative component is necessarily very high. Virtually all the interesting questions we might care to investigate are beyond the range of definite answers, and the same applies even in the instances of many quite trivial matters. It is certain, for instance, that Apollonius wrote the *Conics*; but the question as to how much of the extant text is by Apollonius himself, rather than by his editor Eutocius, has not yet received the attention it demands. In view of this, one's assertions about these materials must always be received within a range of plausibilities and probabilities. A phrase like "one thus sees..." signifies a claim of a rather high degree of likelihood, whereas one like "it may well be..." signifies a low one. An oddity of our language is that ostensible intensifiers like "surely," "certainly," and "doubtless" actually indicate quite the opposite of certitude, by signalling a situation within which doubts *can* be raised; their force is largely suasive, but will at least communicate to the reader the nature of my spiritual commitments to my claims. If I sometimes lapse into incautious use of the indicative mood, the reader should be forewarned to supply the needed qualifications.

In the matter of my interpretive method, I subscribe, however imperfectly, to three simple ideals: to consider all the available elements pertinent to a given question; to advocate that view, among all the viable contenders, which best conforms with the general character of the ancient geometric tradition and with the most straightforward conception of the mathematical field in relation to other disciplines; to prefer that view among viable alternatives which most enhances the interest value of the extant documents. Applying these principles can often be a delicate matter. In the case of the first, the mere survival of a document need not compel us to believe what it says. If two commentators happen to disagree on a certain point, we must of course try to determine which (if either) is correct and to explain why there is a discrepancy. By the same token, if the statement of only one commentator is known, that need not be true. Here the second principle assists in ascertaining whether a testimony might be suspect. If, for instance, Pappus asserts something about the geometric field which flatly disagrees with what we find in the technical literature, surely we would not hypothesize a lost corpus of writings merely to make his view correct, but rather, we would admit his error and try to figure out why he made it.[24] One can hope eventually to gain a sense of the intrinsic credibility of the various authorities in their various contexts. One learns quickly, for instance, that Archimedes commits no technical errors and that any assumption to the contrary on our part would be merely presumptuous.

The second principle serves to introduce an assumption of economy in interpretations. It is of course possible that ancient mathematical thought differed from that of other periods in certain striking and fundamental respects. But I believe we must be forced to such conclusions, rather than concur in them carelessly or even as a matter of principle. It seems to me, for instance, that the technical activity of problem solving in mathematics and other fields is never particularly sensitive to external factors, like the opinions of one or another school of philosophers. To be sure, the line of division may be difficult to draw sharply in some cases, and what start as external considerations might sometimes, through a very subtle process, become part of the internal conditions of research activity. But I do not imagine the ancient geometers as constantly looking over their shoulder. It is surely absurd to suppose that Platonist notions of the perfection of circles induced mathematicians to restrict their solving techniques to compasses; or even that Zeno's paradoxes of motion instilled in Eudoxus a fear of the infinite. The notion that ancient mathematics was somehow a vast exercise in dialectical philosophy must miss a very important point: that geometry is rooted in an essentially practical enterprise of problem solving.[25]

As for the third principle, I believe that scholars have a responsibility to keep alive the interest in the documents they examine. It seems to me quite wrong to *assume* that a certain document is a forgery, unless the reasons are compelling, for that will merely guarantee that no one takes it seriously thereafter.[26] Again, it seems to me wrong to *assume* that a certain mathematical writer is incompetent, unless we are driven to that view on clear grounds. On the other hand, it is equally important to avoid romantic notions about the expertise of our authorities in the face of clearly contradictory indications. The stature of Pappus as an outstanding mathematical intellect, for instance, has prevented the intelligent use of his writing as a *source* for the earlier geometric tradition.[27] Again, if one begins to doubt that Aristotle himself produced the extraordinary mathematical reasonings in the *Meteorologica*, this work might eventually be returned to the mainstream of thinking about ancient mathematical science.[28]

It is perhaps in the interest of scholarly precision to approach general theses about the nature of a discipline with scepticism. But most of the major treatments of the ancient geometry, such as that of Heath, have taken that wise counsel too far. They set before us what purports to be a factual narrative of the ancient field, but which is hardly other than a *mélange* of details without strong signs of interrelation, or even of internal consistency. The most exciting aspect of this material, in my view, is that a closer examination reveals the general lines of a coherent development, and that the extant record, despite its incomplete state, actually makes sense as the remnant of an extraordinary movement in thought whose basic outline is discernible. If the reader remains unconvinced of the details of the argument in the following study, yet nevertheless comes to appreciate the unsatisfactory state of current scholarship and to perceive the possibility of achieving a coherent view of this movement, then my effort shall have attained its principal objective.

NOTES TO CHAPTER 1

[1] One may consult the *Oxford Classical Dictionary* (2nd ed., Oxford, 1970) for a brief account and bibliography of the life and work of Plutarch.

[2] *Moralia*, 579 a–d; cf. the text, translation and annotations on this and related passages by P. H. de Lacy and B. Einarson, *Plutarch's Moralia* (Cambridge, Mass./London: Loeb Classical Library), VII, 1959, pp. 396–399.

[3] The ritual background to mathematics is a prominent concern in several articles by A. Seidenberg; cf. his "Ritual Origin of Geometry," *Archive for History of Exact Sciences*, 1963, pp. 488–527 and his "Origin of Mathematics," *Archive for History of Exact Sciences*, 1976, 18, pp. 301–342. Some of these ideas have been taken up and extended by B. L. van der Waerden, "Pre-Babylonian Mathematics," *Archive for History of Exact Sciences*, 1980, 23, pp. 1–46. The thesis of a philosophic background to pre-Euclidean mathematics, especially in the form of Eleatic influences on the Pythagoreans, has been championed by Á. Szabó; cf. *Anfänge der griechischen Mathematik*, Budapest and Munich/Vienna, 1969 (translated as *The Beginnings of Greek Mathematics*, Dordrecht, 1978). I have presented a detailed critique of Szabó's views in "On the Early History of Axiomatics," in *Pisa Conference (1978) Proceedings*, ed. J. Hintikka *et al.*, Dordrecht, 1981, I, pp. 145–186.

[4] The restriction to given lines A, $2A$ is of course inessential; one may in general posit lines A, B, in any given ratio and obtain $A^3 : X^3 = A : B$, as the ancient treatments of the problem are well aware.

[5] Cf. Eratosthenes, as cited by Eutocius in *Archimedis Opera*, ed. J. L. Heiberg, III, p. 88; cf., also, Proclus, *In Euclidem*, ed. G. Friedlein, Leipzig: Teubner, 1873, p. 213. Hippocrates' work is discussed in Chapter 2.

[6] Theon, *Expositio rerum mathematicarum*, ed. E. Hiller, Leipzig: Teubner, 1878, p. 2.

[7] Plutarch, *Lives: Marcellus*, xiv, 5–6. An alternative version of this anecdote, mentioning the geometers Archytas, Eudoxus, and Menaechmus, appears in Plutarch's "Why Plato said that god always geometrizes" (*Moralia*, 718 e–f), one of several discussions on topics related to Plato's *Timaeus*. Certain theological overtones present in this version of the Delian story must be due to Plutarch's elaborations of his source narrative, just as in the *Marcellus* passage the concern to link this story with historical trends in geometry and mechanics must also be his.

[8] One may compare the text, translation and comments by P. Shorey in *Plato: The Republic* (Cambridge, Mass./London: Loeb Classical Library), 1935, II, pp. 170f, 112f.

[9] *Archimedis Opera*, ed. J. L. Heiberg, III, p. 90.

[10] Notably, von Wilamowitz; cf. Chapter 2.

[11] *Archimedis Opera*, III, pp. 78–88; these are discussed in Chapter 3.

[12] Cf. Chapter 6.

[13] Consider the following bold-sounding claim on the "practical purpose of geometric investigations":

The graphical solution of problems of construction forms a major goal of geometry. Even where this goal seems to a certain degree to have been forgotten through the advancing of theoretical investigation, its guiding influence on this investigation is quite easily perceived.

The statement is made by the Italian geometer, F. Enriques, in the article concluding the anthology of contributions by him and collaborators on advanced questions pertaining to elementary geometry, *Questioni riguardanti la geometria elementare* (2nd ed., 1907; I have followed the second German edition, 1923, II, p. 327).

By contrast, another mathematician has this to say about the ancient studies on the conic sections:

> I believe that one did not in general effect practically this construction [of the cubic root] via conic sections, but that rather it was a theoretical means—as it can also be for us—a means for the extension of understanding [*Erkenntnis*].

(From H. G. Zeuthen, "Die geometrische Construction als 'Existenzbeweis' in der antiken Geometrie," *Mathematische Annalen*, 1896, 47, pp. 222f; this essay will be discussed further in Chapter 8.) One may expect that the same division of opinion on the relative roles of practical constructions and theoretical investigations would still mark modern discussions on the nature of mathematics.

[14] Eutocius, in *Archimedis Opera*, III, pp. 54–106. Pappus, *Collection*, ed. Hultsch, I, pp. 54–68, 270–284. On Menelaus, Geminus, and Sporus, see T. Heath, *History of Greek Mathematics*, I, pp. 226, 229f; II, pp. 222-226, 260f. I will propose further views in the sequel.

[15] Heath, *History of Greek Mathematics*, Oxford, 1921, I, Ch. 7. Böker, "Winkel und Kreisteilung," *Pauly Wissowa*, Ser. II, 9_1, 1961, col. 127–150; "Würfelverdoppelung," *ibid.*, col. 1193–1223; "Neusis," *Pauly Wissowa*, Suppl. IX, 1962, col. 415–461. *Neuses* are familiar in the form of sliding marked ruler constructions; see Chapters 2, 5, and 6. Tannery, "Pour l'histoire des lignes et surfaces courbes dans l'antiquité," *Bull. Sci. Math. Astr.*, 1883, 18, 278–291 and 1884, 19, 19–30, 101–112. Loria, *Ebene Kurven: Theorie und Geschichte*, 2nd ed., Leipzig/Berlin, 1910 (1st Italian ed. 1902; 2nd Italian ed. 1909).

[16] Cf. the discussion of Nicomedes' conchoid in Chapter 6.

[17] Cf. Heath's treatment in *History*, I, pp. 218–220. A distinguished exception to this tendency is the long study by A. D. Steele, "Ueber die Rolle von Zirkel und Lineal in der griechischen Mathematik," *Quellen und Studien*, 1936, 3, pp. 287–369. Steele blends a careful examination of the ancient sources with a good grasp of the modern algebraic view of these issues; he argues persuasively that the restriction to compass and straightedge was only rarely the explicit condition of the ancient study of geometric problems (cf. Chapter 8).

[18] For a discussion of the reducibility of third-order problems together with a detailed survey of methods of cube duplication and angle trisection, one may consult the article by A. Conti in F. Enriques, *Fragen der Elementargeometrie*, II (cited in note 13 above); for a discussion of the transcendentality of π, also with historical background, one may consult B. Calo in the same book. In the anthology *Monographs on Topics of Modern Mathematics*, ed. J. W. A. Young, 1911 (Dover, repr., 1955) one may find an article on "Constructions with ruler and compass" by L. E. Dickson, and another on the "History and transcendence of π" by D. E. Smith.

[19] The pun is possible in Greek, where the word *telos* and derivatives carry the various senses of "end," "end result," and "goal"; cf. Aristotle, *Physics*, II, 2, 194 a 30.

[20] On Eudoxus, see Chapter 3.

[21] The ancient field of the philosophy of mathematics is richer than my statement suggests. But that study seems primarily to have engaged philosophers, interested in accommiodating mathematical questions within their own favored philosophical outlooks, like Platonism, Epicureanism, or Scepticism. The participation of mathematicians within this field appears to have been slight, and its influence on their researches slighter still. This ancient philosophical field is the subject of recent papers and a monograph in progress by I. Mueller.

[22] We have already alluded to an instance of this attitude, in the treatment of the two versions of the Delian story surviving from Eratosthenes.

[23] See, for instance, Carpus' remarks on the nature of geometric problems, cited in Chapter 8. Dissatisfaction with the ancient formal style led several 17th-century geometers to reconsider questions of method and notation; see, in particular, Descartes, *Géométrie*. The method of analysis thus served to stimulate the development of new algebraic approaches in geometry; cf. M. S. Mahoney, *The Mathematical Career of Pierre de Fermat*, Princeton University Press, 1973, Ch. II.1, III, V.

[24] This example is not hypothetical; cf. Chapter 8.

[25] See the view expressed by Enriques, cited in note 13 above.

[26] This is clearly the effect which von Wilamowitz' rejection of the Eratosthenes document has had, discouraging use of its potentially significant testimony to the early history of cube duplication; cf. note 10 above and Chapter 2.

[27] Pappus' repute both as a mathematician and as a spokesman on the nature of mathematics is an issue in Chapter 8 and will be further examined in the sequel volume.

[28] See Chapter 4.

CHAPTER 2

Beginnings and Early Efforts

When did geometers first initiate the quest for solutions of geometric problems? One seems to find some hints toward a form of answer from certain discussions by Proclus, who, as leader of the Academy in Athens in the 5th century A.D., included guidance in the study of elementary geometry as part of his teachings in the Platonic philosophy. His commentary on Euclid's Book I is a rich source not only of ancient views on the philosophical aspects relating to mathematics, but also on details of its historical development. From Proclus we receive several items on pre-Euclidean efforts derived from the history of geometry compiled by Aristotle's disciple Eudemus of Rhodes toward the close of the 4th century B.C. Two of these relate to problems of construction: I, 12 (how to draw the perpendicular to a given line through a given point not on that line) and I, 23 (how to draw an angle equal to a given angle at a given point on a given line). Reportedly, these problems were first solved by Oenopides of Chios, a geometer and astronomer, elsewhere identified as an associate of Hippocrates of Chios and hence datable to after the middle of the 5th century B.C.[1]

This witness to the work of Oenopides has led scholars to wonder how it could be that such elementary efforts emerged so late. Assuming to the contrary that they must have been known much earlier, they infer that it was the specific form of the solutions, as found in Euclid's treatment, which Oenopides originated, so that he would become the one responsible for the formal project of effecting geometric constructions within the restriction of employing compass and straightedge alone, as is characteristic of Euclid's method in Book I. Presumably, then, Oenopides had access to a diverse range of ways for solving these and other problems, and set about the task of regularizing and classifying problems according to the means of construction adopted.[2]

If such a bald statement of this thesis seems unsympathetic, a more subtle representation would merely camouflage its intrinsic implausibility. To be sure, much older precedents for the sort of geometric construction assigned to Oenopides can be detected. For instance, Babylonian tablets from the mid-2nd millennium B.C. display figures of circles and squares drawn with the assistance of instruments, while the Egyptians' use of cords in their practice of geometric mensuration earned for them the name "rope-stretchers" among the later Greeks.[3] But the type of inquiry supposedly engaged in by Oenopides is of an entirely different order of formal sophistication: to effect constructions on the consciously formal restriction to a specified set of means. It is no obvious matter to explain how and why such a distinctly formal move in geometry should appear so early. Attempts to account for it within the context of an alleged formalization of the general geometric field through linkage with pre-Socratic philosophical developments, as with the 5th-century Eleatics and Pythagoreans, are fraught with difficulties of their own.[4]

In the case of Oenopides, such doubts are magnified upon closer consideration of the details communicated by Proclus; specifically, that Oenopides introduced his construction of the perpendicular for its utility in astronomy; and that for the phrase "at right angles" he employed the term "gnomon-wise" (*kata gnōmona*), that is, with reference to the "gnomon" or pointer of a sundial. It thus seems evident that Eudemus, Proclus' source, has gleaned his information from a treatment of astronomical constructions known under the name of Oenopides.[5] But then it hardly follows that these were the earliest constructions of these problems, even under the special manner of construction they employ, but only that these were the earliest such instances of them known to Eudemus. More importantly, Oenopides would surely not have been occupied in a consciously formal geometric effort, for the astronomical context indicates that he was showing how to arrange the construction of astronomical devices, like sundials, via appropriate instruments of construction. The latter surely included compasses and rulers, but we have no reason to suppose that he rejected the use of others. Such devices as set squares, forms of compasses and sectors, angle-measuring devices, plummets, and sliding marked rulers were all in the repertory of techniques available to geometric practitioners in antiquity, and attested in the mathematical literature from the decades before and after Euclid.[6]

It is imperative, I maintain, to raise these doubts about the formal nature of the work of Oenopides and his contemporaries. While *we* might consider it obvious and natural to classify constructions according to the means employed and to assign privileged status to those demanding only compass and straightedge for their execution, our intuitions in such matters are thoroughly conditioned through knowledge and adoption of the objectives of the formal geometric tradition, advanced primarily through the works of Euclid and Apollonius. But in fact this formal restriction on the treatment of problems in itself betokens attainment of a sophisticated theoretical level. Thus, a historical inquiry into the study of problems must recognize the sophistication implicit in this move and ask about the time, manner, and motive of its introduction. Most treatments of ancient geometry, and specifically those dealing with the history of problem solving, assume this formal motive at the outset.[7] This assumption, I believe, presents serious obstacles against gaining insight into the development of the ancient studies as they actually happened. Let us then set aside that assumption. Through the examination of the evidence of the early work, specifically on the two problems of cube duplication and circle quadrature, we may hope to discover the roots of the ancient field of problem solving.

THE DUPLICATION OF THE CUBE

We possess accounts of the early stages of the study of the cube duplication in two reports derived from Eratosthenes of Cyrene, a prominent man of science and letters from the latter part of the 3rd century B.C. One of these is the fragment from a scene in his dialogue, the *Platonicus*, and is preserved by Theon of Smyrna and Plutarch in writings from the 2nd century A.D.[8] The other takes the form of a letter addressed by Eratosthenes to his royal patron Ptolemy III Euergetes in association with a transcript of the description of an actual model of his device, the "mesolabe" ("mean-taker"), for the mechanical solution of this problem and the text of an epigram dedicating it in the temple of the Ptolemies. The latter account is preserved as one of eleven texts on this problem compiled by Eutocius of Askalon, the 6th-century-A.D. commentator on Archimedes' work.[9] The value of such a document for examining the early history of these efforts would seem evident; but it has not been exploited by historians, owing to their general acceptance of the case challenging its authenticity.[10] It is thus important first to consider whether that case is to be sustained.

Eutocius' version runs to five full pages in the standard edition by J. L. Heiberg. We may divide it into five sections: (i) A historical introduction cites a scene from a tragedy about Minos by an unnamed dramatist and a legend telling of Plato's involvement with the oracle of Delos as precedents for an interest in the problem of doubling the cube. Specific note is made of efforts by Hippocrates of Chios and by three associates of Plato: Archytas, Eudoxus, and Menaechmus. (ii) Eratosthenes' mechanism for finding means is cited and praised for its greater practicality in comparison with the earlier methods. Several contexts of potential application are enumerated, among them ship building and military engineering, by way of illustrating the utility of the device. (iii) A full account of the geometric theory of the device is given. (iv) This is followed by a description of certain physical details pertaining to the actual construction of a working model. (v) The text closes with a description of the votive monument raised by Eratosthenes to commemorate its invention: this consisted of a bronze exemplar of the "mesolabe" set atop a pillar with an explanatory inscription engraved below. A full transcript of the latter is given, including (a) a brief account of the geometric theory of the device, parallel to that in (iii), and (b) an epigram of eighteen verses singing the praises of its virtues in contrast to the same prior efforts named in (i), and finally dedicating it to Ptolemy and his son.[11]

Since the penetrating examination of this document by U. von Wilamowitz-Moellendorff in 1894, the authenticity of the closing epigram and of the descriptive text just preceding it [(v-a) and (v-b)] have been granted.[12] The epigram is too expertly crafted, too fine a specimen of the difficult form of Hellenistic elegiac couplets, he argued, to be so easily dismissed as a forgery the way critics before him had contended. The descriptive text on the model of the device would raise no suspicions (save for an occasional trivial interpolation), while the event of an inventor or artisan's dedicating a work sample at a shrine is not without precedents in this period. But von Wilamowitz had no such approval to give to the accompanying letter. First, he argued, this presents an extended account of the device (iii) which would surely be superfluous, given the presence of the dedicatory model itself. Second, on points of detail it conflicts with the "genuine" part of the text (v), he alleged, as well as with the fragment from the *Platonicus* available to us through Theon and Plutarch; on the whole, however, it appears to assert little beyond what one might have read in the inscription or inferred from that. Third, this text was not known to Pappus, who wrote about two centuries before Eutocius, for his own version of Eratosthenes' method differs on points of detail in the construction and proof from the one given in the letter, although it presents a highly condensed version of materials parallel to Eutocius' items (iv) and (iii) or (v-a).[13] Finally, and most damaging of all, the letter is banal: it is written without any sense of style and without any recognition of the occasion of its communication. One can hardly conceive, argued von Wilamowitz, that a man of such literary talents as Eratosthenes, writing to his patron, should frame his account in such a pedestrian manner. Thus, he concluded, the letter must be taken as a late forgery, still unknown at the time of Pappus; its author presumably transcribed from the votive monument the epigram of Eratosthenes and the partially mutilated description of the device and then, misled by the epigram's devotional invocation of Ptolemy as deity of the shrine, mistook the offering as a personal gift and so produced the letter as a companion explanatory document. But, he added, the ineptness of the forger's effort is betrayed on many points, most strikingly in its omission of second person forms of address almost until the very end. On the other hand, the composition has preserved for us the genuine epigram, for which we may be grateful.

Shall one accept this view of the provenance of the letter? To be sure, its style is concise, perhaps a bit abrupt; but it is not an illiterate production. One has yet to display such outright anachronisms which would mark it as impossible for a writer of the 3rd century B.C. Moreover, the historical and technical information it presents is entirely compatible with what we learn from our other sources on the early studies. Indeed, the absence of anachronisms relating to the mathematical content is impressive and would suggest at the very least a degree of thoughtfulness and skill on the part of the alleged forger. A comparison with the "genuine" parts of the document shows that the "forger" has not limited himself to merely what has appeared in the epigram, but goes beyond it on several points in ways which indicate a real familiarity with this material, rather than mere fabrication. For instance, he observes that of the earlier methods only that of Menaechmus admitted any practical implementation "to a certain extent and that with difficulty" (*dyscherōs*); by contrast, the epigram refers only to the constructions by Archytas as "unwieldy" (*dysmēchana*), where in context the charge applies indifferently to the methods of both Menaechmus and Eudoxus as well.[14]

Relative to the argument concerning Pappus, no one has ever presumed that Pappus passed on *all* the sources at his disposal. Even if he did not possess a copy of the source in the form used by Eutocius, that hardly implies the nonexistence of that source at Pappus' time. After all, Eutocius himself presumably passed over the version in Pappus (if he had access to that) in favor of the text he actually reproduces. Surely, Pappus was capable of the same kind of editorial selectivity. As it happens, his text on Eratosthenes is rather more concise than that given by Eutocius, but on the whole not superior to it. Von Wilamowitz makes much of certain discrepancies (e.g., the shape of the sliding plates as triangular according to Pappus, but rectangular according to Eutocius)[15] which have no bearing on the feasibility of the design and its realization. But he is silent on certain insights of Eutocius omitted by Pappus: e.g., that the "mesolabe" can be used for finding not just two, but any number of mean proportionals by the insertion of additional plates.[16] The notion that Eutocius' report could result merely by transcribing and adapting what could be seen on the monument is simply not true, although it might well hold for the version in Pappus. In particular, Eutocius is far more detailed in his recommendations on the physical construction of the device. It is difficult enough to accept that the inscription should remain more or less intact as late as Pappus' time, early in the 4th century A.D.[17]; but that such a memorial of pagan worship could survive the volatile spirit of the later part of that century and the next is truly incredible.[18] Von Wilamowitz' argument, that most of the information in the letter would be superfluous in the presence of the monument itself, is not compelling. One may consider the example of the ornate astrolabe invented and built by Synesius, disciple of Hypatia, statesman, philosopher and bishop; the specimen was forwarded to his noble associate Paeonius accompanied by a letter describing in detail the conception and physical execution of the device. Many of these details would be evident from the device, including the two epigrams engraved on it, which Synesius quotes in full: "let it be set down for such as may read it later, since for you it is enough that it lies on the tablet."[18a] One thus sees in Synesius' dedication a close parallel to Eratosthenes' letter on the "mesolabe."

The strongest indication of the letter's authenticity, however, would appear to be precisely what von Wilamowitz took to be its most questionable feature: its banality. For surely a forger would have let his imagination run wider than the narrow compass of factual materials given here. By contrast, those facts do amount to the kind of information which technical accounts accompanying other mathematical writings of this period contain. The prefaces to the treatises by

Archimedes, for instance, rarely go beyond stating the principal theorems in the work and making brief observations on the methods used.[19] One would suppose that his correspondents Dositheus and Eratosthenes were receiving these communications by virtue of their official positions at Alexandria; but even this inference is not directly affirmed by anything actually said in the prefaces.[20] The *Sand-Reckoner* provides a good parallel to Eratosthenes' letter. It is devoted primarily to the exposition of the geometry and number theory needed for a specific computation dealing with the dimensions of the cosmos.[21] Within it are details of the design of a sighting device by Archimedes himself and a report of particulars, both geometric and observational, relating to its actual use. The discussion elaborates a theme introduced through a general literary reference (i.e., the inadequacy of the account of the infinite by certain unnamed philosophers). In all these respects, then, it follows a pattern like that of Eratosthenes' letter. Indeed, both are framed as communications to royal patrons, yet the *Sand-Reckoner*, despite its much greater length, is as sparing in its use of second-person forms as the letter is.[22] The Gelon, addressed as "king" by Archimedes, is believed to have served as regent in Syracuse for some thirty years, although he never reigned as king in his own right.[23] Was he still a youth or already a mature man when Archimedes wrote to him? Was the writing sent to him at Syracuse, or elsewhere? And where was Archimedes himself at that time: at home or abroad? Did he read the letter orally before an audience or was the writing intended for Gelon's personal study? Was Archimedes a kinsman of the regent or his former tutor? The document provides not a single clue toward the answer to any of these questions of context.

In view of this, how can one fault Eutocius' document for its failure to establish a setting? Von Wilamowitz' critique thus falters through his own failure to reckon with the character of the appropriate literary genre. Without grounds for suspicion, then, we may treat Eutocius' text no less seriously than any other, as a historical source. Indeed, von Wilamowitz' plea for admitting the epigram, falling within a genre of which he is a recognized master, would naturally prepare us for admitting the whole document as genuine. The burden of proof rests upon those who would maintain the contrary.

Under the view that the letter is authentic, its ostensible conflict with the alternative account from the *Platonicus* takes on entirely new significance. With reference to the efforts of the geometers in Plato's Academy, the letter says this:

> [a] After some time they say that certain Delians fell into the same difficulty as they set about to double one of the altars in accordance with an oracle, and so sending out word they asked the geometers with Plato in the Academy to find for them what was sought. These applied themselves diligently and sought (how) to take two means of two given lines, [b] and Archytas of Taras is said to have discovered (a solution) by means of semicylinders, Eudoxus by means of the so-called curved lines. [c] But it happened that these all wrote in demonstrative fashion, none being able to manage it in a practical way or to put it into use, save to a certain small extent Menaechmus and that with difficulty.[24]

It would appear that two or three different sources have been stitched together:

(a) an account of the legend of the Delian oracle and the communication to the Academy; (b) a reference to solutions by Archytas and Eudoxus; and (c) an account somehow acquainted with writings derived from these geometers, in particular, from Menaechmus. Doubtless, communications between Archytas and the Academy were consistently good, through visits by Plato and others to Italy, and through correspondence; but the ancient biographical traditions do not mention an actual residence by Archytas at the Academy.[25] Thus, line (b) reads uncomfortably as a simple continuation of (a). Further, both (a) and (b) are reported at some detachment from primary sources ("they say;" "it is said"). By contrast, (c) seems to speak from a certain familiarity with such sources. When Eutocius elsewhere presents Archytas' method, Eudemus is named as source.[26] One would suppose that Eudemus also treated the methods of Eudoxus and Menaechmus and that his versions were available to Eratosthenes. That being so, the tentative tone of line (b) is puzzling: surely it would be *obvious* that Archytas used semicylinders and Eudoxus used curved lines. Here then is a place where an interpolator may have inserted a line, based on his reading of the allusions to these geometers in the epigram.[27] Without it, line (c) follows as a fully appropriate technical observation after the story in (a).

The other account of this incident is transmitted by Theon of Smyrna in the introductory section of his compilation of mathematical materials pertinent to the study of Plato:

> For Eratosthenes says in his writing the *Platonicus* that when the god pronounced to the Delians in the matter of deliverance from a plague that they construct an altar double of the one that existed, much bewilderment fell upon the builders who sought how one was to make a solid double of a solid. Then there arrived men to inquire of this from Plato. But he said to them that not for want of a double altar did the god prophesy this to the Delians, but to accuse and reproach the Greeks for neglecting mathematics and making little of geometry.[28]

This version of the story is further attested in two passages from Plutarch, for whom the moral is that one needs dialectical skill to maneuver among the ambiguities of oracles.[29] Following von Wilamowitz, many have taken these discrepancies to militate against the authenticity of the version in Eutocius. One is thus to suppose that the forger of the letter diluted Eratosthenes' account in the preparation of his own. But why should he have sought to remove all those details which give the story vividness: the occasion of the plague and Plato's personal intervention to interpret the true meaning of the oracle? The story in the letter seems to have no point beyond explaining why a certain group of geometers came to concern themselves with cube duplication.

The embellishments reported by Theon and Plutarch suggest that their versions derived from the elaboration of the more pedestrian account cited by Eutocius. Doubtless, Eratosthenes himself was responsible for these changes, as he sought in his *Platonicus* to dramatize aspects of Plato's view of mathematics. After all, motives of this very sort led Theon and Plutarch to cite this passage in their own works. Elsewhere, Plutarch exploits the cube duplication to illustrate another aspect of Plato's philosophy: that geometry is not a matter of perceptible things, but only of the eternal and incorporeal. Thus, he has Plato criticize the mechanical solutions of Archytas, Eudoxus, and Menaechmus as spoiling the true good of geometry.[30] This alternative version echoes the second part of Eutocius' account, where the practicability of these geometers' solutions is brought up for comment. But there is a patent difference: according to Eutocius, Eratosthenes criticizes the earlier efforts for their being too abstract; by contrast, Plutarch derives from him a criticism of their overly mechanical character. Of course, this discrepancy need only be one of emphasis. The surviving accounts of these methods are in fact fully geometrical in the formal manner; but they do rely on certain mechanical conceptions, like the generation of curves by moving lines or by the intersection of solids of revolution.[31] The student of Plato's abstract philosophy might well view the latter with disapproval. On the other hand, Eratosthenes' own solution entailed the production of a mechanical device, and in his eyes its practical viability is a major asset; in this context, the practical limitations of the earlier designs could well be a point of note, inducing him to underscore their purely theoretical nature. Here again, the discrepancy is undoubtedly due to Eratosthenes himself. The intent in the letter is historical, but that in the *Platonicus* is dramatic and philosophical. He might thus take certain liberties in the latter, denied to him in the former. To set the scene for his dialectical point, he can elaborate the mechanical elements more explicitly than the original treatments actually did. Such fictionalization would surely be acceptable in a work which made no pretension of being plainly historical.

Accepted as a serious historical source by a man well placed for a knowledge of this period of Greek mathematical history, Eratosthenes' text as reported by Eutocius offers hope of insight into matters often obscured in modern discussions. First, although historians typically wish to leave open the question of the historical validity of the Delian oracle,[32] one now readily perceives the story to be a fabrication, originating from within the Platonic Academy around the middle of the 4th century B.C. By this time the problem of cube duplication was already familiar, for as Eratosthenes himself notes, important advances in its analysis were made by Hippocrates of Chios, that is, almost a half-century before.[33] It would seem odd indeed for the Delian oracle to be concerned with an old problem which then happened to be eluding the efforts of contemporary geometers. But on the other hand, as a dramatic way of affording recognition and motivation to those efforts, the story makes good sense, especially since geometers associated with the Academy were prominent in this activity. But one can only guess what the context and purpose of the story might have been at its first composition. As we have indicated, the specific morals attached to it by Eratosthenes (in the *Platonicus*) and by Plutarch seem to be their own additions, but they are likely to have sensed accurately the ancient Academicians' view of these matters. Any student of the *Republic* can appreciate the strength of Plato's insistence on purity and abstractness of mathematical entities.[34] The kinematic element in the constructions for the cube duplication might thus provoke discussion of the relative status of geometry and mechanics, and indeed this very issue is important within Aristotle's theory of the order of the sciences.[35]

Furthermore, Eutocius' text of Eratosthenes is noteworthy for the sparseness of the historical information it transmits. One should have thought that Eratosthenes would be concerned with producing as full an account as he could of the studies preceding his own, and his text does leave that impression. Yet it is amazing that he turned up with so little. He appears to have discovered nothing worth mentioning from the whole century or so separating him from the efforts of the Academicians. As for the even earlier efforts, he cites only the work of Hippocrates and a scene from an unnamed tragic poet.[36] In the latter, when King Minos is told that the tomb he had ordered for Glaucus will have dimensions of one hundred feet on a side, the king answers that this would be too small, and so, "let it be double; without mistaking its fine form, swiftly double each member of the tomb." The new structure will of course be eight times larger than the original one in volume, not its double. Eratosthenes calls attention to the poet's error. But actually there is none here: the poet intends that each dimension shall be doubled; he is not articulating the mathematicians' problem of the cube duplication. Evidently, Eratosthenes has misconstrued the passage in his desire to find precedents for an interest in this problem.

In effect, Eratosthenes has sought some motivation for Hippocrates' study of the cube duplication. But Hippocrates' effort by its very nature reveals that motive; for in Eratosthenes' account, this is what he does:

> It used to be sought by geometers how to double the given solid while maintaining its shape...After they had all puzzled for a long time, Hippocrates of Chios was first to come up with the idea that if one could take two mean proportionals in continued proportion between two lines, of which the greater is double the smaller, then the cube will be doubled. Thus he turned one puzzle into another one, no less of a puzzle.[37]

Hippocrates' insight is of course not restricted to lines assumed in a 2 : 1 ratio. If for any two given lines, A and B, we can insert the two mean proportionals, X and Y, then $A : X = X : Y = Y : B$. Thus, by compounding the ratios, one has $(A : X)^3 = (A : X)(X : Y)(Y : B)$, that is, $A^3 : X^3 = A : B$. Thus, X will be the side of a cube in the given ratio (B : A) to the given cube (A^3).[38]

In his closing line here, Eratosthenes seems to perceive little merit in Hippocrates' move. But in fact, by this stroke Hippocrates has put the problem into a form permitting the application of a whole new range of geometric techniques, those of proportion theory. In general, this procedure of "reducing" (*apagōgē*) one problem to another from whose solution its own would follow is a powerful technique of problem solving, a forerunner of the fruitful method of geometric "analysis." As Proclus tells us, Hippocrates was the first geometer known to have applied the method of reduction in the investigation of "puzzling diagrams," i.e. difficult problems of construction.[39] We will see later that Hippocrates appears to have adopted a similar strategy in the investigation of the circle quadrature.

Later geometers clearly recognized that such a *reduction* does not in itself amount to a *solution* of the problem posed. But would Hipppocrates already have made this distinction in the case of his treatment of the cube duplication? A passage from Aristotle suggests that he did. In *de Anima* II, Ch. 2, Aristotle observes that one may define a term like "squaring" (*tetragōnismos*) not only by saying *what* it is (i.e., the equality of a square with a rectangle), but also by stating its *cause* (i.e., the finding of a mean proportional).[40] Now, the latter refers to the reduction of the given problem of areas, such as one may read in Euclid's VI, 17, and is the precise plane analogue to Hippocrates' problem of solids. But the construction of this planar problem, whether in the form of a squaring (as in *Elements* II, 14) or in that of finding a mean (as in VI, 13), was certainly already available to Hippocrates. A long fragment of his study of the quadrature of crescent-shaped figures survives, and the techniques of proportions are applied throughout to situations far more demanding than the squaring of rectangles.[41] Since the distinction between the reduction and the solution was clear in the instance of the planar case, the same must surely have been appreciated in the solid case as well. We may suppose, then, that Hippocrates and his followers continued their research into the cube duplication by seeking the construction of the two mean proportionals, where the latter effort achieved its first clear successes through the discoveries of Archytas, Eudoxus, and Menaechmus.

These considerations help clarify the way in which the cube duplication was first articulated as a problem. Within the study of the measures of plane figures, affected via the techniques of proportions, it would be perfectly natural for a geometer like Hippocrates to consider the case of two mean proportionals and so perceive its relation to the problem of volumes, by analogy with the case of the single mean. In this way, the new problem emerges through the natural development of geometric research. Now, it sometimes happens that research moves in new directions in response to external stimuli: practical problems arising from commerce, government, and engineering are a manifest basis of much in the most ancient mathematical traditions; ritual might sometimes play a role; or philosophy might instil a sensitivity to formal questions.[42] The story of the Delian oracle has thus often been used as an example of such external motivation. But one can now see that the ancient evidence on the origins of the cube duplication affirms quite a different view: that the legend arose within Plato's Academy around the middle of the 4th century B.C., long after the problem had attained notoriety through the work of Hippocrates; and that Hippocrates himself articulated this problem as a natural extension of the geometric field of his time. In the absence of direct testimony to earlier work in the 5th century, one has no grounds for supposing a development of any other sort.

THE QUADRATURE OF THE CIRCLE

In connection with the problem of constructing in a given angle a parallelogram equal to a given rectilinear figure (*Elements* I, 45), Proclus remarks:

> Having taken their lead from this problem, I believe, the ancients also sought the quadrature of the circle. For if a parallelogram is found equal to any rectilinear figure, it is worthy of investigation whether one can prove that rectilinear figures are equal to figures bound by circular arcs.[43]

Although the words "I believe" reveal this view to be merely a surmise on Proclus' part, rather than a conclusion based on more or less explicit testimony, it is far from unreasonable. Indeed, there appears to be a certain sense of the problem of circle quadrature in remnants of the ancient Egyptian and Mesopotamian traditions as far back as the middle 2nd millennium B.C. and earlier. Specifically, the scribe of the Rhind Papyrus works out the area of a given circle first by subtracting 1/9 of its diameter and then squaring the result.[44] Thus, in effect, the circle is equated with the square whose side equals 8/9 the diameter of the circle. This rule is illustrated by a figure in which a square is divided into nine equal squares, where the diagonals of the four small squares at the corners are drawn in. In this way one sees at once that the circle might be approximated as 7/9 the enclosing square. One might conjecture that the scribe went on to subdivide each of the squares into nine squares and then tried to estimate how many of the 81 resulting squares combined to approximate the circle. But another approach might have been followed. Knowing that problems of areas often require computing squares and their roots, the scribe might have sought a value near 7/9 which resulted from a squaring. Since $7/9 = 63/81$, where the latter is just short of $64/81 = (8/9)^2$, a quasi-deductive route might be possible for deriving the rule applied by the scribe of the papyrus.

If such was indeed the basis of the rule, the scribe would also have a basis for recognizing the measurement of the circle as a geometric problem. The same would not be true within a tradition relying entirely upon empirical derivations of its geometric rules. It is hardly possible, for instance, that the Egyptians could have known the 1 : 3 ratio of the volumes of the cone and cylinder having equal altitude and base other than by observing that the fluid contents of a conical vessel can be emptied exactly three times into a cylindrical one, or by some comparable physical measure.[45] But inaccuracies of manufacture, spillage, and other such factors would render this procedure approximate. There would thus be no firm demarcation between rules which are exact (as is that for the ratio of volumes) and those which are approximate (e.g., the circle measurement). Certainly, within the early traditions the aim was to obtain practical procedures. The project of somehow producing a construction of the square equal to the circle seems remote from this objective, even if it may be perceived as entailed within the results given. Its explicit formulation as a problem must then be sought elsewhere.

We may gather that Greek mathematicians had hit upon this formulation of the problem before the close of the 5th century B.C. For in his comedy *The Birds* (staged in 414 B.C.) Aristophanes presents this lampoon of the astronomer Meton arriving as self-appointed civic surveyor of Cloud-cuckoo-land:

> If I lay out this curved ruler from above and insert a compass—do you see?—
> ...by laying out I shall measure with a straight ruler, so that the circle becomes square for you.[46]

Now, there is no report of Meton's inquiring into the circle quadrature, and the passage here actually views him as somehow dividing the circle into quadrants.[47] But the force of the jest would seem to require that the audience perceive it to refer to a real problem under investigation by geometers at that time. This conforms with reports we have of certain efforts by Hippocrates, Antiphon, and Anaxagoras.

Relative to Hippocrates, Aristotle tells that he proposed a false proof of the quadrature of the circle "by means of segments" or "by means of lunules."[48] The latter refers to Hippocrates' masterful construction and quadrature of crescent-shaped figures bounded by arcs of circles; we turn later to the question of the bearing this effort had on the circle quadrature. According to the account preserved by Simplicius, the 6th-century Aristotelian commentator, drawing explicitly from a report by Eudemus, Hippocrates prefaced his treatment with demonstrations of certain basic theorems on circles:

> He made his start and set down first some things useful for those [constructions], that segments of circles have to each other the same ratio as their bases in power; and this he proved from having proved that the diameters have the same ratio in power as the circles....[49]

This passage notes explicitly that Hippocrates *proved* these theorems. Moreover, it goes on to supply further details of the derivation of the segment theorem in good agreement with a version given much later by Pappus.[50] We may thus be confident that toward the close of the 5th century, Hippocrates already appreciated the essential concerns of the deductive form entailed by the proofs of theorems such as these. Now, the lines just cited include the statement of a theorem identical to one in Euclid's *Elements* (XII, 2): "Circles are to each other as the squares on their diameters." Simplicius' terminology of "powers" (*dynameis*) instead of "squares" (*tetragōna*) reassures us that here he has followed the archaic Eudemean usage in his text, rather than improvise through his own knowledge of Euclid.[51] But we encounter difficulty in supposing that Hippocrates advanced the proof of this theorem on circles in its Euclidean form. For the limiting method on which that depends is due to Eudoxus, as we learn on the good authority of Archimedes.[52] Then what sort of proof could Hippocrates have provided a half-century or more before Eudoxus?

Aristotle's remarks on the false efforts at circle quadrature link Hippocrates

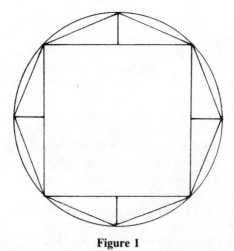

Figure 1

with two sophists from early in the 4th century, Bryson and Antiphon.[53] What Simplicius reports of the latter may help us here. Antiphon proposed to construct a rectilinear figure equal to a given circle by a procedure involving the successive doubling of the number of sides of an initial inscribed polygon. For instance, if one starts with the inscribed square, erects the perpendicular bisector to each side, marks the point where each meets the circle, and connects each of these points to the adjacent vertices of the square, one will have erected an isosceles triangle on each side of the square, and their combination with it will produce the inscribed regular octagon. (See Figure 1.) One may next bisect the sides of this octagon, ultimately resulting in the inscribed 16-gon; and so on.

> Continuing in this way, at the point where the area was exhausted, he said he will have inscribed a certain polygon in the circle in this manner, such that its sides conform to the arc of the circle on account of their smallness. Since we are able to produce the square equal to any polygon,...because the polygon has been produced conforming to the circle, we shall also have produced a square equal to the circle.[54]

Antiphon's step yielding an arc conforming (*epharmozein*) to the sides of the inscribed polygon would appear to be drawn from notions familiar through the ancient atomists. For them, the microscopic world of atomic reality might be qualitatively different from the macroscopic world as perceived by our senses. Specifically, the circle appears to meet its tangent at a single point only, and we insist on this as part of the abstract conception of this configuration; but in their physical manifestation they will meet along a small line segment. This very question was discussed by Democritus and some of the sophists, notably Protagoras; and one of Simplicius' sources on Antiphon, the 3rd-century-A.D. commentator Alexander of Aphrodisias, observes that this principle—the circle's meeting the tangent line at one point only—was the one whose abolition lay at the root of Antiphon's fallacious quadrature.[55] We thus perceive a firm dialectical context for Antiphon's argument.

The invalidity of his circle quadrature is evident, for it relies upon the actual "exhaustion" (*dapanan*) of the whole area lying between the circle and the polygons, while the indicated procedure of inscribing polygons will not do this if applied only a finite number of times.[56] Hence, for Aristotle and the commentators, its refutation is not a matter for the geometer to be concerned about. By contrast, modern scholars have tended to assign to Antiphon credit for the introduction of the important notion of polygonal approximation, essential for the ancient quadratures of curvilinear figures.[57] But is this a plausible position to maintain? Given that Antiphon and Hippocrates were near contemporaries, the latter without doubt being rather the older, shall we suppose that the sophist, in the course of framing a muddled geometric argument to support his dialectical position, lent essential insights to the geometer? Or is it not more likely that he was drawing from and modifying procedures already familiar among geometers? The latter certainly typifies the later philosophical tradition, as represented by Plato, Aristotle, the Stoics, and the Sceptics, for instance, and their stance toward the technical sciences of their times. By contrast, it seems unclear why a geometer like Hippocrates should turn to the sophists for technical assistance in a theorem of this sort. The fact that Antiphon's use of this technical procedure lapses into patent fallacy, while Hippocrates requires the same procedure for the proof of a theorem implemented with full logical precision in his quadratures of lunules, merely confirms one's doubts about Antiphon's originality in this technical context.

If we assign to Hippocrates knowledge of the polygonal procedure preserved in our text of Antiphon's quadrature, then we obtain a good basis for reconstructing a proof he could have used for the theorem on the ratios of circles. For one notes a subtle difference between Antiphon's method and the one used in the Euclidean proof of this theorem (XII, 2). Euclid obtains each polygon in the sequence by bisecting each of the arcs cut off by the vertices of the polygon preceding it. Antiphon, on the other hand, builds each new polygon from its precursor by adding the isosceles triangles onto each of its sides. Now, if one considers two circles, each enclosing similar polygons constructed in this way, since the initial polygons and the added triangles taken pairwise are similar, these figures will be in the same ratio, namely that of the squares of the diameters of the circles. Hence, the *sum* of *all* these figures, extended indefinitely, will be in the same ratio. Since these sums constitute the circles themselves, it follows that they too have the ratio of the squares of their diameters, as Hippocrates asserts.[58]

From the strict logical viewpoint, of course, this argument fails, for it applies to the infinite case a theorem on the ratios of sums which can be established for sums of only a finite number of terms, if even that, given the technique of proportions available to Hippocrates.[59] But this approach is so strongly founded on clear notions of the continuity of magnitude that the difficulty might not necessarily have been appreciated at an early stage of these studies. We must assign to Eudoxus the first full awareness of this difficulty and the discovery of the method of limits by which it could be removed. It is noteworthy that the technique of sums we have just proposed is a standby within Archimedes' demonstrations; that Archimedes provides a proof, but only for the finite case (*Conoids and Spheroids*, Prop. 1); nevertheless, that he applies its infinite case in all of the planar and solid measurements presented in his *Method*.[60]

One cannot say for certain whether such studies of the circle quadrature can be set much before the time of Hippocrates. Writing in the 2nd century A.D., Plutarch mentions that the 5th-century natural philosopher Anaxagoras "drew the quadrature of the circle" during a spell in prison.[61] This would place an awareness of the theoretical character of the problem of circle quadrature to not much later than the middle of the 5th century B.C. It is an especially intriguing attribution, in that Anaxagoras insisted on the continuity of matter as the basis of the nature of things and seemed to perceive some of the mathematical implications of the principle of indefinite divisibility, the very principle later violated in Antiphon's circle quadrature. To be sure, Proclus reports that Anaxagoras "touched on many things relating to geometry."[62] But the ancient tradition otherwise assigns to him no specific interest in mathematics, and Plutarch's testimony is quite casual, inserted merely to illustrate the point that the thinking man finds happiness even under trying circumstances. The subsequent notoriety of the circle quadrature might all too easily color the testimony transmitted by later authorities. Thus, we must forswear using Anaxagoras' alleged contribution as a sign of early interest in the problem.

From Aristotle we may gather that a construction of the circle quadrature was attempted on the basis of Hippocrates' quadrature of lunules. For in his account of the logical method of "reduction" (*apagōgē*) he remarks thus:

> We are the closer to knowing, the fewer the mean terms in the syllogism. For instance, as the circle together with the lunules is equal to a rectilinear figure, we are that much the closer to the squaring of the circle itself.[63]

Viewed in the light of Hippocrates' reduction of the cube duplication to the problem of the two mean proportionals, this passage seems to indicate that Hippocrates attempted a similar strategy for the circle quadrature: to cast it into an alternative form permitting the application of a wider range of the techniques then known. This is fully borne out in the extended discussion of circle quadrature presented by Simplicius commenting on a related passage from the *Physics* where Aristotle mentions the false quadratures.[64] In addition to sketching and criticizing Antiphon's argument on the circle quadrature, as we have already seen, and noting several other efforts on this problem, Simplicius gives a detailed account of Hippocrates' quadrature of lunules. Indeed, he provides two different versions, a simplified treatment known through Alexander's commentaries, followed by an elaborate series of constructions and proofs quoted verbatim from Eudemus' history. This passage from Eudemus is our most important fragment from the pre-Euclidean geometry, providing invaluable insight into the content, methods, and terminology of the early tradition. We shall examine it in detail, after a brief consideration of the shorter version from Alexander.

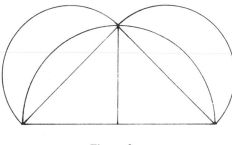

(a)   **Figure 2**

In Alexander's text, two cases of lunules are presented[65]:(a) Starting from a right isosceles triangle, we draw on each of its sides a semicircle. (See Fig. 2a.) Since the square on the hypotenuse equals the sum of the squares on the two legs, and since circles are as the squares on their diameters, it follows that the semicircle on the hypotenuse equals the two semicircles on the legs. If we now remove the portion common to the large semicircle and the two smaller ones, we find that the isosceles triangle which remains is equal to the two lunules which arch over its legs. Thus, by equating the lunules to a rectilinear figure, we have effected their quadrature.[66] (b) In the second case, we begin with the trapezium formed as a bisected regular hexagon and draw semicircles on each of its sides. Since the longest side of the trapezium is double each of its shorter sides, the large semicircle equals four of the smaller ones. As before, subtracting the common portion, we find that "the remaining lunules together with the (small) semicircle are equal to the trapezium." (See Fig. 2b.)

Recalling Aristotle's remark on the reduction of the quadrature of the circle to that of the lunules, cited above, we see that he evidently had this construction of Hippocrates in mind. But precisely what conclusion Hippocrates himself intended to draw from it has been debated among ancient and modern com-

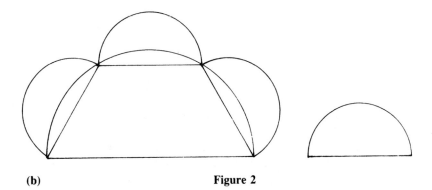

(b)   **Figure 2**

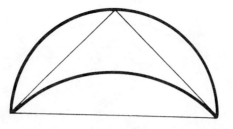

Figure 3

mentators alike.[67] Alexander, for instance, supposed that Hippocrates wished to argue that since the lunule in (a) has been squared, while the sum of the circle and the lunule in (b) has also been squared, then the circle itself has been squared. If so, he would have committed the patent error of assuming that the quadrature of a single case of lunule amounts to the quadrature of all cases. Simplicius, by contrast, found it incredible to conceive that a geometer of Hippocrates' outstanding caliber could have lapsed into such a blunder. Under this view, one would charge to others, for instance, to Aristotle, the attempt to construe Hippocrates' results as a circle quadrature. Doubtless, this is the correct view, for as we saw in the parallel instance of the cube duplication, Hippocrates' reduction of the problem there was not likely to have been regarded as a solution of it. As we shall see later, Simplicius' text provides further insights into Hippocrates' view of the circle quadrature.

The alternative account, which Simplicius quotes "word for word" (*kata lexin*) from Eudemus, presents the full treatment of four cases of lunules. The first is essentially identical to Alexander's case (a), while the fourth bears comparison, if somewhat more loosely, with his case (b). The other two cases are new, constructing lunules which are in the one instance greater and in the other less than the arc of a semicircle. Translations and technical summaries of this text are readily available, so that a brief summary will suffice here.[68] My principal aim now is to retrieve the line of thought underlying Hippocrates' constructions and thus to reveal his sense of the more general class of lunules.

(i) In the first case, Hippocrates begins with the right isosceles triangle and circumscribes the semicircle on its hypotenuse. (See Fig. 3.) This much conforms with case (a). But now, instead of drawing semicircles on the legs also, he draws on the hypotenuse the circular segment similar to those which appear on the legs of the triangle. In this way he obtains a single lunule, the figure bounded by the two circular arcs, and shows that it equals the initial triangle. This follows, since the similar segments are as the squares on the chords which define them, so that the larger segment equals the sum of the two smaller ones. If to the mixtilinear figure inside we add the two small segments, we obtain the lunule; if we add to it, alternatively, the large segment, we obtain the triangle. Thus, the triangle and the lunule are equal to each other. It is clear that this lunule is the same as each of the two lunules constructed in case (a). But the present treatment reveals a greater technical range. For instance, its author operates freely with similar

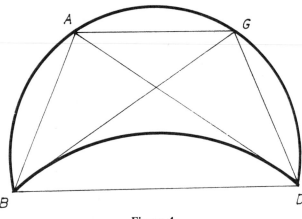

Figure 4

segments, not just semicircles. The pertinent definition, given earlier in the text, stipulates them as subtending equal angles in their respective circles. As we saw above, his proof that similar segments, like similar rectilinear figures, have the ratio of the squares on homologous sides signifies that major progress has already been made toward the Eudoxean manner of measuring curvilinear figures. In all these respects, the treatment of case (i), despite its relative simplicity, prepares us nicely for the more intricate constructions which follow.

(ii) Hippocrates now assumes the construction of a trapezium whose base has to each of its three remaining equal sides the ratio $\sqrt{3} : 1$. (See Fig. 4.) He next draws the circumscribing circle and then draws on the base the circular segment similar to those which have thus been formed on the three other sides. The text establishes in full detail that the arc of the outer bounding circle is greater than a semicircle.[69] The actual proof of the quadrature of the lunule is omitted from Eudemus' text, so Simplicius supplies his own, remarking that Eudemus must have considered it obvious. But in fact the analogous proofs are given in full for each of the following two cases, and textual considerations show these to be in the pre-Euclidean manner of Eudemus.[70] Thus, one may infer that Simplicius' text came to lose portions of the proof through scribal omissions.[71] At any rate, since the one large segment equals the three small ones, the same process of addition as used in (i) shows that the lunule equals the trapezium.

From these two cases we may recognize the condition Hippocrates has in mind for the construction of squarable crescents. If we conceive of a rectilinear figure bounded by two polygonal arcs, the outer consisting of n sides each equal to s_n, the inner consisting of m sides each equal to s_m, and if $s_n^2 : s_m^2 = m : n$, and if furthermore the sides s_n, s_m subtend equal angles in the circles circumscribed about their respective polygonal arcs, then the lunule bounded by these circular arcs will equal the rectilinear figure bounded by the polygonal arcs. Although Hippocrates does not formulate the construction in such general terms,

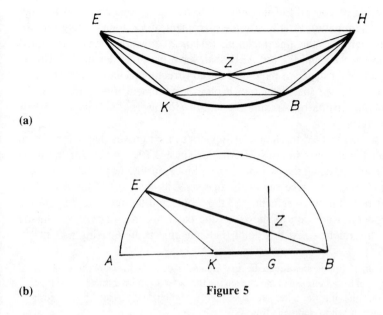

(a)

(b) Figure 5

his treatment of the third case indicates that a conception of this sort did indeed guide his thought. His examination of this next case, like those which preceded it, is entirely in the synthetic mode, and so tends to obscure the manner of its discovery. I shall thus attempt to reconstruct on its basis an alternative version in the analytic mode in order to display its essential idea.[72]

Let an investigation of the lunule of the type just stated be proposed, where the two bounding arcs correspond to polygonal arcs, the one of three equal sides, the other of two. Thus, $s_2 : s_3 = \sqrt{3} : \sqrt{2}$. Let us conceive that the figure has been produced, consisting of the double arc EZH and the triple arc EKBH (see Fig. 5a); if we pass a circle through each of these polygonal arcs, then we must also have that the circular arcs EZ, ZH, EK, KB, BH all subtend the same angle at the center of their respective circles. We join EH and consider the angles ZEH and KEH. As the former intercepts the single arc ZH, while the latter intercepts the double arc KBH, it follows that angle KEH is double ZEH, so that EZ bisects angle KEH and its extension must pass through B. Furthermore, point Z lies on the perpendicular bisector of BK[73], while point E lies on the circle of radius EK and center K. Thus, given lengths in the ratio $\sqrt{3} : \sqrt{2}$, to accomplish the construction we need only take the shorter length as KB, draw the circle having the same length as radius and center at K, draw the perpendicular bisector of KB, and then so place a line passing through B that the portion of it intercepted between the circle and the perpendicular bisector equals the larger of the given lengths (see Fig. 5b). In this way, both the placement and the length of EZ will be determined, and from it the rest of the figure can be completed.

The latter sets out precisely the same form in which Hippocrates accomplishes the construction of case (iii) of the lunules. In particular, one perceives through this analysis how natural it is to use the placement of EZ for effecting this. In Hippocrates' terms, "the line *inclining* (*neuousa*) toward B and intercepted between the line DG and the arc AEB shall have the given length."[74] This is the earliest known instance of the use of the constructing technique called *neusis* by the Greeks. As we shall later see, it plays a prominent role in many of the problem-solving efforts by Archimedes and his followers a century and a half after Hippocrates.[75]

Having so constructed the lunule, Hippocrates next proves that it is equal to the associated rectilinear figure EZHBK; for since $3 \text{ EK}^2 = 2 \text{ EZ}^2$, the three outer segments will equal the two inner ones. Since, furthermore, "the *mēniskos* (lunule) consists of the three segments together with that part of the rectilinear figure apart from the two segments,"[76] the lunule and the rectilinear figure are equal. Hippocrates goes on to show that the outer bounding arc of this lunule is less than a semicircle. At this point Eudemus injects an amazing remark:

> In this manner Hippocrates squared every lunule, since in fact (*eiper*) [he squared] the [lunule] of a semicircle and the one having the outer arc greater than a semicircle and the one [having the outer arc] less than a semicircle.... But he squared a lunule together with a circle as follows.[77]

What are we to make of "*every* lunule"? Does it signify merely the three particular cases just examined? But in context a more general sense seems to be indicated, since one moves at once to the consideration of the case pertaining to the circle quadrature. Then has Eudemus erred in supposing that squaring one lunule or even a few is tantamount to squaring them all? If this is so, then did Hippocrates himself make the same error, thus bringing on himself Aristotle's charges of having produced a fallacious argument on the circle quadrature? But it is hard to imagine how the one responsible for these carefully ordered proofs on the lunules could then have committed such an elementary blunder in reasoning. Now, there is indeed a sense in which Hippocrates might claim to have established the quadratures of a whole class of lunules, namely those for which the *m* similar segments along one bounding arc equal the *n* similar segments along the other arc. Although, of course, he has not produced the general *construction* of these figures, he can assume that their *quadratures* are obvious by virtue of the three cases given. If this is what Hippocrates intended by asserting that "every lunule" had been squared, we would infer that Eudemus, and doubtless others before him, had misconstrued his claim as referring to every possible form of lunule.

As for case (iii), so also for (iv), the synthetic mode adopted by Hippocrates conceals its essential line of thought. But it emerges if one views this case in relation to case (b) in Alexander's version, by analogy with the relation of cases (i) and (a). If we start with the hexagon inscribed in a circle, instead of using either its side or its diameter as base, let us introduce the chord connecting the two nonadjacent vertices H and I. (See Fig. 6.) Thus, HI is the side of the

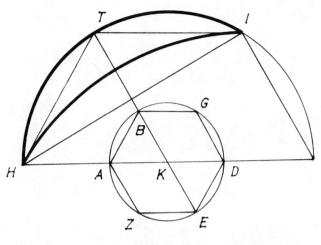

Figure 6

inscribed equilateral triangle, while HT is the side of the hexagon, so that HI : HT = $\sqrt{3}$: 1. If we draw over HI a segment similar to each of those over HT, TI, it will equal three times either of those segments. In the familiar manner, it follows that the triangle HTI equals the lunule together with one of those smaller segments. Thus, squaring the lunule reduces to squaring that segment. Hippocrates conceives of a set of segments similar to it and equal to it in sum; these are the segments on the sides of a hexagon inscribed in a circle whose diameter has to that of the initial circle the ratio 1 : $\sqrt{6}$. The segment over HT thus equals the sum of all six segments on the sides of the hexagon ABGDEZ. Adding to both the area of the hexagon and then combining with the earlier result, we find that the triangle together with the hexagon is equal to the lunule plus the smaller circle. In Eudemus' text the construction begins with the drawing of the two circles, so that the motivation behind the choice of 1 : $\sqrt{6}$ as the ratio of their diameters remains unclear until well into the proof. Having established the equality just stated above, he concludes:

> Thus, if the rectilinear figures [i.e., the triangle and the hexagon] mentioned can be squared, so also can the circle with the lunule.[78]

This ends the quotation from Eudemus; for Simplicius observes in the very next sentence:

> One ought to place greater trust in Eudemus [than in Alexander] to know things pertaining to Hippocrates of Chios, for he was nearer to his times, being a disciple of Aristotle.

It is clear in Eudemus' account that Hippocrates makes no presumption of having squared the circle itself, but only the circle in combination with the lunule. This

36 Ancient Tradition of Geometric Problems

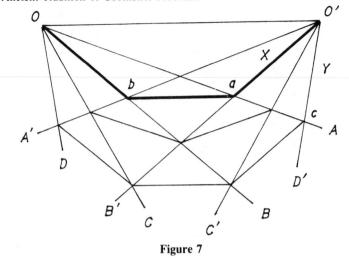

(a)

Figure 7

is so clear, in fact, that Simplicius questions the view that Hippocrates intended this here at all, and so suggests that when Aristotle speaks of his fallacious quadrature of the circle "via segments," he must refer to an entirely different argument.[79] Far more likely, however, is that Hippocrates presented this last case of lunule as a successful *reduction* of the circle quadrature, without its being a *resolution* of that problem. He would thus leave for later research the investigation of this lunule. If some followers rashly asserted that the circle had indeed been squared in this way, we have no grounds at all for charging this error to Hippocrates himself or even for suggesting that he wished slyly to leave this impression.

Aside from Hippocrates' attitude toward the circle quadrature, we should consider what his view of constructions was. It will be helpful to consider first the more general class of lunules and certain modern contributions to their study. At each end of the given line segment OO' let there be drawn a system of equally spaced rays, OA, OB, OC, OD, ..., and O'A', O'B', O'C', O'D', ..., respectively, where in each case the angle θ separating consecutive rays is the same. (See Fig. 7a.) Then we may form a polygonal arc from O to O' by selecting the intersections of appropriate rays from each system. For instance, if point a is the intersection of rays OA, O'B' and point b is that of OB, O'A', then the broken line O'abO will be a polygonal arc consisting of three equal lengths; for if we pass through them the circular arc joining O and O', each length will be a chord subtending angles equal to θ relative to the points O and O'. Thus, the circular segments having these chords as base will be similar to each other. (See Fig. 7b.)[80] In analogous fashion, we can construct between O, O' another polygonal arc having as many equal elements as desired.[81] As before, we obtain a set of circular segments similar to each other and to the segments associated with any other such polygonal arc constructed via the same initial sets of rays. Let two such arcs be taken, the inner having n elements, the outer having m elements. Denoting by X a side of the inner arc (e.g., O'a) and by Y a side of the outer

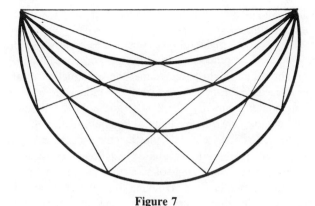

(b)   Figure 7

arc (e.g., 0'c), and applying the identity of sines to the triangles OO'a and OO'c, we have $X : Y = \sin m\theta : \sin n\theta$. The condition that the lunule bounded by the two corresponding circular arcs be squarable in Hippocrates' sense is that $nX^2 = mY^2$, so that $\sin m\theta : \sin n\theta = \sqrt{m} : \sqrt{n}$. If the general case is denoted by the pair (n, m), then the three cases examined by Hippocrates are (1, 2), (1, 3), and (2, 3). These all happen to be constructible in the familiar "Euclidean" sense, that is, by means of compass and straightedge alone. That two other cases, (1, 5) and (3, 5), are of the same sort was noted by Wallenius (1766) and Clausen (1840)[82]; for one can readily see that their defining relations reduce to a quadratic in $\cos 2\theta$. Furthermore, researchers have established that the lunule $(1, p)$ is *not* constructible in this manner when p is a prime not of the Gaussian type (that is, of form $2^{2^k} + 1$) (Landau, 1903), nor is the lunule (n, p) so constructible for $n = 1, 2, \ldots, p - 1$ (Tschakaloff, 1929).[83] Moreover, even for Gaussian prime p (other than 2, 3, and 5), the lunule $(1, p)$ is not constructible (Wegner); furthermore, apart from the cases already known to be constructible, the lunule (n, m) is not constructible when n, m are both odd integers (Tschebotaröw, 1935).[84]

For our purposes now, it is important to realize that this modern conception of the problem of constructing the lunules is anachronistic as far as Hippocrates' study is concerned. For we recall that his construction of lunule (2, 3) depended on a *neusis*, that is, the suitable inclination of a line segment of given length. As it happens, an alternative mode for effecting the *neusis* he requires is possible via planar methods (i.e., compass and straightedge), for it may be expressed as a quadratic relation falling within the class of problems known as the "hyperbolic application of areas," so that a solution can be worked out on the model of Euclid's *Data*, Prop. 59, based on *Elements* II, 6 and VI, 29.[85] While it is a reasonable conjecture that Hippocrates had access to these techniques, nothing in our text directly supports that claim. Certainly, his manner of introducing the *neusis* gives no hint of requiring us to effect this step by any means other than the obvious: the manipulation of the marked line EZB.[86] Indeed, even later geometers like Archimedes freely admit *neuses* in their constructions, where sometimes, as in *Spiral Lines*, Prop. 5, an alternative "Euclidean" method is possible.[87] Thus, I would maintain that Hippocrates' goal was to find any construction he could. If he proceeded from the general conception of the configuration given above, he might perceive that the construction of any lunule (n, m) satisfying the condition $X : Y = \sqrt{m} : \sqrt{n}$ reduces to a determination of the appropriate angle θ. Initially (for $\theta = 0$) the ratio $X : Y$ equals $m : n$. By experience of actually constructing figures, he could realize, if not prove, that as θ increases the ratio decreases, and hence that the specific configuration sought is possible.[88] In principle, his admission of *neuses* as a constructing technique makes accessible to him a wider range of cases than just those of the "Euclidean" type, including, for example, cases reducing to cubic relations.[89] Thus, the fact that he presents only the three squarable cases indicates that neither he nor any other ancient geometer who tried his hand at this problem was able to come up with any cases besides these. In this instance, then, not having recourse to the later algebraic methods in geometry constituted a real barrier against the discovery of solutions to these problems.

As noted above, Simplicius assigns greater weight to Eudemus' text than to the one transmitted by Alexander, believing that the former more closely represents the work of Hippocrates himself. Now, from the purely *textual* point of view, he seems quite correct in maintaining this. In certain contexts, for instance, Eudemus adheres to older terminology discontinued in the tradition after Euclid: he speaks of "powers" (*dynameis*) instead of "squares" (*tetragōna*), and uses expressions like "the line *on which (eph' hēi)* AB" about as frequently as their short forms (e.g., "the line AB," or simply "the AB"), while only the latter are found in the later formal tradition.[90] By contrast, Alexander's text is cast in full conformity with Euclidean usage. On the other hand, from the *technical* viewpoint, Alexander's constructions of lunules (a) and (b) are rather simpler than the analogous cases (i) and (iv) of Eudemus; and the technical refinement evident throughout the proofs of Eudemus' four cases clearly surpasses the treatments in Alexander. It is of course conceivable that Alexander's text originated from an effort to simplify an earlier version in the manner of Eudemus. But we recall that in his passage on "reduction," Aristotle alludes to a quadrature set in the form of Alexander's case (b), rather than Eudemus' case (iv). This affirms that Alexander's version too derives ultimately from a 4th-century source. Its simpler treatment of the lunules well suits what one would expect of the early phases of such a study; and as we have seen, it assists considerably in retrieving the motivations underlying the more elaborate constructions given by Eudemus.

This raises the question of the provenance of Eudemus' text. Much depends on Eudemus' reliability and sophistication as a historian. While we possess far too little of his work to make a secure judgment on this matter, it would seem that the purposes occasioning his account of Hippocrates would be admirably served through a fair transcript of his source materials, while Eudemus' motives for producing a radically modified account, indeed even his competence to do so in such a detailed technical context, would be entirely unclear. Can we then assume that Eudemus' text reflects with reasonable accuracy an actual writing of Hippocrates? There is of course no difficulty in supposing that Hippocrates' initial version of the study of lunules was in the simpler form indicated by Alexander, and that he later produced the more refined version recorded by Eudemus. On the other hand, it is equally plausible that the latter version grew out of the researches of followers inspired by his efforts to pursue this study. Doubtless, Eudemus was in the position to distinguish between a genuine writing of Hippocrates and a later writing in the tradition instigated by him. But would he insist on making this distinction clear in his discussion of the quadratures? We can hope that he would, yet must admit that it is possible he did not.[91]

These considerations affect one's use of the fragment as an index of the form and terminology of 4th-century mathematical writing, for we cannot with certainty locate it closer to the time of Hippocrates, toward the beginning of that century, than to the time of Eudemus, nearer its end. Nevertheless, its reliability as a testimony of Hippocrates' findings seems secure, for we have Eudemus' explicit statement several times in the fragment that the four cases of lunules were all studied by him, and it is hard to suppose that he could have been mistaken in this regard. Furthermore, the proofs of the quadratures, even in the simpler version reported by Alexander, indicate advances in the geometric tradition around Hippocrates' time; access to a wide range of theorems and constructions of rectilinear and circular figures, results of the geometric theory of proportions, and so on, as well as appreciation of the basic concerns in the deductive ordering of proofs. But regardless of whether Hippocrates did indeed work out the impressively formalized synthetic treatments given by Eudemus, rather than, say, a more loosely framed analytic treatment, his contribution to the study of these geometric problems is unquestionable.

PROBLEMS AND METHODS

This survey of the earliest efforts within the ancient tradition of problem solving has revealed the central importance of the work of Hippocrates of Chios. In particular, his studies of the cube duplication and of the quadrature of the circle and related figures laid the foundation for later researches. According to Proclus, he was also responsible for the first compilation of "Elements," that is, of a systematized presentation of propositions and proofs.[92] From the surviving fragments of his work, we may infer that this included substantial materials on the geometry of circles (cf. Euclid's Book III) as well as an exposition of the properties of plane figures dominated by the techniques of proportion theory (cf. Book VI), even if of course there were still lacking the more rigorous methods of limits and proportions (comparable to Books V and XII) due to Eudoxus.

Although stories like the Delian oracle on the doubling of the cube are often cited as suggesting an external motivation for the early studies of such problems, our examination of Eratosthenes' account of them and our inquiry into Hippocrates' work serve to discount this view. On the contrary, the interest in these problems can be seen as a fully natural outgrowth of researches within the geometric field. For instance, having cast the problem of squaring a given rectangle into the form of finding the mean proportional between its length and width, Hippocrates could be led to perceive that the finding of two mean proportionals is equivalent to increasing a given cube in a given ratio. In this way, his principal contribution to the study of the cube-duplication problem follows from the same techniques which characterize his study of plane figures. This study of quadratures leads readily to the further question as to whether a comparable construction for the circle might be produced. Here, of course, the quest for a solution fails, but not before the intriguing detour into the construction and quadrature of the lunules. To be sure, only three cases are actually found, although Hippocrates was surely aware of the defining conditions for an unlimited class of quadrable cases. The tantalizing result of having squared a certain lunule in combination with a circle would of course deceive no one into supposing that the circle itself had been squared. But in all these instances an important principle has been established: that curvilinear figures, specifically those associated with circular arcs, are not different in kind from rectilinear figures as far as their quadrability is concerned. Thus, when the later commentator Ammonius (5th century A.D.) presumes to have affirmed the impossibility of the circle quadrature through the intrinsic dissimilarity of the circular and the rectilinear, by analogy with the noncomparability of curvilinear and rectilinear angles, his disciple Simplicius brings forward the quadratures of the lunules as an immediate counterexample.[93]

In the matter of the lunules, the standard assumption is that Hippocrates consciously restricted the means allowable in their construction to the "Euclidean" devices of compass and straightedge. Although the cases he presents are indeed constructible by these means, the view flies in the face of his unexplicated introduction of a *neusis* in the construction of his third lunule. While of course the compass and straightedge were by this time long familiar as constructing instruments, and while the range of constructions effectible by these means was doubtless already appreciated to be quite extensive, nevertheless other means were then known, and still others would later be invented. The enterprise of discovering the solutions to problems could hardly be well served by the imposition of such a restriction at this early stage. Hippocrates must surely have been interested in finding constructions, using whatever means were available to him. The explicit restriction to one or another mode of construction is by its nature primarily a formal move, motivated by the urge to divide and classify the collected body of established results. Until the geometric corpus had attained a size and diversity meriting such efforts, there could hardly be much sense in engaging in these formal inquiries. While that level surely was reached around the time of Apollonius, it is open to debate whether it had already reached that level at the time of Euclid.[94] Applying this notion to the much earlier time of Hippocrates must be viewed as purely anachronistic.

There is a subtle distinction to be made here. The activity of seeking the solutions to geometric problems is intrinsically formal, to the extent the constructions are to be provided within an implied context that imposes some restrictions on the available means and some measure of deductive ordering in the justifications. Thus, if one can rightly portray the most ancient traditions as empirical in that physical measures might be acceptable to justify certain claimed geometric results, then one cannot assign to these an involvement in problem solving. For in this environment any figure will be squarable, for instance, to within any desired perceptible degree of accuracy. By virtue of addressing the study of geometry in abstraction from overtly empirical measurements, the early Greek geometers would soon come upon configurations whose production proved intractable. For instance, the range of techniques known for the construction and study of similar rectilinear plane figures would be recognized as leading to no evident solution for the cube duplication or the circle quadrature. It is surely no mere coincidence, then, that Hippocrates was the first known systematizer of geometry as well as the first to formulate these constructions as problems, while such an awareness among earlier geometers, even Oenopides, is quite dubious.

On the other hand, the conscious restriction to a specific set of constructing techniques, like the compass and straightedge in sharp separation from others, would be premature at Hippocrates' time. This, I believe, may explain why the problems of rectifying and dividing circular arcs did not seem to attract interest until a century or more later. For a tradition of geometric practice which includes forms of protractors and cords among its tools might readily admit their abstract analogues in the more formalized study of figures, just as the use of sliding rulers gives rise to *neuses* and the use of compasses suggests the Euclidean postulate of circles.[95] Thus, the trisection of the angle became a problem only after a tightening of the restrictions on construction techniques, and hence only within a formally more sophisticated geometric field. By contrast, the problems of the cube and the circle present difficulties at the much earlier level of development. Indeed, one may anticipate that it was through the investigation of these and related problems that the clearer conceptions of the general nature of the problem-solving enterprise were to emerge.

NOTES TO CHAPTER 2

[1] Proclus, *In Euclidem*, ed. Friedlein, pp. 283, 333. On Oenopides and Hippocrates, cf. T. L. Heath, *History of Greek Mathematics*, I, pp. 174–202 and the articles in *Pauly Wissowa* and the *Dictionary of Scientific Biography*.

[2] Heath, *op. cit.*, p. 176; cf. Á. Szabó, *Anfänge der griechischen Mathematik*, III.19. I criticize this view in my "On the Early History of Axiomatics," p. 150.

[3] O. Neugebauer, *Mathematische Keilschrift-Texte*, I (*Quellen und Studien*, 1935, 3 : A), pp. 137–142. On the "rope stretchers" see Heath, *op. cit.*, pp. 121ff, and p. 178, where the term appears in connection with the mathematical talents of Democritus.

[4] This view is advocated with particular zeal by Á. Szabó, as in the work cited in note 2 above.

[5] Oenopides is reputed, for instance, to have discovered the obliquity of the ecliptic;

see Heath, *op. cit.*, p. 174. This is of course merely a testimony of the infancy of astronomical science among the Greeks in the 5th and 4th centuries, for such facts were certainly long familiar among the Mesopotamians. Of course, the specifically geometric approach to astronomy is characteristic of the Greek style, as contrasted with the arithmetic approach of the Mesopotamians. In this sense, the obliquity might indeed have been a new feature with Oenopides.

[6] Some of the instruments used in the practical work of geometry and architecture are named by Aristophanes (e.g., *Birds*, 1001 ff) and Plato (e.g., *Philebus* 51c, 56b, 62b). Among later developments one may note the elaborate sighting device described by Hero (in his *Dioptra*, *Opera* III) and the compass for drawing the conic sections devised by Isidore of Miletus (see the interpolation into Eutocius' commentary on Archimedes' *Sphere and Cylinder*, in Archimedes, *Opera*, ed. Heiberg, 2nd ed., III, p. 84). An instrument of the latter kind is described by al-Qūhī and other Arabic geometers; see the account by F. Woepcke, *L'Algèbre d'Omar*, 1851, p. 56n. In the same class may be mentioned the various mechanisms for the cube duplication used by "Plato," Eratosthenes, and Nicomedes (cf. Eutocius, *op. cit.*, pp. 56 ff, 90 ff, 98 ff; these are discussed in Chapters 3 and 6). A survey of the devices familiar in the ancient practical geometry is given by O. A. W. Dilke, *The Roman Land Surveyors*, Ch. 5. On the ancient architectural instruments of stone dressing, in particular, the set square, compass, ruler, and plumb line, see A. Orlandos, *Les Matériaux de construction et la technique architecturale des anciens grecs*, École française d'Athènes: *Travaux et Mémoires*, 16, pt. 2, 1968, pp. 59–69.

[7] See, for instance, Heath, *op. cit.*, I, pp. 218 ff; Becker, *Mathematisches Denken der Antike*, pp. 74 ff.

[8] Theon, *Expositio rerum utilium ad legendum Platonem*, ed. Hiller, p. 2. Plutarch, *Moralia* 386e, 579b. These passages have already been cited in Chapter 1.

[9] Cf. Archimedes, *Opera*, III, pp. 88–96.

[10] See, for instance, Heiberg's note in *Archimedes*, III, p. 89n; Heath, *History*, I, pp. 244 f; I. Thomas, *Greek Mathematical Works*, I, p. 256n; B. L. van der Waerden, *Science Awakening*, pp. 160 f. They all merely refer to von Wilamowitz as having established the inauthenticity of the document; see note 12 below.

[11] Eutocius in *Archimedes*, III, pp. 88–90 (i), 90 (ii), 90–92 (iii), 92–94 (iv), 94–96 (v).

[12] "Ein Weihgeschenk des Eratosthenes," *Göttinger Nachrichten*, 1894; repr. in *Kleine Schriften* II, 1971, pp. 48–70.

[13] *Collection*, ed. Hultsch (III) I, pp. 56–58.

[14] Eutocius, *op. cit.*, pp. 90, 96.

[15] See Fig. 1 (a-b) in Chapter 6 for Eratosthenes' construction.

[16] Eutocius, *op. cit.*, pp. 90, 96.

[17] For a gauge of the longevity of such a structure, one may consider the grave marker of Archimedes, all but lost within a century and a half of its foundation; the circumstances of Cicero's rediscovery of the monument in the 1st century B.C. are related in his *Tusculan Disputations*, V, 64ff.

[18] Recall that Hypatia, the learned daughter of Theon of Alexandria, died in 415 A.D. at the hands of a mob of rioting monks; cf. Heath, *History*, II, pp. 528 f, and P. Brown, *The World of Late Antiquity*, London, 1971, pp. 103 f.

¹⁸ᵃ See A. Fitzgerald, *The Letters of Synesius*, pp. 258–266 (esp. pp. 262–266); Fitzgerald provides a useful discussion of Synesius' career in the Introduction and notes.

¹⁹ Cf., in particular, the prefaces to *Sphere and Cylinder* II and to *Conoids and Spheroids*; those to *Quadrature of the Parabola*, *Sphere and Cylinder* I and *Spiral Lines* provide a certain small amount of contextual information.

²⁰ Indeed, G. J. Toomer expresses misgivings over the general assumption that Alexandria was a monopolizing center of technical studies in this period; cf. his *Diocles*, p. 2. One notes that the prefaces to works by Diocles and by Apollonius are somewhat more ample in background information than those to the Archimedean writings.

²¹ *Opera*, II, pp. 216–258. For accounts, see the Archimedes monographs by Heath, by Dijksterhuis, and by Schneider.

²² In the 43 pages of the standard edition of the *Sand-Reckoner*, one finds about a dozen second-person forms (e.g., pp. 216, 218, 220, 234, 244, 246, 258); in Eutocius' text of Eratosthenes, a five-page document, there are three (pp. 90, 94, 96), not counting the epigram.

²³ For an account of the general background, see M. I. Finley, *Ancient Sicily*, New York, 1968, Ch. 9. I apply these materials toward the dating of the *Sand-Reckoner* in my "Archimedes and the *Elements*," pp. 234–238.

²⁴ Eutocius, *op. cit.*, pp. 88–90; cf. Heath, *History*, I, p. 245. I have inserted the reference letters.

²⁵ On Archytas, see Heath, *op. cit.*, pp. 213–216.

²⁶ Eutocius, *op. cit.*, p. 84.

²⁷ *Ibid.*, p. 96.

²⁸ *Expositio*, ed. Hiller, p. 2.

²⁹ *Moralia*, 386e, 579b–d. Cf. note 8 above.

³⁰ *Ibid.*, 718e f; *Vita Marcelli* xiv, 5. The latter contains an extended account of the life and exploits of Archimedes; see the Archimedes surveys cited in note 21 above.

³¹ Details appear in Chapter 3.

³² Heath, *History*, I, p. 246; van der Waerden, *op. cit.*, pp. 161–163. The latter seems to treat the story as a dramatization by Eratosthenes; but in a more recent effort, he wishes to treat it as a serious datum reflecting a ritual tradition within early mathematics; see his "On Pre-Babylonian Mathematics (II)," *Archive for History of Exact Sciences*, 1980, 23, pp. 37 f. This thesis is advocated by A. Seidenberg in several articles, particularly, "The Ritual Origin of Geometry," *Archive for History of Exact Sciences*, 1963, 1, pp. 488–527.

³³ Eutocius, *op. cit.*, p. 88; to be discussed further below.

³⁴ Note in particular Plato's criticisms of geometric research and his recommendations for mathematical education in *Republic*, Books VI–VII.

³⁵ Cf. *Post. Ana.* I, 7 and *Meta.* M.3. Passages on the ordering and nature of the sciences are collected and discussed by Heath, *Mathematics in Aristotle*, pp. 4–12, 46, 59 f, 64–67, 225.

³⁶ Eutocius, *op. cit.*, p. 88.

³⁷ *Ibid.*, pp. 88–90.

[38] A form of the proof, applied to the more general case of similar solids, appears in Euclid's *Elements* XI, 33 (porism).

[39] *In Euclidem*, p. 213. On the method of analysis, see Chapter 3.

[40] *De Anima* II, 2, 413 a 16–20. This passage is noted, if briefly, in most editions of the work; cf. W. D. Ross, Aristotle's *De Anima*, Oxford, 1961, p. 217, and also Heath, *Mathematics in Aristotle*, pp. 191–193. Most commentators, however, attempt to construe it as a specific reference to the Euclidean constructions of either or both of the two problems, rather than as a statement of the reduction of the one to the other.

[41] This fragment, preserved in Simplicius' commentary on Aristotle's *Physics*, is discussed further in this and the next chapter.

[42] On the ancient Egyptian and Mesopotamian mathematical traditions, see O. Neugebauer, *Exact Sciences in Antiquity*, Ch. 2, 4; and van der Waerden, *op. cit.*, Ch. 1, 2. On the thesis of the ritual origins of technique, see note 32 above. Establishing a perspective on the relation of geometry and philosophy in the pre-Euclidean period is an aim of Chapter 3 below.

[43] *In Euclidem*, pp. 422 f. Proclus cites as an example of a theorem on the circle quadrature Archimedes' *Dimension of the Circle*, Prop. 1; see Chapter 5.

[44] Problem 48; cf. the edition by A. Chace *et al.*, 1927–29, II, Plate 70. For a discussion, see R. Gillings, *Mathematics in the Time of the Pharaohs*, 1972, pp. 139–146, and H. Engels, "Quadrature of the Circle in Ancient Egypt," *Historia Mathematica*, 1977, 4, pp. 137–140. The rule is equivalent to approximating π as 256/81, that is, 3 13/81 or a bit less than 3 1/6

[45] That this relation was recognized is indicated by the use of the correct rule for the volume of the truncated pyramid in the Moscow Papyrus (c. 1600 B.C.); see van der Waerden, *op. cit.*, pp. 34 f. For this would seem to require knowledge of the 1 : 3 ratio between the pyramid and the corresponding parallelepiped. Physical procedures for the measurement of irregular surfaces and solids are found in the Greek metrical tradition; cf. Hero, *Metrica* I, 39 and II, 20.

[46] *The Birds*, lin. 1001–1005; cf. Heath, *History*, I, pp. 220 f and I. Thomas, *op. cit.*, pp. 308 f. Thomas translates *kampylos kanōn* (literally, "curved ruler") as "flexible rod," and this may be supported in the light of Aristotle's mention of the "leaden rule" used by architects (*Nic. Eth.* 1137 b 30) and by Diocles' use of a flexible ruler of horn for the purposes of curve plotting; see Toomer, *Diocles*, pp. 159 f. H. Mendell informs me of an alternative rendering of the passage which conforms better both to sense and to the standard punctuation: "If I lay out the ruler and insert this curved compass from above—do you see?...I shall measure with the straight ruler by laying (it) out..." In this way, "curved" modifies "compass" (cf. Aristophanes, *Clouds*, 178) rather than "ruler". This of course does not affect the later testimonia to the use of flexible rulers in geometric practice.

[47] Cf. Heath, *loc. cit.* The passage itself gives no fully clear notion of just *what* Meton is up to, and perhaps this is part of the joke.

[48] *Physics* 185 a 16; *Sophistical Refutations* 171 b 15; cf. Heath, *Mathematics in Aristotle*, pp. 33–36.

[49] Simplicius, *In Physica*, ed. Diels, I, p. 60. The fragment from Eudemus occupies pp. 61–68, followed by an assessment on pp. 68 f; Simplicius takes up other versions of the circle quadrature on pp. 53–60.

[50] *Collection* (V), I, pp. 340–342. See my discussion in "Infinity and Continuity," pp. 127 ff.

[51] On this terminology, see my "Archimedes and the *Elements*," pp. 240n, 254n, 264; and my "Archimedes and the Pre-Euclidean Proportion Theory," p. 196.

[52] On the Eudoxean methods and their attribution via Archimedes' *Quadrature of the Parabola* and *Sphere and Cylinder* I, see my "Archimedes and the Pre-Euclidean Proportion Theory," pp. 194 ff.

[53] For a general account of Antiphon's thought, see W. K. C. Guthrie, *A History of Greek Philosophy*, III, pp. 285–294. Simplicius' passage on his circle quadrature (*In Physica*, pp. 54 f) would appear to derive from Eudemus, for his specific critique of this argument is cited (*ibid.*, p. 55). The same critique, if without express citation of Eudemus, is given also by Themistius (4th century A.D., *In Physica*, ed. H. Schenkl, *CAG* 5, Pt. II, 1900, pp. 3 f), namely, that it violates the principle of the unlimited divisibility of continuous magnitude. But Simplicius also cites an alternative critique by Alexander, so that he may well have derived his account of Antiphon via Alexander, rather than directly from Eudemus.

[54] Simplicius, *op. cit.*, pp. 54 f.

[55] *Ibid.*, p. 55 (see note 53 above). On Protagoras' treatment of this issue, see Aristotle's *Metaphysics* B, 2, 997 b 32 and its discussion by Guthrie, *op. cit.*, II, p. 486 and III, p. 267. On Democritus' work, see Heath, *History*, I, pp. 176–181 and Guthrie, *op. cit.*, II, Ch. viii, especially pp. 487 f.

[56] Cf. Eudemus' view, cited in note 53 above.

[57] See, for instance, Heath, *History*, I, p. 224.

[58] See my "Infinity and Continuity," pp. 133 f. The summation theorem for the finite case is given in *Elements* V, 12.

[59] Hippocrates seems to appeal to a concept of proper "parts" as the basis of his notion of ratio; cf. Simplicius, *op. cit.*, p. 61 and the discussion by Heath, *History*, I, pp. 187–191. This could provide the precedent for a theory of proportions for commensurable magnitudes; but the extension to incommensurable magnitudes, as in the Eudoxean theory, had not yet been worked out at Hippocrates' time. See my "Archimedes and the Pre-Euclidean Proportion Theory," and Chapter 3.

[60] Further examples are found in Pappus' measurements of figures bounded by spirals; see Chapter 5.

[61] Plutarch, *Moralia* 607 f (cf. Diels and Kranz, A 38, II, p. 14). The view that this might signify a *writing* by Anaxagoras on circle quadrature is discounted by J. Burnet (*Early Greek Philosophy*, 4th ed., p. 257) and by Guthrie (*op. cit.*, II, p. 270).

[62] *In Euclidem*, pp. 65 f. Vitruvius also mentions Anaxagoras for a study of perspective; cf. Heath, *History*, I, pp. 172-174. For a general account of Anaxagoras, see Guthrie, *op. cit.*, II, Ch. iv.

[63] *Prior Analytics* II, 25; cf. the passages cited in note 48 above.

[64] See note 49 above.

[65] Simplicius, *op. cit.*, pp. 56 f.

[66] This readily suggests an alternative construction in which the right triangle may be scalene rather than isosceles. Here too the two lunules will sum to the area of the triangle.

46   Ancient Tradition of Geometric Problems

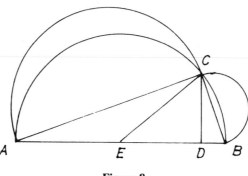

Figure 8

Simplicius' text does not mention this variant, but it has slipped into modern accounts as if by Hippocrates himself. M. Simon notes having failed to discover this variant in works prior to the late 17th century ("Lunulae Hippocratis," *Archiv für Mathematik und Physik*, 1904–05, 8, p. 269). But a medieval precedent may be found in treatments by ibn al-Haytham and Joannes de Muris; cf. M. Clagett, *Archimedes in the Middle Ages*, III, pp. 1315n, 1321f. One notes that since the small segments over the sides AC, CB are not similar, neither are the lunules; hence the lunules will not be in the ratio of the similar triangles ACD, DCB, respectively, but rather, lunule AC will be less and lunule CB greater than its corresponding triangle. Ibn al-Haytham observes that the lunules will be equal to the respective triangles ACE, ECB (where E is the center of the circle) when the arcs AC and CB are equal. The relation is of course not general, for in the scalene case the lunule over the greater leg will be greater than that over the smaller leg; the maximal discrepancy occurs for central angle CEB having sine equal to $2/\pi$ (approx. 39½°), where the lunules will be in the ratio 3.1616... : 1. [This is the configuration shown in Fig. 8.]

[67] For a review of modern interpretations, particularly those of Tannery and Björnbo, see Heath, *History*, I, p. 196n and Thomas, *op. cit.*, I, p. 310n. The opinion of Alexander is cited by Simplicius, *op. cit.*, pp. 57 f, 60, while his own view appears on pp. 60, 69.

[68] For full accounts, see Heath, *History*, I, pp. 191–200 and Thomas, *op. cit.*, I, pp. 234–253. A major interpretive issue here centers on the segregation of Eudemus' text from Simplicius' interpolations. Proposals have been offered by Allman, Diels (1882), Tannery (1883), Rudio (1907), and Becker (1936). For references, see Thomas, *op. cit.*, p. 237n. Rudio's translation and commentary contain many useful insights ("Bericht des Simplicius...," *Bibliotheca Mathematica*, 3_3, 1903, pp. 7–62); see also my "Infinity and Continuity," pp. 127–130. Thomas follows Rudio's abridgment, so that the reader interested in judging for himself must go back to Diels' text.

[69] This is done by applying the principle that in right triangles the square on the longest side equals the sum of those on the other two sides, but in obtuse triangles it is greater and in acute ones less; cf. *Elements* I, 47 and II, 12–13. In the present case, since angle BAG is obtuse, $BG^2 > BA^2 + AG^2 = 2 GD^2$, whence $BD^2 (= 3 GD^2) < BG^2 + GD^2$, so that angle BGD is acute. Hence, arc BGD is greater than a semicircle, since the angles inscribed in segments are proportional to the corresponding supplementary arcs (cf. Simplicius, *op. cit.*, p. 60).

[70] Specifically, this text includes archaic usage of the form "the line *on which* AB," where later writers would invariably say "the line AB"; see below, note 90.

[71] Signs of such minor textual corruption appear elsewhere, for instance, in the treatment of case (iii) where certain steps in the construction are out of the proper order (Simplicius, *op. cit.*, p. 65); cf. Thomas, *op. cit.*, p. 244n.

[72] On the method of "analysis and synthesis," standard in the later problem-solving tradition, see Chapter 3.

[73] For if one introduces line HZK, the angles at K and B will be equal; hence, Z is the vertex of the isosceles triangle ZKB.

[74] Simplicius, *op. cit.*, p. 64.

[75] See, in particular, Chapters 5, 6, and 7. A survey of ancient *neuses* is given by R. Böker in *Pauly Wissowa*, Suppl. ix, 1962, col. 415–461. Cf. note 86 below.

[76] Simplicius, *op. cit.*, p. 66. The text adds, superfluously, "and the rectilinear figure is [the meniscus] with the two segments, but without the three."

[77] *Ibid.*, p. 67. Writing "[he squared]" after "seeing that" (or "since in fact," in my rendering of *eiper kai*), Heath assumes an ellipsis, and he is well supported in view of the grammatical role of *eiper* as a subordinating conjunction. This entails, however, the fallacy of concluding the general from the particular. Wishing to avoid this, Tannery and Björnbo have proposed excising "every" and "seeing that" as interpolations (see Heath, *loc. cit.*).

[78] Simplicius, *op. cit.*, p. 68.

[79] *Ibid.*, p. 69; cf. Aristotle, *Physics* 185 a 16 (note 48 above).

[80] Note that lines OC and O'C' will be tangent to the circular arc.

[81] This is under the restriction that θ be less than $180/m$, for m the number of elements in the polygonal arc.

[82] Cf. Heath, *History*, I, p. 200.

[83] E. Landau, "Ueber quadrierbare Kreisbogenzweiecke," *Sitzungsberichte*, Berlin Math. Ges., 1902–03, pp. 1–6 (suppl. to *Archiv für Math. und Phys.*, 4, 1902–03); cited by F. Enriques (ed.), *Fragen der Elementargeometrie*, 1923, II, pp. 304–308. Tschakaloff, "Beitrag zum Problem der quadrierbaren Kreisbogenzweiecke," *Math. Zeitschrift*, 1929, 30, pp. 552–559; this and other efforts are cited by L. Bieberbach (cf. note 89 below), pp. 140 f, 159.

[84] References are given by A. D. Steele, "Zirkel und Lineal," *Quellen und Studien*, 1936, 3 : B, p. 317. The algebraic condition to which the case (1, 9) reduces has some constructible roots, but these are imaginary; its real roots are nonconstructible.

[85] See Heath, *Euclid*, I, pp. 386 f and Chapter 3 below.

[86] Steele provides an alternative planar construction of this *neusis*, following the model of a result from Apollonius' *Neuses* as discussed by Pappus (*op. cit.*, pp. 319–322). But Steele goes on to insist, on textual grounds, that the Hippocrates–Eudemus passage, as extant, could not have intended any such alternative method in substitution for the *neusis*.

[87] See Heath, *Archimedes*, pp. ci–ciii, who follows Zeuthen in supposing that Archimedes must have understood alternative constructions of the *neuses*. But Dijksterhuis (*Archimedes*, pp. 138 f) insists, I believe correctly, that Archimedes accepted the *neuses* in their own right; see Chapter 5.

[88] As θ continually decreases from $180/(m + n)$ to 0, the ratio X : Y continually increases from 1 : 1 to $m : n$. Since the ratio $\sqrt{m} : \sqrt{n}$ lies within this range, X : Y will assume this value for some angle within the corresponding domain of θ. One may observe that Hippocrates' *neusis* for the case (2, 3), when X = Y, yields a construction for the regular pentagon. Some have wished to assign a method of this sort to the ancient Pythagoreans; but, unfortunately, there is no documentary support for this claim (cf. R. Böker, "Winkelteilung," pp. 137 f).

[89] In this connection, Steele notes Vieta's treatment of the case (1, 4); see *op. cit.*, p. 318. *Neuses* may be used for the solution of third-order problems like the cube duplication and the angle trisection; see Chapters 5, 6, and 7. For an account of this construction technique, see L. Bieberbach, *Theorie der geometrischen Konstruktionen*, Basel, 1952, Ch. 16, 17.

[90] See note 70 above. The language of proportions here is also archaic. Eudemus' text speaks of the proportion A : B = G : D by the expression "A is greater than B by the same part that G is greater than D" (Simplicius, *op. cit.*, p. 60). This may be compared with the phrasing in Archytas (fr. 2; Diels–Kranz, 47 B 2, I, pp. 435 f) and in the pseudo-Aristotelian *Mechanics*, Ch. 20. By contrast, Euclid, Archimedes, and other late writers will say "A has the same ratio to B that G has to D," or more simply, "A is to B as G is to D."

[91] Note, for instance, that the commentators commonly speak of "those in the circle of..." (*hoi peri*), thus leaving it unclear whether they refer to the master of the tradition or to his followers. Eudemus himself was evidently willing sometimes to make historical claims on the basis of inference. He says of Thales, for instance, that he knew a certain theorem on congruent triangles (*Elem.* I, 26), for he "must have used this" for determining the distance of ships at sea, as he is reported to have done (cf. Proclus, *In Euclidem*, p. 352). The difficulty of discerning the actual basis of this and other claims by Eudemus has thus led to a spectrum of widely divergent views on the achievements of Thales. For references, see W. Burkert, *Lore and Science in Ancient Pythagoreanism*, 1972, pp. 415–417.

[92] Proclus, *In Euclidem*, p. 66.

[93] Simplicius, *op. cit.*, pp. 59 f. Ammonius here refers to the "horn angle" defined by the space between the circle and its tangent (cf. *Elem.* III, 16 and Heath, *Euclid*, II, pp. 39 ff). Proclus makes several references to this and related mixed and curvilinear angles (cf. *In Eucl.*, pp. 122, 127, 134). These figure within Philoponus' discussion of Bryson's circle quadrature (see Chapter 3).

[94] This will be discussed in more detail in Chapters 7 and 8.

[95] In particular, the availability of "curved (flexible) rulers" suggests an extremely simple practical procedure for these constructions; see note 46 above and Chapter 3.

CHAPTER 3

The Geometers in Plato's Academy

Through Hippocrates' efforts a start was made toward the investigation of geometric problems. But the first discovery of actual solutions for more difficult problems, like the circle quadrature and the cube duplication, was made by geometers in the generations after him. Hippocrates' reduction of the latter problem, for instance, to the form of finding two mean proportionals between two given lines served as the basis for the successful constructions by Archytas, Eudoxus, and Menaechmus. At the same time, new geometric methods were introduced, in particular the use of special curves generated through the sectioning of solids or through the geometric conception of mechanical motions. Furthermore, Hippocrates' precedent for the "reduction" of one problem to another could lead others to the articulation of the versatile technique of geometric "analysis."

During the 4th century B.C., Plato's Academy in Athens became a center for geometric studies. This is reflected, as we have seen, in such legendary accounts as that of the Delian oracle, portraying Plato's interaction with mathematicians. Although later authorities sometimes assign to him the discovery of specific technical results, like a method for the duplication of the cube, the veracity of such reports is dubious and generally discounted. In the reasonable assessment by Proclus, it was rather through the special position he afforded mathematical study within his program of philosophical education, elaborating a plan initiated by the older Pythagoreans, that Plato encouraged mathematical research, as well as through his incorporation of technical examples into his writings to illustrate points of method.[1] Plato appears to have been much impressed by the technical rigor of some of the older geometers, notably Theodorus of Cyrene, a contemporary of Socrates, and by the genial insights of geometers of his own generation, in particular the Pythagorean Archytas of Tarentum and Theaetetus of Athens.[2] Of other geometers, like Leon and Leodamas, we know little. But towards the middle of the century, Eudoxus of Cnidus became affiliated with the school, and he and his immediate disciples, like the brothers Menaechmus and Dinostratus, profoundly stimulated research into mathematics and mathematical science. It is natural to suppose that the formal precision characteristic of the geometric methods of Eudoxus developed in part as a response to the intellectual emphases of this philosophical environment. Interestingly, Eudoxus' tenure at the Academy coincided with that of the young Aristotle, so that the work of this geometer can hardly but have had its influence on the elaboration of the philosopher's views on logic and the formal structure of science.[3]

Recognizing these aspects of the context of research in this period will prepare us for its distinctive character. Even as the geometric field expanded impressively in the range of established technical results, there advanced parallel to this a rigorization of the forms of mathematical exposition. The latter has often been exploited as evidence of the influence of philosophical developments on mathematical research and has induced many scholars to propose dialectical reasons for explaining technical issues, like the ancient treatment of infinity or the selection of methods of construction. But such views are too readily exaggerated. Surely, the technical disciplines were on the whole autonomous in setting the directions of research. If in this period geometry and philosophy interacted to a greater degree than at other times in antiquity, the effect on geometry was subtle and not specific: an enhanced sensitivity to questions of formal proof. By contrast, the reciprocal effect on philosophy could often be specific, as mathematics offered clear models for inquiries into the general structure of knowledge. Failure to appreciate this distinction must inevitably hinder the effort to capture the nature and motives of technical work in the pre-Euclidean generations.

SOLUTIONS TO THE CUBE DUPLICATION

According to Eratosthenes, three of Plato's associates worked out solutions for the Delian problem: Archytas "via semicylinders," Eudoxus "via curved lines," and Menaechmus "via the triads cut out from the cone."[4] Among the dozen methods in his survey, Eutocius presents some remarks on a text of Eudoxus' method, along with complete details of the construction and proof of those of Archytas and Menaechmus, the former based on an account by Eudemus. In the present section we review these, adding a fourth method transmitted by Eutocius under the name of Plato.

The construction of Archytas is a stunning *tour de force* of stereometric insight.[5] Its essential idea, following the form of the reduction of the problem by Hippocrates, is to find the two mean proportional lines AI, AK between two given lines AM, AD via an arrangement of similar right triangles as shown in Fig. 1. Archytas determines point K as the intersection of three solids. Starting with a semicircle whose diameter AD' equals the larger given length and in

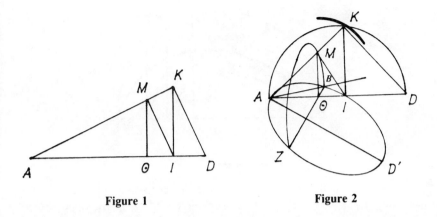

Figure 1   **Figure 2**

which the chord AB equals the smaller, he conceives an equal semicircle erected in the plane perpendicular to the given semicircle; as the diameter AD' turns on the pivot A, this circle will generate the surface of a torus (see Fig. 2). If a right semicylinder has been erected over semicircle ABD' as its base, then the torus will meet it along a certain curve. At the same time, he conceives that the given semicircle rotates about its diameter AD' as an axis, so that the extended chord AB generates the surface of a cone. Let this intersect the "cylindric curve" at K. Archytas then shows that this effects the construction: if one draws triangle AKD, the line KI perpendicular to AD, the line AM equal to AB, and the line Mθ perpendicular to AD, he can prove that the angle AMI is a right angle. Hence, the three triangles AMI, AIK, AKD are similar, so that the four lines AM, AI, AK, AD are in continued proportion.

The synthetic expository mode here adopted by Eutocius inevitably obscures the underlying pattern of thought. But an analysis along the following lines may suggest an approach available to Archytas. Since the objective is to find an arrangement of the triangle AKD such that AM has the given length, let the chord AK be conceived to rotate on A, and for each position of K let there be drawn KI perpendicular to AD and IM perpendicular to AK (see Fig. 3). Then M will trace a curve such that the ray AM continually decreases from its initial value AD, assuming arbitrarily small values in the vicinity of A. At some position, then, AM will equal the given line AB, and from this the solution of the problem follows.[6] Instead of effecting this via a pointwise construction of the locus of M, one can seek the common intersection of the loci associated with the points K, I, and M.[7] The locus of K, as the vertex of a right triangle of hypotenuse AD, is secured by keeping it on the circumference of the circle AKD (see Fig. 4). The length AM can be kept equal to AB by setting M on the trace of B as its semicircle rotates on its axis AD. As for the locus of I, this can be solved as follows: in the desired terminal position, we require that IM be perpendicular to AK, where I is figured as the foot of the perpendicular from K to AD. Now, M lies on the circle ZMB of diameter ZB; setting θ as the

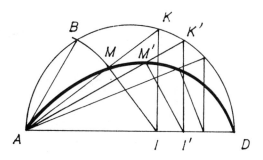

Figure 3

intersection of ZB with AI, Mθ is then perpendicular to both of those lines (for it lies along the intersection of the two semicircles AKD, ZMB, each perpendicular to the base plane which contains the lines AD, ZB). Since angle ZMB is inscribed in a semicircle, $M\theta^2 = Z\theta \cdot \theta B$; since it is required by hypothesis that angle AMI be a right angle, $M\theta^2 = A\theta \cdot \theta I$. Thus, $A\theta \cdot \theta I = Z\theta \cdot \theta B$, so that ZθB and AθI will be intersecting chords in the same circle (cf. *Elements* III, 35); that is, I lies on the circle ZABD′. Hence, one may take the semicircle ABD′ as the locus of I, so that the vertical KI will lie in the right semicylinder erected over ABD′ as its base.

Through an analysis of the sort just given, one sees that the desired configuration of similar right triangles can be arranged through the determination of K as the intersection of the torus, semicylinder, and cone. The construction and proof by Archytas, as transmitted by Eutocius, follows precisely as the synthesis answering to this analysis.

Eutocius possessed a text purporting to set out Eudoxus' construction of the problem, but expresses reservations concerning its authenticity:

> We have happened upon writings of many clever men reporting on this problem [of finding the two mean proportionals]. Of these, we deprecated the writing of

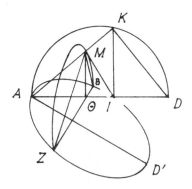

Figure 4

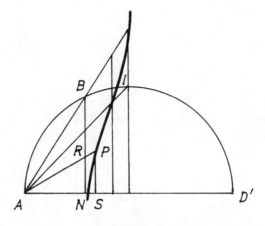

Figure 5

Eudoxus of Cnidus, since he says in the introduction that he discovered it by means of curved (*kampylai*) lines, but in the proof, in addition to not using curved lines, he even finds a discrete proportion and then uses it as if it were continuous. This would be impossible to imagine in the instance of anyone even moderately versed in geometry, let alone in the instance of Eudoxus.[8]

Scholars have understandably deplored Eutocius' decision to omit any further details on this method, for one would surely expect that even this defective text preserved the basis for a satisfactory reconstruction.[9] Despite this omission, several proposals have been made. One makes use of the curve we mentioned above in connection with the analysis leading to Archytas' solution (cf. Fig. 3). A second suggestion, advocated by Tannery, also starts from Archytas' figure.[10] Tannery proposes that Eudoxus considered the orthogonal projection of the curves of intersection onto the base plane. The section of the cylinder by the torus projects onto the semicircle ABD'. The section of the cone and the torus gives rise to a plane curve of the following description (see Fig. 5): in the semicircle ABD' draw BN perpendicular to AD' and let an arbitrary ray from A meet BN at R; mark S on AD' such that AS = AR and let the perpendicular to AD' at S meet AR extended at P. Then the locus of P associated with all such rays is the second curve of projection, and its intersection with the semicircle ABD' is the point I which solves the problem. Although Tannery depends on the use of modern coordinate representations to display this curve, it can be worked out according to ancient geometrical methods without undue difficulty [see Fig. 6]. For consider the point Q on the intersection of the torus and the cone. From the torus, it follows that AQ is the mean proportional of AP, AD, since it is a chord in the semicircle AQD. From the cone, it follows that AQ = AT. Now, AB is a chord in the semicircle ABD', so that it is the mean proportional of AN, AD'. Thus, $AT^2 : AB^2 = AP \cdot AD : AN \cdot AD' = AP : AN$ (since AD = AD'). By similar triangles, AT : AB = AS : AN, whence $AS^2 : AN^2 = AP : AN$, or AS

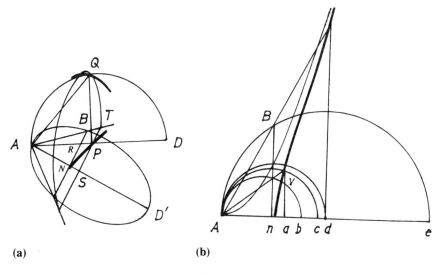

Figure 6

: AN = AP : AS. But by similar triangles, AP : AS = AR : AN, so that AR : AN = AS : AN, or AR = AS. This last result leads to the construction of the curve as specified above, where one must introduce some form of pointwise procedure actually to produce the curve and find its intersection with the semicircle ABD'.[11]

A proposal of quite a different sort has been made by R. Riddell, who links the recursive configuration of right triangles essential to Archytas' method with a kinematic configuration derived from Eudoxus' work in geometric astronomy.[12] Addressing the problems of describing the motion of planets by means of uniform circular motions, Eudoxus hit upon the conception of the "hippopede" curve: if one considers a point P set on the equator of a rotating sphere, and one embeds this sphere within a second sphere rotating with an equal speed in the opposite direction and communicating its own motion to the inner sphere, then if the two spheres have the same poles, the point P will merely remain fixed; but if the spheres are inclined at an angle, P will trace a curve in the form of a double loop, or figure-8, for each complete revolution of the two spheres. (See Fig. 7a.) It was named "hippopede," or "horse fetter," for its shape, and may be alternatively conceived as the intersection of the inner sphere with a right cylinder.[13] Projected onto the equatorial plane of the outer sphere, the curve becomes the circle ARH [see Fig. 7(b)]. The effect of the rotation of the inner sphere is to raise P above this plane from its initial position at A, so that its projected position is on the circle QR'G; the effect of the rotation of the outer sphere is to transfer the whole figure back over its initial position ARH, thus leaving the projected position of P at R.[14] If this looping path of P is now superposed over the slow rotation of a point on the equator of yet another enclosing sphere (corresponding to the ecliptic circle on the celestial sphere), then the resultant

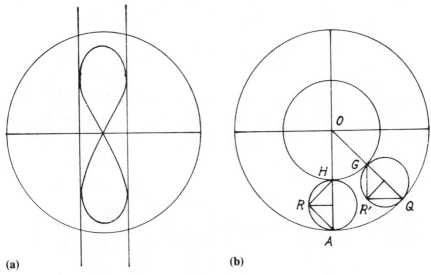

Figure 7

motion of P gives a fair semblance of the path which planets are seen to mark off in the heavens. Unfortunately, this model proves remarkably inept at saving the observed appearances of planetary motions: it fails to account for the variations in orientation, size, and shape of successive loops; in some instances, like Venus, it does not yield a retrograde motion at all; and it places the planet always at the same distance from the observer and so leaves unexplained the differences in brightness.[15] Surely, such difficulties as these induced later astronomers to try entirely different approaches, like the heliocentric model of Aristarchus, setting the planets each in its own circular orbit around the sun and each with its own period of revolution.

But if the Eudoxean scheme of homocentric spheres held limited interest for astronomers beyond a few decades after Eudoxus, this need not signify that the curve he introduced failed to draw the attention of geometers. Indeed, the hippopede is mentioned several times by later authors for one or another of its features.[16] Riddell thus invites us to consider another curve generated via a modification of Eudoxus' arrangement. Instead of having the outer sphere rotate in reverse with the same speed as the inner, let it rotate with twice that speed. This will have the effect of returning the projected position of QR'G through an equal angle to the opposite side of its initial position, so.assuming the position JRH [see Fig. 8(a)]. The space curve traced by P will thus arch to an upward maximum corresponding to the first quarter-turn of the inner sphere, then move downward, arriving at its minimum at the end of the third quarter-turn.[17] The curve of projection will accordingly pass through two inward-facing arches, as it weaves within the space bounded by an outer circle and an inner ellipse [see Fig. 8(b)]. Now consider the projection R of an arbitrary position of P [see Fig.

56   Ancient Tradition of Geometric Problems

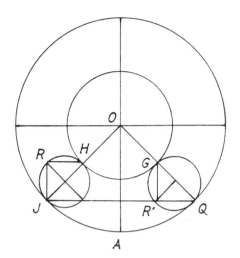

(a)

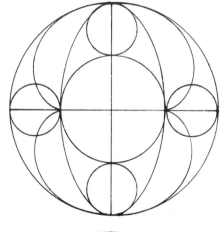

(b)

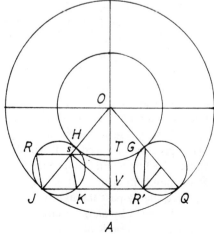

(c)

Figure 8

8(c)]. Draw the line RT perpendicular to the initial radius 0A, and meeting the terminal radius 0J at S. Since the angles at 0 are equal, JQ is perpendicular to 0A, meeting it at V. Drawing SV, Riddell shows that VS will be perpendicular to 0J on the condition that 0H is one-half 0A.[18] For this makes R' the bisector of QV, whence JK (side of the rhombus RSKJ, where RJ = R'Q) will equal one-half JV; hence, VK = KJ = KS, so that angle JSV is a right angle. Thus, the triangles 0TS, 0SV, 0VJ fall into the configuration central for Archytas' duplication of the cube: that is, the lines 0S, 0V are two mean proportionals between 0T, 0J. Thus, to find the two means between two given lines, one produces the projection of this modified Eudoxean curve, where 0A equals the larger given line and 0H equals one-half of the line; marking off the shorter line as 0T, one then draws the perpendicular to 0A at T, meeting the curve at R, and one locates the point R' as the projection of the point P' where the circle through P parallel to the equator of the outer sphere meets the equator of the inner sphere. Then, the perpendicular to 0A from R' meets the outer circle at J and 0A at V, from which the diagram may be completed to yield 0S, 0V as the required mean proportionals.[19]

Relative to these proposals by Tannery and Riddell, it is a mere quibble to point out that Archytas does not actually introduce the section of the torus and the cone, nor does the ancient tradition ever signify such a variant on Eudoxus' hippopede.[20] These are both ingenious constructions, quite in the spirit of the ancient geometrical methods. It will emerge from what follows, however, that further considerations point to a different alternative.

The construction "according to Plato" is the first of Eutocius' texts on the cube duplication, and thus follows immediately his brief comment on the omitted Eudoxean method.[21] The "Platonic" procedure employs a mechanical device to produce a configuration consisting of three consecutive similar right triangles GBD, DBE, EBA, where the sides AB, BG are given [see Fig. 9(a)]. Its base is the beam Hθ pivoting on the given point G such that its endpoint H always lies along the line of AB [see Fig. 9(b)]. A second beam HZ is fixed at a right angle to the first. A third beam KL is free to slide along HZ, but is always perpendicular to it. If K is set along the line of GB, then KL will meet BA (or its extension), say at A', and when this point comes to coincide with A in the course of the device's manipulation, the construction will have been effected, so that DB, BE will be two mean proportional lines between the given lengths GB, BA [see Fig. 9(c)].

The Platonic provenance of this method is open to serious question. First, Eratosthenes makes no reference to it among his remarks on the early efforts at cube duplication. This is all the more noteworthy since his interest in the Platonic philosophy is evident through his writing of the *Platonicus* and his elaboration there of the story of Plato's involvement with the Delian oracle.[22] Surely, then, Eratosthenes would have been particularly keen to indicate an actual solution by Plato of the Delian problem, had he known of one. Furthermore, this method depends on the conception of a mechanical device. Indeed, Eutocius' text goes into considerable detail on the fashioning of the grooves and the pivots ensuring that the beams maintain their perpendicular orientations throughout their motion.

58   Ancient Tradition of Geometric Problems

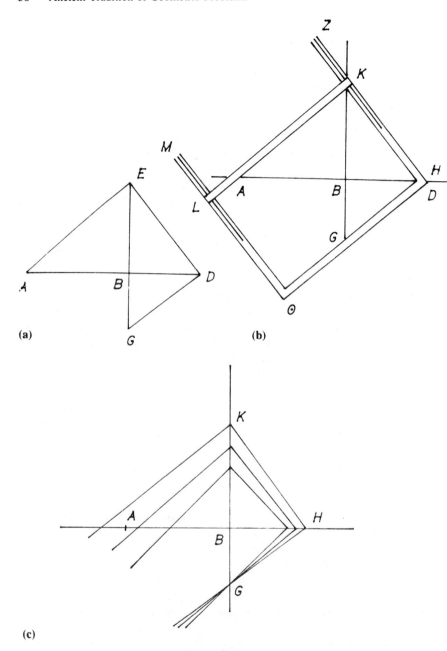

Figure 9

One is astounded at the flexibility of the traditions which on the one hand attribute such a mechanism to Plato, yet on the other hand portray him as the defender of the purity of geometry and the sharp critic of his colleagues for their use of mechanical procedures in geometric studies. But surely the latter tradition, as typified in Plutarch's version of the Delian story, is closer to the mark in capturing Plato's own view of the nature of geometry. For one can find strong expressions of the abstract character of the entities and methods of geometry in the Platonic writings.[23] The careless habit of assigning feats of insight to such heroes as Plato and Pythagoras might easily tempt later writers to overlook issues of simple consistency.

But if we acknowledge, as most scholars now do, that this method cannot be attributed to Plato, we then face the question of who its originator was. Its absence from Eratosthenes' report might suggest a later dating. But as we shall later see, it appears to be implied as the background to Diocles' method only a short time after Eratosthenes. The Platonic association now assumes new significance, for it suggests that Eratosthenes himself played a role in the transmission of this method through its inclusion in a work like the *Platonicus*. This does not lead one to suppose that Eratosthenes invented this device in addition to the other mechanical procedure transmitted under his name, but rather that he might have introduced it as the centerpiece of a discussion on the nature of geometry between Plato and one of his cube-duplicating colleagues.[24] Then which colleague? Only Archytas, Menaechmus, and Eudoxus are named in this connection, and of these the first two used methods which would not bring the pseudo-Platonic device into mind, as we may gather from Eutocius' accounts of them. This leaves the Eudoxean method, of which we know only that it employed some form of "curved lines." Can one try to identify his solution with the pseudonymous one?

The manipulation of the pseudo-Platonic device, in the manner described by Eutocius, does not give rise to any curve. But instead of adjusting the sliding beam KL such that K is held to the line GB (extended), let us place KL so that it will always pass through A [see Fig. 10(a)]. Then K will trace out a curve, and where it meets GB (at E) the configuration for solving the problem is obtained. To view its generation a bit more abstractly, we conceive of two parallel rays AK, GH, turning on the given pivots A and G. From the intersection H of GH and AB (extended) draw the perpendicular meeting AK in K; then K describes our curve. Alternatively, we may draw from the intersection K of AK and GB (extended) the perpendicular meeting GH in H, and H will then describe a curve of the same type, differently oriented [see Fig. 10(b)]. In modern works these curves are called "ophiurides," a family of the third order, and the curve used by Diocles for solving the cube duplication arises in the limiting case where B coincides with either A or G.[25] One may note that a procedure equivalent to the pseudo-Platonic method occurs in an Arabic version based, it would appear, on an account by the early 2nd-century-A.D. geometer Menelaus.[26] Here, an overtly mechanical description in the style of Eutocius, replete with grooved beams and pivots, accompanies a parallel description framed in the more abstract

60   Ancient Tradition of Geometric Problems

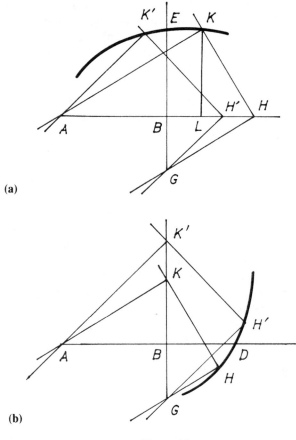

Figure 10

geometric manner just given. Neither actually generates a curve, although the method for doing so is evident. The omission might well be due to Menelaus, for he does the same thing in his treatment of Archytas' method.[27] We thus have a possible explanation of how a text of Eudoxus' solution "by means of curved lines" might come into Eutocius' hands, now bereft of any mention of curves. Moreover, if the text was at all unclear on the positioning of K along line GB, one would not see how the discrete proportion GB : BH = HL : LK becomes the continuous proportion GB : BD = BD : BK. These two defects marred Eutocius' source on the Eudoxean method and so suggest an association with the pseudo-Platonic method.

Under this view of the identity of the two methods, we would have to assign to Eudoxus a treatment in the more abstract manner suggested above, for Eratosthenes criticizes Eudoxus, no less than Archytas and Menaechmus, for having addressed the problem in a strictly theoretical manner without thought of practical execution. As indicated above, Eratosthenes himself emerges as a likely candidate to bear responsibility for casting this method into the form of an actual

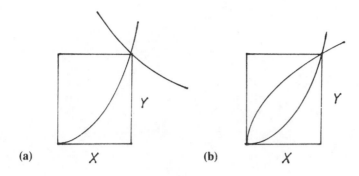

Figure 11

mechanical device. His writing of the *Platonicus* would offer a natural context for this and through this association induce the later writers mistakenly to attribute the method to Plato. One can detect in Eutocius' version of the pseudo-Platonic method certain coincidences of phrasing which link it to other texts he presents.[28] Such editorial interventions on his part would be necessary in the event, as we propose here, that his source ultimately derived from the relatively informal context of a discussion of the Platonic philosophy by Eratosthenes.

The third of the early solutions mentioned by Eratosthenes entailed the "sectioning off from the cone the triads of Menaechmus." Eutocius presents a text of Menaechmus' method in which the desired lines are obtained first from the intersection of a hyperbola with either of two parabolas, then, in a second version, via the intersection of the two parabolas.[29] More specifically, if we wish to find for two given lines A, B the two lines X, Y which are the mean proportionals between them, then we may construct the parabola satisfying the condition $X^2 = AY$ (or, indifferently, the parabola satisfying $Y^2 = BX$) and the hyperbola satisfying $XY = AB$. The intersection of these two curves yields the lines X, Y which are the desired mean proportionals [see Fig. 11(a)]. Alternatively, we may consider the intersection of the two parabolas; for this gives rise to the same two lines X, Y [see Fig. 11(b)].

The essential line of thought, which we have paraphrased here following Eutocius, is entirely clear, especially since Eutocius precedes his formal synthesis of the problem with an analysis. Assuming the four lines have been set out, such that $A : X = X : Y = Y : B$, if X is set at a right angle to Y, the proportion $A : X = X : Y$ associates X and Y as coordinates of a given parabola; similarly in the instance of the hyperbola and the second parabola. The notion of setting the lines at right angles is quite naturally suggested by the configuration of the pseudo-Platonic method which would surely be familiar to Menaechmus, if that was indeed the method used by Eudoxus [see Fig. 11(c)]. What has puzzled scholars is the manner by which Menaechmus was able to construe these relations as properties of sections of cones. Eutocius' text is of limited use in this regard, for it is framed entirely in conformity with the Apollonian form of the theory

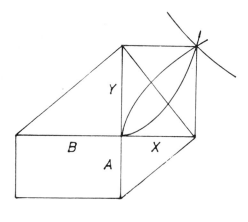

(c)

Figure 11

of conic sections. A text adhering closely to the treatment of this problem by a 4th-century geometer like Menaechmus could hardly employ terminology like "latus rectum," "ordinate," "asymptote," and perhaps not even the terms "parabola" and "hyperbola." Thus, modern commentators have engaged in ambitious reconstructions of the rudimentary theory of conic curves which Menaechmus might have developed for the purposes of his solution of the cube duplication.[30] As one can well anticipate, any such theory must have embraced a substantial portion of the content, if not of course the form, of Apollonius' Books I and II.

This effort has provoked surprisingly little uneasiness among historians of this period.[31] Yet apart from this testimonium of the work of Menaechmus, there are but two passages which might reflect the very early study of conics, the one in the pseudo-Aristotelian *Problems*, the other in Euclid's *Phaenomena*, and hence both datable to around 300 B.C.[32] At about that same time Euclid and his contemporaries were in the position to make a first compilation of the theory of conics, of which we possess a good, if indirect, indicator in the works of Archimedes.[33] But the tradition of an early theory instituted by Menaechmus nearly a half-century before Euclid depends on but a single line of Eratosthenes' verse: "Do not thou seek to section off from the cone (*kōnotomein*) the Menaechmean triads."[34] Reading this line in its context, we recognize at once that the "Menaechmean triads" (note the plural) refer to the three specific conic curves (the hyperbola and two parabolas) associated with each choice of the two given lines in the problem of finding the two mean proportionals. Contrary to the familiar view, these can hardly refer to the three forms of conic section: parabola, hyperbola, and ellipse. For at no known stage of the ancient theory were these curves generated together as triplets, while the ellipse has no bearing on the ancient solutions of the cube duplication, as transmitted to us, whether from Menaechmus or from anyone else. Eratosthenes' term *kōnotomein* is no guarantee that Menaechmus actually produced these curves as sections of cones, for anachronistic applications of terminology have been known to occur among historians of mathematics, while Eratosthenes' epigram was not composed as a historical effort anyway. Indeed, strictly speaking, Eratosthenes does not even claim that Menaechmus sectioned the cone, but only that "you," the contemporary reader of the epigram, need not do so. We thus have good cause to consider whether Menaechmus might have constructed his solving curves in some other manner.

An appropriate method is readily conceived on the basis of the pointwise construction of the curves answering to the three relations derived from the assumption of the continued proportionality of the lines A, X, Y, B.[35] In the operation of the means-finding apparatus attributed to Plato (Fig. 9b), we maneuver the right angle GHZ so that the leg GH always passes through G while H always lies on the extension of AB [Fig. 12(a)]; if for each position H we complete the rectangle HBKN, the vertex N will trace a curve whose coordinates are such that BH ($= X$) is the mean proportional of GB ($= A$) and BK ($= Y$). Similarly, if the right angle AKH is turned so that the leg AK always passes through A while K lies on the extension of GB [Fig. 12(b)], completing the rectangle HBKX, we trace the curve whose coordinates are such that BK is mean proportional of HB and BA ($= B$). Thus, where these two curves intersect, the coordinates X, Y will be the two mean proportionals between the given lines A, B, as required by the problem. In an alternative construction, we can figure Y as the half-chord perpendicular to the diameter of a circle such that Y divides that diameter into segments of length B and X (cf. *Elements* VI, 13 or II, 14 [see Fig. 12(c)]. To produce the curve associated with the relation $Y^2 = BX$, then, one sets out as many lengths for X as one wishes, determines Y for each in the above manner, locates a point for each such pair of coordinates, and connects the points by means of a flexible ruler or other suitable drawing aid. The construction of the curve associated with the relation $X^2 = AY$ is of course the same [see Fig. 12(d)]. In the case of the relation $XY = AB$, one may find Y coordinate to X via the "parabolic application of area," based on the fact that the diagonal of the rectangle of sides $B + X$ and $A + Y$ passes through the origin (cf. *Elements* I, 44) [see Fig. 12(e)]. Otherwise, the generation of a curve through a selection of values for X follows the same pattern. Having drawn two of these curves over the given lines A, B, one determines from their point of intersection the lines X, Y which solve the problem of the mean proportionals.

Pointwise procedures of this sort are familiar among later writers. Diocles, for instance, constructs his own curve for solving the cube duplication by determining a set of points and then connecting them via a flexible ruler; his alternative construction of the parabola, framed in effect around its focus-directrix property, is also in the pointwise manner.[36] Eutocius describes analogous procedures for drawing all three kinds of conics, and notes that writers on mechanics often adopt such methods for the conics where other methods do not work well.[37] Similarly, Anthemius marks out the contours of burning mirrors by orienting the tangent lines at selected points.[38] No source identifies the originator of these

64 Ancient Tradition of Geometric Problems

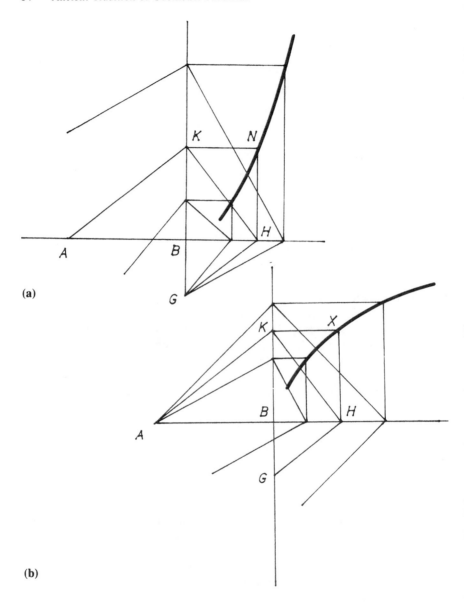

Figure 12

pointwise methods. But Diocles' dependence on Menaechmus is evident in his presentation of a form of the two-parabola solution for the cube duplication, and one can readily assume that he also found with Menaechmus a precedent for his use of a pointwise construction of those curves.[39] It is right, I believe, to view this method of constructing curves to be an insight of significance in the development of geometry, yet hardly overreaching the abilities of pre-Euclidean geom-

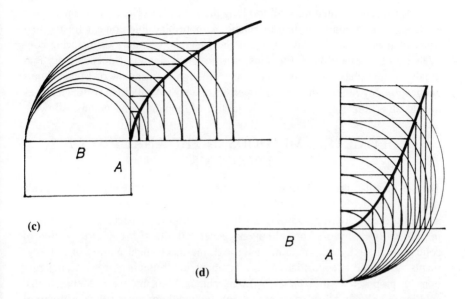

(c)

(d)

eters. Indeed, its invention prior to Menaechmus, or perhaps by him, is rendered plausible by the association with Diocles.

I propose, then, that Menaechmus based his solution on curves defined with respect to the second-order relations among the mean proportional lines. This does away with the warrant to speculate on how Menaechmus might have formed these curves as sections of cones and which of their properties he might have

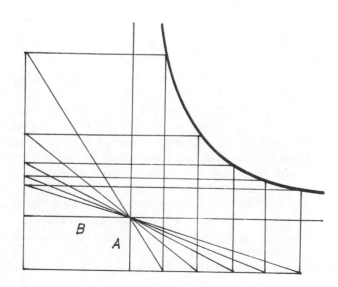

(e)

proved, exercises which have all too long distracted historians from considering evidence from around the time of Euclid, a more plausible context for the first study of these sections. This view has the welcome corollary of restoring confidence in the reliability of Eutocius, our principal source of documentary materials on this phase of early geometry. Although his account of Menaechmus unquestionably reflects the reworking effort of a post-Apollonian editor, it serves now as an accurate guide to the essential line of thought Menaechmus followed.

GEOMETRIC METHODS IN THE ANALYSIS OF PROBLEMS

The set of 4th-century solutions to the cube duplication already exploit a wide range of geometric methods which would remain characteristic features of the ancient problem-solving tradition. Each of the methods considered, for instance, introduces one or more special curves, generated either through the section of solids, as in Archytas' solution, or through the mechanical motion of line segments, as in the pseudo-Platonic arrangement, or through pointwise constructions, as proposed for the solution by Menaechmus. Eutocius' text of Menaechmus' solution highlights two further aspects of methodology not displayed in any of his other ten texts. One is its use of the second-order relations of lines associated with the pre-Euclidean technique of "application of areas," on which was built the later theory of the conic sections. The second is its adoption of the two-part format of "analysis and synthesis" which would become, in the hands of Apollonius and his contemporaries, a powerful instrument for the discovery and exposition of the solutions to problems.

The "application of areas," as we learn from Proclus on the authority of Eudemus, owes its origins to the ancient Pythagoreans, although Plutarch knows of a tradition assigning its invention to Pythagoras himself.[40] It relates to a series of problems now extant in Euclid's Book VI and supporting lemmas to be found in his Books I and II. The objective is to "apply" (*paraballein*) a given area to a given line segment (I, 44) or, alternatively, to construct a parallelogram similar to a given figure and equal to a second given figure (VI, 25). One may also seek to apply the given area to a given line segment such that the resulting parallelogram either "exceeds" (*hyperballein*) or "falls short" (*elleipein*) by a figure similar to a given parallelogram (VI, 29, 28, respectively; cf. II, 6, 5). In the earlier stages of this technique, one would take the figures of application to be rectangles and the figures of excess or defect to be squares. Thus, for the first, or "parabolic" case, one seeks to construct the line X such that $AX = M$ for a given line segment A and a given rectilinear area M, e.g., the square of side B. In the "hyperbolic" case, one seeks X such that $X(X + A) = B^2$, and in the "elliptic" case, the X such that $X(A - X) = B^2$ [see Figs. 13(a), (b)].[41] The attribution to the Pythagoreans merely confirms what one infers from Euclid's *Elements*: that these techniques were familiar in the pre-Euclidean geometry, that is, in the 4th century B.C. and perhaps the latter part of the 5th century B.C. But it is now well known that the same problems, treated in the form of numerical

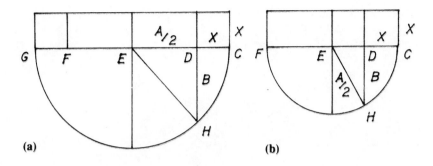

Figure 13

solving procedures, were a prominent part of the Babylonian tradition more than a millennium earlier.[42] One would thus infer that the Greeks first learned these techniques through contact, direct or indirect, with the Babylonian tradition.[43] The same would also have provided them a precedent for working out ordered procedures for the solution of *problems*; but the specifically *geometric* form into which the Greeks cast their own version of this technique is not manifest in the Babylonian texts now extant.

A most notable example of the "application" technique is in the division of a line into "extreme and mean ratio" (VI, 30; cf. II, 11), through which one can effect the inscription of a regular pentagon in a given circle (IV, 10, 11) and the inscription of the regular icosahedron and dodecahedron in a given sphere (XIII, 16, 17). In all these instances, Euclid's treatments are in the synthetic manner. But Eutocius' text of Menaechmus' solution of the cube-duplication suggests through its adoption of the double mode that these other problems might also have owed their initial solutions to analytic treatments. Indeed, Pappus preserves a set of alternative constructions of the inscribed regular solids effected by means of analysis and synthesis.[44] Proclus relates that Plato taught the method of analysis to Leodamas for use in his geometric researches and that Eudoxus used analyses in his studies of the "section" (*tomē*).[45]

It will be useful to reconstruct an analysis along the lines suggested by Euclid's synthesis of the problem of inscribing the regular pentagon in a given circle. As we shall later see, a comparable analysis is discernible as the basis of Archimedes' solution of the inscription of the regular heptagon (Chapter 5). Assume then that the regular pentagon ABCDE has been inscribed in the circle [see Fig. 13(c)] and draw the diagonals AD, BE meeting in F. In the triangle AFE the angles at A and E are equal, intercepting the equal arcs DE, AB, respectively. Since angle BAD intercepts the double arc BCD, it will be double the angle FAE. Since, furthermore, the angle BFA is external to triangle FAE, it will equal the sum of angles FAE, AEF, that is, twice angle FAE. Thus, angles BAF, BFA are equal, so that AB = BF. Since F is the vertex of the isosceles triangle FAE, it lies on FG, the perpendicular bisector of the given side AE. Its position may thus be given via *neusis*, in which the intercept FB from a line drawn from the

68   Ancient Tradition of Geometric Problems

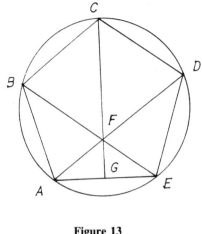

(c)

Figure 13

given point E shall have a given length (= AB) and lie between a given line (FG) and a given circle (that with center A and radius AB). The corresponding synthesis is evident, and in fact is nothing other than a variant of Hippocrates' third lunule construction.[46]

Euclid's method in IV, 11 does not employ a *neusis*, however, so that the implied analysis must have followed a different course. If we adopt the same figure as before, we have that the triangles ABE, FAE are similar, since their base angles (at B, E, and A) intercept the equal arcs AE, AB, ED, respectively. Thus, AB : BE = EF : AE, and since AB = AE = BF, it follows that BF^2 = EF • EB. We have thus reduced the problem of inscription to one of dividing a line into segments satisfying this second-order relation. The latter problem, familiar to the ancients as the division into "extreme and mean ratio" (cf. *Elements* VI, Def. 3), and to us as the "golden section," can be effected via the methods of the "application of areas," as Euclid does in II, 11 and again in VI, 30.[47] In the latter, the given area BE^2 is applied hyperbolically to the given line BE; that is, BE^2 = X(BE + X) [see Fig. 13(d)]. Setting BF = X, it follows that BF^2 = BE • EF, as required. But in II, 11 an independent method is used where BF is constructed as the difference of ½ BE and the hypotenuse of the right triangle of legs BE, ½ BE [see Fig. 13(e)].[48] The alternative method would seem to indicate an origin at an earlier stage of the development of these techniques.[49]

If Euclid's procedure drew from the *neusis*-related analysis just given, we would expect a synthesis like this: divide the given line BE at F as specified in the analysis; complete the isosceles triangle with BE as base and legs BA, AE each equal to BF; circumscribe the circle about BAE; then BA, BE are the side and diagonal, respectively, of the regular pentagon inscribed in this circle. If we then inscribe in the given circle an angle equal to ABE, we shall have determined an arc A'E' subtended by a side of the regular pentagon inscribed

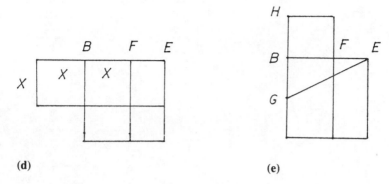

(d)   (e)

in it, from which the figure is readily completed. This approach is of course entirely within the scope of the early techniques; it is sufficiently straightforward in conception to make one uncomfortable in denying that the ancients knew it. But the synthesis given by Euclid in IV, 11 is quite different. We thus receive no encouragement from him in supposing the ancients followed this method,[50] and must frame another analysis as the basis of his method.

Assume again the regular pentagon ABCDE inscribed in the circle and now draw the diagonals AC, CE [see Fig. 13(f)]. It is evident that the triangle ACE is isosceles and that each of the angles at its base is twice the angle at its vertex. The inscription thus reduces to the construction of an isosceles triangle of this form; the synthesis of the latter problem is given by Euclid in IV, 10 in order to effect the inscription in IV, 11. To construct the triangle, we can produce the following analysis: assume ACE is isosceles, where the angle at A is twice that at C [see Fig. 13(g)]. Let AF bisect the angle at A, whence triangles AFE, CAE are similar. Thus, EF : EA = EA : AC, or EA^2 = EF • AC. Now, EA = FC,

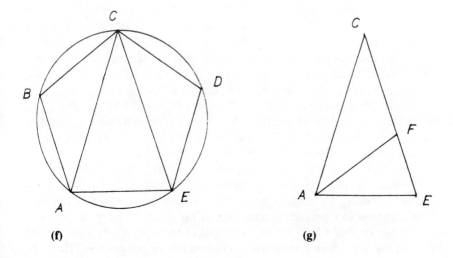

(f)   (g)

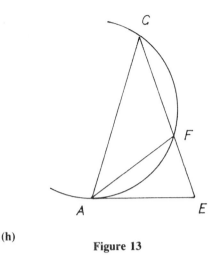

(h)

Figure 13

each being equal to line AF in the isosceles triangles AFE, FAC, respectively. Furthermore, AC = EC. Thus, FC^2 = FE • EC. In this way, the construction of ACE reduces to the division of a given line CE in accordance with the extreme and mean ratio, as given in II, 11 (or VI, 30).

I have no doubt that the analysis underlying IV, 10—by which the construction of CAE is reduced to finding F such that FC^2 = FE • EC—took this form. The techniques of proportions and similar figures were already prominent in the ancient geometry quite early, as we have seen through the problem-solving efforts of Hippocrates. Moreover, treatments survive of closely related problems which adopt precisely this configuration of similar triangles.[51] Nevertheless, the synthesis Euclid presents in IV, 10 does not merely retrace the steps of this analysis. Euclid assumes the division of CE such that CF^2 = FE • EC and completes the triangle CEA such that EC = CA and EA = FC [see Fig. 13(h)]; it is to be shown that the angles at A and E are each twice that at C. Draw AF and introduce the circle AFC. Since EA^2 = FE • EC, EA is tangent to the circle (*Elements* III, 37), whence angles FAE and FCA are equal (III, 32). Euclid next proves that angles EFA and EAC are equal;[52] since the triangle ECA is isosceles, the latter angle equals FEA, so that the triangle FEA is also isosceles, whence EA = AF. But by construction, EA = CF; angles FCA and FAC are thus equal base angles in the isosceles triangle FAC. The exterior angle EFA equals their sum, or twice the angle at C. Thus, in the triangle CEA, the base angle EAC (which equals angle EFA) is twice the angle at the vertex C.

Our reconstruction of the analysis has not prepared us for Euclid's introduction of the circle and the angles formed by its tangent. But these moves are quite naturally motivated by Euclid's pedagogical decision to defer the presentation of the techniques of proportion until later in his treatment of plane geometry (i.e., in Books V and VI). The constructions of the regular polygons in Book IV must thus be effected exclusively via methods of congruence. Thus, the equality of angles EAF and ACE, which could have followed at once from the similarity of the corresponding triangles through the proportionality of their sides, must now be secured via the tangent EA to the circle AFC. This discrepancy raises a subtle point for the interpreter. Euclid's formalized organization of the *Elements* need not have followed the order of discovery; although it is conceivable that the whole body of plane geometry in Books I–IV was worked out prior to the adoption of proportion techniques by the ancient geometers, I find that only remotely plausible as a position on the historical development of the field. Since the purpose of an analysis of a problem is fundamentally heuristic, it is entirely feasible that this part utilize methods formally excluded at the stage of synthesis. For it is the synthesis alone which has formal status. In the case of the construction in IV, 10 at issue here, one could fashion an analysis which conforms step for step with Euclid's synthesis.[53] But that would be an exercise without either pedagogical or heuristic interest. In the light of this, Euclid's omission of analyses in his treatments of construction problems in the *Elements* follows naturally from the editorial decision to formalize the exposition of this material.

To the extent that the *Elements* do reflect the earlier procedures—certainly in more general respects, if not always on points of specific detail—we may submit the reconstructions above as a specimen of the problem-solving field among the 4th-century geometers. Characteristic of the method are the exploitation of analyses and the reduction of the desired configurations to conditions relating to the application of areas. From this period we may cite two further witnesses to the adoption of this approach: one is a passage from Plato's *Meno*, the other a set of remarks by Proclus on a geometer named Amphinomus.

The *Meno* (86 e – 87 b) introduces a method of reasoning "from hypothesis" which, but for its name, is identical to that of "reduction" as used, for instance, by Hippocrates in his attack on the cube duplication. The issue in the dialogue is to establish whether virtue is teachable. In the absence of a satisfactory definition of what virtue is, Socrates proposes that one consider a hypothesis from which to pursue the examination of the main question.

> I say "from hypothesis" in the manner that the geometers often make inquiry, whenever someone has asked them, for instance about an area, whether this area here can be stretched out as a triangle in this circle here, one would say 'I don't yet know whether this is of such a sort, but I think that as a certain hypothesis the following will assist in the matter. If this area is such that the one who has stretched (it) along its given line (makes it) fall short by an area such as is the stretched (area) itself, then it seems to me that a certain result follows, but a different result, if it is impossible that these things be done. Having hypothesized, then, I wish to say to you whether the result about its stretching in the circle is impossible or not.'[54]

Like all too many of the mathematical passages in the dialogues, this one leaves much to be desired in terms of clarity. Plato's phrasing here is quite free, whether by reason of his deliberate choice to suit the colloquial context of the dialogue, or perhaps through imprecision in the mathematical terminology of his day. For instance, the procedure of "stretching *in*" (*enteinein*) is clearly the inscription

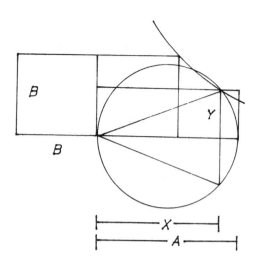

Figure 14

of one figure within another. But can one assume that the procedure of "stretching *along*" (*parateinein*) refers to the application (*paraballein*) of area, as would appear to be the case here? Is the area which falls short (*elleipein*) "by such as is" (*toioutōi, hoion ēi*) the applied area similar to it, or *equal* to it? Does "its given line" signify the diameter of the given circle, or perhaps some other line serving as a side for the figure of application?

These obscurities have occasioned dozens of efforts of interpretation. But one of the oldest attempts still remains the most plausible and widely accepted.[55] The problem as initially posed requires the inscription of a given area as a triangle in a given circle. We may figure the given area as a square of side B and denote by A the diameter of the given circle [see Fig. 14]. There is no loss of generality in requiring that the inscribed figure take the form of an *isosceles* triangle. In accordance with the mode of analysis, assume the figure has been inscribed as specified; let the altitude X of the triangle lie along the diameter, and let Y be half its base, set at right angles to the diameter. The condition on the area thus stipulates that $XY = B^2$. Furthermore, since Y is a half-chord in the circle, Y is the mean proportional between X and $A - X$, the two segments of the diameter it cuts off; that is, $X : Y = Y : A - X$. In this way the problem is reduced to another one, corresponding to that stated in the second part of Plato's passage: the given area B^2 is now to be applied to the given line A such that the area by which it falls short (namely, the rectangle of sides Y and $A - X$) be similar to the area applied (the rectangle of sides X and Y). Thus, the passage may be viewed as based on this reduction or analysis of the problem of inscription.

If we complete the analysis, in the manner suggested by Menaechmus' treatment of the cube duplication, we will observe that the condition on areas ($XY = B^2$) entails that X and Y are coordinates in a given curve, namely, the hyperbola having as its asymptotes the diameter of the circle and the tangent drawn at its endpoint, where the hyperbola passes through the point whose coordinates both equal B. We have proposed that Menaechmus introduced such curves in the form of pointwise constructions corresponding to the defining area relation. But whatever construction he used, it would perfectly suit the purposes of the problem discussed in the *Meno*. This effectively derails the principal objection raised by some against this interpretation of the passage: that it supposes a problem incapable of solution by the methods available to pre-Euclidean geometers.[56]

A remarkable feature of the *Meno* passage is that it expresses the mathematical project not as the actual *solution* of a problem, but rather as the determination of the *possibility* of its solution. This has led many to view the passage as discussing a "diorism," that is, the statement of the necessary condition for the solvability of a problem. But in the mathematical literature diorisms have the form of explicit conditions on the givens of the problem. In the present case, this might be the statement that the given area must be less than the area of an equilateral triangle inscribed in the given circle; having verified this relation to hold for particular values of the givens, one would know that the problem is solvable in this case, even before one has begun the solution of the problem as such. Although it is often the case that the analysis of a problem reveals the appropriate form of its diorism, nevertheless, the articulation of the diorism is quite different from the analysis or reduction of the corresponding problem. We thus have to explain why Plato here frames this example of problem reduction as if it were equivalent to the determination of possibility.

According to the usual chronology of Plato's works, the *Meno* is transitional between the earlier and the middle dialogues, so that its composition is placed around 385 B.C.[57] This is two decades or more before the activity of Menaechmus, a disciple of Eudoxus whose association with the Academy falls in the period after 370 B.C. Some of the geometric techniques available to Menaechmus would not yet have been introduced at the earlier time of the *Meno*, and these might include the methods of curve drawing discussed above. Thus, Plato's emphasis on the possibility of the inscription problem might be taken to signify that geometers had then discovered the diorism, but not the actual solution of this problem. For a proof of the diorism can be effected via elementary methods of construction,[58] but this is not true of the inscription problem itself. With the introduction of the methods affiliated with Menaechmus' cube duplication, not only could the problem now be solved, but one also would obtain an alternative proof of the diorism; for the problem of application is possible only when the auxiliary hyperbola (in the derived problem) intersects the circle, and in the limiting case, where the curves meet at a single point of mutual tangency, one finds that $X = \frac{3}{4} A$, equivalent to the case of the equilateral triangle.[59]

Supporting this view is one's intuition that the successful solution of the inscription problem could not long have antedated Menaechmus' solution of the cube duplication. But the form of the *Meno* passage must surely have been affected by certain dramatic and dialectical considerations at least as much as by these historical and technical ones. Plato's principal concern in the dialogue is over a question of possibility: the teachability of virtue. It thus suits his purposes to frame the geometric example in like manner, as a determination of the possibility of a construction, even though the geometric method illustrated is that of reduction for the solution of problems. Further, Plato's account here is likely to reflect his own interpretation of what the true goal of such geometric activities is: not the finding of constructions, but rather the establishment of their possibility, that is, the demonstration of the existence of the solving entities. In the later mathematical writings problems are often asserted in the form, "it is possible (*dynaton estin*) to construct...," and taken literally, this might indicate an existential motivation underlying the activity of problem solving. I argue below (cf. also Ch. 8) that this provides a poor account of the objectives of the geometric research in this period. But geometers' actual motives need not much have affected Plato's view of them. I think that the special concerns of his general philosophy could well have encouraged Plato to represent, or misrepresent, the geometers' search for constructions as essentially an investigation into their possibility. This would account for the discrepancy between the *Meno* passage and the standard format in mathematical writings.

Proclus reports that Leon, a younger contemporary of Plato, "found diorisms," so that we may infer additional evidence of interest in such studies of problems (*In Euclidem*, pp. 66 f). Proclus elsewhere provides an example further revealing the use of techniques of area application in the analysis of problems.[60] A certain geometer named Amphinomus is said to have distinguished problems as "ordered," "intermediate," or "unordered" according to whether they permit of one, several, or an unlimited number of solutions. By way of illustrating this last kind, Proclus cites the problem of dividing a given line segment into three parts in continued proportion. If one divides the given line into segments in the ratio of 2 : 1 and one applies the square of the smaller segment to the larger segment as to fall short by a square, then the given line will be divided into three equal parts which are thus in continued proportion. If the given line is then divided into segments in any ratio greater than the double, say the triple or any larger multiple, and if in similar fashion the square on the smaller segment is applied to the larger as to fall short by a square, then again the line will have been divided into three parts in continued proportion. Let us designate by X the larger of the initial segments, by Y the smaller; then the procedure of application produces Z such that $Y^2 = Z(X - Z)$. It thus follows that the segments Z, Y, X − Z are in continued proportion, while their sum equals the given line. Since the application is possible only when Z is less than or equal to one-half X, whence the maximal value for Y is seen to be one-half X, Proclus' initial choice of the division $X : Y = 2 : 1$ represents the limiting condition for the possibility of solving the stated problem. One would suppose that a "diorism" to this effect had been worked out in the original treatment. But as the number of larger multiples is unlimited, each yielding a solution, the problem has an infinity of solutions and so falls under Amphinomus' class of "irregular" (or "unordered," *atakta*) problems.[61]

This is a nice specimen of the use of application techniques for problem-solving. A 4th-century provenance is indicated, but not entirely secured. One would suppose that Amphinomus himself cited this problem as an instance of his "unordered" class, although its appearance in this context might possibly be due to a later editor like Proclus. What we know of Amphinomus otherwise is contained in three additional references by Proclus, all of which associate him with 4th-century figures. Amphinomus is said to have upheld the view of Aristotle, that geometry does not inquire into the causes of things (*In Eucl.*, p. 202).[62] He is linked with Speusippus, Plato's nephew and successor, as maintaining that all propositions in geometry should be framed as theorems, for this form best captures the eternal and nongenerable nature of geometric entities; by contrast, Proclus continues, the mathematicians allied with Menaechmus emphasized the problematic element in geometric research, whether this seeks the production of an entity or the determination of its properties (*In Eucl.*, pp. 77 f).[63] Furthermore, when Proclus observes that not every proposition has a true converse, but that a proposition will be convertible when it refers to the primary or essential attribute of a class, he adds that "these things did not escape the notice of the mathematicians around Menaechmus and Amphinomus" (*In Eucl.*, p. 254). This passage recalls a discussion by Aristotle, to the effect that a proper definition should pick out attributes which apply to all and only the members of the class defined (*Posterior Analytics* I, 4, 5). The special interest which geometers might have in the convertibility of propositions is clear from another comment by Aristotle:

> If it were possible to prove something true from something false, the procedure of analysis (*analyein*) would be easy; for it would convert of necessity. For let us posit that A obtains; from this supposition these things follow which I know do obtain, namely B. Then from these things I can show that the former (A) obtains. Now the things in mathematics tend to be convertible, for they posit nothing contingent, but only definitions—and in this they differ from the things in dialectical philosophy (*ibid.*, I, 12).[64]

Aristotle here portrays analysis as a deductive procedure leading from the desired premise (or construction), hypothesized to be true (or constructible), to another one known to be true (or constructible). This accords with the examples we have already discussed, as well as with those in the later mathematical literature. Moreover, Aristotle's awareness of the relevance of the issue of convertibility reveals that current mathematical practice had recognized the need for the corresponding synthesis to effect the demonstration; where steps in the analysis are not simply convertible, one would have to identify the conditions under which conversion is valid, that is, the conditions of the diorism. Aristotle thus indirectly attests to the maturity of this method among the geometers of his generation.[65]

We thus see that Menaechmus' method for solving the cube duplication typifies an important field of the pre-Euclidean geometry. It is but one of several known instances employing the technique of application of areas, a versatile instrument for the analysis of problems. Indeed, the field of analysis had been sufficiently advanced by around the middle of the 4th century B.C. that certain special features of the logical structure of analysis began to attract the attention of philosophers. For instance, the "diorism" expressing the conditions for the possibility of solution results from consideration of the conditions permitting the steps in the analysis to be converted, so to produce the formal synthesis. Among some there already appears to be emerging a self-conscious adherence to the formalist ideal. In the question as to whether one ought to frame geometric propositions as theorems or as problems, Proclus finds a spokesman for the theorematic view, as one might predict, in the person of a 4th-century Platonist philosopher. But he draws from the geometer Menaechmus a sense of the priority of problems. This division is surely ominous. For a preoccupation with the formal exposition of theorems, at the expense of the analysis of problems, could hardly avoid retarding the effort to discover new solutions. On the other hand, the formal work of synthesis can sometimes raise difficulties demanding geometric insight for their resolution. The efforts of Eudoxus, to which we now turn, illustrate this impressively.

EFFORTS TOWARD THE QUADRATURE OF THE CIRCLE

As geometers sought ways actually to construct the square equal to a given circle, they came to recognize that the underlying theorems on circle measurement posed difficulties of their own. Hippocrates' measurement of the lunules, for instance, must assume that circles have the ratio of the squares on their diameters. Any proof which Hippocrates could have offered for this theorem, such as the one we proposed above on the model of Antiphon's sophism on the circle quadrature, must inevitably have adopted a naive manner of limits, for the formally correct method, as presented in Euclid's Book XII, owed its first introduction to Eudoxus.

Aristotle names the 4th-century Sophist Bryson, along with Hippocrates and Antiphon, as having attacked the problem of circle quadrature.[66] But the later commentators Alexander (3rd century A.D.) and Proclus, Ammonius, and Philoponus (5th and early 6th centuries), who inform us of the details of Bryson's argument, seem to know only this much[67]: having circumscribed a *square* about the circle and then inscribed another within it, and having taken a third *square* intermediate between the first two, Bryson claimed that this *square* equalled the circle, "for things greater than and less than the same things are equal."[68] On the meaning and criticism of this argument, the commentators disagree. The simplistic reading leads at once to absurdity: one might claim on the same basis that 9 and 10 are equal, since they are both greater than 8 and less than 12.[69] A more sophisticated interpretation takes into consideration the sequence of *all* circumscribing and inscribed regular polygons: then "the rectilinear figure" which is less than all the circumscribing polygons and greater than all the inscribed polygons will equal the circle, since the circle satisfies the same inequalities.

Philoponus dismissed this latter effort by Proclus to refurbish Bryson's principle in the form "to that than which there is a greater and a lesser there is also an equal."[70] It is hard to see the force of Philoponus' attempted counterexample, however. To be sure, the "horn angle" contained between a circle and its tangent is less than every rectilinear acute angle, while the corresponding mixed angle between the circle and one of its diameters is greater than every rectilinear acute angle. But when he observes that this has not established the existence of an equal angle,[71] we might well ask what angle he is referring to. Through further elaboration of this class of mixed angles, Philoponus shows that he would deny what the modern theory of curves would affirm: that the angle at which two curves intersect each other is taken equal to the angle between their tangents at the point of intersection. Despite the fact that these angles are obviously *different* from each other geometrically, it is important to accept them as quantitatively *equal* if quantitative statements about tangents are to be possible at all. It is interesting that writers in the ancient tradition of geometric optics adopt this principle quite freely, manifesting none of the qualms expressed here by Philoponus.[72]

Philoponus further objects to Proclus' interpretation of Bryson on the grounds that it misses the point of the problem of circle quadrature:

> Those squaring the circle did not seek (to establish) whether it is possible that there exists a square equal to the circle, but rather thinking that it can be so, they tried to produce a square equal to the circle.[73]

But in accepting Proclus' argument as an existence proof, he is perhaps too kind. For Proclus has far too easily slipped in the "rectilinear figure" presumed at once greater than all the inscribed figures and less than all the circumscribed ones. If, for instance, this intermediate figure is being generated along with the sequence of bounding polygons, then it will effectively turn out to be an infinite-sided regular polygon, namely the circle itself. But a subtle modification of the argument can avoid this difficulty. For each bounding polygon, let us construct the equal square. As the number of sides increases, the sequence of inscribed polygons gives rise to a sequence of squares of increasing area, while the sequence of circumscribing polygons gives rise to a sequence of squares of decreasing area. Since, furthermore, all the squares in the one sequence are less than all the squares in the other, there will be a square intermediate between the two sequences. That the difference between the circumscribed and the inscribed polygons can be made arbitrarily small is geometrically obvious and can be secured by a relatively straightforward bisection argument.[74] Hence, the intermediate square is unique, from which its equality with the circle follows from Bryson's principle. In this way, a Weierstrassian procedure for *defining* the magnitude of the circle might be foreshadowed, if dimly, in Bryson's approach. This is the interpretation adopted by some modern writers.[75] But we do well to note that the ancient commentators on Bryson seem to have no firm account of the details of his argument and do not themselves give a precise statement of his principle and a valid form of its application, despite their own acquaintance with the limiting methods in the tradition after Eudoxus. Indeed, the issue at the heart of Bryson's sophism, the articulation of the conditions for equality between

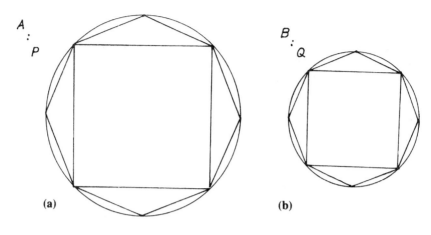

Figure 15

continuous magnitudes, receives no fully satisfactory explication either among the philosophers or among the geometers of antiquity. One encounters instance upon instance in the writings of Euclid, Archimedes, and others where notions of continuity enter, based on intuitive assumptions rather than on explicit formulations.[76]

Bryson's argument must reflect an interest among mid-4th-century geometers in the formal questions entailed by the assumption that problems like the circle quadrature do indeed have solutions. In this context Eudoxus devised a special technique, called by modern writers the "method of exhaustion" or the "indirect method of limits," which could secure the demonstration of theorems like those required by Hippocrates, even if it left in abeyance some of the deeper foundational issues implicit in Bryson's argument. The Eudoxean method, as we learn principally from Archimedes, is the basis of the measurements of the circle, pyramid, cone, and sphere in the manner now extant in Euclid's Book XII. Archimedes himself, whom we may perceive to be working from Eudoxus' treatments in preparing such early works of his own as the *Dimension of the Circle*, applies essentially the same technique one finds in the *Elements*.[77]

Let us then sketch Euclid's theorem on the area of the circle (*Elements* XII, 2) as a representative instance of this method. One proposes to demonstrate that circles have the ratio of the squares on their diameters. Let the circles be A, B, and the squares be C, D [see Fig. 15]. If it is not true that $A : B = C : D$, then for X such that $A : X = C : D$, X will either be less than or greater than B. Assume first that X is less than B. We now inscribe in B a regular polygon, form from this the inscribed regular polygon having twice as many sides, and continually repeat this procedure on the newly formed polygons. Since the difference between the circle and the polygon is reduced by more than half at each step in this procedure,[78] we will eventually obtain a polygon Q such that $B - Q$ is less than $B - X$[79]; that is, Q is greater than X. Now inscribe in A the regular polygon P similar to Q. Since P : Q = C : D = A : X and P is less than A, it follows that Q is less than X. This contradicts our having found Q greater than X. Hence, X cannot be less than B. One might conceive an exactly analogous procedure utilizing circumscribed polygons whereby the alternative assumption that X is greater than B is reduced to a contradiction. But Euclid perceives a much simpler alternative. Using the same notation as before, we posit the quantity Y such that Y : B = A : X = C : D. Since X is greater than B, Y is less than A, so that we are in precisely the same situation as before: the existence of such Y leads to a contradiction, so that X cannot be greater than B either. Hence X equals B, and the stated proportion of the circles and the squares of the diameters is thus affirmed.[80]

The procedure of inscribing regular polygons and successively doubling the number of sides has an evident precedent in Antiphon's circle quadrature, although Eudoxus surely drew it from Hippocrates' proof of the theorem on the ratios of circles,[81] for Eudoxus' proof had the effect of rigorizing the treatment of this very same theorem. The key insight in the revised version is the appeal to successive bisection in the context of an indirect proof. The lemma that any finite magnitude can be reduced by repeated bisection so as to become less than any assigned finite magnitude of the same kind is characteristic of all the limiting arguments in *Elements* XII and is retained in the majority of limiting theorems of Archimedes. It is proved in *Elements* X, 1; but the method employed there does not conform to the style in Book XII.[82] One detects that this lemma could be deemed sufficiently obvious as not to require an explicit proof at the initial stages of Eudoxus' theory, but that later disciples sought to ground it on a more fundamental axiom, comparable to the "Archimedean axiom" on continuous magnitudes.

It is worth noting certain features which do *not* characterize Eudoxus' method. Convergence is always one-sided, usually via a sequence of inscribed figures. The device of two-sided convergence, via sequences of paired inscribed and circumscribed figures approaching each other arbitrarily closely and hence approaching a common limit, is typical of Archimedes' more advanced measurements, but appears to have no precedent in earlier technical writings.[83] It thus seems implausible that Bryson's sophism on circle quadrature included a clear representation of the two-sided manner of limits, so that the relatively unsophisticated form of the argument suggested by the report of Alexander (where the inscribed rectilinear figure continuously increases toward the circumscribed figure, and so at some time passes through a figure equal to the circle in magnitude) seems more likely to capture Bryson's attempt than the more elaborate version proposed by Proclus. Another striking feature of Eudoxus' method is its failure to utilize the forms of the theory of proportion given in *Elements* V. For comparison, we may reconstruct an alternative treatment in the Euclidean manner[84]:

To prove A : B = C : D, let us suppose the contrary. Then, first, let A : B be greater than C : D. There exist integers m, n such that $mA > nB$, while $mC \le nD$.[85] Since $A > n/mB$, we may inscribe in A a regular polygon P such that $P > n/mB$.[86] Let Q be the polygon similar to P inscribed in B. It follows that $mP >$

nQ. Now P and Q are similar, so that P : Q = C : D; since mC ≤ nD, one has mP ≤ nQ. This contradicts the previous inequality relating P and Q. If, then, A : B is less than C : D, the inverted ratios satisfy the opposite inequality, so that the same contradiction results. Thus, A : B = C : D.

The most straightforward reason for Euclid's failure to adopt such a procedure in the case of this and comparable theorems in Book XII is that Book XII owed its origin to earlier work than that which yielded the proportion theory presented in Book V. While the latter is usually ascribed to Eudoxus, an alternative proportion theory conforming quite closely to the style adopted in Book XII is implicit in certain proofs in Archimedes and other later writers.[87] One would thus infer that Book V reflects modifications in Eudoxus' theory by his followers, but that Euclid's treatment of the limiting theorems in Book XII adheres to the methods actually used by Eudoxus.

The principal advantage which Euclid could have obtained by employing this reconstructed alternative method for the theorem on circles is its elimination of the need to hypothesize the existence of the magnitudes X and Y satisfying the proportionalities A : X = C : D = Y : B. This hypothesis of the existence of the fourth proportional is a fixture in all applications of the Eudoxean-Archimedean method of limits, as well as in the reconstructed Eudoxean proportion theory mentioned above. It is one of several striking instances of the appeal to nonconstructive assumptions within the ancient geometry.[88] Like Bryson's intermediate square, these magnitudes can be assumed to exist via appeal to some intuition of the nature of continuous magnitude in general. Here one perceives no concern whatever to ground such appeals on explicit constructions of the hypothesized terms. Indeed, Philoponus' remark on the circle quadrature entirely separates the issue of the *existence* of the square equal to the circle from that of the *construction* of that square, in that the former is assumed from the start as the basis for the latter. Ever since Zeuthen, the view that constructions serve the purpose of existence proofs has been a favorite interpretation among writers on ancient geometry.[89] The present counterinstances should give one pause in subscribing to this view as if it were one the ancients would have held of their own efforts.

Among the ancient constructions of the circle quadrature, the method utilizing the special curve called the "quadratrix" is often assigned a pre-Euclidean origin. Pappus cites its use by Menaechmus' brother Dinostratus and by Nicomedes (late 3rd century B.C.), and the latter is affirmed by Iamblichus.[90] But in two places Proclus associates this curve with a geometer named Hippias.[91] If this is indeed Hippias of Elis, the Sophist, a contemporary of Socrates, the quadratrix would be the earliest known instance of a special curve among the ancients, and Hippias would deserve to be ranked foremost in mathematical expertise not only among the Sophists of his generation, but among the geometers as well.[92] But the technical sophistication of this would then so strongly foreshadow discoveries by Eudoxus and Archimedes that one is immediately put on guard. A too hasty acceptance of its attribution to such an early figure threatens to distort one's natural view of the chronology of the ancient technical methods by as much as a century or two.

Left to a vote count among modern authorities, the attribution would win acceptance comfortably. Tannery and Björnbo defended the case quite insistently, and it has been adopted by Cantor, Heiberg, Heath, and many others since.[93] Dissident voices include Allman and Hankel among the older writers and Guthrie more recently.[94] The case rests entirely on the fact that Proclus does mention Hippias *of Elis* in one other passage, as an authority to support the claim that "Mamercus, the brother of the poet Stesichorus... had a reputation in geometry" in the time before Pythagoras.[95] Clearly, the context has nothing to do with the quadratrix and signifies nothing of the mathematical expertise of Hippias himself. Does this single passage then justify shifting the burden of proof onto those who doubt the identification of Hippias the sophist with Hippias the geometer? The substantial body of fragments from the Sophist give no evidence of interest or competence in technical matters.[96] To be sure, Plato has Socrates, in the dialogue *Hippias Minor*, address him as "a skillful calculator and arithmetician... the wisest and ablest of men in these matters," to which Hippias assents quite unequivocally; he is later noted as adept in geometry and astronomy as well, and indeed boasts of his own skill in all the arts, including the making of rings, seals, the very shoes, cloak, and tunic he was then wearing, not to omit the arts of memory, of writing prose and poetry of all forms, of music and so on.[97] The point is surely clear: Hippias is being ribbed as a self-styled jack of all trades, but an actual master of none in particular. This is slippery ground to base a claim of his accomplishments in the field of geometry. Can one so readily disregard the negative testimony of Aristotle, who cites the circle quadratures by Hippocrates, Antiphon, and Bryson, but none by Hippias? The later Aristotelians, for all their ample commentary on these passages on the circle quadrature, likewise omit any mention of Hippias. Yet we may infer from Proclus that this man's contribution was hardly a casual one:

> Apollonius proved the properties of each of the conic lines, and Nicomedes (did the same) for the conchoids, and Hippias for the quadratrices, and Perseus for the spirics [i.e., the sections of the torus].[98]

Thus named in the company of late 3rd-century figures like Apollonius and Nicomedes, known for entire treatises devoted to the properties of the curves associated with them, Hippias would appear to be credited here for a comparable treatment of a *class* of curves, the "quadratrices" (note the plural). Yet several writers, including Proclus himself, assign a share in the study of this curve to Nicomedes also; indeed, it was Nicomedes, according to Iamblichus, who appears to be responsible for naming the curve "quadratrix" *(tetragōnizousa)* in recognition of its role in the quadrature of the circle. One wonders what was left for Nicomedes to do, however, if Hippias the Sophist had so advanced the study of this curve more than two centuries earlier. When we consider further that the name Hippias was quite common,[99] how can we deny that the Hippias

82   Ancient Tradition of Geometric Problems

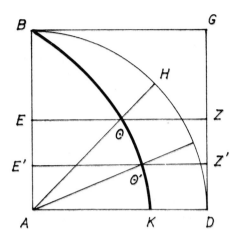

Figure 16

of the quadratrix was not the 5th-century sophist, but rather a 3rd- or 2nd-century geometer extending the work of Nicomedes?

Thus, strictly on the basis of textual evidence, the case for assigning this curve to the sophist is exposed as implausible. But if the technical implications of this attribution had been considered by its advocates, it would never have been entertained seriously in the first place. We shall now find it useful to provide a brief account of this curve, in order to show how it relates to 3rd-century elaborations of the Eudoxean methods.[100] With reference to the square ABGD, let us conceive that the ray AB rotates on A from an initial position at AB to a terminal position at AD, while in the same time BG moves parallel to itself from BG to AD (see Fig. 16). The moving lines will thus intersect at a point θ which traces a curve running from B to a point K lying on AD. One may show that the given length AB is the mean proportional between the terminal length AK and the length of the arc of the quadrant BD. Since one can construct the line which is the third proportional of the lines AK, AB,[101] this line will equal the quadrant BD. Having thus rectified the arc of the quadrant, one can effect its quadrature, since its area is equal to one-half the product of its radius AB and its arc BD.[102]

As one can well anticipate, the full proof of this construction, such as one finds it in Pappus, depends heavily on the use of limiting methods of the Eudoxean type.[103] It is set out in the indirect form and assumes properties of similar arcs whose formal proofs may be associated with Archimedes.[104] A certain subtlety of method is required even to secure the conception of the terminal position at K; for at this position the defining rays coincide. This need for an appeal to continuity to specify K occasioned misgivings over the method among later commentators like Sporus.[105] One can hardly imagine that it would have failed to spark discussion among 4th-century geometers and philosophers and that some reflection of this would emerge in the comments by Aristotle and his interpreters. One further notes that the method actually effects a *rectification* of the arc BD. Yet Aristotle is quite ignorant of any such method, when in the *Physics* (VII, 4) he criticizes a more primitive method of the following sort:

> What then is the situation (with respect to motions) in the case of the circle and the straight line? For it would be absurd to suppose that this cannot be moved in a straight line similarly to that in a circle, but that the straight must necessarily move either faster or slower .... Nor does it matter at all if one were to say that the straight does in fact necessarily move faster or slower. For then the arc shall be greater and less than the straight line, so that also (it shall be) equal. For if in the time A the one traversed the (line) B, the other the (arc) G, then B would be greater than G. For that is how "faster" was defined. Then also the faster (shall traverse) the equal in a lesser (time), so that there shall be some part of (time) A in which that (moving along) B shall traverse the equal of the circle, while that (moving along) G (shall traverse) G in the whole time A. To be sure, then, if these are compatible (*symblēta*),[106] what has just been said will follow, namely, a straight line shall be equal to a circle. But they are not compatible; hence, neither are the motions.[107]

At the heart of this argument is the Brysonian principle: that if a varying magnitude can become now greater, now less than a specified value, at some time in the interval it will equal that value. Here, the inequalities are arranged through the assumption that circular and rectilinear motions can be compared to each other via the relational terms "faster" and "slower." It is hardly an advertisement of Aristotle's geometric insight that he dismisses so perfunctorily this perfectly plausible rationale for the existence of the line equal to the circle. But one may note that Aristotle invariably treats with great respect the technical findings of such colleagues as Eudoxus and his followers.[108] He would hardly have set aside so casually the present argument, were its conclusion and the kinematic conceptions supporting it already secured through the quadratrix construction by Dinostratus, let alone by Hippias over a half-century earlier.

The lemma associating the rectification of a circular arc with the area of its sector is indispensable if one is to derive from the quadratrix a solution of the circle quadrature. A proof is extant as the first proposition of Archimedes' *Dimension of the Circle*, and Pappus cites Archimedes explicitly both in this passage on the quadratrix and in numerous other places in the *Collection* where this same lemma is introduced. Other writers like Hero, Theon, Proclus, and Eutocius invariably assign this result to Archimedes, often with specific reference to the *Dimension of the Circle*.[109] We have no evidence whatsoever to justify assuming an awareness of this result by an earlier Greek geometer. This must certainly rule out any idea of its use by Hippias a century and a half before Archimedes. But it also discourages assigning its use by Dinostratus, around three-quarters of a century before Archimedes, for the manner of proof adopted by Archimedes involves only such techniques as were available to any geometer in the Eudoxean tradition.[110] For instance, he uses the familiar bisection method in the context of two independent applications of one-sided convergence, first

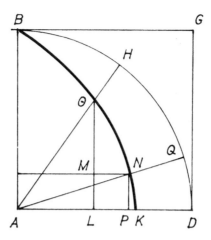

Figure 17

by inscribed regular polygons, then by circumscribed polygons. None of the more sophisticated limiting methods, such as appear in his *Sphere and Cylinder*, are brought to bear on the circle measurement. Can one comfortably accept the implication that pre-Euclidean geometers freely admitted this lemma in advanced constructions, like that involving the quadratrix, but that its elementary proof was overlooked for decades until Archimedes?

These considerations thus reveal the utter implausibility of assigning the use of the quadratrix to a Presocratic like Hippias of Elis. But they raise the question of reconciling a post-Archimedean dating with the testimony of Pappus that "for the quadrature of the circle a certain curve has been received by Dinostratus and Nicomedes and some other more recent figures; it takes its name from the property relating to it, for it is called by them 'quadratrix'."[111] As we have already noted, Iamblichus associated this naming of the curve with Nicomedes alone, but assigns no involvement on the part of Dinostratus. This suggests that Dinostratus might have introduced this curve in the context of a problem other than the circle quadrature. Pappus shows in detail how the quadratrix may be used for the division of an angle.[112] Let us propose to divide a given angle into three equal parts. We introduce the quadratrix BθK and draw the ray AH making with AD an angle equal to the given (see Fig. 17). Let AH meet the curve at θ and draw the ordinate θL; next divide θL at M such that ML is its third part. If a perpendicular to θL is drawn through M, it will meet the curve at a point, say N. Then the ray AN makes with AD an angle one-third that of the given angle HAD. This is evident at once from the motions which define the curve; for they entail that the ordinates have the same ratio as the associated arcs, e.g., that θL : NP = arc HD : arc QD, hence, also, as the associated angles (HAD, QAD).

It is not difficult to imagine how a curve of the above sort might be introduced for the purposes of the angle division. Aristotle's critique of the attempted rectification of the arc suggests an extremely simple practical procedure: figure the given angle as HAD in a circle, and with a cord or flexible ruler lay out the arc HD; next straighten the ruler and divide the segment HD in the desired ratio (say, QD : DH = 1 : 3); now lay the ruler back against the circle to find the arc QD, whence angle QAD will be the desired part of HAD. Thus, the angle division brings forward a natural association of curvilinear and rectilinear magnitudes. Further consideration might lead one to try to correlate circular and linear motions, the one traversing arc BD uniformly in the same time that the other traverses the straight line equal to the arc BD, and then to replace the latter with the precisely similar motion traversing the straight line BA. Although it would hardly be a trivial step to contrive in this way a curve answering these motions and work out its application for the angle division, the way has been well prepared through the kinematic constructions of curves by Archytas and Eudoxus. In this way Dinostratus was in the position to convert the simple procedure, too hastily dismissed by Aristotle, into an ingenious method based on a new curve. A slightly different orientation of the same motions yields the spiral studied somewhat later by Archimedes and his colleague Conon.[113] One would thus suppose that their discoveries relating to the spiral led Nicomedes to discover the use of the former curve for the circle quadrature, and so to dub it the "quadratrix." Only after him, around the beginning of the 2nd century B.C. at the earliest, could the geometer Hippias compile these findings into a writing on this curve.

This view of the origins and study of the quadratrix charges Pappus with an erroneous, or at least misleading, statement in his giving to Dinostratus a share in the use of this curve for circle quadrature. But he is hardly free of similar inaccuracies elsewhere. For instance, he writes thus of the early efforts on angle trisection[114]:

> Having set out to divide the given rectilinear angle into three equal parts, the ancient geometers were at a loss for the following reason. We say that there are three kinds of problems in geometry: some are called planar, some solid, and some linear. [Pappus now explicates this classification in detail.] Since there is this difference among problems, the former geometers were not able to discover (a solution for) the cited problem of the angle, being by nature of the solid class, when they sought it by means of planes.[115] For the sections of the cone were not yet familiar to them, and so they were at a loss. Later, however, they trisected the angle by means of the conics, using for its discovery the following *neusis*.

Pappus' thinking here is apparent. The angle trisection is assuredly a solid problem, that is, one solvable by means of conic sections. Hence, its *proper* solution would elude geometers until they had access to the requisite body of theorems on conics. The method he goes on to discuss betrays strong signs of a provenance near the time of Apollonius, however, so that Pappus might be construed to mean that no earlier methods of angle trisection had been worked out. This, of course, is an inference which no one would accept. Methods using the "quadratrix," the spirals or the *neuses* of Archimedes, or the conchoids of Nicomedes would all be assignable to periods earlier than the end of the 3rd century. Pappus has allowed his use of the conceptions and terminology of the later development of geometry to distort his characterization of the earlier efforts, much as we have argued Eratosthenes did in alluding to the "cone sectioning" of Menaechmus. But if indeed Pappus supposed that the mere introduction of a certain curve, as by Dinostratus, carried with it a knowledge of the whole range of its applications, embracing both the solution of the angle trisection and that of the circle quadrature, we need hardly imitate his mistake. For it is surely plain that the latter effort at least depended on advances in geometry due to Archimedes.

GEOMETRY AND PHILOSOPHY IN THE 4TH CENTURY

The technical methods of geometry experienced significant advances at the hands of the group of geometers affiliated with Plato's Academy. Problems like the cube duplication and the angle trisection were attacked and solved in a variety of ways. Special curves were devised for these ends via ingenious intuitions of mechanical motions. Archytas, for instance, used the curve formed as the intersection of a torus and a cylinder for doubling the cube; of the same kind was the "curved line" of Eudoxus, arguably the ophiuride generated by the moving lines in the pseudo-Platonic solution; so also was Dinostratus' quadratrix, conceived for trisecting the angle. At the same time the techniques of the application of areas were being implemented for the solution of a range of problems, including the construction of regular polygons and solids. Menaechmus perceived how these give rise to second-order curves by which the cube duplication can be solved, although his method for producing the parabolas and hyperbola so defined is more likely to have been a pointwise procedure than the conception of sectioning the cone familiar to Euclid and his contemporaries several decades later. Furthermore, the method of analysis received considerable development in this period, exploited not only by geometers like Leodamas, Leon, Eudoxus, and Menaechmus, but even by a philosopher like Aristotle in his discussions of problem solving in general.

Responding to the philosophical environment of their efforts, these geometers markedly improved the formal basis of the organization and proof of geometric theorems. Proclus names about a dozen figures, most having an association with the Academy, who contributed to the composition of *Elements* of geometry, precursors of the various parts of Euclid's treatise.[116] Particularly noteworthy were the insights by Eudoxus, who instituted techniques of limits and of proportions in a manner exemplary for its formal precision. The former filled gaps in Hippocrates' treatment of the measurement of the circle, showing how his use of infinite limiting procedures could be effected through finite sequences in the context of indirect proofs. Eudoxus' theory of proportions permitted the extension of theorems on proportions to the cases of incommensurable magnitudes, and so secured the technical advances in the study of irrationals, most notably by Theaetetus in the preceding generation.[117] One thus sees that Eudoxus, however imbued with the spirit of formalism encouraged by his philosopher colleagues in the Academy, nevertheless drew the principal inspiration for his technical efforts from the earlier geometers like Hippocrates, Archytas, and Theaetetus.

This convergence of geometric and philosophical interests has sorely tempted many scholars engaged in the study of these ancient efforts. Accordingly, no other period of the ancient geometry has been so thoroughly laced with inappropriate interpretive positions as this pre-Euclidean period.[118] Concerning the early circle quadratures, one commonly attributes to Sophists like Antiphon and Bryson pioneering insights into the nature of limiting methods, while to Hippocrates one imputes blatant oversights and elementary logical errors.[119] Democritus emerges as a precursor of Archimedes in the devising of a geometric method of indivisibles,[120] while Hippias of Elis is alleged to anticipate his discoveries on the measurement of the circle through the use of the quadratrix. Eudoxus is motivated not by the ambition to advance the technical achievements of earlier geometers, but rather by the concern to accommodate the arguments of Zeno and Aristotle against the use of infinites.[121] Indeed, dialectical motives like the search for existence proofs are read into the whole ancient activity of solving geometric problems of construction.

Whence this tendency to distort the interpretation of the ancient geometry? Reconstructing the motives of others, even of near contemporaries, is always an uncertain enterprise. But I suspect, in this instance, that the particular intellectual background of the principal commentators is a significant factor. Tannery and Björnbo, for instance, those most responsible for the wide acceptance of the case relating to Hippias of Elis, are remembered primarily for their competence in philosophy and philology. How natural that they should seize upon the notion that a man of their own mind, like this Sophist, could have influenced the development of geometry! By contrast, no such presumption marked the view of Hankel, by profession a mathematician. One observes, furthermore, that the period of the late 19th and early 20th centuries witnessed a phenomenal shift in perspective on the nature of mathematics.[122] Developments in analysis, geometry, number theory, and the theory of sets, for instance, revealed that certain formal issues, long ignored or unperceived, lay at the heart of important technical difficulties, and one might even maintain that the whole of mathematics was subsumed within logic. Hard on the heels of the satisfactory resolution of foundational questions in analysis and geometry, however, came a new "foundations crisis" in the field of logic itself. In all this, the close interaction of mathematical and philosophical inquiries was evident as a stimulating area for research. From this vantage point one could readily project a comparable form of interaction between the philosophical paradoxes raised by Parmenides and Zeno and the geometry of the early Pythagoreans; or between the philosophical repercussions of the discovery of irrationals and the choices of technical methods by the 5th- and 4th-century geometers. One hypothesized a paralysis of research, a renunciation of the use of proportions in geometry, and so on, until this impasse was finally broken through the efforts of Eudoxus.[123] One seemed not to care that the extant evidence, fragmentary as it is, indicates no signs of such paralysis or renunciation, but rather a remarkable degree of continuity in the development of technical methods.

The alternative view advocated here acknowledges the significant interaction between geometry and philosophy over the special field dealing with the formal nature of proof, but that, on the whole, the two disciplines developed autonomously. Is that remarkably different from the course of modern mathematics? Save for those actively engaged in the study of foundational questions, how many mathematicians accept the general implications of those studies as an immediate factor in their own researches? It is interesting to consider the case of O. Becker, a prominent interpreter of pre-Euclidean geometry. As an adherent to the Husserlian phenomenology in the 1930's, Becker was especially sensitive to the questions of method raised by the earlier mathematical intuitionists. He thus probed the work of Eudoxus for signs of a similar concern.[124] If he did indeed discern formal aspects of the development of proportion theory never so clearly perceived by scholars before him, he nevertheless failed to find evidence of other concerns of the modern philosophy of mathematics, like the stricture against nonconstructive assumptions in proofs. But how many of Becker's own contemporaries shared these concerns in their researches?

This habit of reading present views into the past is hardly a recent phenomenon. The later Pythagoreans told the tale of one Hippasus doomed to shipwreck for his sacrilege of having divulged the secret of the irrational to the uninitiated.[125] The story of the Delian oracle seeks to link Plato directly with the early researches on cube duplication, and Eratosthenes converts it into a parable of Plato's philosophy of mathematics.[126] Here is material for theories of "foundations crises" and philosophers' influence on the progress of geometry, should anyone seek it. Let the ancients have their fun, spinning yarns to illustrate their favorite notions in philosophy. But we need not take these too seriously as guidelines for tracing the history of the ancient technical researches.

NOTES TO CHAPTER 3

[1] *In Euclidem*, p. 66.

[2] On these geometers, see my *Evolution of the Euclidean Elements*, Ch. 3, 7, 8. Short accounts may be found in the *Dictionary of Scientific Biography*.

[3] Perhaps the most ambitious attempts to assess the mathematical component of Aristotle's thought are T. Heath, *Mathematics in Aristotle* and H. J. Waschkiess, *Von Eudoxos zu Aristoteles*.

[4] Cited by Eutocius in Archimedes, *Opera*, ed. Heiberg (2nd ed.), III, p. 96.

[5] *Ibid.*, pp. 84–88. An extremely lucid account is given by B. L. van der Waerden, *Science Awakening*, pp. 150 f. See also O. Becker, *Mathematisches Denken der Antike*, pp. 76–80 and T. L. Heath, *History of Greek Mathematics*, I, pp. 246–249.

[6] Becker (*op. cit.*, p. 79) presents this curve via a somewhat modified construction and cites its use by Villapaudo, as reported by Viviani (1647); see also G. Loria, *Ebene Kurven*, 1902, p. 317. Kepler's use of the same curve is noted by W. Breidenbach, *Das delische Problem*, Stuttgart, 1953, pp. 31 f.

⁷ For this part of the analysis, one may compare the account by van der Waerden. The principal difference lies in that I seek a determination of the locus of point I as a result of the analysis, while van der Waerden introduces it as an unexplained intuitive insight at the start.

⁸ Eutocius, *op. cit.*, p. 56.

⁹ See, for instance, Heath, *History*, I, p. 249; and I. Thomas, *Greek Mathematical Works*, I, p. 260n.

¹⁰ P. Tannery, "Lignes courbes," *Bulletin des sciences mathématiques*, 1884, 19, p. 101; cf. also *Mémoires de la Société des sciences physiques et naturelles de Bordeaux*, 2_2, pp. 277–283. Accounts appear in Heath, *History*, I, pp. 249–251; and Becker, *op. cit.*, pp. 78–80. Breidenbach also notes Descartes' use of this curve (*op. cit.*, pp. 29 f); cf. the *Géométrie* (ed. Smith and Latham), pp. 154–157.

¹¹ One may note that the curve generated via the condition AR = AS [Fig. 6(a)] is extendable beyond the range of the intersection of the torus and the cone (i.e., up to the point where AR = AB). Setting PS = y, AS = x, and AN = a, one obtains $y : x = \sqrt{x^2 - a^2} : a$, or $a^2y^2 = x^2(x^2 - a^2)$. The curve extends to infinity with increasing slope as x increases, becoming asymptotic to the parabola $ay = x^2$. Its inflection point V lies at the intersection of the line $x = $ Aa $= \sqrt{3/2}\, a$ and the circle of diameter Ac $= 1/2\sqrt{1/2}\, a$, so that AV = $3/2\, a$. (Fig. 6(b)) V here corresponds to I in the construction of the two mean proportionals AK, AI between AD, AB, for AD : AB $= \sqrt{3\sqrt{3}} : \sqrt{2\sqrt{2}}$. For Ab : AN = 3 : 2, V will correspond to T, the terminus of intersection of cone and torus. The problem of doubling the cube corresponds to Ae : AN = 4 : 1.

¹² R.C. Riddell, "Eudoxan Mathematics and the Eudoxan Spheres," *Archive for History of Exact Sciences*, 1979, 20, pp. 1–19.

¹³ In the modern theory of curves, this is termed the "spherical lemniscate"; projected onto the plane tangent to the cylinder at the double point, it becomes the Bernoullian lemniscate; cf. C. Zwikker, *Advanced Plane Geometry*, Ch. XVII. For its relation to the sections of the torus, see Chapter 6.

¹⁴ This sketch is not intended to be self-sufficient. For details, consult Riddell. Other treatments are given by O. Neugebauer, *Exact Sciences in Antiquity*, pp. 153 f, 182 f; van der Waerden, *op. cit.*, pp. 180–182; and Heath, *History*, I, pp. 329–335.

¹⁵ Detailed critiques of Eudoxus' astronomical theory are given by Neugebauer, *History of Ancient Mathematical Astronomy*, pp. 677–685; Heath, *Aristarchus*, Ch. 16; and L. E. J. Dreyer, *History of Astronomy...*, Ch. IV.

¹⁶ Cf. Proclus, *In Euclidem*, pp. 112, 127 f.

¹⁷ Drawings of this curve are provided by Riddell, *op. cit.*, Figs. 12, 14.

¹⁸ Under this condition, the projection curve is a two-cusped epicycloid, as shown in Fig. 8(b), and is related to the epicyclic curves discussed in Chapter 7. When 0G is less than GQ, the curve will have two loops in place of the cusps, and when 0G is greater than GQ, the curve will arch smoothly inward at these positions. Riddell's Fig. 11 illustrates a curve of the latter form.

¹⁹ This is Riddell's solution method. But it leads to an alternative which dispenses with reference to the rotating spheres. If we conceive of the plane curve as generated by the epicyclic motion of circle JRH, then the perpendicular from T will meet it in R, while the generating circle will have assumed the position JRH; thus, 0J is determined, from

which the construction of 0S, 0V follows. If we admit the epicyclic curve alone, without reference to the continually changing positions of circle JRH, nevertheless, the specific position of that circle corresponding to the point of intersection R can be produced by a planar construction; for the circle will pass through a given point and be tangent to two given circles. The latter problem falls under one of the classes of planar constructions treated by Apollonius in his *Tangencies*; cf. Chapter 7.

[20] But one may note that the modified Eudoxean scheme which Schiaparelli would assign to Callippus introduces the sphere of double speed, as in Riddell's construction. The three-sphere arrangement, however, produces a more complicated curve in which the figure-8 obtains additional double loops at each extremity. For a discussion, see Heath, *Aristarchus*, pp. 212–216.

[21] Eutocius, *op. cit.*, pp. 56–58.

[22] See the discussion in Chapter 2. According to the lexicon of Suidas, Eratosthenes was known as "a second (*deuteros*) or new (*neos*) Plato"; cf. I. Thomas, *op. cit.*, II, p. 260.

[23] See, in particular, *Republic* 510 c–e, 525 ff.

[24] This is comparable to the view proposed by van der Waerden, *op cit.*, pp. 159–165. He does not actually attribute the pseudo-Platonic construction to Eratosthenes; but that appears to be implied in his account.

[25] Cf. Loria, *Ebene Kurven*, I, p. 50. A more detailed discussion appears in Chapter 6 (note 107).

[26] Banū Mūsā, *Verba filiorum*, Prop. 17; cf. M. Clagett, *Archimedes in the Middle Ages*, I, pp. 340–344. Further discussion will appear in the sequel. Only the geometric form of the construction is given in the Latin version of this work (made by Gerard of Cremona in the 12th century). But the Arabic version, extant in the recension by al-Ṭūsī (13th century), adds to this the details of the physical mechanism by which the construction can be produced.

[27] *Verba filiorum*, Prop. 16; cf. Clagett, *op. cit.*, pp. 334–340.

[28] Note, for instance, the expressions "to the extent that..." (*epi tosouton mechris an*) and "the beam has the position such as has..." (*ton kanona thesin echein hoian echei...*) used here by Eutocius (*op. cit.*, p. 58) and their reappearance in his versions of the methods of Philo and Apollonius (*ibid.*, pp. 62, 64, 66). His mechanical terminology parallels that in his texts of Nicomedes (*ibid.*, pp. 98–100) and of Pappus (*ibid.*, p. 70; cf. *Collection* VIII, p. 1070), less that in his text of Eratosthenes (*ibid.*, pp. 92–94).

[29] Eutocius, *op. cit.*, pp. 78–84.

[30] See, for instance, H. G. Zeuthen, *Lehre von den Kegelschnitten*, Ch. 21, especially pp. 457–466 and the synopsis of "Menaechmus' probable procedure" by Heath, *History*, I, pp. 251–255 and II, pp. 110–116 and his *Apollonius*, Ch. 1.

[31] After sketching Zeuthen's reconstruction of a Menaechmean derivation of the asymptote property of the hyperbola, Coolidge observes: "I confess to finding it a bit difficult to believe that Menaechmus could work out all this. But we are forced to accept something of the sort if we are to credit him with the discovery of the conic sections in connexion with the problem of duplicating the cube" (*History of the Conic Sections*, p. 5). Such rare, but well-founded misgivings are set at rest in the view proposed below.

[32] Both relate to the elliptical appearance of circles viewed obliquely. For discussion, see Chapter 4.

[33] For accounts of the early theory, see Zeuthen, *op. cit.*, Ch. 2, 19, 20, and Heath, *Apollonius*, Ch. 3. A compilation of Archimedean theorems on conics is given by Heiberg in "Die Kenntnisse des Archimedes über die Kegelschnitte," *Zeitschrift für Mathematik und Physik*, hist.-litt. Abth., 1880, pp. 41–67.

[34] Eutocius, *op. cit.*, p. 96.

[35] Note that the curves discussed above as reconstructions proposed for Eudoxus' method rely on pointwise procedures, as do several of the accounts of other curves, like the quadratrix; cf. van der Waerden, *op. cit.*, p. 192.

[36] See *Diocles*, ed. Toomer, Props. 12 and 4, respectively. Eutocius (*op. cit.*, pp. 66–70) presents a version of the former in which the points are joined by rectilinear elements, rather than by means of a flexible ruler. But the flexible ruler is clearly indicated in Diocles' own account; cf. Toomer, pp. 159 f.

[37] Eutocius in Apollonius, *Opera*, ed. Heiberg, II, pp. 230–234.

[38] See Heiberg, *Mathematici Graeci Minores*, 1927, pp. 78 ff, 85 ff. These constructions are discussed in Chapter 6.

[39] Diocles, *op. cit.*, Prop. 10.

[40] Proclus, *In Euclidem*, p. 419. Plutarch, *Moralia*, 720 a, 1094 b; see Heath, *History*, I, pp. 144 f. For accounts of this technique, see Heath, *ibid.*, pp. 150–154; van der Waerden, *op. cit.*, pp. 118–126; and Becker, *op. cit.*, pp. 60–64.

[41] In Fig. 13(a), if CD = X, DF = A, E bisects DF, and DH = B, then the "hyperbolic" case makes DH the mean proportional between CD and DG ($= A + X$). This gives rise to the right triangle EDH having two given sides ED ($= \frac{1}{2} A$) and DH; thus, EH ($= \frac{1}{2} A + X$) is also given, from which CD can be solved. Similarly, in Fig 13(b) the "elliptic" case makes DH the mean proportional between CD and DF, whence ED ($= \frac{1}{2} A - X$) is given from the triangle EDH. For the relation between these analyses and the syntheses given by Euclid, see Heath, *Euclid's Elements*, I, pp. 382–388 and II, pp. 260–267; and Becker, *op. cit.*, pp. 62 f.

[42] See Neugebauer, *Mathematical Cuneiform Texts*, 1945, and his *Exact Sciences in Antiquity*, pp. 40–42; examples are given by van der Waerden, *op. cit.*, pp. 63–71, and their relation to the Greek technique is elaborated, *ibid.*, pp. 122–124.

[43] On the manner of this transmission, see my "Techniques of Fractions." Neugebauer remains vague (*Exact Sciences*, pp. 149–151, 181 f), although he insists on a real link between the two traditions.

[44] *Collection* (III), I, pp. 132–162. This is not to suggest that Pappus' treatment has a pre-Euclidean provenance. These constructions may well have their origin with Aristaeus in the time of Apollonius; cf. Chapter 7.

[45] *In Euclidem*, p. 67. For an interpretation of this reference to Eudoxus, see my *Evolution*, Ch. VIII.

[46] See Chapter 2, especially note 88.

[47] For further discussion, see my *Evolution*, Ch. II/II, V-I/IV.

[48] This results easily from an analysis based on "completion of the square." Assuming $BF^2 = FE \cdot BE$, one has $BF^2 + BF \cdot BE = BE^2$, or $(BF + \frac{1}{2} BE)^2 = BE^2 + (\frac{1}{2} BE)^2$, so that $BF + \frac{1}{2} BE$ is the hypotenuse of a right triangle of sides BE, $\frac{1}{2}$ BE. Setting BG = $\frac{1}{2}$ BE and HG = EG, it follows that BF = HB is the required length. The corresponding synthesis is given in Euclid's II, 11. This form of analysis has been

proposed as underlying some of the computations preserved in Mesopotamian tablets from the mid-2nd millennium B.C.; cf. van der Waerden, *Science Awakening*, pp. 69 f. It thus contributes to the view of the transmission of the older technique to the Greeks during the pre-Euclidean period cited in note 43 above.

[49] R. Fischler proposes some further views on these early studies in "A Remark on Euclid II, 11," *Historia Mathematica*, 1979, 6, pp. 418–422.

[50] One must avoid assigning too much weight to this form of negative testimony, however. Euclid's work is not an exhaustive repository of the geometric knowledge of his time, and one must allow for the prior availability of alternative methods from among which Euclid could choose for the proofs in the *Elements* (cf. my *Evolution*, Ch. I/II). Nevertheless, it is always preferable, when engaging in reconstruction, to have some positive evidence from the surviving documents.

[51] See the discussion of the Aristotelian and Apollonian locus problems in Chapter 4; the related problem preserved by Pappus (Chapter 4, note 16); and the *neuses* from Heraclitus and Apollonius (Chapter 7).

[52] Euclid adopts an oddly roundabout method here, taking angle EFA as an exterior angle to triangle CFA. One could proceed more directly by considering the triangles AFE, CAE which have equal angles at C and A and share the angle at E; thus, the remaining angles EFA, EAC are equal.

[53] Heath reports just such a reconstruction, taken from Todhunter; cf. *Euclid's Elements*, II, pp. 99 f.

[54] See also the translations by Heath, *History*, I, p. 299; and Thomas, *op. cit.*, I, pp. 394–397.

[55] This was proposed by August (1829) and Butcher (1888) and adopted by Heath, *History*, I, pp. 299–303. For a detailed survey of these and many other proposals, see R. S. Bluck, *Plato's Meno*, 1961, pp. 441–461. Bluck discusses the general method in this passage in relation to geometrical analysis, *ibid.*, pp. 76–85.

[56] So Heijboer, cited by Bluck, *op. cit.*, p. 448. Cf. Heath's response to the objections by Benecke, *History*, I, p. 302.

[57] Bluck, *op. cit.*, Introduction, Ch. E, esp. pp. 118-120 reviews the arguments for the dating of the *Meno*.

[58] We provide an elementary proof that the equilateral triangle is the greatest of all triangles inscribed in the same circle. Since of all triangles on a given chord AB [Fig. 19(a)] the isosceles triangle CAB is the greatest, we have only to compare the equilateral with the isosceles cases. First consider the isosceles triangle ADE [Fig. 19(b)] whose base DE is parallel to and shorter than a side BC of the inscribed equilateral triangle ABC. Since the bisector H of arc ADE lies on arc AB, triangle ABE will be greater than ADE; for the altitude BF of the former will be greater than the altitude DG of the latter. From the previous case, the equilateral triangle ABC will be greater than ABE, since the former is isosceles with respect to the common base AB. Hence, *a fortiori*, ABC is greater than ADE. Similarly, if side DE is greater than BC [Fig. 19(c)], ABE will again be greater than ADE, for here the bisector of arc ADE will lie on arc BC. It follows as before that the equilateral triangle ABC is greater than ADE. We have thus established that the equilateral triangle is greater than any other triangle inscribed in the same circle. It seems to me that no feature of this proof of the diorism would have posed any difficulty for a Greek geometer. Thus, the technical effort from which Plato draws here must surely

The Geometers in Plato's Academy 93

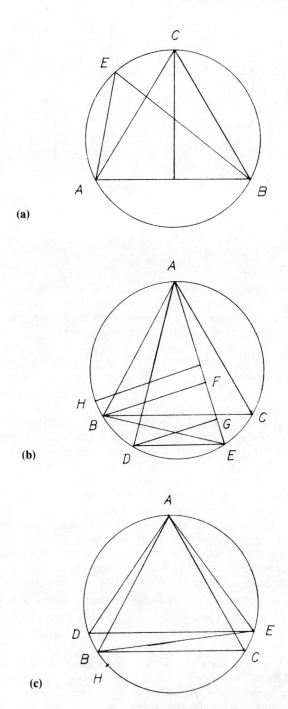

Figure 19

94 Ancient Tradition of Geometric Problems

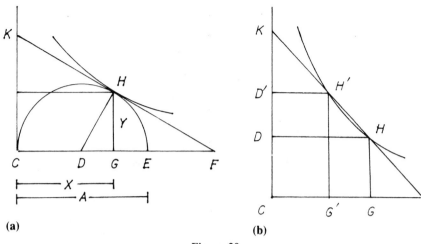

Figure 20

have been directed toward the explicit construction of the inscribed triangle of given area. Our good fortune in learning of this effort through this passage thus owes to Plato's slip in phrasing the diorism in accordance with the terms of the construction problem itself, thus losing sight of its independent treatment via elementary methods.

[59] In Fig. 20(a), setting HG = Y, CE = A, and CG = X, we have, from the property of the tangent to the hyperbola, FH = HK, whence FG = GC (cf. Chapters 5, 6, 7). Since HG is the mean proportional of FG, GD and also of EG, GC (= FG), it follows that DG = GE, so that X = ¾ A. The property of the tangent can readily be deduced from the constant-product property of the curve. Let the curve be figured with respect to the lines FC, CK such that for any points H, H' on it, GH · HD = G'H' · H'D' [Fig. 20(b)], and let the chord HH' meet CF, CK at F, K, respectively; it is claimed that FG = D'H'. For HD : H'D' = G'H' : GH; by similar triangles, G'H' : GH = FG' : FG. Thus, HD − H'D' : H'D' = FG' − FG : FG. Since HD − H'D' = FG' − FG (= GG'), H'D' = FG, as claimed. If the secant is now adjusted so that H' approaches H, in the limiting position of the tangent one has FG = GC, as required above. One may note that this derivation effectively reverses the order followed in Apollonius' *Conics*, for there the constant-product property of the hyperbola (II, 12) follows after the equality of the segments of the secant, FH = H'K (II, 8). As it happens, Apollonius does not use II, 8 for the proof of the lemma II, 10 (that FH · HK = FH' · H'K) on which II, 12 depends. This odd isolation of II, 8 (and its corollary tangent theorem, II, 9) doubtless signifies stages in the composition and transmission of the *Conics*; see Chapter 7.

[60] *In Euclidem*, pp. 220 f. This passage is discussed by Heath, *Euclid*, I, p. 128.

[61] A similar instance is cited by Pappus as an example of an "indeterminate" (*adioriston*) problem: the application with square excess over a given line segment yields widths arbitrarily large or small, depending on the magnitude of the area applied (*Collection* (VI), II, p. 542). Thus, the "unordered" problems of Amphinomus are parallel to the "indeterminate" of Pappus, in that each refers to problems having separate solutions for each choice of a parameter from a specified range. Menaechmus' curves for the cube duplication correspond precisely to this class of problems.

[62] *In Euclidem*, p. 202. A puzzle here is that Aristotle *does* maintain that demon-

strations explain the causes of propositions; cf. *Posterior Analytics* I, 2, 71 b 8 - 32; and 24, 85 b 23 - 28. The difficulty is noted by G. Morrow, *Proclus: A Commentary ...*, 1970, p. 158 n.

[63] *In Euclidem*, pp. 77 f. There would seem to be a discrepancy with Proclus' ascription to Amphinomus of a classification of problems (see note 61 above). But perhaps a commitment to the priority of theorems over problems need not entail the view that the format of problems should be eliminated altogether.

[64] 78 a 7 - 14; cf. also *Prior Analytics* II, 2–4. Other discussions of analytic method in the more general dialectical context appear in *Nic. Eth.* 1112 b 11 - 24 and *Post. Ana.* I, 22, 84 a 6 ff. Such passages are examined by J. Hintikka and U. Remes, *The Method of Analysis*, Ch. 2, in the context of Pappus' account of this method (*Collection*, VII, Preface).

[65] This passage correctly understands analysis as the examination of the deductive consequents of the hypothesis; other passages in Aristotle and later writers (see the references in note 64) attempt to construe it alternatively as a search for appropriate antecedents, but this confusing feature conflicts with the unanimous testimony of the mathematical evidence. It would also render superfluous any concern over convertibility, since the analysis (of antecedents) would of itself have produced the deductive sequence of the synthesis. I shall not attempt to press further these early passages on the nature of analysis, but the reader who wishes to do so may consult J. Barnes, "Aristotle, Menaechmus, and Circular Proof," *Classical Quarterly*, 1976, 26, pp. 278–292 for a more ambitious interpretation.

[66] *Soph. Ref.* xi and *Prior Ana.* I, 9. For discussion see Heath, *Mathematics in Aristotle*, pp. 47–50.

[67] Philoponus, *In Ana. Post.*, ed. Wallies, pp. 111–114 cites views of Alexander, Ammonius, and Proclus concerning Bryson's argument. Simplicius' long passage on the circle quadrature does not consider Bryson (*In Physica*, ed. Diels, I, pp. 53–69).

[68] The specification of *squares* appears only in the comment by pseudo-Alexander (*In Soph. El.*, ed. Wallies, p. 90); on this basis one has been led to speculate whether Bryson intended that the square formed as the arithmetic or the geometric mean of the constructed squares ought to equal the circle. But in the paraphrase by Philoponus, Alexander considers an unrestricted "rectilinear figure," thus indicating a more sophisticated view.

[69] So Proclus, as cited by Philoponus, *op. cit.*, pp. 111 f. Proclus objects to Alexander's interpretation on the grounds that it would make the fallacy of Bryson's argument the same as that of Antiphon's: that the circle will coincide with the straight line. Proclus thus seems to have in mind the case where the inscribed and circumscribed figures are so close to each other that the intermediate polygon becomes indistinguishable from either and hence coincides with the circle. This need not be rejected as a possible view of Bryson's argument, even if Proclus himself goes on to propose an alternative.

[70] *Op. cit.*, pp. 112 f. The principle already appears in the Aristotelian *Mechanics* (847 b 25): that whatever changes from one extreme to the other (e.g., from convex to concave, or from great to small) must sometime be in the mean state (e.g., straight or equal). It may well be rooted in Academic discussions about the greater and lesser, the limit and the unlimited, such as in Plato's *Philebus* (23 c - 27 c).

[71] *Ibid.*, p. 112. Cf. Proclus, *op. cit.*, p. 234, who observes that in the case of the

horned angle, one does not pass from the lesser to the greater through the equal; that is, no rectilinear angle equals the curvilinear.

[72] Mixed angles are introduced in the pseudo-Euclidean *Catoptrics*, Hero's *Catoptrics*, Euclid's *Optics* in the recension by Theon, and the Bobbio mathematical fragment (discussed in Chapter 6, note 99). For references, see Toomer, *Diocles*, pp. 156 f and my "Geometry of Burning-Mirrors in Antiquity."

[73] *Op. cit.*, p. 112.

[74] Archimedes proves this in a more sophisticated manner in *Sphere and Cylinder* I, 5, 6.

[75] So Heath, *Mathematics in Aristotle*, pp. 48 f. See also Becker, "Eudoxos Studien II" and "III" (*Quellen und Studien*, 1933 and 1936, respectively) and Ian Mueller, "Aristotle and the Quadrature of the Circle" in *Infinity and Continuity in Ancient and Medieval Thought*, ed. N. Kretzmann. Mueller quite plausibly injects a kinematic element into the conception of the sequence of squares.

[76] For instance, the existence of the fourth proportional is regularly assumed in the limiting theorems of Euclid and Archimedes, doubtless in agreement with Eudoxus' procedure (see discussion below). See also Becker, "Eudoxos-Studien II" and my "Archimedes and the Pre-Euclidean Proportion Theory." Mueller points to this as a characteristic nonconstructive feature in Euclid's geometric methods; see his *Philosophy of Mathematics . . . in Euclid's Elements*, 1981, Ch. 3.2, 6.3.

[77] On Archimedes' use of Eudoxean techniques, see my "Archimedes and the *Elements*" and "Archimedes and the Pre-Euclidean Proportion Theory." For accounts of Eudoxus' method of limits, see van der Waerden, *Science Awakening*, pp. 184–187 and Mueller, *Philosophy of Mathematics*, Ch. 6.3.

[78] Euclid proves this by considering the rectangle enclosing each segment on the sides of the inscribed polygon. Since the inscribed triangle equals exactly one-half the rectangle, while it, in its turn, is greater than the segment, the triangle is greater than one-half the segment. Hence, the passage from one polygon to the next will remove more than one-half the difference in area between the circle and the initial polygon.

[79] This follows from the lemma (X, 1) on the indefinite diminution of a finite magnitude through successive bisection.

[80] Via this ploy, Euclid eliminates the need for circumscribed figures in all the limiting theorems in Book XII (i.e., Props. 2, 5, 10, 11, 12, 18). This feature may originate with Eudoxus' treatment of these theorems. But it is just the sort of refinement which one would expect from a subsequent rigorizing editor, for instance, a follower of Eudoxus or Euclid himself. One need not deny then that Eudoxus might have employed circumscribed figures in his treatment. In this case, Archimedes could have seen in Eudoxus' work a precedent for his own use of such figures, as in the *Dimension of the Circle*; see Chapter 5.

[81] See Chapter 2.

[82] This was noted by Becker in his "Eudoxos-Studien IV" (*Quellen und Studien*, 1936). For a discussion of its implications for understanding Eudoxus' form of proportion theory, see my "Archimedes and the Pre-Euclidean Proportion Theory."

[83] See my "Archimedes and the *Elements*" and Chapter 5.

[84] A more elaborate reconstruction is proposed by Becker in his "Eudoxos-Studien

IV." For accounts of the Euclidean technique of proportions, see Heath, *Euclid*, II, pp. 116–131; van der Waerden, *op. cit.*, pp. 187–189; Mueller, *op. cit.*, Ch. 3; and my *Evolution of the Euclidean Elements*, Ch. 8 and Appendix B.

[85] One can readily provide a constructive test for the inequality $A : B > n : m$. If a sequence of regular polygons P is inscribed in circle A, while a sequence of similar regular polygons Q' is circumscribed about B, then one will eventually have $P : Q' > n : m$, whence *à fortiori* $A : B > n : m$. A similar test applies in the case of the opposite inequality.

[86] This may be done by the bisection method, as in XII, 2, or by the alternative method of concentric circles used by Euclid for the volume of the sphere; cf. XII, 16.

[87] This view is elaborated in my "Archimedes and the Pre-Euclidean Proportion Theory."

[88] See Becker, "Eudoxos-Studien II." For other instances one may note Archimedes' *neuses* in *Spiral Lines*, and indeed *all* applications of *neuses*; also, the assumption of points of intersection of given lines and curves (cf. Heath, *Euclid*, I, pp. 234–240).

[89] "Die geometrische Construction als 'Existenzbeweis' in der antiken Geometrie," *Mathematische Annalen*, 1896, 47, pp. 222–228. For a criticism of this view in relation to the Euclidean geometry, see Mueller, *op. cit.*, pp. 15, 28.

[90] *Collection* (IV), I, pp. 250–252; Iamblichus is cited by Simplicius, *In Physica*, I, p. 60.

[91] *In Euclidem*, pp. 272, 356.

[92] This is a view actually maintained by some; cf. K. Freeman, *Pre-Socratic Philosophy*, Oxford, 1946, p. 355; and W. D. Ross, "Hippias of Elis" in the *Oxford Classical Dictionary*, 2nd ed., p. 517.

[93] Tannery, "Lignes courbes," 1883, pp. 278–281; Björnbo, "Hippias," in *Pauly Wissowa*, 8, 1913, col. 1708 f; Heiberg, *Geschichte der Mathematik . . . im Altertum*, p. 5; Heath, *History*, I, pp. 23, 225 f. Apparently, these writers maintained this as the "traditional" view, and so felt little need actually to argue a textual case for it. Cantor's line is typical: this Hippias must be Hippias of Elis, for the ancient tradition preserves no mention of any *other* geometer of this name. But, of course, the tradition preserves merely single references to many geometers (as a glance at the name-indices to Pappus, Proclus, and Eutocius reveals); while we may easily estimate that dozens have been forgotten altogether. Cantor also begs the extremely important question as to whether the Sophist *was* a geometer, and so must invent reasons for his being omitted from Proclus' list of Euclid's precursors (*In Euclidem*, pp. 65–67).

[94] Hankel, *Geschichte der Mathematik*, 1874, p. 151n. Allman initially accepts Hankel's view (*Greek Geometry*, 1889, pp. 92–94), but is later persuaded by Cantor's line and so adopts a position midway between Hankel and Tannery: that Hippias did indeed introduce the curve, but for angle trisection rather than circle quadrature (*ibid.*, pp. 93n, 189–193). This is the view held by van der Waerden (*op. cit.*, pp. 146, 191) and by Becker (*op. cit.*, pp. 95 ff); but Guthrie seems unconvinced (*History of Greek Philosophy*, III, pp. 283 f).

[95] *In Euclidem*, p. 65.

[96] For a survey, see Guthrie, *op. cit.*, pp. 280–285.

[97] *Hippias Minor* 367 c–368 e. In the *Protagoras* (e.g., 315 c, 318 e) Hippias appears

as a teacher of the mathematical sciences and natural philosophy, but no specific indication is given of his geometric expertise.

[98] *In Euclidem*, p. 356.

[99] Eighteen figures bearing this name are noted in *Pauly Wissowa*, Vol. 8.

[100] A more detailed discussion is given in Chapter 6. For other technical accounts, see Heath, *History*, I, pp. 226–230; van der Waerden (on Dinostratus), *op. cit.*, pp. 191–193; and Becker, *Denken*, pp. 95–98.

[101] Cf. *Elements* VI, 11.

[102] Cf. Archimedes, *Dimension of the Circle*, Prop. 1.

[103] *Collection* (IV), I, pp. 250–262; the proof is given in pp. 256–258.

[104] E.g., that similar arcs have the ratio of the diameters of their respective circles. For a discussion of this theorem and its relation to the extant *Dimension of the Circle*, see my "Archimedes and the *Elements*" and "Archimedes and the Pre-Euclidean Proportion Theory." Note that this use of the quadratrix also requires the equivalent of the inequality $\sin x < x < \tan x$; the first known instance of this is in Archimedes' *Dimension of the Circle*, Prop. 1.

[105] Cited by Pappus, *Collection* (IV), I, pp. 252–256. The difficulty is examined in the accounts by Heath and van der Waerden (see note 100 above).

[106] From *symballein*, *asymblēta* is often rendered as "incommensurable" (for which, however, the usual term is *asymmetros*) or as "incomparable," that is, literally, "unable to meet." Aristotle doubtless has in mind the fact that circles and lines cannot be made to coincide with each other. This would bring this passage into the context of Antiphon's and Bryson's circle quadratures; cf. note 69 above and Chapter 2.

[107] *Physics* 248 a 19 - b 7; cf. Heath, *Mathematics in Aristotle*, pp. 140–142. The elliptical phrasing is quite typical of Aristotelian writing, but can be supplied with little fear of ambiguity.

[108] See, for instance, Aristotle's discussions of the Eudoxean and Callippean planetary systems in *Metaphysics* XII, 8.

[109] See, for instance, Pappus, *Collection* (IV), I, p. 258, (V) pp. 312 ff, and (VIII), p. 1106; also *In Ptolemaeum*, ed. Rome, 1931, pp. 254 ff. Hero, *Metrica* I, 26 (cf. I, 37). Theon, *In Ptolemaeum*, ed. Rome, 1936, pp. 359 ff (here following Zenodorus; see Chapter 6). Proclus, *In Euclidem*, p. 423. One may of course add Eutocius' commentary on *Dimension of the Circle*, *op. cit.*, pp. 228 ff. The writers and scholiasts in the metrical tradition growing out of Hero's work concur in the Archimedean attribution; cf. Hero, *Opera*, ed. Heiberg, IV-V.

[110] The elementary character of Archimedes' method here is one of the indications I employ for arguing an early dating for this work; see my "Archimedes and the *Elements*."

[111] *Collection* (IV), I, pp. 250–252. Note that the word for "received" (from *paralambanein*) would regularly mean "received *(from)*", as in the manner of inheritance. I suggest reading "*apo*" ("from") for the manuscript's "*hypo*" ("by"), so to obtain "a curve received from Dinostratus *et al.*" This would have the effect of removing the implication made in the passage as it now stands, that Dinostratus introduced the curve in the context of circle quadrature.

[112] *Collection* (IV), I, pp. 284–286; the spiral is used for the same purpose, *ibid.*, pp. 286–288.

[113] See Archimedes' *Spiral Lines* and Pappus, *Collection* (IV), I, pp. 234 ff. This is discussed in Chapter 5.

[114] *Collection* (IV), I, pp. 270–272.

[115] That is, by means of the "Euclidean" constructions using compass and straightedge alone.

[116] *In Euclidem*, pp. 66 f.

[117] For a discussion of these studies, see my *Evolution*, Ch. 8.

[118] I examine several views of this sort in my "On the Early History of Axiomatics" and "Infinity and Continuity."

[119] One perceives such a tendency in Heath's accounts and those he cites; see his *History*, I, pp. 222, 224, 196n.

[120] The chief proponent of this view is S. Luria, "Die Infinitesimaltheorie der antiken Atomisten," *Quellen und Studien*, 1933, B : 2, pp. 106–185; it is elaborated by J. Mau, *Zum Problem des Infinitesimalen bei den antiken Atomisten*, Berlin, 1954. Brief accounts appear in van der Waerden, *op. cit.*, pp. 137 f; and Guthrie, *op. cit.*, II, pp. 487 f. See also Heath, *Euclid's Elements*, III, pp. 366–368.

[121] The thorough study by H. J. Waschkiess of the roots of Aristotle's continuum theory (*Von Eudoxos zu Aristoteles*), I believe, pays insufficient attention to the technical sources of Eudoxus' geometric methods. By contrast, Mueller maintains that no specific philosophic motive need be invoked for understanding the methods of Eudoxus and Euclid; cf. *op. cit.*, pp. 10, 234.

[122] For surveys of the background of the modern studies of foundations, one may consult N. Bourbaki, *Éléments d'histoire des mathématiques*, "Fondements des mathématiques," pp. 9–63; and M. Kline, *Mathematical Thought from Ancient to Modern Times*, 1972, Ch. 40–43, 51.

[123] For references, see my *Evolution*, Ch. 9 and the papers cited in note 118 above.

[124] See his "Eudoxos-Studien I–IV," *Quellen und Studien*, 1933–36.

[125] See my discussion in *Evolution*, Ch. 2.

[126] See Chapter 2.

CHAPTER 4

The Generation of Euclid

With specific reference to the three "classical" problems our sources preserve no solving efforts from the half-century or so separating the successors of Eudoxus from Archimedes. Nevertheless, this intermediate period, the generation of Euclid in Alexandria,[1] witnessed important advances in the field of geometric problem solving. At least five works by Euclid alone bear directly on the implementation of analytic efforts, while many of the more advanced problems which would concern geometers later in the 3rd century were defined and their analysis initiated at this time.

The *Elements* of Euclid are justly described as the most widely known and used technical textbook ever written.[2] In this work Euclid gathered the separate fields of elementary plane and solid geometry and number theory and edited prior treatments to produce a comprehensive and generally uniform compilation. Here the efforts of Eudoxus and his disciples in the field of "exhaustion" measures of curvilinear figures are gathered, as well as their development of the theory of proportions of magnitudes; the researches of Theaetetus and his successors in the fields of irrational lines and the construction of the regular solids are presented, together with the treatment of number theory on which they depend. These materials, which occupy somewhat more than half of the *Elements*, constitute fairly well circumscribed subjects whose study is of interest in its own right. But the remainder of the work is more elementary in character, intended to provide the basis of theorems necessary for the further study of geometry. Throughout, the treatment is synthetic, so that it sometimes is not immediately evident what the significance of a particular theorem might be. This is true, for instance, of much of Book II, where one meets theorems having the appearance of auxiliary lemmas relating to a field often called "geometric algebra."[3] These contain the proofs of theorems related to the techniques of the geometric application of areas which we discussed in connection with the cube duplication by Menaechmus.

Yet for all the sophistication of its logical structure and the intricacy of some of its constructions, especially with respect to the irrationals and the regular solids, the *Elements* is predominantly a treatise of an introductory sort, as its title implies. The researches assembled in it were initiated decades before Euclid by Hippocrates, Theaetetus, and Eudoxus and advanced by their successors. Even the organization into treatises comparable to the *Elements* had been pursued by several scholars in the middle and latter parts of the 4th century.[4] Thus, not only is one hard put to identify any specific theorem in the *Elements* as the discovery of Euclid, but even his responsibility for the form of the proofs there presented and for the selection and ordering of the theorems in the separate books is quite in doubt. Clearly, the "Pythagorean theorem" was familiar a millennium before Euclid among the mathematicians in ancient Mesopotamia, for instance; but its proof in I, 47, which Proclus ascribes to Euclid himself, is patently a modification of that in VI, 31, and the result in this latter form can hardly be denied to Hippocrates, for it is fundamental for his quadrature of the lunules.[5] Again, the Euclidean proportion theory in Book V without doubt incorporates modifications on the form of the theory initiated by Eudoxus; but several passages from Aristotle's discussions of the continuum make clear that such changes were already well under way three or four decades before Euclid.[6] We may say the same for the theory of irrationals in Book X, or even that Euclidean principle *par excellence*, the postulate of parallels.[7] The key conceptual insights as well as the major portion, if not the whole, of the work of securing them through formal proof had already been achieved by Euclid's predecessors.

Thus, the appropriate measure of the geometric researches conducted by Euclid and his contemporaries is to be sought not in the *Elements*, but in the *Data* and the lost *Porisms* and *Conics*, conceived for abetting the solution of geometric problems. But before describing these works, let us turn to the examination of a striking problem-solving effort representative of the new developments very near Euclid's time.

A LOCUS PROBLEM IN THE ARISTOTELIAN CORPUS

An odd stroke of good fortune has preserved for us a fine example of the synthesis of a locus problem from late in the 4th century: it is embedded as a lemma within the discussion of the shape of the rainbow in the Aristotelian *Meteorologica* (III, 5) and reads thus:

> since the points K, H have been given, so also would the (line) KH be given, and also the (line) MH (is given), so that also the ratio of MH to MK. Thus M touches a given circumference.[8]

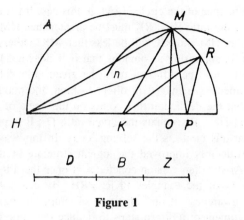

Figure 1

In the terminology of the "givens" (*data*) standard in the analytic literature, this expresses the result that the locus of points whose distances from two given points are in a given ratio (other than unity) is a given circle. This passage is unique in the Aristotelian corpus in that it provides a full geometric demonstration, and all the more remarkable in that the manner of proof is identical to that employed a century later by Apollonius in his second book *On Plane Loci*, by virtue of which the construction now goes under the name of the "Apollonian circles."[9] Its proof should thus inform us of the technique adopted at an early phase of the problem-solving tradition.

The writer introduces a line segment and divides it into parts D and B such that D : B = MH : MK, the given ratio (see Fig. 1).[10] He then requires that there be produced (*prospeporisthō*) an addition Z to B such that D : B = B + Z : D, and that accordingly, KH be extended to P such that Z : KH = B : KP. He joins MP and claims that "P will be the center (*polos*) of the circle toward which the lines from K fall," that is, the broken lines KMH satisfying the condition of the locus. Let the line PR be taken such that Z : B : D = KH : KP : PR, so that PR : KP = PH : PR. This means that the triangles PKR and PRH are similar, for they have the angle KPR in common. Thus, PR : KP = HP : PR = D : B , the given ratio. Hence, on the one hand, R is a point on the locus, while on the other, PR is the mean proportional of KP, PH whose magnitudes are given (that is, independent of the choice of R) so that PR is also given in magnitude. Hence, PR = PM and the locus required is the circle of center P and radius PM.

This passage is studded with interpretive puzzles. First, there is the question of the relation between the Aristotelian and Apollonian versions of the proof. The constructions are evidently the same, but there are differences in the demonstrations. The Aristotelian form adopts an indirect proof, in which the auxiliary line PR is hypothesized to be greater or less than PM, and it is deduced that both HR : RK and HM : MK equal the given ratio D : B. But why this should be an apparent contradiction is not quite clear. Perhaps the writer has assumed the special context of his construction, where the trial point R is to be taken along the circumference of the circle HAM; in this case it is indeed true that for PR less than PM, the ratio HR : RK must be greater than HM : MK, since HR will be greater than HM and RK will be less than MK. Otherwise, he seems to beg the question in supposing as obvious that if R does not lie on the arc Mn the ratio HR : RK must be different from the given ratio.[11] In the Apollonian version the argument is managed in direct fashion: the magnitude PM is not introduced, so that the deduction that points on the circle of radius PR satisfy the condition that HR : RK equals the given ratio D : B is precisely what has been sought, that this circle is the solving locus. In this way, the Apollonian version has considerably improved the logical structure of the proof. On the other hand, the Aristotelian version concludes at once that PR : KP = HP : PR from the similarity of the triangles PKR, PRH. By contrast, the Apollonian version, despite awareness of the same pair of similar triangles, launches into an incredible argument of a dozen steps to deduce the same proportionality of lines. I find it hard to conceive how Apollonius could have missed the transparency of this step, and thus I would assign the longer detour to a later editor, possibly Eutocius, momentarily puzzled by a step assumed as obvious in Apollonius' proof.[12]

A second puzzle is that this construction seems to be more complicated than it need be. Simson has proposed a very neat alternative form which develops from the condition of the locus in a more natural manner.[13] Consider what an analysis of the problem might be: given the points H, K, we choose any point M such that HM : MK equals the given ratio. (See Fig 2.) Divide HK at L so that the ratio of the segments HL : LK equals the same; then by *Elements* VI, 3, the line ML bisects the angle HMK. Now, extend HM beyond M to S and draw MT to bisect the angle SMK; one then has TH : TK = HM : MK.[14] Thus, points L and T, as well as M, lie on the locus. Since ML and MT are the bisectors of supplementary angles, angle LMT is a right angle. We may thus draw the circle through L, M, T of diameter LT. As the points L, T are given via the given points and ratio which define the locus, and as L, T in their turn specify the circle of diameter LT, it follows that all points M of the locus lie on that circle. Thus, to construct the locus, one need merely determine the points L, T and draw the circle. This form of the solution is manifestly simpler than the

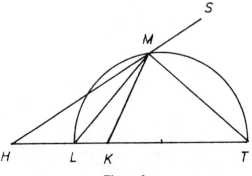

Figure 2

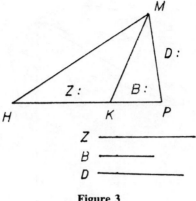

Figure 3

Aristotelian/Apollonian form. The contrast is sharpened when one observes that the points L, T do not enter into the latter form at all. Perhaps the originator of the ancient version did not have at hand the theorem on the bisector of the external angle, so critical for Simson's alternative. But one gains nothing in attempting to second-guess the ancient writer. What is striking is that Apollonius has sustained this relatively cumbersome approach to the problem. Not only is his dependence on a model essentially identical to the Aristotelian version thus affirmed, but also his treatment of earlier results is revealed. Apparently, his aim in the *Plane Loci* was not to redo the older solutions, but rather to collect these and use them as preliminary to the presentation of his own findings.

This also shows that the analysis implicit in the ancient construction of this locus could not have proceeded along the lines suggested by Simson.[15] The implied analysis would appear to have taken the following form: given the points H, K, we assume the point M on the locus, so that HM : MK equals the given ratio. [See Fig. 3] If we now extend HK and draw PM such that angle PMK equals angle KHM, then the triangles PMH, PKM will be similar, since they share the angle at P, while the angles at M and at H are equal. Thus, HM : MK = PH : PM = PM : PK, so that each equals the given ratio, and PM is the mean proportional of PH, PK. Consider lines Z, B, D such that Z : B : D = HK : KP : PM. Since D : B is a given ratio, and D is the mean proportional between B and B + Z , it follows that Z is given from D and B. Since HK is also given, KP and PM are given in magnitude, for they are each in a given ratio to HK. Thus, the locus is seen to be the circle of center P and radius equal to PM. The construction and synthesis answering to this analysis are evidently those given in the Aristotelian/Apollonian solution; and in the presence of such an analysis, that solution loses all of the complexity which marks the synthesis when it stands by itself. Indeed, the construction is seen to hinge on a single insight, a rather clever one, of introducing the angle at M equal to that at H. In this way, Apollonius' decision to retain the older form, rather than search for a handier alternative, becomes easy to understand.[16]

106   Ancient Tradition of Geometric Problems

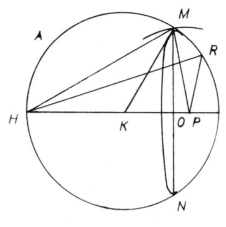

Figure 4

Our third puzzle is, why does this construction appear in the *Meteorologica* at all? For his explanation of the circular shape of the rainbow, the Aristotelian writer requires that the point M, marking the elevation of the bow above the horizon, be determinate. Upon rotation of the figure about the axis HKP, the point M will trace a circle, of which the half above the horizon defines the position of the whole bow (see Fig. 4).[17] His stipulation for M is that it be the point on the given circle HAM for which the distances HM, MK are in a given ratio. This reasonably opaque assumption is not motivated through any physical considerations, such as the nature of the optical rays and their reflection; nor is any attempt made actually to specify the value of this ratio. Thus, we are hardly any better off than if he had simply stipulated that the angle PKM be a constant. Furthermore, he applies the condition in an unnecessarily complicated way. For, as we have seen, if M is taken on the circumference such that HM : MK equals the given ratio, then for any point lower than M, the corresponding ratio will be greater than the given, while for any point higher than M, the ratio will be less. This is the view adopted by the later Aristotelian commentators, Alexander and Olympiodorus, who paraphrase in minute detail the entire construction and proof of the locus problem, although they have utterly failed to perceive its significance.[18] These considerations make plain the derivative character of this context of the locus problem. The author of the *Meteorologica* has recognized that this construction offers one method for fixing the elevation of the rainbow. But the absence of the analysis leaves the construction without clear geometric motivation, while the stipulation of lines in given ratio stands without physical motivation. Moreover, the unnatural manner of deploying this condition within the special context of the passage—that is, presenting the general solution of the locus, rather than merely arguing the uniqueness of the point on the given circle satisfying the condition on ratios—seems to indicate that the writer was sufficiently impressed by this particular result to create a place for it within his meteorological discussion. That is, it was taken to be a significant discovery, and that, we should suppose, was by reason of its being recent and representative of advanced study in geometry at the time.[19]

In speaking of the "Aristotelian writer," I betray a reluctance to take Aristotle as the author of this solution of the locus problem. It ranks among the finest samples of geometry we know of from the pre-Euclidean period, and nothing in the Aristotelian corpus prepares us for the notion that Aristotle himself was capable of such a technical feat.[20] The passage is singular on several accounts: only here and in a passage two chapters earlier in the same book of the *Meteorologica* does one come upon the presentation with full proof of a geometric proposition. The nearest parallels are a set of theorems on proportions involving infinite magnitudes given at several places in the *Physics* and *De Caelo* and effected with no remarkable skill or insight.[21] Otherwise, Aristotle's mathematical passages either merely allude to aspects of proofs or take up the discussion of metamathematical issues. These are valuable as reflections of mathematical work of the period, but they provide no evidence of Aristotle's active participation in that work. Furthermore, one may well entertain suspicions about other features of the passage in the *Meteorologica*: the optical ray is conceived as emanating from the eye and outward to the object seen, here from K to H via reflection at M; this conforms with standard usage in ancient optical studies, but directly contradicts the theory of vision Aristotle favors against this view in *De Anima* (II, 7), in which seeing is an effect of the transparent medium acting on the eye, so that the account of vision harmonizes with the accounts of the other senses. Why should Aristotle not have adopted the same view in the *Meteorologica*? It is geometrically indifferent, and, indeed, optical writers sometimes do adopt it in a context, as here, where the object seen is a luminous body like the sun or a star.[22] This discrepancy has long been noted by commentators, both ancient and modern; but it seems not to have provoked the uneasiness one should have expected.[23] Again, the analysis of the rainbow introduces the bizarre device of placing the sun (H) and the cloud producing the rainbow (M) at equal distances from the observer (K). This fiction, sometimes called the "meteorological hemisphere,"[24] is patently absurd from the cosmological standpoint, as Aristotle must have known, in view of his familiarity with the Eudoxean astronomy (in *Metaphysics* XII, 8) and his consistent distinction between the sublunary sphere of the elements and the higher domain of the heavenly bodies. Why should Aristotle here so casually evade the plain principles of cosmology? If one supposes that he merely adopts a standard meteorological convention, just as we saw before he followed the optical convention in the conception of the visual ray, then one must admit that Aristotle's thought has had no impact on the formation of those conventions among the mathematical scientists of his day. Yet the adoption of this convention seems odder still in that it provides no particular advantage for the account of the rainbow. Conceiving the sun in its true position merely renders the ray HM effectively parallel to the horizon,[25] and so would require an alternative condition for determining the altitude of M, for the condition on the ratio of the lines HM, MK becomes meaningless. This seems, then, to be another indication that the account of the rainbow has been fashioned to serve as the vehicle for this problem, rather than to provide the context for its original statement and solution.

If these considerations do not quite rule out the authenticity of *Meteorologica* Book III, they surely deserve more serious attention than Aristotle scholars tend to afford them. At the least, we recognize that the geometric locus problem in the passage on the rainbow is an importation from geometric practice, not Aristotle's own discovery. But the suggestion of its nonauthenticity at once raises the problem of dating. It has already been observed that the *Meteorologica* appears to be a composite document, different sections reflecting times ranging from the mid-350's to the late 340's.[26] Conceivably, then, some of its parts might be later still, and some might even be the work of disciples after Aristotle's death in 322. If so, how much later? If the passage on the locus problem derives from the 3rd century, near or even after the time of Apollonius, then its significance for the history of ancient geometry diminishes almost to nil.

In this instance we possess a valuable index of early provenance: the frequent appearances of archaic terminology in expressions like "the line *on which* (are) AB" (*hē grammē* eph' hēi *AB*), where standard usage after Euclid would read "the right line AB" (*hē eutheia AB*).[27] We have seen that the older form is characteristic of usage in the Eudemus fragment on Hippocrates cited by Simplicius, and an analogous form for logical terms is prevalent in Aristotle's *Prior Analytics*. By contrast, the shorter form is exclusive in Euclid, Archimedes, and Apollonius and the writers who followed them.[28] One must suppose that the author of the solution of the locus problem was in touch with the regular tradition of geometry of his time, so that if that happened to be much after the beginning of the 3rd century, the use of such nonstandard terminology would be difficult to explain.

Thus, I view the account of the rainbow in the *Meteorologica*, and the version of the locus problem to which it has been specially accommodated, as a product of the late 4th century, written more likely by a disciple of Aristotle than by the philosopher himself. The solution of that problem is not original in this context, as one sees from the absence of the analysis which would motivate the construction, but more clearly, from the partial garbling of the proof, needlessly encumbered by an indirect format of demonstration.[29] What, then, was the nature of the author's source and of the geometric activity which produced it? We have little means for determining this. But certainly when Apollonius reproduced this construction in his *Plane Loci* a century later, he was drawing not from the *Meteorologica*, but from its mathematical sources.[30] The *Meteorologica* passage thus reveals to us the manner in which this result was received by a mathematically oriented writer, late in the 4th century, and so indicates both the topicality and the significance of its discovery in the field of research then current.

EUCLID'S ANALYTIC WORKS

The problem from the *Meteorologica* manifests several features standard in the classical form of problem solving. It is a problem of locus, that is, the construction of the set of points (typically a line or curve) satisfying a stated condition. But here it is actually expressed not as a problem, but as a theorem about "givens": that if two points are *given* and a ratio is *given*, then the circle is *given* which is such that the lines drawn from each of its points to the given points are in the given ratio. That Euclid devoted an entire treatise, the *Data* ("givens"), to theorems framed in this manner reveals that the form was important within geometric studies of the time. We recall, furthermore, that the Aristotelian problem assumes an auxiliary construction: that the line Z may be produced (*prospeporisthō*) such that for given lines B, D, D : B = B + Z : D. Here too Euclid composed a massive treatise of *Porisms*, now lost, so called perhaps only for their role as auxiliaries to other constructions, although the stronger sense of "production," that is, of terms of a stipulated description, might also apply.[31] Most important, we have seen that the Aristotelian problem, while presented in strictly synthetic fashion, is best understood as framed in answer to a preliminary analysis. Clearly, this analytic method had attained prominence by Euclid's time, for Pappus includes these and other treatises by Euclid among the corpus of works, the *Topos analyomenos* (frequently translated as "Treasury of Analysis"), designed in the interests of analytic investigations.[32]

The *Data* is a complement to the *Elements*, recast in a form more serviceable for the analysis of problems.[33] As in the Aristotelian example, each of its theorems demonstrates that a stated term will be given on the assumption that certain other terms are given. The subject matter overlaps that of the *Elements*, dealing with ratios and with configurations of lines and of plane figures, both rectilinear and circular. Indeed, only in rare instances does the *Data* present a result without a parallel in the *Elements*.[34] The proofs follow in a deductive sequence, so that one might refer back to previously proved theorems; in the Aristotelian problem, for instance, one assumes that the ratio of given terms is given, or that a term in given ratio to a given term is itself given—such propositions appear in the *Data* (Prop. 1, 2) and must have been available to the Aristotelian writer in a comparable form. Again, some proofs may assume results proved in the *Elements*; for instance, properties of similar triangles are needed in the Aristotelian problem. In view of this, one might suppose the *Data* was effectively superfluous in light of the *Elements*.[35]

But the analysis we sketched for the locus problem in the preceding part reveals the special utility of the "data" format for this type of investigation. There, the analysis started from the supposition that the construction has been done, namely, that the point M is such that HM : MK equals the given ratio. The construction of P then led to a set of ratios linking the lines HK, KP, PM. We next assumed an arbitrary line D as *given*; hence B, the line in given ratio to D, is also *given* (so that D : B = PM : KP). Furthermore, if Z is the line such that D is mean proportional between B and B + Z, then it too is *given*, and it has to HK the same ratio that B, D have to KP, PM, respectively. But HK is *given* (for the points H, K are given as basis of the locus to be constructed), so that the ratio HK : Z is *given*; and B, D are *given*, so that KP, PM are *given* (the latter in magnitude only), whence the circle of center P and radius PM is *given*. Thus, the locus of M is *given*.

This ends the analysis and serves as the basis for the synthetic construction of the solution, as we meet it in the Aristotelian and Apollonian texts. One proceeds generally in the reverse order: starting from the given points H, K and the given ratio, one introduces auxiliary lines D, B in that ratio and produces the line Z, as described in the analysis; from this the lines KP, PM are determined and the circle is drawn. The proof consists of verifying that HM : MK equals the given ratio. In this instance, each step of the analysis was simply convertible. But it sometimes happens that the reversal of the logical order is possible only when certain additional conditions are satisfied; the determination of the appropriate conditions would then require a separate treatment, called the "*diorismos*," in effect specifying when the problem as stated is in fact constructible.[36] The investigation of diorisms seems to have been advanced already by the middle of the 4th century, so that the procedure of analysis and synthesis was familiar as an instrument for the solution of problems several decades before Euclid.[37]

Referring back to our sample analysis, one observes how the concluding section is shaped by a series of "givens," each determined from preceding ones, until one has arrived at the desideratum of the construction as given. The adoption of this terminology is not merely a formalism; it serves a critical purpose by keeping separate two sets of terms having quite different logical status. Some terms are known only conditionally, in that they follow upon the hypothesis of the analysis that the construction has been effected; but other terms are known by virtue of the definition of the construction which one is to produce. The latter are the "givens." When these two sequences of terms arrive at a common term, one may consider the analysis to be complete. Without a terminological ploy like that of the "givens," the logical status of these sequences would soon become hopelessly confused. Thus, far from being superfluous, Euclid's *Data* accords formal recognition to the advanced stage of analytic researches at his time.

The *Meteorologica* passage already employs the form of sequential "givens," so that Euclid cannot be taken as having originated this style, although it is possible that his *Data* represents the first major effort to organize the materials of elementary geometry in this form. Indeed, some parts of the *Data* are modeled after their analogues in the *Elements*, indicating that these parts are not likely to have had an independent provenance.[38] We lack earlier instances of the sequential usage, doubtless owing to the general loss, but for fragments, of documentation from the pre-Euclidean period, so that one cannot specify the origin of this usage. But for the simple notion of the "given" in geometry, one has evidence from early in the 4th century in the "hypothesis" passage from the *Meno* (86e–87b), where the project is to apply a certain area "along the *given* line of this circle here...." The phrasing is somewhat loose, as the dialogue context would recommend, but the formal basis can be seen: the conditions to be satisfied in the construction of a problem are framed in terms of certain "given" magnitudes. This is invariably the form in the problems in Euclid's *Elements*.[39] It is perhaps best to view this as a natural development, formalizing usage in common language. For one comes upon comparable expressions in dialectical passages, for instance, without geometric connotations. Aristotle describes the good legislator as "capable of taking the *given* constitution into account," that is, the one actually existing, rather than some hypothetical ideal (*Politics* IV, 1, 1288 b 28); or, again, he dismisses the notion that an eternal thing might be subject to generation or passing away, as this "conflicts with one of the *givens*," that is, the generally accepted principles of natural philosophy.[40] In view of such parallels, however, one should not rule out the possibility that geometry and dialectic interacted in the development of this usage, at a time when both fields were seeking to formalize the elements of research.

A further contribution by Euclid to the study of advanced problems lies in the field of the conic sections. In the production of his treatise *On the Conics*, Euclid was able to draw on the efforts of an earlier contemporary, named Aristaeus "the elder" by Pappus, for he is said also to have written a treatise on this subject.[41] As neither work survives, doubtless owing to their having been superseded by Apollonius' far more extensive treatment about a century later, our knowledge of their nature is indirect and severely limited. One must suppose that whenever Archimedes assumes without proof a theorem on conics, "for that has been demonstrated in the *Conic Elements*," he refers to the work either of Aristaeus or of Euclid.[42] From Apollonius himself and commentary by Pappus and Eutocius, we learn that most of the materials on the general properties of the curves in Apollonius' Books I and II were familiar within the older theory, and that his third book extended a body of theorems useful for the solution of locus problems initiated by Euclid. As for the fourth book, investigating the manner of intersections of conics, Apollonius assigns its origins to Conon of Samos, a geometer in the generation just after Euclid and an important influence on Archimedes' early work. The remaining four books on normals, similar conics, and conjugate diameters were the product of Apollonius' own more advanced researches.[43] Thus, the theory at Euclid's time may be viewed as consisting essentially of the materials of the first two books and some portions of the third.[44]

As we have seen, the standard view that the theory of the conics was initiated and strongly advanced by Menaechmus around the middle of the 4th century is open to serious question. His treatment of the cube-duplication problem indicates an interest in the family of curves specified via second-order relations in accordance with the techniques of the application of areas, and he and his followers surely were able to discover properties of such curves and use them in the solution of other problems.[45] But the original context of the discovery that such curves resulted from the sectioning of cones is a matter of conjecture. O. Neugebauer, for instance, has suggested the study of the shadows cast in sundials as a possible context.[46] Alternatively, the field of optics seems a likely prospect. For instance, Euclid demonstrates that "the wheels of chariots appear sometimes circular, sometimes *drawn in* (*parespasmenoi*)" (*Optics*, Prop. 36), and Pappus includes among his lemmas to the *Optics* the construction of the ellipse whose appearance the circle assumes in the latter case.[47] Moreover, the earliest extant references to the conic section have to do with the same situation: for instance, that the

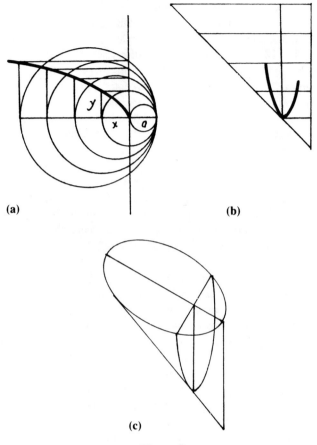

Figure 5

gibbous moon looks like a "shield" (*thyreos*) or that a circle viewed obliquely takes the shape of "the section of an acute-angled cone."[48]

Even if my view of the earliest phase of the study is correct, we should still recognize that the Greek geometers could not long have remained content with the type of pointwise constructions we have proposed. Curves arose as abstractions of motions (e.g., circles and spirals) or through the intersection of solids (e.g., the curve of Archytas). The sort of algebraic relation by which one defines a curve in the Cartesian manner was admitted by the Greek as a fundamental *property* (*symptōma*) of the curve, but not as the basis of its conceptualization.[49] One can imagine how the search for a natural mode of generation might have progressed from the pointwise construction to the sectioning of cones. For instance, the curve introduced via the relation $Y^2 = AX$ might entail for its pointwise construction a diagram consisting of a system of circles sharing a vertex and having their diameter lying along the same line [Fig. 5; cf. Ch. 3, Fig. 12(c)]; for each circle the chord is drawn perpendicular to that diameter, as to divide the diameter into segments equal to A and X, so that the chord itself will be twice Y.[50] One might then see, if the lines X are drawn in elevation at right angles to the plane, that the diagram becomes the plane projection of a cone, one of its generators being the line perpendicular to the plane and passing through the common vertex of the circles, and the curve being formed as the cone's intersection with a plane parallel to that generator.

Under this view, the study of these curves, as defined via the application of areas, had progressed significantly before it was discovered that the curves formed as sections of the cone satisfy the same second-order relations. The work of producing a full-fledged treatise on the conic curves, like those by Aristaeus and Euclid, could proceed quite rapidly, for most of the properties of these curves are deduced from their *symptōmata*, the second-order relations, rather than from their generation as sections. It is important to realize that the role of the latter aspect of the curves was to provide a natural conceptual view of their generation. It did not serve for their practical construction, for later writers found pointwise procedures far more convenient than any solid construction.[51] On the other hand, it was largely superfluous for the work of proving theorems, at least in the more advanced stage of the theory at the hands of Apollonius. Realizing this helps to explain another rather puzzling feature of the early theory of the conics. We learn from Pappus and Eutocius that before Apollonius the conics were specified such that the sectioning plane was taken perpendicular to the side of the cone.[52] Thus, parabolas were formed only via the perpendicular sectioning of the right cone with a right angle at its vertex, and were accordingly named "sections of the right-angled cone." Similarly, ellipses and hyperbolas were formed, respectively, from cones with acute and obtuse angles at the vertices. The report of the commentators is confirmed by usage in Archimedes and Diocles, and the older names persist in some accounts well after Apollonius.[53] Yet why this restriction? It seems to belie the fact that specimens of all three types can be formed as sections from any cone—a consideration which surely must have led Apollonius to remove this restriction on sections, although the nature of oblique sections was already well appreciated by Euclid and Archimedes before him. Odder still, it prohibits the formation of a circle as a "conic section." Clearly, then, the retention of this mode for so long was due to its introduction as a deliberately artificial aspect of the theory; that is, it reflected not the primitiveness, but the sophistication of that theory around the time of Euclid.

Our account of the initial phase of these studies revealed that the second-order curves at first lacked a natural mode for their generation. The sectioning of the cone could provide a suitable mode, and indeed under the restricted manner of sectioning there will be a unique section corresponding to each second-order curve.[54] To see this, first consider the parabola: in the application mode it is specified via a single parameter, the line segment over which the application is performed (line A in the relation $Y^2 = AX$). Correspondingly, when it is formed from the perpendicular section of a right-angled cone, a single term, the distance D of the cutting plane from the vertex of the cone, will designate the curve. [See Fig. 6(a).] It follows that $A = 2D$. Since the distance along the diameter

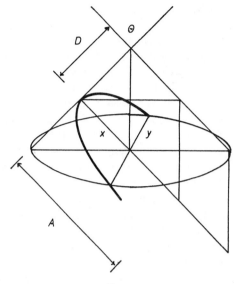

(a)

Figure 6

of the section from its vertex to where it meets the axis of the cone also equals D, we obtain a straightforward sense for Archimedes' expression of the parameter as "double the line up to the axis."[55] Turning now to the hyperbola and the ellipse, we may specify each form via two lines such that an area similar to the rectangle defined by the two lines is applied to one of those lines as to exceed or fall short by a square (i.e., the lines A, C in the relations

$$\frac{A}{C} Y \cdot Y = AX \pm X^2$$

where the positive sign corresponds to a hyperbola, the negative sign to an ellipse). We have two important indices supporting this view of the initial forms of the curves. First, it leads immediately to the form implied as the standard in Archimedes' theorems, namely as $Y^2 : X(A \pm X) = C : A$.[56] Second, Pappus' statement of the definitions explicitly figures the excess or defect as a *square*. But in Apollonius' definitions there is a subtle difference; for the "latus rectum" C and "diameter" A give rise to the relation

$$Y^2 = CX \pm \frac{C}{A} X \cdot X$$

so that the excess or defect is similar to the figure defined by the lines C, A. Pappus has not erred, as some have charged; he has merely followed a source which conforms with the pre-Apollonian designations.[57] Like the parabola, each hyperbola and ellipse may be associated uniquely with a perpendicular section of a cone [see Fig. 6(b) and 6(c)], where the distance D between the vertex of the section and the vertex of the cone and the angle θ at the vertex of the cone

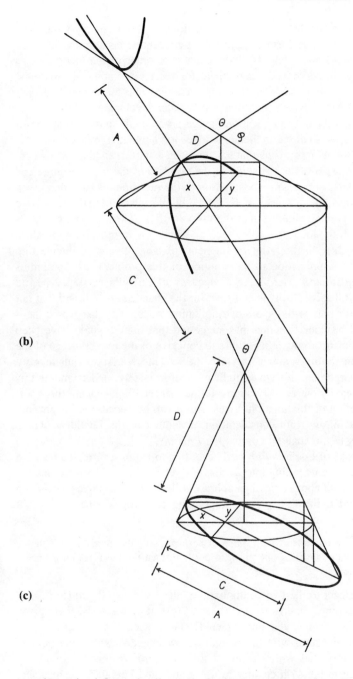

(b)

(c)

are determined from the lines A, C via the relations $C = 2D \tan(\frac{1}{2})\theta$ and $A = D \tan\theta$ (for the ellipse, θ acute) or $A = D \tan\varphi$ (for the hyperbola, θ obtuse and φ its supplement).[58] Note that just as for the parabola, the latus rectum may be styled as "double the line up to the axis."

This effort to relate the second-order relations to the sections of the cone produced constructions to form an appropriate conceptual background for their study. But it would not much affect the content or order of the theory, or its primary purpose, to serve as an instrument for the investigation of problems. From the start, it was in their form as area relations that the curves had their application, and this remained so throughout the later development of the theory. We have seen how the cube duplication figured prominently in the initial phases of this work, and we shall see in the angle trisection another significant example of its use. Apollonius informs us that Euclid made a start in another area where the conics are indispensable: the study of the problem of the locus of three and four lines.[59] In this, one seeks a construction of the set of points such that given three (or four) lines, the distances A, B, C (and D) of each point to the respective given lines satisfy the condition that the ratio $A \cdot B : C^2$ (or $A \cdot B : C \cdot D$) is given. In general, the locus is a conic section. Theorems from Apollonius' third book (Prop. 16–22), on the ratios of segments of tangents and of intersecting chords, have their application in the solution of this problem, as Apollonius observes, noting in strong terms the inadequacy of Euclid's version: "he did not effect the synthesis [of this locus] save for the chance part of it and that not successfully," for only with the aid of Apollonius' own newly discovered theorems could that be done. Zeuthen has suggested that Euclid would have been hindered in this effort through his failure to conceive of the two branches of the hyperbola as constituting a single curve.[60] But it seems that Apollonius levels a more serious charge than this would indicate. Unfortunately, understanding this issue depends heavily on reconstructions, for neither the Euclidean nor the Apollonian investigation of this locus survives. But it can be hoped that insight into the former may follow from consideration of another of the Euclidean works Pappus includes in the analytic corpus, the *Porisms*.

The three books of Euclid's *Porisms*, according to Pappus, formed a massive treatise of 171 theorems falling into 29 classes, yet without claim of exhausting the subject matter.[61] Pappus gives a paraphrase collecting into a single proposition several of Euclid's; the first part, incorporating ten propositions of the first species, is this:

> if three points on one line of the *hyption* or *parhyption* figure are given, and the others save for one lie on a line given in position, then that also lies on a line given in position.

That is, with respect to the two configurations, the "supine" figure [Fig. 7(a)] and the "hypersupine" figures [Figs. 7(b) and 7(c)], if points A, B, F are given and points C, D lie on given lines, then E also lies on a given line.[62] One recognizes the connection with those studies now included within the field of projective geometry and which received their modern treatment through researches initiated in the 17th century by Desargues and Pascal.[63] For instance, the converse of the lemma just cited from Pappus provides a proof of the theorem of Desargues on two projectively related triangles: that if the lines joining corresponding vertices of the triangles CED, C'E'D' meet at a point, then the points of intersection A, B, F of corresponding sides are collinear [see Fig. 7(d)].[64] A

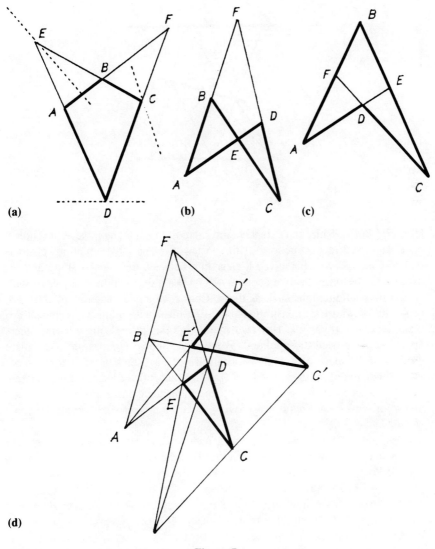

Figure 7

perusal of the 38 lemmas which Pappus presents on the *Porisms* confirms this sense of the affinities between the subject matter of the *Porisms* and that of the more modern field of projective geometry. In addition to the connection with Desargues' theorem just cited, one finds configurations relating to the complete quadrilateral, the preservation of cross ratio through projective transformations, and results on involutions. Most noteworthy is the famous theorem of Pappus on the hexagon inscribed between two lines:

if AB, GD are parallel lines and there fall on them certain lines AD, AZ, BG, BZ and ED, EG are joined, then a straight line arises through H, M, K.[65]

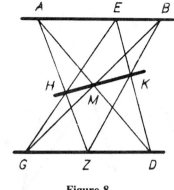

(a)

Figure 8

[See Fig. 8(a).] With such extensive information to guide them, several scholars have attempted reconstructions of this lost work. The most ambitious effort is that of Chasles, who perceived a clear link between the ancient study and his own researches in projective geometry.[66] Working through the many cases suggested by configurations such as these, Chasles compiled a series of over 200 propositions whose demonstrations could be managed within the technical domain available to Pappus. However impressive this project might seem, others appear to have been disappointed. For surely Pappus' lemmas are but a mere shadow of the scope of Euclid's work itself.[67] Thus, Zeuthen refers to the more general context of these studies in the modern theory—for instance, the theorem of Pappus is a special case of Pascal's theorem on hexagons inscribed in conics—and so proposes that Euclid was moving toward a projective theory of the conic sections.[68]

It would appear, however, to be a questionable procedure on our part to extrapolate in this way and insist that because the *potential* for such a theory is foreshadowed in these lemmas, the work to which they relate, Euclid's *Porisms*, must *actually* have advanced such a theory. It is an interpretive issue of no small subtlety to determine in what sense and to what degree Euclid and later geometers developed a certain body of geometric materials which have been subsumed in more recent times within the field of projective geometry. This caution is especially well advised because the ancient manner of effecting these results entirely lacks the conceptions and methods characteristic of the modern field. Consider, for instance, Pappus' treatment of the theorem on the inscribed hexagon. He establishes this in two cases, as the bounding lines are parallel (Lemma 12) and as they intersect (Lemma 13)[69]; in this he already betrays the absence of an essential feature of projective methods: the equivalence of parallel and nonparallel configurations through the introduction of points and lines at infinity. In the nonparallel case, with reference to the lines GHTE and NGZD, Pappus has, from a prior lemma (3), GE · HT : GH · TE = GN · ZD : ND · GZ; similarly, with reference to NGZD and DKLE, he has GN · ZD : ND · GZ = DK · EL : DE · KL [see Fig. 8(b)].[70] It thus follows that GE · HT : GH · TE = DK · EL :

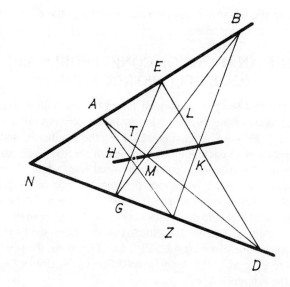
(b)

DE · KL. He can thus conclude from another lemma (10) that HMK is a straight line. Pappus' proof is structurally identical to the standard projective one: the perspectivity centered on A projects GHTE onto GZDN, while that centered on B projects GZDN onto LKDE; the projection GHTE onto LKDE is the composition of these two perspectivities, and since it has the fixed point E, it is itself a perspectivity. The center of projection must be the intersection of GL, TD, namely, the point M; hence the line HK passes through M.[71] From a comparison of these two treatments, one sees that Pappus' Lemma 3 amounts to the specification of a projective transformation (here, a perspectivity) via its cross-ratio-preserving property; its converse (Lemma 10) plays the role which the modern approach casts to the fundamental theorems on the composition and determination of projectivities and perspectivities.

There emerges from this a tantalizing puzzle: in view of the utter absence of projective techniques from the ancient treatment, how did it happen that the author of the *Porisms* came to investigate the same configurations and properties which are prominent in the modern theory? Just what was Euclid up to? What we know of this work indicates that it was ancillary to a larger field of research.[72] Indeed, Pappus says so explicitly: "the *Porisms* of Euclid are a most ingenious set mustered toward the analysis of the more heavy-laden problems," and notes their utility for locus problems of the sort found in abundance in the *topos analyomenos*.[73] Now, as we have noted, results comparable to some in the *Porisms* have been applied with stunning success toward the study of conics since the 17th century. In the light of Euclid's contributions to the early theory of conics, one may well conjecture that Euclid compiled the *Porisms* to support a problem-solving activity related to the conics. This view would be a somewhat weakened form of that advocated by Zeuthen. Let us then consider one of the problems of this sort expressly assigned to Euclid, the locus with respect to three

120 Ancient Tradition of Geometric Problems

and four lines, to see whether it can provide insight into the nature and the role of the *Porisms*.

THE ANALYSIS OF CONIC PROBLEMS: SOME RECONSTRUCTIONS

One may observe that the pre-Apollonian forms for the central conics express them as special cases of the three-line locus. In the defining relations $Y^2 : X \cdot (PP' \pm X) = L : PP'$, for given lines L, PP', the lengths X and $PP' \pm X$ represent the distances of each point of the conic from two given parallel lines, as measured in the direction of a given transversal line, while Y represents its distance from that transversal, as measured in the direction of the parallels (see Fig. 9). One thus sees that the ellipse (for $PP' - X$) and the hyperbola (for $PP' + X$) which are tangent to the parallel lines at P, P' and have PP' as diameter and L as latus rectum will satisfy the property of the three-line locus relative to these given lines and the given ratio L : PP'. In effect, then, the three-line locus problem seeks to generalize the defining condition of the conics for other configurations of the reference lines.

Solutions of the general problem can be constructed on the basis of two propositions from Apollonius' *Conics*:

> III, 16 : if OP, OQ are tangent to a conic and KK' is a secant line parallel to OQ which meets the conic in R, then $OP^2 : OQ^2 = PK^2 : KR \cdot RK'$ [see Fig. 10(a)];
>
> III, 17: if RR', SS' are secants parallel to OP, OQ, respectively, and meet each other in J, then $OP^2 : OQ^2 = RJ \cdot JR' : SJ \cdot JS'$. [See Fig. 10(b)][74]

The latter theorem holds not only when J is an internal point (as shown in Fig. 10b), but also when the secants intersect externally. The theorems are related

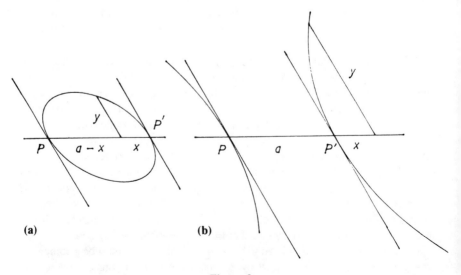

Figure 9

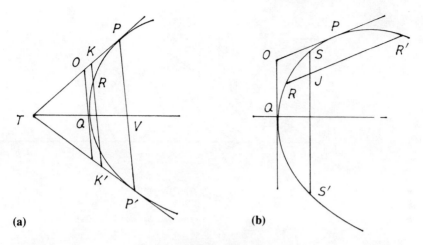

Figure 10

in that they each generalize familiar properties of tangents and secants to the circle, where $OP = OQ$, $PK^2 = KR \cdot RK'$ and $RJ \cdot JR' = SJ \cdot JS'$ (cf. Euclid's *Elements* III, 35, 36). Since Archimedes asserts as "proved in the *Conic Elements*" this same property of the intersecting chords in a conic (*Conoids and Spheroids*, Prop. 3),[75] we thus perceive that this property was established within the earlier theory of conics.

The fact that a conic has the property of the three-line locus follows at once from III, 16,[76] for if we draw tangents TP, TP′ to the conic, bisect chord PP′ in V, and join TV meeting the conic in Q, then TV will be a diameter corresponding to the ordinates parallel to PP′ [see Fig. 11(a)].[77] Moreover, if 0Q is

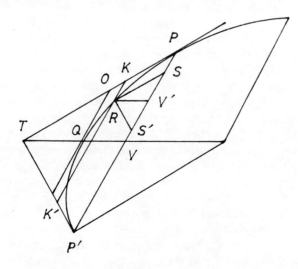

Figure 11

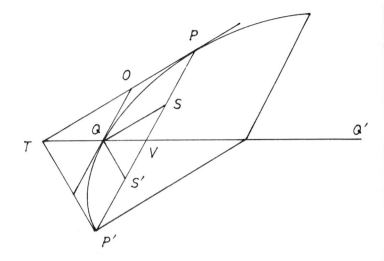

(b)

drawn parallel to PP', it will be tangent to the conic at Q. If next we draw lines from R parallel to TP, TV, TP' meeting PP', respectively, in S, V', S', then KR = PS, RK' = S'P' and PK : RV' = PT : TV. We thus have $OP^2 : OQ^2 = PK^2 : KR \cdot RK' = (RV'^2 : PS \cdot S'P') \cdot (PT^2 : TV^2)$, so that $RV'^2 : PS \cdot S'P'$ has a given value, independent of the choice of point R on the conic. As the lengths PS, S'P', RV' are the respective distances of R from the lines TP, TP', PP' as measured in the given directions of PP', TV, one sees that R lies on the three-line locus relative to the lines TP, TP', PP', and the given value of the ratio.

Conversely, to find the curve answering the condition of the three-line locus, we note that if Q is the intersection of the curve with TV, then $QV^2 : PS \cdot S'P' = M^2 : N^2$, a given ratio [see Fig. 11(b)]. Since PS = S'P' = 0Q, and 0Q : TQ = PV : TV, one has QV : TQ = (M : N) (PV : TV), a given ratio. We may thus determine Q on the given line VT by means of this proportion and then introduce the conic of vertex Q, diameter TV and tangents TP, TP'. That is, when Q bisects VT, the conic is the parabola whose latus rectum L satisfies the relation $PV^2 = QV \cdot L$. Otherwise, we find Q' via the harmonic relation QV : VQ' = QT : TQ'; then QQ' will be the diameter of the conic and L its latus rectum, for $PV^2 : QV \cdot VQ' = L : QQ'$. One notes that this yields an ellipse when QV is less than TQ and a hyperbola when it is greater.

In this way, the solution of the three-line problem, where the solving locus is a parabola, an ellipse, or a single branch of a hyperbola, was available to anyone knowing the property proved in *Conics* III, 16. An aspect of the problem not within the range of the theory of conics before Apollonius relates to the cases where both branches of the hyperbola are considered.[78] We may recognize two additional configurations analogous to III, 16: one in which the tangents are drawn to P, P' on one branch and R lies on the second; another in which the tangents are drawn to different branches. Apollonius presents the analogues in III, 18, 19 from which one can complete the construction of the locus.

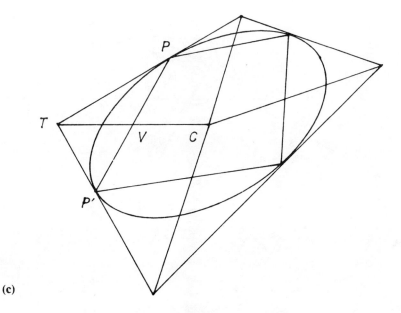

(c)

Having established that the conic satisfies the property of the three-line locus, one can derive that it satisfies the property of the four-line locus by considering the tangents drawn to the four vertices of an inscribed quadrilateral and applying the three-line result in succession to each of the four sides [see Fig. 11(c)].[79] But the conversion to a construction of the four-line locus is not as straightforward as that of the three-line locus. Zeuthen has thus proposed a different construction developing from the relation of the segments of intersecting chords (III, 17).[80] In the first instance, let the given four lines intersect in the trapezium ABCD, sides AD and BC being parallel; we wish then to determine the locus of points R such that RI · RK : AI · IB has a given value, the four segments here being the distances from R to the respective sides of ABCD as measured in the directions AD, AB (see Fig. 12). Let us conceive the conic curve about ABCD whose diameter is the bisector of the parallel sides AD, BC and whose ordinates lie in the direction of AD. Then by III, 17, the ratio RI · IS : AI · IB has a constant value for all chords drawn parallel to fixed directions, while RK = IS. Hence, this conic will be a solution for the locus of R if it can be contrived that this ratio has the value specified in the locus problem when the chords are parallel to AD, AB, respectively. To construct this conic, Zeuthen proposes the introduction of an auxiliary conic which will turn out to be similar to that required. Since the tangents parallel to AD, AB meet in segments such that the ratio of their squares $OP^2 : OQ^2$ equals that of the products of the segments of the intersecting chords, we may assign segments $0'P'$, $0'Q'$ respectively parallel to AD, AB, such that $0'P'^2 : 0'Q'^2$ has the given value, and draw P'T' parallel to the bisector of AD, BC. In this way, $0'P'$, $0'Q'$ will be tangents and P'T' the direction of the diameter of the auxiliary conic similar to the one sought. Extend

124 Ancient Tradition of Geometric Problems

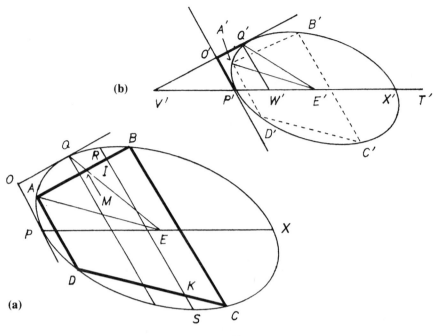

Figure 12

0'Q' to meet P'T' at V', and draw from Q' a line parallel to 0'P' meeting P'T' at W'. We now produce the point X' on P'T' in harmonic relation with V', P', W'; that is, P'V' : P'W' = X'V' : X'W'. Thus, P'X' is the diameter of the conic (via *Conics* I, 36) and its latus rectum L' is determined from the relation Q'W'^{2} : P'W' · W'X' = L' : P'X' (via I, 21). To complete the construction, one must circumscribe about ABCD the conic similar to that in the auxiliary figure. For instance, since the bisector E' of P'X' is the center of the auxiliary conic, the direction E'Q' is known; the analogous line in the required figure must meet the diameter at the same angle and pass through the midpoint M of AB, a chord parallel to the tangent corresponding to 0Q. This determines the center E of the required conic. If we join E, A, the direction of the corresponding line in the auxiliary conic is the same, whence A' is found as the correlate of A. We can now inscribe the trapezium A'B'C'D', similar to ABCD, and from it determine the points 0, P, Q, X which specify the conic. Alternatively, once the ratio EQ : EP of conjugate diameters has been found, Zeuthen proposes an application of *Conics* III, 27 from which the diameter of the required conic can be determined, thence the conic itself.[81]

In Zeuthen's account, not only could Euclid produce a solution of this sort for the case of the four-line locus where the reference figure is a trapezium, but also he could extend this to the general quadrilateral. For this Zeuthen uses a lemma based on one of the locus propositions Pappus cites from the *Porisms*.[82]

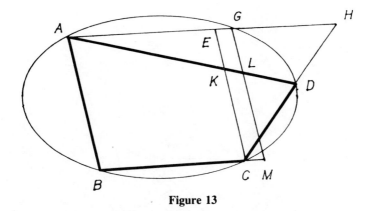

Figure 13

Unfortunately, Zeuthen must assume that Euclid had access to an extension of the cited proposition in which the locus of a given point can be taken to be not only a line, but also a conic (that is, a curve described by the four-line locus). But one may also see that Zeuthen's method is unnecessarily complicated; for the transition from the trapezium to the general quadrilateral can be effected without recourse to the lemma he employs. Let the quadrilateral ABCD be the reference figure for a four-line locus, such that distances are to be measured in directions parallel to AB, BC (see Fig. 13). We complete the parallelogram ABCE and extend AE to meet the solution curve of the locus in G; let AG meet CD in H and let AD meet EC in K; and let the line from G parallel to EC meet AD in L and BC in M. Since G is on the locus, $GH \cdot GA : GL \cdot GM = r : s$, the given ratio; while $GM = EC$ and $GA : GL = AE : EK$ via similar triangles. This determines GH, so that the given figure ABCD determines G on the locus. Since the five points A, B, C, D, G determine the conic uniquely, one is thus permitted to solve the locus via the trapezium ABCG.[83]

As in the case of the three-line locus, the complete solution of the four-line locus would be out of reach before Apollonius' introduction of the two-branched conception of the hyperbola. Securing the requisite lemmas is the project in III, 20–23. One would thus suppose that only the partial solution, perhaps along the lines of the method reconstructed by Zeuthen, was worked out by Euclid and his followers, then later extended by Apollonius.

A puzzle attaches to the set of theorems which close Apollonius' Book III. Like the propositions (16–23) just considered, the later set (Prop. 53–56) deal with relations of segments of secants and tangents drawn to conics. Consider III, 54 (see Fig. 14): if TQ, TQ' are tangents to a conic, V the bisector of chord QQ', and if to any point R on the curve lines are drawn from Q and Q' to meet in S' and S, respectively, the line drawn from Q' parallel to TQ and the line from Q parallel to TQ', then the segments QS, Q'S' contain a given area; that is, $QS \cdot Q'S' : QQ'^2 = (PV^2 : PT^2)(TQ \cdot TQ' : QV^2)$, a given ratio.[84] Zeuthen suggests that this theorem might play a role in Apollonius' own solution of the

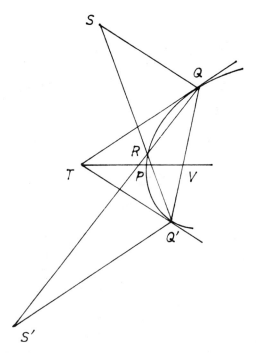

Figure 14

three-line locus problem, and this has been accepted by Heath and others.[85] But this view can hardly be correct, since III, 54 is merely an elaboration of III, 16, while the latter itself suffices for a straightforward solution of this problem, as we have seen. Elsewhere, Zeuthen notes the patent projective character of III, 54: for the constancy of the product $QS \cdot Q'S'$ enables one to assign to each line in the pencil at Q a unique line from the pencil at Q', and this correspondence specifies a conic projectivity in the manner of Steiner.[86] To be sure, this element of the theorem is implicit in it. But even if one can now perceive here the "germ of the projective generation of the conic sections",[87] one can well doubt whether any ancient geometer would have conceived the same result in this sense.[88] Indeed, Apollonius' theorem immediately suggests his intent to solve a specific problem: to determine a conic passing through three given points and being tangent at two of these to given lines. For the givens specify the value of $QS \cdot Q'S' : QQ'^2$, so that the ratio $PV : PT$ is given from III, 54. This determines P and also the harmonic conjugate point P' (from $PV : PT = P'V : P'T$), so that the diameter PP' is known. From the relation $QV^2 : PV \cdot VP' = L : PP'$, the latus rectum L is also known. We have thus found the required conic.[89] As before, the other propositions (55 and 56) permit solutions for other configurations of the problem, where both branches of a hyperbola are to be considered. In view of this straightforward interpretation of these theorems, Zeuthen's projective hypothesis becomes less persuasive.

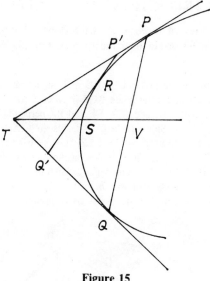

Figure 15

An interest in constructing conics in answer to such initial conditions of incidence and tangency may be detected in other parts of the *Conics*. For instance, one learns in III, 41 that three lines tangent to a parabola cut each other in proportion; that is, PP' : P'T = P'R : RQ' = TQ' : Q'Q (see Fig. 15). The corresponding problem would be to determine a parabola tangent to three given lines and passing through a given point on one of them. Knowing any one of the points P, Q, R, the other two can be found via the relation in III, 41. If V bisects PQ, then TV will be the diameter of the parabola, and if S bisects TV, S will be its vertex (for PT is tangent, so VS = ST); thus $PV^2 = SV \cdot L$ determines its latus rectum L. In our discussion of Apollonius in Chapter 7, we shall find that the detailed examination of problems of this sort formed a major area of interest within his work.[90]

Euclid's efforts on problems like that of the three- and four-line locus thus initiated a fruitful activity of problem solving within the range of the developing theory of conics. The fragmentary character of the extant evidence limits one's ability to distinguish Euclid's own findings and methods from those of the later geometers who continued his work. But our brief consideration here of samples from this field should make an important aspect of this work clear: that its concerns and methods were decidedly *not* projective. Zeuthen's perception of correspondences with the objectives of modern projective geometry seems rather to mistake technical coincidences for essential insights into motivation. Our hope to find within this material the context providing the rationale for the Euclidean *Porisms* has thus been frustrated; for it is the markedly projective appearance of its subject matter which most requires explication. The key we require will emerge, as we consider yet another of the lost Euclidean treatises, the *Surface Loci*.

128 Ancient Tradition of Geometric Problems

AN ANGLE TRISECTION VIA "SURFACE LOCUS"

Pappus presents a series of three angle trisections, of which the third raises certain suggestions of an origin near the time of Euclid. We may sketch it as follows (see Fig. 16):[91]

> Let it be required to cut the given arc ABG at B such that arc BG is its third. (Analysis:) let it be done, so that angle BGZ is twice BAZ. Draw GD to bisect angle BGZ, meeting BA at D, and draw DE and BZ perpendicular to AG. Thus, AD = DG and AE = EG. Since angle BGA is bisected, AG : BG = AD : DB (*Elements* VI, 3); but AD:DB = AE : EZ (via similar triangles), so that AG : AE = BG : EZ = 2 : 1. Thus, B lies on the locus such that BG = 2EZ, for given points G, E; that is, the locus which is such that $BZ^2 + ZG^2 : EZ^2 = 4 : 1$. This is a hyperbola, given in position, so that its intersection with the given arc ABG solves the problem. And the synthesis is obvious.

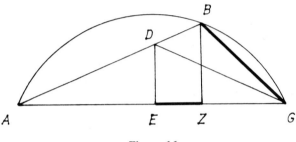

Figure 16

One sees that the angle trisection has been reduced to the specification of a hyperbola via focus and directrix, that is, as the locus of points whose distances from a given point and a given line are in a given ratio greater than 1 : 1. But we note that the locus is first transformed into one involving the *squares* of those distances, rather than the distances themselves. Elsewhere, Pappus presents a proof of this locus construction, and there too the problem is expressed in terms of the squares of the distances.[92] It is part of a general account of the focus-directrix specification of all three conics which Pappus provides as lemmas to Euclid's *Loci on Surfaces*.[93] Since the angle trisection is incomplete without this lemma, and since it conforms with the special formulation of the problem appearing in it, we must suppose that both bore some relation to this lost Euclidean analytic work.

We face the problem, however, that nearly a century after Euclid, the parabolic case of this locus is constructed by Diocles without indication of awareness of a prior treatment.[94] Although Diocles might merely have been ignorant of Euclid's treatment, it seems better to suppose that he did indeed play a part in the discovery of this locus. But what part? Certainly, if the treatment of the parabolic case given by Pappus were known earlier, the solution of Diocles, which is essentially identical to it, could hardly have been original. Is it possible, then, that an alternative method was modified by Diocles, thence to serve as a basis for the version given by Pappus?

A consideration of the Euclidean work will assist us. What was a "surface locus"? It is natural to suppose that it had to do with the tracing of curves on given surfaces, as via the intersection of two surfaces, and this view is in accord with remarks by Pappus and Proclus.[95] If so, we may perceive in the cube duplication of Archytas and the hippopedes of Eudoxus instances of such curves, and the former enters as a locus directed toward the solution of a problem. Other instances arise in Pappus: one may trace spirals on the surface of cylinders, cones, and spheres via projections of the plane spiral. While one usually assigns the definition of the latter to Archimedes, it is clear from Pappus that this curve was known to Conon somewhat earlier, so that its introduction and first study might well have occurred in Euclid's time.[96] To illustrate surface locus, Pappus describes the curve formed by intersecting a plectoid surface (i.e., a spiralling ramp) by an oblique plane; the orthogonal projection of the plane curve so formed is the quadratrix.[97] Pappus' method is analytic: assuming E on the quadratrix AK, the ratio EZ : arc DG (= AB : arc AG) is given; for θ on the spiral, the ratio θD : arc DG (= AM : arc AG) is given (see Fig. 17). Setting EI = θD, the ratio of EI : EZ is thus given (Pappus does not state its actual value, AM : AB), so that I is given as on the intersection of the surface with a given plane through ZI. One may note that this alternative form, like the "mechanical" mode it would supplant, still requires appeal to continuity for determining its terminal position, the key to its use for the circle quadrature.[98] Doubtless, the projective form arose in connection with other applications of the curve, such as its use for the division of angles. Euclid himself is likely to have known of such uses through the efforts of predecessors like Dinostratus.

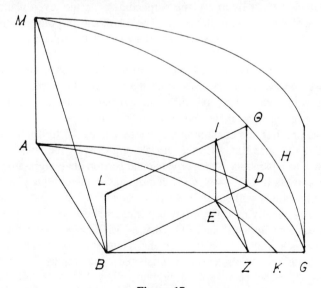

Figure 17

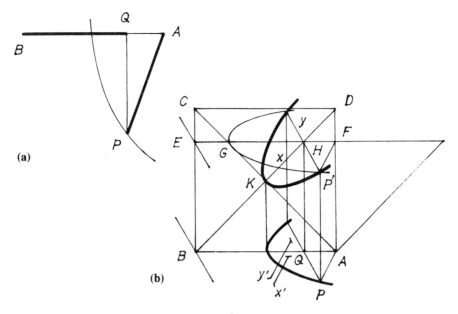

Figure 18

Using a method directly suggested by the solid projections presented by Pappus for the quadratrix, we can construct a solution to the focus-directrix problem in the following manner: first consider the case where the given ratio is 1: 1, the given points A and B; we wish to find the locus of P such that AP = BQ, for Q, the foot of the perpendicular to AB from P (see Fig. 18). Let the square ABCD lie in a plane perpendicular to the given plane, and the line EF in a plane parallel to the given. Then the locus of points at distance AF from F is a circle which meets AC, the diagonal of the square at G; while the locus of points at distance EB (= AF) from the line through E perpendicular to BC in the elevated plane is a line parallel to that line meeting the diagonal BD at H. If we now consider all positions of the horizontal EF, the first locus will be a cone with vertex at A, axis AD, one generator coinciding with AC, and the vertex angle a right angle; while the second locus will be a plane passing through the line BD and containing the line perpendicular to it in each elevated plane. Since the lines BD, AC meet at right angles at K, the intersection of the surfaces will be a parabola, "the perpendicular section of the right-angled cone," and if we express it as $Y^2 = N \cdot X$ via coordinates in the oblique plane, then the line of application N will equal 2DK, as we saw above. Now, if we project this parabola orthogonally onto the original plane, lines AP and BQ will be equal, for the associated lines FP' and EQ' (that is, EH), respectively equal to them are equal to each other in the elevated plane. Thus, P lies on the required locus.[99] Moreover, the projection leaves Y unaltered, but shrinks X by the factor $1 : \sqrt{2}$, so that the curve will be expressed by $Y'^2 = N' \cdot X'$, for $N' = 2\sqrt{2}$ AK or 2AB. Thus, the required locus is a parabola of parameter twice the given line.

The Generation of Euclid 131

The cases of the hyperbola and the ellipse are analogous, each utilizing a rectangle ABCD in place of the square used for the parabola. Let AL : LB be in the given ratio, and draw the right triangle ABK such that KL is the altitude to the hypotenuse AB (see Fig. 19). Then if we complete the rectangle ABCD

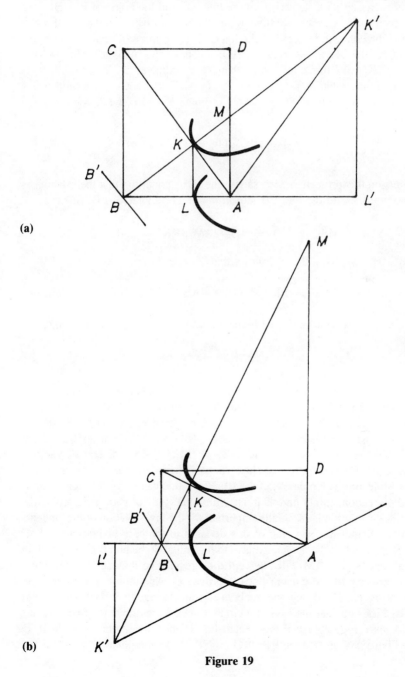

Figure 19

such that AKC is its diagonal, $(AB : BC)^2 = AL : LB$, the given ratio. We designate the line BB' perpendicular to AB in the base plane perpendicular to that of ABCD. If we now introduce the cone of axis AD and generator AC, it will intersect the plane of the lines BK, BB' in a conic section, each of whose points is such that its distances from the line AD and the plane of BC, BB' are in the given ratio AL : LB. Setting K' as the point where BK extended again intercepts the cone (in the case of the hyperbola, its opposed branch) and by M the point where BK meets the axis AD, we have the diameter KK' and the parameter 2KM for this conic, such that the coordinates X, Y for each of its points satisfy the relation $Y^2 : X (KK' \pm X) = 2KM : KK'$. If we project this conic onto the base plane, the projected ordinate $Y' = Y$, while the projected abscissa

$$X' = \left(\frac{LB}{BK}\right) X.$$

The points on this projected curve satisfy the locus condition that their distances from the point A and the line BB' are in the given ratio AL : LB. Since

$$Y'^2 : X' (LL' \pm X') \left(\frac{KB}{BL}\right)^2 = 2KM : KK' = 2AL : LL',$$

and $KB^2 : BL^2 = AB : BL$, it follows that

$$Y'^2 : X' (LL' \pm X') = 2AB \left(\frac{AL}{LB}\right) : LL'.$$

Thus, the curve solving the locus is the conic of diameter LL' and parameter

$$2AB \left(\frac{AL}{LB}\right).$$

The view that Euclid's *Surface Loci* took up the solution of problems via the intersections of solids, in a manner like that given here, raises interesting possibilities concerning the nature of the *Porisms*. Consider, for instance, that in the above construction each of the lines BL', BK' is divided harmonically; that is, BL' : BL = AL' : AL and BK' : BK = MK' : MK. Several of Pappus' lemmas to the *Porisms* intend to show that this harmonic relation is preserved through projection (Lemma 19; cf. 5, 6). Thus, one can see how such lemmas on solid projections might find their use within the study of geometric problems. The harmonic relation is especially important for the investigation of the tangents to the conics. Indeed, by virtue of this division in the case of lines BL', BK' above, it follows that the tangents drawn to the conics at the points corresponding to the abscissas AL, KM will intercept their respective diameters at the point B. This property of the tangent lines is proved by Apollonius in *Conics* I, 34 (cf. 33 for the parabola), but one can conceive an alternative derivation via solid methods. First consider the case of a circle to which are drawn the two tangents KA, KA' [see Fig. 20(a)]; if any secant line from K meets the circle in P, Q and the chord AA' in N, then KP : KQ = NP : NQ (or equivalently, KN is the

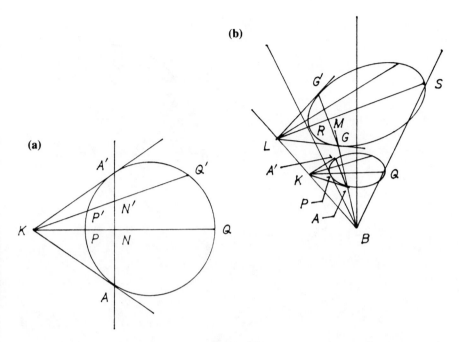

Figure 20

harmonic mean of PK, KQ).[100] Pappus includes a proof of this among the lemmas to the *Porisms* (Lemma 28).[101] Let us next consider the case where PQ is a diameter of the circle, and let us view the circle as the section of a right cone of vertex B [see Fig. 20(b)]. Then if the figure is projected onto an oblique plane from the center B to yield a conic of diameter RS, the projected lines GL, G'L, corresponding to KA, KA', respectively, will be tangents to this curve (for incidence relations are preserved), and the line LRMS will be divided harmonically in correspondence with KPNQ.[102] Hence, the harmonic property of the tangents to conics follows as a projective consequence of the property for the circle. At the same time, the harmonic division of any secant to a conic likewise follows from the case of the circle.

With reference to the same circle, we may ask for the division of secants corresponding to the geometric and arithmetic means [see Fig. 21(a)]. Since AK is the geometric mean of the segments PK, KQ of any secant, it follows that the locus of points J on the secants such that KJ is the geometric mean of PK, KQ will be the circle of center K and radius KA. The points I which mark off KI as arithmetic mean of PK, KQ will be the midpoints of the chords PQ, and hence lie on the circle of diameter OK (for 0 the center of the given circle), since the angle OIK will always be a right angle. If now, as before, the circles are conceived as sections of a solid, they can be projected onto oblique planes and so yield determinations of the conics whose points divide the secants to given conics according to the specified conditions of harmonic, geometric, and arith-

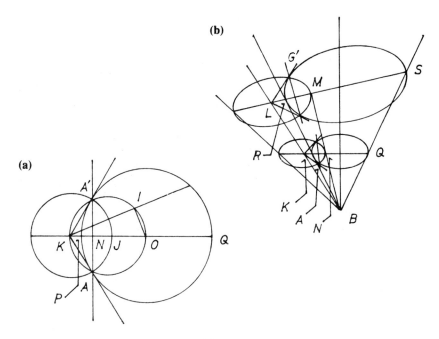

Figure 21

metic division [see Fig. 26].[103] Later we shall encounter a speculation by Zeuthen assigning to Eratosthenes an inquiry into locus problems of this type. The present remarks show how naturally the investigation of solid projections could develop in this direction.

Of these three locus propositions with reference to the circle, Pappus gives only that one relating to the harmonic property, but not those relating to the geometric and arithmetic means. Nevertheless, he does provide, in two other lemmas, certain other properties of a comparable kind (see Fig. 22):

> Lemma 33: If points N, K lying in the line of the diameter of a circle are such that NK is the geometric mean of PK, KQ, and K' is any point on the line

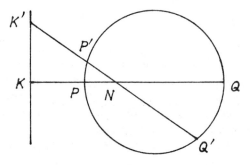

Figure 22

perpendicular to QK at K, then any secant K'P'Q' passing through N is divided such that NK' is the geometric mean of P'K', K'Q'.

Lemma 35: If in the same figure it is assumed that NK is the harmonic mean of PK, KQ, then also NK' will be the harmonic mean of P'K', K'Q'.[104]

One notes that just as the earlier lemma (28) on the harmonic mean produced the construction of the polar line with respect to the pole K lying outside the circle, so in Lemma 35 one has the construction when the pole N lies inside the circle. In both cases one can obtain via solid projection in the cone an alternative proof of Apollonius' corresponding propositions for conics (III, 37–40).

Again consider the section of a right cone by a plane perpendicular to a generator (see Fig. 23). We have seen that if the diameter of the section PQ meets the axis of the cone in N and the base plane in K, then the line KPNQ is divided harmonically; and that if AN is an ordinate, the tangent at A passes through K. Now let any other plane containing line KK' perpendicular to KB in the base plane intersect the cone in the section of diameter P'Q'. Using B as the center of projection, we find that KP'N'Q' is also divided harmonically. Thus, if A' is the projection of A, the line KA' is tangent to the projected conic. The converse suggests a problem: to find the locus of points of tangency A

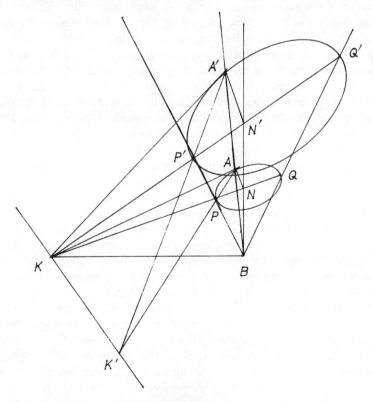

Figure 23

drawn from a given point K to the sections of a given cone made by planes passing through a given line KK'. One approach might consider the projection of one section PQ onto another P'Q' via the center B. As we have seen, since KPNQ is divided harmonically, so also is KP'N'Q'. Since KA, KA' are tangents, the lines AN, A'N' are each parallel to KK' and hence to each other. Since, furthermore, AN, A'N' are radii of the circular sections of the cone made by planes parallel to the base through points N, N', respectively, it follows that AN : A'N' = BN : BN'. Hence, the triangles BAN, BA'N' are similar, so that BAA' is straight. Thus, the sought locus of points of tangency A is the generator of the cone BAA' extended. One may note the corollary, that if the lines AP, A'P' are drawn and extended, they will meet at a point on KK'; for the triangles APN, A'P'N' are in projective correspondence in accordance with the configuration of Desargues.[105] Thus, for all curves in this family of sections, the line AP will pass through the same point K" on line KK'.

These observations on the possible nature of Euclid's investigation of surface loci suggest a route toward the resolution of several puzzles surrounding his *Porisms*. In view of Pappus' lemmas to this work, we would gather that Euclid either assumed them without proof or else used a different method of proof from that given by Pappus. The context of stereometrical problems offers an appropriate set of techniques for such alternative proofs in contrast with the metrical procedures of plane geometry adopted by Pappus. Furthermore, the lemmas reveal the affinity of the subject matter of the *Porisms* with that commonly encountered now in the field of projective geometry. Yet it is clear that the ancients never introduced the conceptions and techniques of projectivity fundamental to the modern field, and so could never have undertaken the systematic exploration of the relations organized within that field. What, then, were the initiating interests which led them into these materials? We have seen how the study of problems relating to conics, as for instance the drawing of tangents, could give rise to configurations significant within the projective field. Among these are the two-triangle figure of Desargues, the complete quadrilateral, the harmonic division of a line and the preservation of this division under projection. For these have natural interpretations in the context of problems involving the formation of conic curves as different planar sections of the same cone.

If through his *Porisms* and *Surface Loci*, Euclid did indeed attempt to inaugurate a stereometrical study of the conics, his approach was soon modified drastically. This three-dimensional aspect is of course still in evidence in Archimedes' studies of the conics of revolution; but they have all but disappeared from Apollonius' theory of conics. For here the formation of these curves as conic sections serves only to fix their conception and to permit the derivation of the planimetric second-order relations of their ordinates. Thereafter, all properties and applications of the curves follow from these relations via the manipulation of proportions of lines in the plane.[106] For instance, the problem of the locus relative to a point and a line, which we solved above through the sections of a cone, is solved planimetrically in an alternative method preserved by Pappus; here the condition of the locus is transformed so that either $Y^2 = 2AB \cdot X$ or the ratios $Y^2 : X$ (LL' $\pm$ X) have a given value with reference to the given line AB and the given ratio AL : LB. It is not difficult to perceive how a stereometric treatment comparable to that reconstructed above might have served as a precedent for the extant planimetric treatment. If Euclid had solved the locus problem in the former manner, perhaps even omitting the parametric expressions for the solving conics, there would have been ample cause for Diocles and for the author of the lemmas to the *Surface Loci* to produce their alternative solutions. In this way, we can reconcile Euclid's awareness of the focus-directrix property of the conics with their re-examination by these later writers.

EUCLID'S CONTRIBUTIONS TO THE STUDY OF PROBLEMS

None of the lost mathematical writings alluded to in ancient sources seems more enigmatic than Euclid's *Porisms*. Pappus' extensive lemmas and commentary on the work have permitted detailed and persuasive reconstructions, especially that by Chasles. But its underlying motive and the scope of the wider geometric field toward which it was compiled remain unclear. The ambitious view advocated by Zeuthen seizes upon the striking correspondences with the modern field of projective geometry to assign to Euclid and his followers the elaboration of a projective theory of the conics, culminating in the projective generation of conics with reference to five given points, a result equivalent to Pascal's "mystic hexagon." Unfortunately, the ancients never introduce the general conceptions essential for the rationale of projective geometry. I have attempted an intermediate position, suggesting that Euclid experimented with a stereometric approach to the study of conics. His own *Optics* provides a ready context for such an inquiry; for instance, his observation that obliquely viewed circles appear "pressed in" (Prop. 36) at once raises the question of the precise determination of the apparent curve, whose solution as an ellipse actually appears among Pappus' lemmas to this work.[107] My view would thus assign to Euclid a form of the study of conics not unlike that later adopted by Desargues and Pascal, in which propositions established for circles are extended to conics via solid projection.[108] This does not intend to deny that the projective *conceptions* were for the most part absent from the ancient work and that these were thus truly novel elements of the 17th-century efforts. Nor do the several reconstructions of conic theorems given above presume to exhibit the method actually adopted by Euclid and his followers, but rather serve as specimens of a field of results readily accessible to them upon the adoption of such a stereometrical approach. The fact that Desargues and Pascal, firmly grounded in the study of Apollonius and Pappus, moved in this same direction suggests that they might in part have been rediscovering forgotten aspects of the ancient work in the process of advancing their own. Our task of interpreting the ancient field will surely have a more appropriate instrument in these 17th-century efforts than in the more sophisticated studies by the 19th-century projective geometers.

It is thus clear that Euclid's study of the conics, his *Porisms* and *Surface*

Loci, together with his *Data* and *Elements*, had special utility for the analytic investigation of geometric problems. The study of locus problems appears to have attracted special interest at this time. We have seen examples in the two-point locus of the Aristotelian *Meteorologica*, the point-line locus (or the specification of a conic via focus and directrix, as in the lemmas to the *Surface Loci*), and the locus with reference to three and four lines; doubtless many more of comparable type were then examined, although our sources no longer preserve clear information on the wider scope of this activity.

As for Euclid himself, one surely does better to view him as an effective teacher and compiler than as a phenomenally gifted mathematical intellect in his own right. The *Elements*, for instance, drew on the discoveries by Theaetetus and Eudoxus and their followers, while its individual books appear to have already received a systematic organization in the decades before Euclid. The *Data* was novel in form only, its content largely duplicating that of the *Elements*, but cast specifically for application in analyses. In the area of the conics, Euclid had the work done by Aristaeus as a precedent. We must regret not knowing more about the *Surface Loci* and its precursors; but, presumably, work extending as far back as Archytas and Eudoxus, in their studies of the cube duplication and other solid constructions, could serve as background. As for the *Porisms*, however rich the potential field might have been relating to it, Euclid's treatment had its weaknesses. For instance, Pappus was able to formulate into single propositions what Euclid presented in as many as ten and could see how some of Euclid's results extend into more complicated configurations in a way not noted by Euclid.[109] The proliferation of special cases which seems to have marked Euclid's approach must surely have obscured the general patterns in this material. But Apollonius charges Euclid not only for having taken up merely "the chance portion" of the locus of three and four lines, but also for having done so "unsuccessfully." These criticisms are implicitly admitted by Euclid's apologists, in a passage cited by Pappus, in that they insist he made no presumption of having exhausted these fields, but that his effort proved to be an indispensable basis for further research, including that of Apollonius himself.[110]

We should keep in mind, then, that in Euclid's time the theory and application of the conics were still relatively new, constituting a field undergoing its first consolidation. His contributions were seminal for the development of problem solving in the 3rd century. But it is entirely unfair of us to demand that his own efforts here matched the refinement of the *Elements* or of the Apollonian *Conics*.

NOTES TO CHAPTER 4

[1] The biographical data on Euclid are meager. His activity at Alexandria is attested in passages by Proclus and Pappus; but the former is anecdotal ("no royal road to geometry" addressed to King Ptolemy), while the latter refers to Euclid's *students* at Alexandria (cf. T. L. Heath, *Euclid's Elements*, I, Ch. I for references and discussion). On the other hand, the nature of Euclid's work well suits the environment of Alexandria: his several textbooks on geometry and mathematical science would find a welcome place in the research and teaching at the Museum and use to the fullest advantage the textual

resources of the Library. One usually sets his activity near the time of the founding of the Museum, thus centering on the first two decades of the 3rd century B.C. Note, however, that Archimedes' earliest recognizable citation of Euclid's *Elements* is in *Spiral Lines*, well after the middle of the century, so that a later dating may be indicated for Euclid (see my "Archimedes and the Elements" and "Archimedes and the pre-Euclidean Proportion Theory"). Recent authors wish to deemphasize the role of Alexandria as a center for mathematical and astronomical study; cf. G. J. Toomer, *Diocles*, p. 2 and O. Neugebauer, *History of Ancient Mathematical Astronomy*, pp. 571 f.

[2] Heath, *History of Greek Mathematics*, I, pp. 357 f. See also his survey of major editions of the *Elements* from antiquity to the early 20th century in *Euclid's Elements*, I, Ch. VIII. In addition to Heath's ample notes on the *Elements*, one may consult the studies by E. Neuenschwander, "Die ersten vier Bücher der *Elemente* Euklids," *Archive for History of Exact Sciences*, 1973, 9, pp. 325–380 and "Die stereometrischen Bücher der *Elemente* Euklids," *ibid.*, 1974–75, pp. 91–125; and I. Mueller, *Philosophy of Mathematics and Deductive Structure in Euclid's Elements*, Cambridge, Mass., 1981. The background to the Euclidean theory of incommensurables is examined in my *Evolution of the Euclidean Elements*, Dordrecht, 1975, while a survey of Book X of the *Elements* is provided in my "La Croix des Mathématiciens."

[3] On this technique, see my *Evolution of the Euclidean Elements*, Ch. VI, Pt. IV. The standard account derives from H. G. Zeuthen; see also Heath, *History*, I, pp. 150–154 and *Apollonius of Perga*, pp. cii–xi.

[4] An account of the relation of the *Elements* to prior efforts appears in my *Evolution*, Ch. IX, with references there to other discussions.

[5] Proclus, *In Euclidem*, ed. Friedlein, p. 426. See Heath's account of the theorem in *Euclid*, I, pp. 350–354. On its Babylonian appearances, in particular in the "Pythagorean triplets" on the cuneiform tablet Plimpton 322, see O. Neugebauer, *Exact Sciences in Antiquity*, pp. 35–40 and B. L. van der Waerden, *Science Awakening*, pp. 76–80.

[6] See my "Archimedes and the Pre-Euclidean Proportion Theory" and *Evolution*, Ch. VIII, Pt. II and Appendix B.

[7] On irrationals see my *Evolution*, Ch. VIII, Pt. IV. On the pre-Euclidean study of the postulate, see I. Toth, "Das Parallelenproblem im Corpus Aristotelicum," *Archive for History of Exact Sciences*, 3, 1967, pp. 249–422.

[8] Aristotle, *Opera* (Bekker), 376 a 4–6. I have altered the punctuation slightly. Note that the writer tacitly assumes that MK = KH (being radii of the same circle) in order to infer that the ratio HM : MK is given.

[9] Apollonius' proof, to be discussed below, is given in full by Eutocius in his commentary on the *Conics*; see Apollonius, *Opera*, ed. Heiberg, II, pp. 180–184. That this result was proved by Apollonius in the second book of the *Plane Loci* is asserted by Pappus, *Collection* (Book VII), ed. Hultsch, II, p. 666. Two detailed accounts, establishing the identity of the two methods and giving full technical and textual comparisons, are presented by Heath, the first in his *Euclid*, II, pp. 197–200 (commenting on VI, 3), the second in his *Mathematics in Aristotle* (1949), pp. 181–190. The latter account was motivated by Heath's regret that writers on the Aristotelian meteorology had overlooked his earlier analysis and so erred in their discussion of this passage. It is to be all the more regretted that the Loeb translator of the *Meteorologica* repeated the oversight a few years later (1952). F. Solmsen's hesitation to accept the accuracy of Heath's position on this passage is likewise regrettable (*Aristotle's System of the Physical World*, 1960, p. 420n).

140   Ancient Tradition of Geometric Problems

[10] I follow the construction in the *Meteorologica*, but have partially adapted the latter part of the proof where it seems muddled and have avoided its use of an indirect mode of reasoning.

[11] It is possible that this slip, together with the adoption of the indirect mode of proof, might indicate the author's having confused this construction with the proof of its converse, the completeness of the solution. For the proof that *all* points satisfying the condition of the locus lie on this circle is indeed well suited to an indirect demonstration, as one sees from the treatment given by Apollonius. Note further that the point n here cannot be the same as N in Fig. 4, even though the author seems to view it so. Apparently, he has confused two senses of the phrase "the arc through M."

[12] Heath notes this detour (*Elements*, II, p. 200) without registering surprise at the monumental lapse implied in the Apollonian treatment.

[13] Heath summarizes Simson's method in both of his accounts.

[14] This relation, not in the *Elements*, is assumed by Pappus in the *Collection* (Book VII), II, p. 730.

[15] Oddly, Heath persists in following Simson's approach in attempting his own reconstruction of the analysis underlying the *Meteorologica* passage. Furthermore, he seems to assume that one knows beforehand that the locus will be a circle. But surely, this fact too must be revealed by the analysis, not just its center and radius. These lapses detract from the effectiveness of his account for displaying the line of thought actually underlying the Aristotelian/Apollonian approach.

[16] It is remarkable that Pappus adopts much the same method in the solution of a related problem. In Lemma 29 to Euclid's *Porisms* (*Collection* (VII), II, pp. 904–906), he seeks to inscribe a broken line AGB within a given segment of a circle, such that AG : GB is a given ratio (Fig. 24). Thus, he draws the tangent line GD, whose length is the mean proportional between BD, DA, whence AD : DB = AG^2 : GB^2. Since AG : GB is given, so also is AD : DB; since D is thus given, so also is G. Here, the tangent line corresponds to PM in the Aristotelian/Apollonian locus problem. In effect, Pappus' method finds the solution via the intersection of two loci: that answering to the given ratio and that answering to the given angle of separation of AG, GB (i.e., the given circular arc). As it happens, al-Bīrūnī solves this same problem in his *Book of Chords*

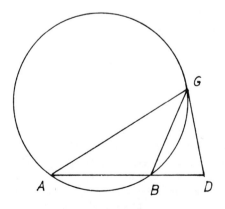

Figure 24

(cf. H. Suter, "Buch der Auffindung der Sehnen im Kreise von...el-Bīrūnī," *Bibliotheca Mathematica*, 1910–1911, 11, p. 36); since the angle is given, he shows one need only construct a triangle with this angle at its vertex, such that the adjacent sides are in the given ratio, and then inscribe the similar triangle in the given segment. That Pappus follows such an elaborate alternative method here must indicate that he has lifted this solution from a context which treated it as a corollary to the two-point locus problem, the latter presented in the Apollonian manner. This may indicate that Euclid himself in the *Porisms* presented the locus problem in this same manner.

[17] The proof that this locus is a circle appears earlier in the account of the rainbow, *Meteor.* III, Ch. 3.

[18] Alexander of Aphrodisias (2nd century A.D.), *In Meteorologica*, ed. Hayduck (*Commentaria in Aristotelem Graeca*, III), pp. 162–172; Olympiodorus (6th century A.D.), *In Meteor.*, ed. Stüve (*CAG*, XII), pp. 243–259. Note that both commentators assume the simpler version of the uniqueness proof just given as Aristotle's intent (Olympiodorus, Lesson 38, pp. 245 ff). They should thus view the entire remaining portion of the geometric proof as superfluous. Yet each goes on to an elaborate line-by-line exegesis of that proof, without seeming to grasp what that proof is intended to establish (*ibid.*, Lesson 39, pp. 253–258). Note that some modern translators and commentators appear to be in the same situation (see note 9 above).

[19] This aspect of the passage was brought to my attention by H. Mendell.

[20] *Pace* Solmsen, who appears content with assigning the proof to Aristotle himself (*op. cit.*, p. 420).

[21] I touch on these passages in "Archimedes and the Pre-Euclidean Proportion Theory" and in an unpublished essay, "Zeno, Aristotle and Eudoxus." But they have not received serious attention among Aristotle scholars, including Ross (*Aristotle's Physics*) and Heath (*Mathematics in Aristotle*). Cf. my review of H. J. Waschkiess, *Von Eudoxos zu Aristoteles*, in *Isis*, 1980, 71, p. 507.

[22] Cf. the pseudo-Euclidean *Catoptrica*, Prop. 30 (Euclid, *Opera*, ed. Heiberg, VII) and Diocles' *Burning Mirrors*, ed. Toomer, Prop. 1–3; here one investigates burning mirrors by which rays from the sun are reflected to a point or a line.

[23] Cf. Solmsen, *op. cit.*, p. 419.

[24] C. Boyer, *The Rainbow: From Myth to Mathematics*, 1959, pp. 39–54.

[25] That is, when the sun is rising or setting at the time the rainbow is seen. Generally, one must take ray MH parallel to the direction of the line from the observer at K to the sun.

[26] See H. D. P. Lee's introduction to the Loeb edition, pp. xxiii–xxv. Note that the authenticity of Book IV, on the nature of chemical processes, has long been doubted, although Lee defends it (*ibid.*, pp. xiii–xxi).

[27] Euclid also standardizes "right line" (*eutheia*) in preference to "line" (*grammē*) for rectilinear segments, and is followed in this by later authors.

[28] On Hippocrates, see Chap. 2 above. Besides the *Prior Analytics*, one may consult Aristotle's *Physics* (IV, ch. 8; VI, ch. 2, 4–7, 9) and *de Caelo* (I, 12; II, 4; III, 2) for this usage. Among later authors, only Philo of Byzantium adopts the same, to my knowledge. A mechanical writer late in the 3rd century B.C., his account of the cube-duplication bears several indications of his distance from the field of geometric research at that time; the passage in Marsden (*Greek Artillery*, p. 110) may be compared with

accounts by Hero, a mechanical writer, like Philo, but much closer to the standard forms of geometrical writing (cf. Pappus, *Collection*, I, pp. 62-64).

[29] See note 11 above.

[30] Although we of course cannot know for certain the identity of the geometer who discovered this solution of the locus problem, it may well be pertinent to this question that Proclus says of Hermotimus of Colophon, a bit before the time of Euclid, that he "composed some things on loci" (*In Eucl.*, p. 67).

[31] For references to Pappus and Proclus on the meaning of "porism," see Heath, *History of Greek Mathematics*, I, pp. 431 f, 434 f.

[32] Pappus' commentary on this corpus occupies the entirety of the 238 propositions of the seventh book of his *Collection*, and is our principal, often unique source as to its contents.

[33] The *Data* has been edited in Greek by Menge (Euclid, *Opera*, VI), and its medieval Latin translation has recently been edited with English translation by S. Ito.

[34] See, for instance, Heath's discussion of Prop. 93 (*History*, I, pp. 424 f).

[35] For authorities espousing this view, see Ito, *op. cit.*, pp. 11 f. But most commentators appear to have had a sense of the relevance of the *Data* for analyses; cf. Heath, *History*, I, p. 422.

[36] Aristotle notes that propositions are not always simply convertible, and sees that as a cause why analytic methods are more useful in geometry (whose propositions tend regularly to be convertible) than in dialectic (*Post. An.* I, 12). For a logical and historical account of this method, see J. Hintikka and U. Remes, *The Method of Analysis*, 1974 and M. Mahoney, "Another Look at Greek Geometrical Analysis," *Archive for History of Exact Sciences*, 1968/69, 5, pp. 319-348. See also Chapter 3.

[37] Proclus informs that Leon, a contemporary of Plato, investigated diorisms to determine the limits of the solvability of problems (*In Euclidem*, p. 66). This bears on the *Meno* passage discussed in Chapter 3.

[38] See, for instance, the striking similarity between *Data*, Prop. 58 and *Elements* VI, 26 on the application of areas. On the other hand, some parts of the *Data* adopt a mode at variance with the *Elements*. The section on parallels, for instance, conceives of a "displacement" (*metapiptein*) of given lines (*Data*, Prop. 28) in contrast with the strictly static view in the *Elements* (I, 30). It may thus be that some portions of the *Data* had already been compiled before Euclid and were incorporated by him into a larger treatment, with his own additions based on the *Elements*.

[39] Cf. the problems in *Elements* I, 1-3, 9-12, 22-23, 31, 44-46; all of IV; further instances in VI, XI, XIII. A comparable usage occurs in the Aristotelian tract *On Indivisible Lines* (970 a 8): to construct a triangle whose three sides are "given" (cf. *Elements* I, 22).

[40] *De Caelo* I, 2, 283 a 6; for other instances, see Bonitz' index in Aristotle, *Opera* (Berlin edition), V, s.v. *didōmi*.

[41] *Collection* (Book VII), II, p. 672; cf. p. 676 and below, note 57. We return to Aristaeus in Chapter 7.

[42] Cf. *Quadrature of the Parabola*, Prop. 1-3. A survey of such instances is given by Heath, *Apollonius*, Ch. III. An effort to compile the whole of the Archimedean theory of conics was made by J. L. Heiberg, "Die Kenntnisse des Archimedes über die Kegel-

schnitte," *Zeitschrift für Mathematik und Physik* (hist.-litt. Abth.), 1880, pp. 41–67. But its comprehensiveness has recently been set into question (G. J. Toomer, *Diocles*, p. 4).

⁴³ See Apollonius, *Conics* I, preface and Heath's discussion, *Apollonius*, Ch. I.

⁴⁴ Note that some parts of the earlier theory might happen not to appear in Apollonius. Such seems to be the case for the theorems on the focus-directrix construction of the conics (see Sect. iv below); cf. also Diocles' assumption of the constancy of the subnormal of the parabola, only found in Apollonius' Book V in somewhat different form (cf. Toomer, *op. cit.*, pp. 17, 151).

⁴⁵ For instances of such inquiries, see Chapter 3.

⁴⁶ "On the Astronomical Origin of the Theory of Conic Sections," *Proceedings of the American Philosophical Society*, 1948, 92, pp. 136–138.

⁴⁷ *Collection* (Book VI), II, pp. 588 ff; see Heath, *History*, II, pp. 397 ff.

⁴⁸ [Aristotle], *Problems* XV, 7, XVI, 6; discussed by Heath, *Mathematics in Aristotle*, pp. 264–267. Note that *thyreos* was still admitted as a name for the ellipse by Geminus (1st century A.D.), as cited by Proclus (*In Euclidem*, pp. 111, 126). In his work on spherical astronomy, the *Phaenomena*, Euclid argues the sphericity of the realm of the stars on the basis that it always appears to produce a circle when sectioned by a plane, while in the case of cylindrical and conical figures the result would be an ellipse, that is, "the section of an acute-angled cone" (Preface, *Opera*, VIII, pp. 4–6). Euclid is surely mistaken in alleging that the appearances of these three sections will be different; for the observer will view them from within the plane of their formation, so that the sections of the cylinder and the cone will appear precisely superimposed over the section of the sphere, giving all three the appearance of circles. It is surely significant that Ptolemy does not use this argument in his account of the sphericity of the cosmos (*Syntaxis* I, 3).

⁴⁹ Cf. Proclus, *In Euclidem*, p. 356: one deduces the "symptom" (property) of the curve as the basis of its study. This characterizes Archimedes' treatment of spirals (*Spiral Lines*, Prop. 2, 12; Pappus, *Collection* (IV), I, p. 234), as also Nicomedes' of the conchoids (Pappus, *ibid.*, p. 244) and that of the quadratrix (*ibid.*, p. 252). This distinction between generating mode and property has been exploited by Zeuthen in his argument that the ancients intended problems of construction to serve as existence proofs ("Die geometrische Construction als 'Existenzbeweis' in der antiken Geometrie," *Mathematische Annalen*, 1896, 47, pp. 222–228); I raise reservations concerning this view in Chapter 8.

⁵⁰ Such a construction was used by J. Werner in his *Elements of Conics* (1522), Prop. XI; see M. Clagett, *Archimedes in the Middle Ages*, IV, Pt. I, pp. 254, 277, 298.

⁵¹ Diocles and Eutocius describe such pointwise procedures; see Ch. 3, notes 36–37.

⁵² Pappus, *Collection* (VII), II, pp. 672–674; Eutocius, *In Apollonium*, ed. Heiberg, II, pp. 168–170. Note that even those writers like Euclid and Archimedes who subscribed to the older mode were still quite aware of the nature of oblique sections; cf. Heath, *Apollonius*, pp. xxxvi, xlv, and also note 48 above.

⁵³ See Toomer, *Diocles*, pp. 9, 14. Geminus, and after him Proclus, sometimes use the older terminology, perhaps a suggestion of their reliance on early sources; cf. *In Euclidem*, p. 111.

⁵⁴ For a similar intuition of the role of the solid construction within the early theory of conics, see van der Waerden, *Science Awakening*, p. 245 and H. G. Zeuthen, *Lehre von den Kegelschnitten*, pp. 467–469. The view is also sketched in Zeuthen's "Con-

struction als Existenzbeweis'' (see note 49 above). One should note that all these allegedly aim to provide a rationale for the *Menaechmean* conception of the conics.

⁵⁵ This interpretation of the expression conforms with the view of Zeuthen and Heath; cf. Heath, *Archimedes*, pp. clxvii–iii; and Toomer, *op. cit.*, p. 13.

⁵⁶ Cf. Toomer, *op. cit.*, pp. 5–7 and Heath, *Apollonius*, Ch. III and pp. lxxix ff.

⁵⁷ Pappus, *Collection* (VII), II, p. 674; cf. Heath, Apollonius, pp. lxxxiii–iv.

⁵⁸ One may consult the derivations given by Toomer (*op. cit.*, pp. 10–13) and E. J. Dijksterhuis (*Archimedes*, pp. 58, 59) for a proof that line C, as shown in Figs. 6(b), 6(c), is the line of application (the "latus rectum") in the parametric expression of the conics.

⁵⁹ Apollonius, *Conics* I, Preface; cf. Pappus, *Collection* (VII), II, pp. 672 ff and Heath, *Apollonius*, pp. xxxi–ii. Eutocius misconstrues this problem as referring to the finding of the two mean proportionals (Apollonius, *Opera*, II, p. 186).

⁶⁰ For a discussion and reconstruction of the Apollonian investigation of this locus, see Heath, *Apollonius*, Ch. V and Zeuthen, *Kegelschnitte*, Ch. 7, 8.

⁶¹ Pappus, *Collection* (VII), II, pp. 648 ff; discussed by Heath, *History*, I, pp. 431 ff.

⁶² This follows Simson's account, as reported by I. Thomas, *History of Greek Mathematics*, I, pp. 482–484. See also van der Waerden, *op. cit.*, pp. 287–290.

⁶³ For a compact survey of the history of projective geometry, see L. Cremona, *Elements of Projective Geometry*, 1893, pp. v–xii. The classic treatment is by Chasles, *Aperçu historique des méthodes en géométrie*, 1837 (1875; 1889).

⁶⁴ Van der Waerden reconstructs a proof, *op. cit.*, pp. 287 f.

⁶⁵ Pappus, *Collection* (VII), II, p. 884 (Lemma 12; Lemma 13 treats the nonparallel case). For the entire set of lemmas, see *ibid.*, pp. 866–918. Brief accounts appear in Heath, *History*, II, pp. 419–424 and van der Waerden, *Science Awakening*, pp. 287–290.

⁶⁶ *Les trois livres des Porismes d'Euclide*, 1860. The plan of this reconstruction is adumbrated in his *Aperçu historique*, pp. 274–284. Heath's assessment of his effort, in light of those of Simson (1776), Zeuthen (1886), and others, appears in *History*, I, pp. 431–438.

⁶⁷ Cf. Heath, *History*, I, p. 437. Heath's principal reservation against accepting Chasles' reconstruction has hardly any standing at all: it is that Chasles makes out the propositions in Euclid's *Porisms* to be on a par with Pappus' lemmas, while one should expect that the lemmas were pitched at a far lower level than the theorems in the associated treatise. But if such is *often* the case with Pappus' lemmas, it seems not *always* to have been so. For instance, the *neusis* problem from Heraclitus (Prop. 72) is surely comparable to the *neusis* constructions in Apollonius' work; a hyperbola construction among the lemmas to the *Conics* (Prop. 204) is the effective equivalent of a construction now to be found in *Conics* II, 4 (see Chapter 7, note 33); the focus-directrix constructions for the conics (Props. 236–238) must match or surpass their correlates in the Euclidean *Surface Loci* (see below). In several instances, Pappus provides problems with full analysis and synthesis: Props. 85, 87, 105, 107–109, 117; and he sometimes offers not only one, but sometimes two alternative proofs of lemmas (e.g., Props. 35, 36, 39). This must surely indicate that the lemma had a certain interest in its own right. Finally, one may consider

the construction of the ellipse conforming to the obliquely viewed circle (*Collection* VI, Prop. 53); this is in fact far *beyond* the scope of the technical methods employed in the associated treatise, Euclid's *Optics*. We thus have no reason to be uncomfortable with the view that in the instance of an early work like Euclid's *Porisms*, Pappus might provide in his lemmas a modification, or even an improvement, on materials in the original.

[68] *Kegelschnitte*, Ch. 8, especially pp. 173–184; cf. Heath, *History*, I, pp. 437 f. Zeuthen shows how a certain porism enunciated by Pappus can lead to the specification of a conic projectively, via its property as a four-line locus (*op. cit.*, pp. 169 f; we discuss this locus below). In this way, one comes to the point of a verification of Pascal's theorem on the hexagon inscribed in a conic; for that can be seen as a consequence of the generation of a conic as the locus of the vertex of a variable triangle whose other two vertices move along given lines, while its sides turn about given points. In noting this connection with the porism, Zeuthen nevertheless acknowledges that the ancients were not likely to have recognized the Pascal property; for Pappus would surely have mentioned it were it known (*op. cit.*, pp. 495 f).

[69] *Collection* (VII), II, pp. 884–886.

[70] In the parallel case, the ratio becomes ZD : GZ in each instance, as Pappus establishes in his Lemma 11.

[71] H. S. M. Coxeter, *Projective Geometry*, New York, 1964, pp. 38 f; W. T. Fishback, *Projective and Euclidean Geometry*, 2nd ed., New York, 1969, pp. 67–69; D. J. Struik, *Analytic and Projective Geometry*, Reading, Mass., 1953, p. 66; B. E. Meserve, *Fundamental Concepts of Geometry*, Reading, Mass., 1955, pp. 61–63.

[72] For instance, the lemma on the ellipse in *Collection* VI, Prop. 53 (see note 67 above) assumes a result proved in Pappus' Lemma 28 to the *Porisms* (VII, Prop. 154). Heath discusses a remarkable coincidence between Lemma 31 (VII, Prop. 157) and Apollonius' III, 45 on the foci of a central conic (*Apollonius*, pp. xxxix f).

[73] *Collection* (VII), II, pp. 648–652.

[74] Cf. Heath, *Apollonius*, pp. 95 f, whose lettering I adopt in part.

[75] Cf. *ibid.*, pp. xxv, xlviii.

[76] Zeuthen does not propose a reconstruction of the three-line locus; Heath develops an unnecessarily complicated form based on *Conics* III, 54–56 (*ibid.*, pp. 122f). Thus, the present version is my own.

[77] That is, QV will bisect all chords parallel to PP'.

[78] This limitation within the pre-Apollonian researches is stressed by Zeuthen; cf. Heath, *op. cit.*, pp. lxxxiv ff, cxli.

[79] For details, see Heath, *op. cit.*, pp. 123–125. Although he derives his form of the three-line locus from III, 54–56 (see note 76 above), this is immaterial to the subsequent derivation of the four-line locus from the three-line locus.

[80] *Lehre von den Kegelschnitten*, Ch. 7, 8; summarized by Heath, *op. cit.*, Ch. v.

[81] Cf. Heath, *op. cit.*, pp. cxlv f.

[82] Cf. *ibid.*, pp. cxlvii–ix.

[83] That five points determine a conic is established by Apollonius in *Conics* IV, 25 using the harmonic division of secants drawn to conics; results of the latter type are likely to have been well known in the earlier theory (see below). To effect the locus in the

146   Ancient Tradition of Geometric Problems

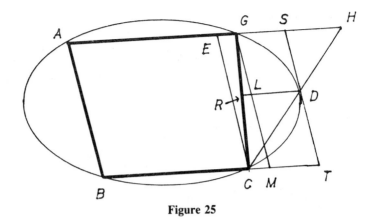

Figure 25

manner presented here, we require that if G is on the locus relative to ABCD, then D is on the locus relative to ABCG; that is, AS · DR : DS · DT = AG · GH : GL · GM (where DR is parallel to BC, and SDT is parallel and equal to AB). By similar triangles, AS : DS = AG : GL, while DR : GH = CR : CG = DT : ST, and ST = GM. Hence, as claimed, (AS : DS)(DR : DT) = (AG : GL)(GH : GM).

[84] Here, P is the intersection of the conic with the given diameter TV.

[85] Cf. note 76 above; Heath designates his Section III, 53–56 by the heading, "The Locus with respect to 3 Lines &c" (*op. cit.*, p. 119). Cf. Zeuthen, *op. cit.*, pp. 132 ff, 162 f.

[86] Cf. Zeuthen, *op. cit.*, pp. 122–126, 159 f.

[87] H. Balsam on these theorems in *Apollonius von Perga*, p. 134; cf. Zeuthen, *op. cit.*, p. 163: "The ancients knew completely the generation of conic sections via projective pencils, apart from their comprehension under the generic concept of projectivity."

[88] Strong reservations are voiced by J. L. Coolidge, *History of the Conic Sections*, 1945, p. 21: "Most surprising is Zeuthen's statement in this connexion (p. 163) that the Greeks were familiar with the Chasles–Steiner theorem.... I confess that it seems to me that there is very little basis for such a general statement."

[89] Note that Beaugrand, in a letter of 1639 criticizing Desargues for his nonclassical approach to the study of conics, sees at once this very application of this Apollonian theorem; cf. R. Taton, *Desargues*, 1951, pp. 187 f.

[90] Cf. Heath, Ch. iv; Zeuthen, Ch. 15.

[91] *Collection* (IV), I, p. 284. His first method of trisection employs a *neusis*, effected via a hyperbola (*ibid.*, pp. 272–276), while the second is another solution via hyperbola, identical to the third given here (*ibid.*, pp. 282–284). On these alternative constructions, see Chapters 5–7. Arabic and Greek texts related to these items will be presented in the sequel to the present volume.

[92] *Collection* (VII) Props. 236–238, II, pp. 1004–1014.

[93] Pappus gives one other lemma to the *Surface Loci* (*ibid.*, Prop. 235, p. 1004): the specification of a surface via the condition that its parallel sections are conics of a given relation. The locus reduces either to a cylinder with a conic at its base, or to a cone. See Heath's discussion, *History*, I, p. 440; II, pp. 425–426; and *Archimedes*, pp.

lxii–v. Note that the text bears this meaning without emendation, despite the feeling by Tannery and Heath that there might be some textual problem here.

[94] See Toomer, *Diocles*, Props. 4, 5 and his discussion, p. 17. The view that Euclid knew some form of the focus-directrix property was generally accepted (cf. Heath, *Apollonius*, pp. xxxviii–ix) even by Toomer himself, until his recent edition of Diocles. But Toomer's present view that the appearance of this construction in Diocles must reserve to him credit for the first discovery of this property might, I believe, be an overreaction beyond what the Diocles text requires.

[95] For references, see Heath, *History*, I, pp. 439 f.

[96] Pappus, *Collection* (IV), I, p. 234. For an account of these passages on the various forms of the spirals, see my "Archimedes and the Spirals."

[97] Pappus, *op. cit.*, pp. 258–262. Pappus gives another solid construction for the quadratrix via the plane spiral; the spiral is related to the conical spiral via orthogonal projection, the latter in its turn being projected radially onto the cylindrical spiral. Cf. *ibid.*, pp. 262–264, and Chapter 5 below.

[98] See Chapter 6 on Nicomedes and the quadratrix.

[99] Note that the focus of the oblique parabola will fall at the midpoint of KD, not at D. Thus, neither here nor in the cases of the ellipse and hyperbola given below will the orthogonal projection map a focus onto a focus.

[100] For definitions, see the discussion of Eratosthenes' loci with respect to means in Chapter 6.

[101] The extension of this result to any conic is given by Apollonius in *Conics* III, 37.

[102] This is proved by Pappus in Lemma 19 to the *Porisms*.

[103] If the circles [Fig. 21(a)] are projected onto conics, as from the horizontal to the oblique sections of cones [Fig. 21(b)], the polar line (locus of the harmonic division of secants) for the former will indeed be projected onto that for the latter. But the two conics resulting from the projection of the two circles will not be the locus of the arithmetic and geometric means; for unlike the harmonic division, the proportional divisions corresponding to these means are not preserved under projection in general. (Note, for instance, that the center of the given circle 0 is not projected onto the midpoint of the diameter RS.) A suitable configuration can be obtained if we introduce instead the sections of a cylinder [Fig. 26]. Let the given circle PQ and the loci of its three means be formed as the horizontal section of three intersecting right cylinders, and let the secant KPQ be drawn, where KH, KG, KA are, respectively, the harmonic, geometric, and arithmetic means of the segments KP, KQ. If an oblique section of the cylinders is now made, a configuration of intersecting ellipses will result, and the points H', G', A' will divide K'P'Q' into segments in the same proportion as do H, G, A the line KPQ (for the lines of projection PP', HH', etc., are parallel). Thus, the line and the two ellipses which are the respective loci of H', G', A' will now be the required loci of the harmonic, geometric, and arithmetic means of K'P', K'Q'. This method of parallel projection will thus generalize the locus problem for ellipses; but I have not found a comparable procedure which might yield the analogous result for parabolas and hyperbolas. Zeuthen takes up these locus problems in his effort to reconstruct Eratosthenes' study of "loci with respect to means." While Zeuthen alludes to the stereometrical procedure for ellipses (*Kegelschnitte*, p. 323), he proposes that Eratosthenes adopted a planimetric method in order to cover the cases of parabola and hyperbola as well; cf. the discussion in Chapter 6.

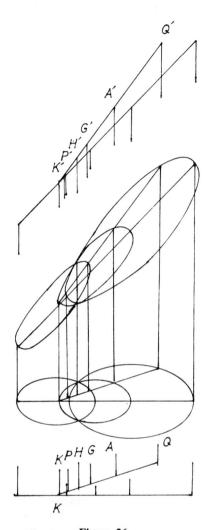

Figure 26

[104] The analogous result for the arithmetic means is trivial, since in this case N is the center of the given circle.

[105] As the points A, A′, P, P′ are coplanar, the lines AP, A′P′ lie in the same plane. They cannot be parallel; for then the plane of APN would be parallel to that of A′P′N′ (since AN, A′N′ are parallel), against the assumption of their intersection along the line KK′. Hence, AP, A′P′ intersect, namely at a point on the intersection of the respective planes of the two sections, the line KK′. As noted above, the Desargues theorem in the plane is entailed by one of Euclid's porisms; cf. note 64.

[106] The exception in the *Conics* is the set of problems in I, 52–60 producing the cone and its sectioning plane in answer to conics specified via given diameter and latus rectum. These, in effect, are converses of the definitions of the curves established via sectionings

of the cone in I, 11–14. Otherwise, the study and application of these curves is always effected via planimetric means in the *Conics*.

[107] See note 47 above.

[108] For texts and discussions of the work of Desargues and Pascal, see R. Taton, *L'Oeuvre mathématique de G. Desargues*, 1951. An extremely brief, but useful, account may be found in Coolidge, *op. cit.*, pp. 28–33. See also the accounts cited in note 63 above.

[109] *Collection* (VII), II, pp. 652–654; cf. Heath, *History of Greek Mathematics*, I, pp. 432 f.

[110] *Collection* (VII), II, pp. 676–678.

CHAPTER 5

Archimedes: The Perfect Eudoxean Geometer

Among the ancient scientists none has attained greater fame, either in his own time or in ours, than Archimedes of Syracuse. His immense achievements in pure geometry extended into the field of mechanics where his theoretical findings had consequences for the design of mechanisms. Siege engines constructed under his supervision held the Roman fleet at bay for three years during the ill-fated campaign which despite his efforts ended in the city's fall and his own murder in 212 B.C. In the popular view Archimedes thus became the prototype of the scientific wonder-worker, an image sustained by the ancient historians and elaborated beyond the limits of credibility by the legendizers.[1] But the Archimedes we meet in his own writings is, with but rare exceptions, exclusively the pure geometer. It is his place within the development of the ancient geometry which shall concern us here.

Archimedes was born near the time that Euclid died, around the second decade of the 3rd century B.C.[2] One might thus suppose that Euclid's work was a dominant influence in Archimedes' training as a geometer and that Archimedes would eventually enter and extend the field of study opened up by Euclid. But this turns out not to be the case. Points of detail relating to Archimedes' early studies and his technique of proportions reveal him to depend on a tradition alternative to Euclid's *Elements*, doubtless closer to the manner of the original Eudoxean treatments.[3] In the area of mechanics which so absorbed his attention, Archimedes could find no precedent in Euclid's work.[4] Yet, as we shall see, the area which Euclid did advance, the application of the method of analysis for the solution of geometric problems, is one which drew relatively little interest from

Archimedes. This split is striking, but not entirely surprising. Archimedes' father was an astronomer, hence competent in the Eudoxean geometric methods and quite capable of rearing his son on the older geometric treatises.[5] Moreover, the separation of Syracuse from Alexandria might well sustain differences in research interests. It would thus appear that Archimedes was too close in time, but too far in distance to come under the spell of Euclid's particular brand of geometry.

Characteristic of the Eudoxean geometry is the application of an indirect method of limits for the measurement of curvilinear plane and solid figures. The centerpiece of this tradition is the issue of the quadrature of the circle. One perceives this problem to be an important factor in Archimedes' early geometric efforts, particularly the *Dimension of the Circle*, while it leads naturally into the surface and volume measurements carried through in his *Sphere and Cylinder* I and the studies of areas and tangents in his *Spiral Lines*.[6] The same Eudoxean approaches find their application in the demonstration of the principles of equilibrium and center of gravity in *Plane Equilibria*. From this a powerful heuristic method emerges by which the content and centers of gravity of figures can be determined through the conceptual weighing of their constituent elements. Archimedes neatly summarizes the manner of its use in the *Method* for obtaining the results whose formal demonstrations he had already communicated in the *Quadrature of the Parabola, Conoids and Spheroids*, and the lost *Equilibria*.[7] These formal proofs adhere meticulously to the style of limits Archimedes developed in refinement of the Eudoxean method. One thus perceives that the techniques and concerns of Eudoxean geometry run through the entire Archimedean corpus, as extant in Greek, and indeed define its subject matter at virtually every point.

By comparison, the investigation of geometric problems assumes a low profile at best in Archimedes' works. It is the concern of only one of the extant Greek writings, *Sphere and Cylinder* II, where six of the nine propositions are problems seeking the construction of segments of spheres according to specifications of size and proportion, and where the double method of analysis and synthesis is applied in each instance. Of course, Archimedes shows himself to be complete master of the technique. Moreover, works preserved now only through their medieval Arabic translations include interesting problem-solving efforts, for instance, an angle trisection implied in one of the propositions in the *Book of Lemmas* and a seven-section of the circle worked out in the *Inscription of the Regular Heptagon*. Both solutions depend on *neusis*, that is, conceptual manipulation of an idealized sliding ruler, and through this have a natural connection with certain constructions, also effected via *neuses*, introduced in the investigation of the tangents in *Spiral Lines*. Few in number perhaps, these constructions of problems set a firm precedent for the solution of problems via special curves, *neuses*, and other means by geometers in the following generation.

Thus, through a survey of Archimedes' contributions to the solution of problems we cannot hope to reveal the scope, or even capture the essence of his geometric achievement. For this must omit discussion of his advanced efforts in the measurement of conics and solids of revolution and of his inquiries into geometric mechanics and hydrostatics.[8] But in the context of our other discussions, the work of Archimedes ought to emerge as the product of a powerful geometric intellect, at once influential for the development of the field, yet oddly out of sympathy with its principal objectives.

CIRCLE QUADRATURE AND SPIRALS

No geometric results discovered by Archimedes were more widely known and used in antiquity than those derived from his study of the measurement of the circle. Later writers cited his *Dimension of the Circle* for the rule that the area of the circle is one-half its perimeter times its radius, analogous to the area rule for triangles; they cited it also for approximations: that the circle is $11/14$ times the square of its diameter, and that its circumference is $3\ 1/7$ times its diameter. While these three results are established in the three propositions, respectively, of the extant *Dimension of the Circle*, the ancients knew this work in a different, more substantial version; for, in particular, they could draw other results from it, such as the analogous area rules for sectors and perhaps also an inequality appropriate for estimating segments of circles.[9] Nevertheless, the extant version clearly reveals the source of Archimedes' formal technique and his skill at adopting this for computational purposes.

Most accounts of Archimedes' works assign this writing to a time relatively late in his career. But this view is the consequence of a plain misunderstanding. When Heiberg proposed his chronological ordering of the Archimedean writings, he saw no clear indications of the placement of the *Dimension of the Circle*, save its obvious affinity in content with the measurements of spheres in *Sphere and Cylinder* I. He thus set it at the end of his list, followed by a question mark to indicate his uncertainty. In the frequently consulted survey of Greek mathematics by Heath, this list was reproduced, but without the question mark for this entry.[10] Thus, the circle measurement was transformed into a late writing, reversing Heiberg's tentative association of it with the relatively early sphere measurement. Further consideration supports Heiberg's intuition, although he appears to have underestimated the early dating of the circle measurement.

What marks the *Dimension of the Circle* as especially early is its adherence to the elementary Eudoxean form of the method of limits. For in his other writings Archimedes introduces a refinement on this method in which the curvilinear figure to be measured is bounded above and below by figures of known measure; as one can arrange that the bounding figures differ from each other by less than an arbitrary preassigned quantity, an indirect argument establishes that the intermediate figure must equal a specified figure also known to lie between these bounds.[11] For instance, Archimedes proves in *Sphere and Cylinder* I, 5 that to a given circle there may be circumscribed and inscribed similar regular polygons whose areas have a ratio less than any preassigned ratio greater than unity. It follows that such bounding figures may be found to differ by less than a preassigned area. To establish the result in *Dimension of the Circle*, Prop. 1, one would need only to note that the area determined as one-half the circumference times the radius of the circle is greater than any polygon inscribed in the circle, but less than any polygon circumscribed about it. For the area of a regular polygon is one-half its perimeter times its in-radius; while the perimeters of the circumscribed and inscribed polygons are respectively greater than and less than that of the circle, their in-radii respectively equal to and less than the radius of the circle. Thus, the area of the circle and the product of one-half its perimeter times its radius must differ by less than any preassigned quantity. Hence, as claimed, they are equal.[12]

Instead of the expected two-sided convergence method, Archimedes follows the Eudoxean one-sided method here, in which the lower bounding figures are considered entirely separate from the upper bounds. In each instance, doubling the number of sides of the polygon reduces at least by one-half the difference of its area from that of the circle.[13] Through the same area rules for regular polygons stated above, it follows that the circle is neither less than nor greater than the specified product; hence it is equal to it. This procedure conforms precisely to the Eudoxean form of the circle measurement, as we know it from Euclid's *Elements* XII, 2, with but one important exception. The Eudoxean theorem has to do with the *ratios* of circles. In the indirect proof, the assumption of a lesser inequality can be reduced to a contradiction through consideration of inscribed polygons. But instead of introducing the circumscribed figures to eliminate the assumption of greater inequality, one may reduce this case to the prior case merely by inverting the ratios. This ploy is not available to Archimedes, since his theorem asserts an equality of areas rather than of ratios of areas. Introducing the circumscribed figures entails a further assumption on the magnitudes of arcs. Just as he had assumed that the perimeter of the inscribed polygon is less than that of the circle, so now he must assume that the perimeter of the circumscribed polygon is greater than that of the circle. The former assumption might be considered obvious by virtue of the conception of the straight line as the shortest distance between two given points. This notion was surely familiar in the pre-Euclidean period, but it is not articulated as an axiom by Euclid.[14] To the contrary, an equivalent for the rectilinear case is established as a *theorem*: that in any triangle the combined lengths of any two sides must be greater than the third side (I, 20). Later philosophers criticized Euclid for belaboring the proof of what is obvious to any ass.[15] But surely Euclid, and doubtless geometers before him, intuited the generality of this result. It is hard to suppose, moreover, that no attempt was made around the time of Eudoxus to establish the proportionality of circumferences and diameters in view of the analogues for circular area and spherical volume in XII, 2 and 18. As far as we know, the requisite axioms receive their first explicit formulation by Archimedes among the postulates prefacing *Sphere and Cylinder* I: that the shortest distance between two points is the straight line joining them; and that for two curves, convex in the same direction and joining the same two points, the one which contains the other has the greater length.[16] In the *Dimension of the Circle*, however, one finds no specific indication that an axiom of this form underlies the steps that the perimeter of the inscribed polygon is less than that of the circle, while the perimeter of the circumscribed polygon is greater. Indeed, in this work no justification for these steps is given at all. Presumably, then, Archimedes could accept them as obvious. In view of his strong dependence here on the earlier techniques of convergence, one need not doubt that in these other regards he could find certain precedents in attempts by earlier geometers to prove theorems on arc length.[17] The formulation of these principles as explicit axioms in *Sphere and Cylinder* I would thus result from Archimedes' own later reflections on the formal requirements of such demonstrations.

The Archimedean area rule in conjunction with the Eudoxean theorem on the ratios of circles (*Elements* XII, 2) yields the result that the circumferences of circles have the same ratio as their diameters. Pappus provides such a proof within a series of lemmas on the ratios of circles and their sectors and segments,[18] and we should expect to find it in the *Dimension of the Circle*, for Archimedes goes on in its third proposition to estimate the ratio of the circumference to the diameter, *not* the ratio of the areas of the circle and the square on its diameter. In the extant Prop. 2, the latter ratio is asserted to equal 11 : 14, but its "proof" actually depends on the result of Prop. 3. Clearly, the writing as we know it has been altered to its detriment through editorial revisions and scribal confusions.[19] The lemmas on related properties of the circle which we find in Pappus and Hero, for instance, may thus derive from Archimedes' work before it had suffered these textual corruptions.

Despite such difficulties the essential skill and genius responsible for the computations in the third proposition are unmistakeable. Archimedes seeks to establish not merely an approximation for the ratio of circumference to diameter, but upper and lower bounds on that ratio. As in the first proposition, the procedure is divided into two separate sections, one dealing with circumscribed figures (leading to the upper bound), the other with inscribed figures (for the lower bounds). This division masks the fact that the computational procedures are identical, each requiring the same property of the bisector of the angle of a triangle (cf. *Elements* VI, 3). Indeed, complete derivations for both procedures are provided, despite their equivalence. Such intermediate proofs often turn out to be the work of interpolators. But even if that is the case here, the editor has only adopted and supplemented the separation of cases fixed in the computation itself. The failure to exploit the potential for refining this procedure by means of the two-sided notion of convergence thus serves to affirm the view that the *Dimension of the Circle* was an early Archimedean effort.[20]

The procedure adopted is as follows. First consider the upper bounds, for which we take the circumscribed regular hexagon as the initial bounding figure. Form the right triangle whose leg A is one-half of one of the sides of this hexagon, whose other leg B is the radius of the circle, and whose hypotenuse is C, so that the central angle formed by B and C is $\frac{1}{12}$ the full circle [see Fig. 1(a)]. If the corresponding element of the polygon having twice as many sides is denoted by A', B', and C', Archimedes' rule establishes that $A' : B' = A : B + C$, while $C'^2 = A'^2 + B'^2$. Starting from the hexagon, he can then proceed in succession to the polygons of 12, 24, and 48 sides, ending with that of 96

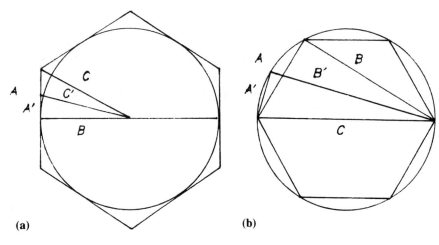

Figure 1

sides. In each case, rounding off due to the approximation of irrational square roots must give an upper bound on A (that is, a lower bound on B and C), thus ensuring that the final value of the ratio of A : B will yield an upper bound for π, namely 96 • 2A : 2B. For the initial value of A : B = 1 : $\sqrt{3}$, he uses the upper bound 153 : 265. The manner of its derivation is unexplained, and has given rise to many attempts at reconstruction by scholars tantalized by the fact that this is one of the convergent fractions in the continued-fraction development of the root.[21] After four applications of the rule, Archimedes obtains 153 : 4673½ as an upper bound on A : B, from which he produces 96 • 153 : 4673½ = 14688 : 4673½ as an upper bound on π. The final value 3 1/7 follows from this through rounding off upward. Here again the effort to find good fractional approximations in the smallest possible terms appears to be assisted by procedures comparable to the development in continued fractions.[22] The configuration for the lower bounds is altered, in that A is the side of the inscribed polygon, while C—rather than 2B—is the diameter of the circle [Fig. 1(b)]. Otherwise, the computation proceeds as before, where rounding off must always be downward so that the ratio A : C will produce 96 A : C as a lower bound on π. The initial value for A : B is 780 : 1351 here (another convergent in the continued fraction for 1 : $\sqrt{3}$), and after four applications of the rule, one obtains 66 : 2017¼ as lower bound for A : C.[23] Hence, 96 • 66 : 2017¼ = 6336 : 2017¼ is a lower bound for π. This is simplified to 3 10/71 via another downward rounding off.[24]

In this way Archimedes has established 3 1/7 as an upper bound and 3 10/71 as a lower bound on π. Although the computation requires that the error due to rounding off must be cumulative, the final values differ by less than one part in 2400 and one part in 4200, respectively, from the actual value. The Greek tradition is aware of no better estimate before the time of Archimedes, while the value 3 1/7 is employed without exception within the tradition of metrical geometry after him.[25] But closer estimates were derived through further applications of Archimedes' method. Hero reports that Archimedes himself produced a different set of bounds in a work *On Plinthides and Cylinders*. Eutocius cites a work called the *Easy Delivery* (*Ōkytokion*) in which Apollonius improved on the Archimedean estimates.[26] Eutocius does not elaborate, but Hero does. In the alternative estimate Archimedes is said to have established 211875 : 67441 as a lower bound and 197888 : 62351 as an upper bound on π. On first inspection these values are disappointing, for the former term is in fact an upper bound on π, while the latter, despite the large terms of the ratio, is a far poorer upper bound than 3$\frac{1}{7}$. Thus, either Archimedes has blundered woefully in the execution of this computation, or else the text has suffered corruption. On the latter hypothesis several attempts at restoration have been made.[27] My own accepts the lower bound as given, but proposes that it has been mistakenly cited as a lower bound when it was actually derived as an upper bound; for one would not wish to suppose that a good value like 211875 : 67441 arose merely by chance through a clumsily managed computation, while the accuracy is precisely of the expected order, about one part of the denominator.[28] Comparable accuracy for the lower bound suggests the value 197888 : 62991, where the manuscript reading of 62351 is easily accountable through scribal error.[29] The correctness of the received numerator is indicated by the factorization 197888 = $2^8 \cdot 773$, where the factors of 2 correspond to the successive doubling of the number of sides of the bounding polygons. The ratios as emended can both be obtained through an Archimedean procedure commencing with decagons and ending with the 640-gons. But as in the *Dimension of the Circle*, the computed bounds could be reduced to lower terms with negligible loss of accuracy, yielding 333 : 106 as lower bound and 377 : 120 as upper bound, from the continued fractions 3 + $\frac{1}{7}$ + $\frac{1}{15}$) and 3 + $\frac{1}{7}$ + $\frac{1}{17}$), respectively. This at once suggests an alternative intermediate value 355 : 113 = 3 + $\frac{1}{7}$ + $\frac{1}{16}$). This same value was established by Chinese geometers in the 5th century A.D.; it was derived again by Western computers in the 16th century. Since it is accurate to about one part in twelve million, however, the ancient computation implied by Hero's figures could not have established it rigorously save through some essential modifications in procedure.[30]

Since the value 3 $\frac{1}{7}$ suffices for the practical purposes of metrical geometry, ancient documentary evidence of such more refined computations is sparse. Only in astronomy was a closer estimate demanded. For the trigonometric studies by Hipparchus around the mid-2nd century B.C., Toomer has reconstructed the value 3438 : 21600 as effective estimate for the ratio 1 : 2π.[31] Since this might be derived simply as the arithmetic mean of the inverses of Archimedes' values in the *Dimension of the Circle*, we cannot cite it in support of Hero's testimony to the refined Archimedean values. Indeed, the larger terms of the latter indicate a degree of accuracy about ten times that required for Hipparchus' uses. In the great system of mathematical astronomy compiled by Ptolemy three centuries later the value adopted for π is 3^p 8' 30" in the sexagesimal notation, that is, $377/120$.[32] This is precisely the value we derived from the upper bound reported by Hero. Ptolemy provides no derivation of the value, but observes merely that it lies between the Archimedean values $3^{10}/_{71}$ and $3^{1}/_{7}$. Although he might well have depended on refined computations such as that cited by Hero, an alternative account is equally possible on the basis of the trigonometric table in Ptolemy's Book I. There the value for chord $1°$ is $1^P\ 2'\ 50''$ for a circle of diameter 120^P; one might thus produce for π the estimate $360 \cdot$ chord $1° : 120 = 3^P\ 8'\ 30'' : 1$.[33] Ptolemy sets out in full detail the computation leading to his value for chord $1°$, and both this value and the estimate for π corresponding to it are correct to the nearest digit in the second sexagesimal place after the units. Thus, his results do not of themselves indicate prior knowledge of this computed estimate for π. This does not rule out that such knowledge might have helped steer the course of the computation of the chords. For instance, the value of chord $1°$ is found via the inequality $^2/_3$ chord $1^1/_2° <$ chord $1° < ^4/_3$ chord $^3/_4°$, where successive bisections of the angle $24°$ have led to the results chord $1^1/_2° = 1^P\ 34'\ 15''$ and chord $^3/_4° = 0^P\ 47'\ 8''$. This leads to chord $1° = 1^P\ 2'\ 50''$, but only because Ptolemy has chosen to adopt an upward rounding off for chord $^3/_4°$ ($= 0^P\ 47'\ 7''\ 24'''\ldots$) instead of the expected downward rounding off. Had he instead adopted the lower figure, the value for chord $1°$ would become $1^P\ 2'\ 49''$ and give rise to a lower estimate for π (e.g., $3^P\ 8'\ 27''$). These difficulties are resolved by carrying through the computations to the third place after the units, as the commentator Theon shows.[34] But awareness in advance that $377 : 120$ was a good approximation for π could have served as a reliable guide.

The juxtaposition of Archimedes and Ptolemy is interesting in that it reveals the difference between the theoretical and the practical approaches to this computation. Ptolemy seeks good approximations for the practical needs of calculation. Expressing results to the nearest digit gives him the advantage of cancellation of errors, so that he obtains values rather closer than those attainable via a strict Archimedean procedure. Now his value $3^P\ 8'\ 30''$ turns out to be slightly greater than π, despite its association with the table of chords, that is, a derivation based on inscribed polygons. The Archimedean procedure, by contrast, controls the direction of rounding off, so that only lower bounds can result from the inscribed figures. Moreover, the production of both upper and lower bounds provides a direct indication of the degree of accuracy of the approximation. In this way, Archimedes reveals a greater concern for the theory of this computation than for the particular numbers it happens to yield. Thus, one cannot be satisfied by the view, already voiced in antiquity, that Archimedes' aim in this work was essentially practical.[35] This might indeed characterize the older Mesopotamian efforts, doubtless dependent on empirical measurements, which resulted in the value $3^1/_8$.[36] But it fails entirely to capture the subtlety of Archimedes' effort. His value of $3^1/_7$ is certainly more accurate than the older figure. But how is it more "practical" when one considers the manipulation of fractions in ordinary contexts? Furthermore, what is "practical" about his lower bound $3^{10}/_{71}$? If, then, Archimedes' purpose in the *Dimension of the Circle* was to deploy geometric theory for the presentation of an algorithm and then to illustrate it through a cleverly managed calculation, he would in effect be demonstrating the greater effectiveness of *theoretical* over purely practical procedures in geometry. But this makes puzzling the motive behind the extended computation implied in the figures reported by Hero. For there is little if any theoretical insight to be gained merely by carrying out the same computational procedure to one or two more decimal orders of accuracy. Neither would this be expected to produce results of greater utility for practice. One thus has some cause to question Hero's assignment of these numbers to Archimedes, and to suspect that perhaps another geometer, like Apollonius in his *Okytokion*, was responsible for them. We shall return to this question later.[37]

The results in the *Dimension of the Circle* cannot be viewed as attempts at a direct solution of the circle quadrature. But ancient and medieval commentators recognized that the proof of the area rule in the first proposition depends on the postulate that one can produce a straight line equal to the circumference of the circle.[38] In effect, then, Archimedes has not solved the circle quadrature, but rather revealed its equivalence to the problem of producing that straight line. Many of Archimedes' theorems have the same force: they demonstrate that the measurement of a specified curvilinear figure can be reduced to that of a certain circle. For instance, in the version of the *Dimension of the Circle* known to Hero, Archimedes proves that the area of a sector is one-half the product of its radius times its arc.[39] Presumably, the same work included the proof that the surface of a truncated right cone equals one-half its slant height times the circumference of its base, as one may see by rolling it out into a plane circular sector. In *Sphere and Cylinder* I, 14 the same result is expressed in the form that the conical surface equals the circle whose radius is the mean proportional between the slant height and the radius of the base circle. Similarly, the principal result of that work proves that the surface of the sphere equals four times its greatest circle, while the area of any segment equals the circle whose radius equals the line drawn from the vertex of the segment to a point on its circular base.[40] In like manner, the volumes of conical and spherical segments are referred to the volumes of cylinders. In *Conoids and Spheroids* the area of the ellipse is shown to be equal to that of the circle whose diameter is the mean proportional between its major and minor axes; the volumes of conoids of revolution are expressed in terms of associated cones.[41] In all these instances, then, the measurement of the figure has been reduced to the fundamental result of the measurement of the circle.

Two such quadratures appear in the *Book of Lemmas*, a work preserved in Arabic and whose translator Thābit ibn Qurra reports opinions of its Archimedean provenance.[42] In the fourth proposition the figure "which Archimedes calls *arbelos* (leather cutter) is drawn, a trigram bounded by three semicircles tangent at the points A, D, C [Fig. 2(a)]. If BD is the perpendicular half-chord in the large semicircle, then the area of the circle of diameter BD equals the *arbelos* ADCBA. The proof substitutes the relation $BD^2 = AD \cdot DC$ into $AC^2 = AD^2 + DC^2 + 2 AD \cdot DC$ to obtain $BD^2 = \frac{1}{2} AC^2 - \frac{1}{2} AD^2 - \frac{1}{2} DC^2$; since circles have the ratio of the squares on their diameters, this establishes that the circle of diameter BD equals the difference between the semicircle of diameter AC and the two semicircles of diameters AD, DC, respectively, where

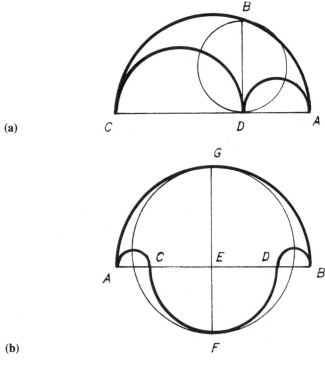

Figure 2

the latter difference is the *arbelos*. A similar quadrature appears in proposition 14: the *salinon* (salt cellar)[43] is bounded by the semicircle ABG, the two equal semicircles on the diameters AC and BD, and the semicircle DFC [see Fig. 2(b)]. By a proof like that for the *arbelos*, it is shown to equal the circle of diameter GF. One observes that the proofs for both figures are very much in the style of the lunule quadratures by Hippocrates of Chios. But there is a forward-looking aspect implicit in two additional lemmas to the figure of the *arbelos*. In Prop. 5 it is proved that the circles EFG, LMN are equal, where each is tangent to the large semicircle, to one or the other of the small semicircles, and to the line which separates them [see Fig. 3(a)]. In Prop. 6 the diameter EF of the circle drawn tangent to all three semicircles is $5/19$ diameter AC when AD : DC = 3 : 2; that is, CP : PO : OA = 4 : 6 : 9 [see Fig. 3(b)].[44] The latter result finds its natural extension in the "ancient proposition" which Pappus presents in Book IV of the *Collection*: if a succession of such tangent circles is inscribed in the *arbelos*, the elevation of the center of the first above the base diameter equals its own diameter, the elevation of the center of the second equals twice its own diameter, the elevation of the center of the third equals three times its own diameter, and so on [see Fig. 3(c)].[45] The theorem given by Pappus may in its turn be viewed as an application of special cases of the problem of constructing circles tangent to three given circles, the complete solution for which

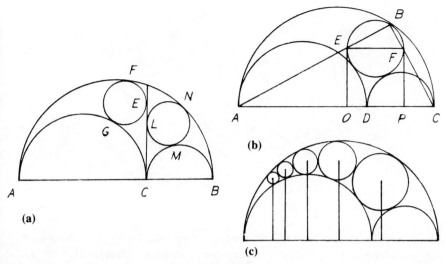

Figure 3

was worked out by Apollonius in his work *On Tangencies*. In this way one can perceive a link between studies initiated by Archimedes and an inquiry pursued later by Apollonius. We shall later discuss comparable links between their studies of circle measurement and *neuses*. Unfortunately, the confused state of the manuscript tradition of the *Lemmas* available to the Arabic translator prevents our delineating precisely those parts actually due to Archimedes. The quadratures of the two figures are examples, if quite simple ones, of the class of curvilinear measurements which dominate the known work of Archimedes. But the theorems on the tangent circles fall within a field of problems not otherwise represented in the Archimedean corpus, but quite characteristic of efforts in the generation after him. We might thus well prefer to assign these results to a follower of Archimedes, rather than to Archimedes himself.

The determination of the areas of figures bounded by spirals further illustrates Archimedes' methods of quadrature. The Archimedean plane spiral is traced out by a point moving uniformly along a line as that line rotates uniformly about one of its endpoints. The latter portion of the treatise *On Spiral Lines* is devoted to the proof that the area under the segment of the spiral equals one-third the corresponding circular sector. For instance, the area lying between the spiral and the radius vector BE is one-third the sector BEH; the area between the whole spiral and the terminal radius BA is one-third the whole circle of radius BA (see Fig. 4). The proofs are managed in full formal detail in accordance with the indirect method of limits. The spirals are bounded above and below by summations of narrow sectors converging to the same limit of one-third the entire enclosing sector, for the sectors follow the progression of the square integers.[46] This method remains standard to this very day for the evaluation of definite integrals as the limits of summations. It is applied throughout the *Conoids and Spheroids* as well for the measurement of the volumes of solids of revolution.

162   Ancient Tradition of Geometric Problems

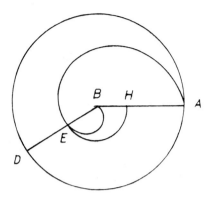

Figure 4

But Pappus reports an alternative treatment, "a remarkable strategy" (*thaumastē epibolē*), by which Archimedes effected the quadrature of the spiral.[47] As this occurs in the context of communications between Archimedes and Conon of Samos, the latter having been deceased for "many years" at the time of the writing of the extant *Spiral Lines*, Pappus here must be preserving for us a much earlier treatment of this theorem. As in the later method, Archimedes here introduces a set of inscribed sectors bounding the spiral from below [see Fig. 5(a)]. At the same time he considers the parallel slices in a cone, that is, the thin cylinders inscribed in it, whose sum bounds the cone from below [see Fig. 5(b)]. Since in both instances the constituent elements follow the progression of the square integers, it follows that their sums bear the same ratio to their respective containing figures. Thus, the lower bounding figure for the spiral has the same ratio to the whole circle that the lower bounding figure for the cone has to the cylinder. As a comparable proportion obtains for the circumscribed bounding figures, it follows that the spiral is to its circle as the cone is to its cylinder, namely, one-third. By a closely related technique, Pappus later establishes that the spherical spiral separates the surface of the hemisphere into two portions

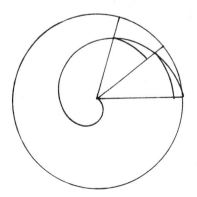 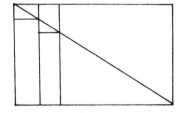

Figure 5

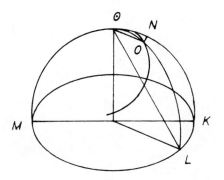
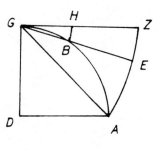

Figure 6

equal, respectively, to eight times the segment ABG and eight times the triangle ADG, where the sector ADGB is a quadrant of a great circle in the sphere (see Fig. 6).[48] As for the plane spiral, the surface of the figure bounded by the spherical spiral toward the initial position θNK is approximated above and below by surface elements corresponding to the elements in a second diagram, namely, the small sectors drawn to the segment ABG. This depends on Archimedes' theorem from *Sphere and Cylinder* I that the area of the surface element θN0 is proportional to the square of θ0, just as the area of the sector BHG is proportional to the square of BG, where BG = θ0.[49] Hence, the sums have the same ratio to their respective enclosing figures. It follows (details of the limiting argument are omitted by Pappus) that the figure bounded by the spiral is to the hemisphere as the segment ABG is to the sector AGZ. The sectors AGZ and ABGD equal each other, however, while the latter is one-eighth the surface of the hemisphere.[50] Thus, as claimed, the figure bounded by the spiral is eight times the segment ABG. Taking the differences, we have that the other portion of the hemisphere bounded by the spiral equals eight times triangle ADG. The extremely close similarity in method used for evaluating the area of this form of the spiral to that of the plane spiral (as in Pappus) and its heavy dependence on results from Archimedes' *Sphere and Cylinder* encourage assigning it an Archimedean provenance. One may note that the spherical spiral is a striking example of a curvilinear figure, like the lunules of Hippocrates, found to be equal to a constructed rectilinear figure.

It is Archimedes' results on the tangents to the spirals which move toward the actual solution of the circle quadrature. Given the spiral whose initial ray is 0B, the tangent drawn to any point C on the curve meets the line drawn from 0 perpendicular to the ray 0C such that the intercept 0E equals the arc CF in the circle of radius 0C which subtends the angle C0B (Fig. 7). Archimedes devotes the major portion of *Spiral Lines* to the formal demonstration of this result for the three cases of arcs equal to, less than, or greater than the whole circumference of the circle (Props. 18, 19, and 20, respectively), together with the requisite lemmas on the general properties of the curve and its tangents. The treatment is entirely formal in the synthetic manner, so that the underlying conception of

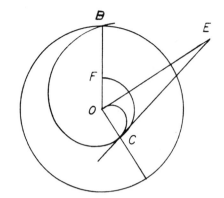

Figure 7

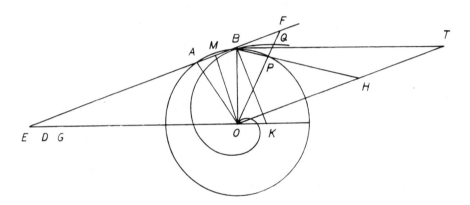

Figure 8

the argument is rather difficult to discern. I thus propose to reconstruct an analysis, closely modelled on the synthesis Archimedes gives, to reveal how its essential features emerged. One should then be able to work through the details of Archimedes' formal treatment with greater ease.[51]

We may frame this analysis as a problem: to draw the tangent to a point on the spiral, for example, the point completing its first full turn. This will be done by specifying the length of the subtangent OE where the tangent drawn to B meets the line drawn perpendicular to the ray OB (Fig. 8). Let a point Q be taken on the spiral on the forward side of B and let the ray OQ be drawn, meeting the circle of radius OB at P. By the definition of the spiral, the rays are in proportion to the corresponding angles, that is, to the arcs of this circle; thus OQ : OB = arc PB + C : C, where C equals the circumference of the circle of radius OB. Since OB = OP, we obtain OQ − OB : OB = QP : OB = arc PB : C. Now extend OQ to meet the tangent at F, and draw the line BP extended to meet in H the line drawn from O parallel to the tangent. Draw from B the line parallel to OE meeting OH extended in T. This makes OEBT a parallelogram, so that OE = BT. Now by similar triangles, FP : BP = OP : PH = OB : PH. Since F is the point on the tangent corresponding to Q, FP > QP ; since, also, PB < arc PB, it follows that FP : BP > QP : arc BP. From the above, QP : arc BP = OB : C. Thus, OB : PH > OB : C, or PH is less than C. Furthermore, PH is less than BT, since BT subtends the largest angle in the triangle BHT. Since BT = OE, it follows that PH is less than both OE and C. This does not establish a relation between OE and C as such. But it is evident that as P tends to B, Q and F converge, so that FP : PB tends toward equality with QP: arc PB; that is, PH converges toward C from below. At the same time PH tends toward equality with BT. We thus have that the subtangent OE equals C, the circumference of the circle of radius OB, the common limiting values of the line PH. One might investigate in comparable fashion the relations arising for P situated on the rearward side of B. Moreover, the result generalizes for portions other than the full turn of the spiral. When the terminal ray OC does not coincide with the initial ray OB, the subtangent OE will be the length of the arc corresponding to the sector angle BOC in the circle of radius OC, where OE is perpendicular to OC (cf. Fig. 7).

Clearly, a treatment of this sort lacks the precision demanded of a formal exposition. But it reveals the length of the subtangent one requires for drawing the tangent, as well as the general line of the argument one might follow to obtain the formal synthesis. In particular, one perceives the role of the insertion (*neusis*) of a length shorter than the subtangent, that is, the line PH between the circle and the line OH parallel to the tangent.[52] One might thus try to show that any line other than BE as drawn here will meet the spiral at another point besides B; that is, that if OD is less than OE, the line BD will cut the spiral at some point R on the rearward side of B, while if OD' is greater than OE, the line BD' will cut the spiral at Q on the forward side of B. This will establish that BE is the tangent, in accordance with the conception of the tangent in the ancient geometry.[53] Alternatively, one might consider the hypothesis that the tangent meets the perpendicular at E such that OE is not equal to C. The case where OE is greater than C (= OG) gives rise to the diagram we have been considering; for any point D lying between E and G, one can insert the length PH = OD and so locate a point Q on the forward side of the spiral from B, where Q will be found to lie beyond F, the point on the extension of the assumed tangent EB; this contradicts the fact (proved in Prop. 13) that Q must lie below the tangent. Similarly, one may derive a contradiction from the assumption that the subtangent is less than C. This approach to the synthesis is in fact the one adopted by Archimedes in *Spiral Lines*, Prop. 18.[54]

The informal analysis proposed here utilizes a somewhat dynamic conception of the limiting process in which as P approaches B, the triangle FPB becomes similar to BOE, while triangle OPH becomes congruent with OBT. To be sure, such conceptions are alien to the formal style of the ancient geometry. Yet one would not like to deny to Archimedes the availability of intuitions of this sort, especially in view of the importance of mechanical heuristic procedures for his other work. Moreover, it is difficult to imagine how one would approach the

166   Ancient Tradition of Geometric Problems

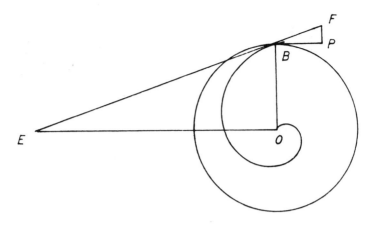

Figure 9

problem of the tangents to motion-generated curves like the spiral without some essentially kinematic intuition. One may note that the present form partly disguises the role of the differential triangle as one would introduce it in a more modern treatment. The triangle FPB would assume that position in the limit, where its side FP will be proportional to the radial velocity of the point tracing out the spiral as it arrives at B, while BP will be proportional to its velocity along the tangent to the circle at B (Fig. 9). Since the motions are uniform, the velocities will be proportional to the total distances traversed by B in these two senses, so that FP : PB = OB : C. But by similar triangles, FP : PB = OB : OE. Hence, OE = C. The procedure actually adopted by Archimedes is far removed from this simpler method, so that one must suppose that its conception of the tangent as the direction of the instantaneous motion composed from the motions which generate the curve, while fundamental for studies of tangents since the 17th century, was still unrecognized by the ancients.[55]

The plane spiral is but one of several forms of spiral studied by the ancient geometers. We have already mentioned the spiral on the sphere, whose quadrature paralleled that of the plane spiral in Pappus' account. Pappus also reports of a construction relating the plane spiral to the spirals drawn to the cone and the cylinder.[56] Given the plane spiral BHA, if one erects KH = HB, perpendicular to the plane of the spiral, the point K will lie on the right cone of vertex B and axis BL (Fig. 10). Thus, as H traces the plane spiral, K will trace a "surface locus," the spiral on the cone.[57] If the ray LK is now extended to meet at θ the right cylindrical surface drawn over the circular base GDA, this point θ will trace another surface locus, the spiral on the cylinder. The latter was familiar to the ancients by the name *cochlias*. Apollonius devoted a writing to its study, while Archimedes is said to have invented the water-lifting device which shares both its name and its figure.[58] We should thus suppose that the ancient geometers discovered a variety of properties of these curves. It happens that the projective relationship just cited provides an alternative way to find the tangent to the plane spiral. First consider the cylindrical spiral: since the ratio θD : arc DG is a

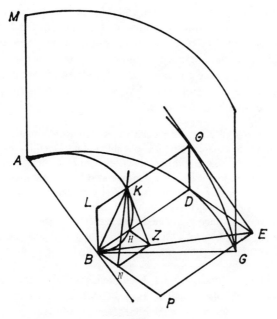

Figure 10

constant for all points θ, the curve may be viewed as the result of wrapping a plane triangle GAM against the cylinder,[59] so that the tangent to θ will be the line θE where DE is perpendicular to DB and equal to the arc DG. Since the cylindrical spiral is the radial projection of the conical spiral, the tangent to the latter at K will project onto the tangent to the former at θ via lines radiating from LB, and hence onto the line KZ, meeting the base plane in Z, such that DE : HZ = θL : KL. HZ will equal the arc of the circle of radius KL corresponding to the arc DG (= DE) of the circle of radius BD. Now the tangent to the conical spiral at K lies in the plane tangent to the cone at K, that is, to the plane through BK and perpendicular to the plane of BLK. This tangent plane meets the base plane in the line BP perpendicular to BH, while the tangent line to the conical spiral will meet BP in N, where ZN is parallel to HB. Since HZ = BN, BN will equal the arc of the circle of radius BH corresponding to the angle GBH. The plane spiral is the plane projection of the conical spiral, so that its tangent at H is the plane projection of the line KN, that is, the line HN. We may thus conclude that the tangent to the plane spiral at H meets the line BP drawn perpendicular to the ray BH such that the subtangent BN equals the arc of the circle of radius BH over the angle GBH. In this way, the conception of the spirals in the manner of surface loci has the potential to provide a heuristic for the results on the tangents to the plane spiral presented by Archimedes in *Spiral Lines*.

Did Archimedes view his findings on the spirals as a construction of the circle quadrature? A sign that he did seems to lie in his manner of expressing them: they are not given as *problems* in the drawing of tangents to spirals, but rather as *theorems* on the properties of the tangents. Moreover, the later tradition did indeed include Archimedes' spirals, along with the quadratrix and the cylindrical spiral, among the known efforts at circle quadrature.[60] What, then, of its success? Certainly his followers did not consider the matter closed, for the studies of the quadratrix, for instance, appear to have been stimulated by Archimedes' efforts, not cut off by them. As for Archimedes' view, a remark from the preface to his *Quadrature of the Parabola* is suggestive:

> Now of those who earlier applied themselves in geometry, some tried to prove [*graphein*: confirm by diagram] that it is possible to find a rectilinear space which is equal to the given circle and to the given segment of a circle, and after these things they attempted to square the space bounded by the section of the whole cone and a line by making assumptions not easily conceded, whence it was not acknowledged by the majority that these things had been discovered [*heuriskomena*] by them. But we understand that none of the earlier ones has tried to square the segment bounded by a line and a section of the right-angled cone [sc. a parabolic segment], which is what has now been discovered by us.[61]

At first it would seem unclear why Archimedes should view the successful quadrature of the parabolic segment as a distinctive achievement within this series of studies of curvilinear areas. For, after all, the lunules of Hippocrates and his own measurement of the spherical spiral, if the above view of its Archimedean origin and dating is accepted, would be familiar as instances of curvilinear figures found to be equal to certain rectilinear figures. Of the study of the ellipse (if that is what the obscure phrase "section of the whole cone" refers to)[62] we know nothing other than the Archimedean quadrature given in *Conoids and Spheroids*. Still, there would appear to be a natural connection with the problem of circle quadrature. In this respect, the allusion to studies of segments of circles is of interest. For Hero preserves the full demonstration of the theorem that the segment of the circle is greater than four-thirds the triangle inscribed in it, that is, having the same base and altitude as the segment.[63] The treatment is in the same manner of Archimedes' *Dimension of the Circle*, and depends on the fact that the triangles AZB, BHG are together greater than one-fourth triangle ABG, where Z, H bisect the arcs AB, BG, respectively [Fig. 11(a)]; applying this result, in turn, to the triangles inscribed in the remaining segments of the circle, one obtains that the segment is greater than triangle ABG plus its fourth plus the fourth of its fourth and so on, where the sequence of parts converges to four-thirds. If one adapts this very procedure for a parabolic arc ABG and takes E, K to be the bisectors of the lines AD, DG, then one finds that the triangles AZB, BHG are together exactly one-fourth the triangle ABG, so that the summation yields that the parabolic segment is exactly four-thirds triangle ABG [Fig. 11(b)]. This is the manner of Archimedes' proof of this theorem in *Quadrature of the Parabola*, Props. 18–24.[64] Appreciating how an approximate quadrature of the circular segment so gives rise to an exact quadrature of the parabolic segment, Archimedes might well feel encouraged in the search for a circle quadrature.

a)
b)

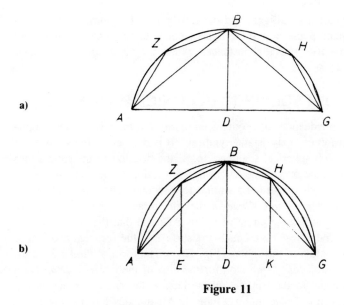

Figure 11

Here Archimedes says nothing about the spirals. But the view that he had already made his basic discoveries relating to these curves is supported by internal evidence. In the preface to *Spiral Lines*, communicated several years after *Quadrature of the Parabola*, Archimedes speaks of a list of theorems he had sent off to Conon years earlier. The promise to provide their formal proofs was being fulfilled in a writing already sent (*Sphere and Cylinder* II), together with the present work (*Spiral Lines*) and another yet to follow (*Conoids and Spheroids*). He states the theorems at issue, and among them are the principal results on the areas and the tangents of the spirals. Now Conon had died many years before the writing of the treatise *On Spiral Lines*; indeed, his recent death had been the occasion for Archimedes to address the *Quadrature of the Parabola* to the new correspondent Dositheus.[65] Thus, before this Archimedes had communicated to Conon some form of these results on spirals, doubtless in a manner like that preserved by Pappus.

It thus appears that Archimedes' solution of the circle quadrature via the tangents to the spiral might fall into that class of efforts "making assumptions not easily conceded," so that the problem remained open. This lends a certain poignance to a remark he makes in the preface to one of his latest writings, *The Method*, addressed to Eratosthenes:

> It happens that these theorems [on the volumes of certain sections of cylinders] differ from those discovered earlier [on the volumes of conics of revolution]. For the latter figures, namely the conoids and the spheroids and their segments, were compared by us in magnitude to figures of cones and cylinders, but none of them was discovered to be equal to a solid figure bounded by planes. But each of those figures [newly communicated] bounded by two planes and surfaces of cylinders is discovered to equal one of those solid figures bounded by planes.[66]

Despite an entire career of diligent and masterly effort, the quest for measurements of curvilinear plane and solid figures equal to rectilinear figures had produced but a handful of actual instances, while the circle quadrature itself remained as elusive as ever.

PROBLEM SOLVING VIA CONIC SECTIONS

In the extant Archimedean corpus, only *Sphere and Cylinder* II includes applications of the method of analysis and synthesis. It is devoted to the solution of problems dealing with spherical segments: e.g., the division of the sphere into segments whose surface or volume are in a given ratio; or the construction of segments equal to a given figure and similar to a second. These depend on the principal results of *Sphere and Cylinder* I: that the surface of the segment ABG (excluding its base) equals the circle of radius AB; and that the volume of the sector equals the cone whose base equals the surface of the segment and whose altitude equals the radius of the sphere (Fig. 12).[67] The form of the latter result brings to mind Archimedes' expression for the area of the circle in *Dimension of the Circle*, as equal to the triangle whose base is the circumference of the circle and whose altitude is its radius. Indeed, Archimedes himself notes this analogy elsewhere as significant for his discovery of the measurement of the sphere.[68]

It is quite straightforward, then, to divide a given sphere into segments whose surfaces have a given ratio (*Sphere and Cylinder* II, 3). Since these segments have the ratio $AB^2 : AE^2 = BX : XE$, one need only divide the diameter of the sphere into segments having the given ratio and then divide the sphere into parts having these line segments as their respective altitudes. The problem of dividing the sphere into segments where the volumes are in the given ratio, however, turns out to be a major undertaking (Prop. 4). Here Archimedes makes use of an expression for the volume of the segment established in Prop. 2: if on the diameter extended, one marks off the point P such that $KD + DX : DX = PX : XB$ (for K the center and DB the diameter of the sphere) (Fig. 13), then the cone whose base is the circle of diameter AG and whose altitude is PX will equal the segment of the same base with vertex at B. If one now determines the analogous point L producing the cone LAG equal to the segment DAG, then the analysis of the problem of dividing the sphere reduces to that of determining the three points X, P, L such that (a) PX : XL is the given ratio; for the cones are as their altitudes and by construction they are equal, respectively, to the segments of the sphere; (b) $KD + DX : DX = PX : XB$; and (c) $KB + BX : BX = LX : XD$, from the determinations of P, L as just stated. Through a considerable exercise in the manipulation of proportions, Archimedes reduces this in its turn to the problem of dividing BD at X such that $BD^2 : DX^2 = XZ:Z\theta$, where $ZB = BK$ and $BZ : Z\theta = PL : LX$. Since PX : XL is given and BK is the radius of the sphere, Z and θ are given, so that only the position of X remains to be solved with reference to the derived condition. One perceives that it entails the solution of a third-order relation in DX ($= 3BK - ZX$). For the present

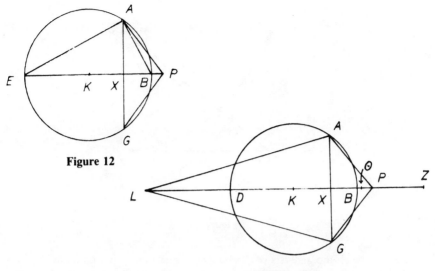

Figure 12

Figure 13

Archimedes merely assumes the solution of this problem for the purposes of producing the synthesis of the problem of dividing the sphere, but promises that "this shall be analyzed and synthesized at the end."[69]

Through some textual mishap suffered by the work not long after its composition, the construction of this auxiliary problem became dislodged from the rest. Already by the close of the 3rd century B.C., the geometers Diocles and Dionysodorus could find no such construction in the copies of *Sphere and Cylinder* available to them for study, and so each worked out his own form of the solution. These are reproduced by Eutocius, who reports that his own researches among the library shelves had brought to light an old book in deplorable condition which presented a treatment of these matters.[70] Since it was written in the Doric dialect, and as it referred to the conic sections by their archaic names ("section of the right-angled cone," rather than "parabola"), he deduced that this was an Archimedean work, for both features are characteristic of his writings. Eutocius goes on to present this text in a restored version devoid of archaisms and replete with appropriate citations of Apollonius' *Conics*. To this he adds versions of the methods used by Dionysodorus and by Diocles. We possess an alternative text of the latter, to which Eutocius' version is reasonably faithful (the principal discrepancy being that Eutocius presents not only the analysis, but also the synthesis, while Diocles omits the latter as "obvious").[71] Thus, only the hypothesis of outright forgery on Eutocius' part could lead us to question that his version of the Archimedean method represents reliably the one originally promised by Archimedes. Not only are the motives for such a forgery entirely unclear, but also there is good cause to doubt that the expertise of a commentator like Eutocius was such as to produce an original treatment of comparable insight, yet so different from the alternatives which might serve for models.[72]

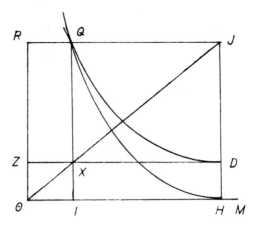

Figure 14

Let us sketch the analysis of Archimedes' solution of this problem. We are to divide the given lines BD at X such that $BD^2 : DX^2 = XZ : Z\theta$ for given points Z and θ. But it is now posed in the more general form in which the given area E may be any value (not necessarily BD^2), while the given line segment F is likewise arbitrary (not necessarily $Z\theta$). Assume then, that the given line ZD $(= \frac{1}{2}BD)$ has been cut at X such that $ZX : F = E : XD^2$ (Fig. 14). Draw $Z\theta$ equal to F at right angles to ZD and complete the parallelogram θHJR with diagonal θXJ. We now determine M on θH extended by the relation $E = \theta H \cdot HM$. Thus, (a) $XZ : Z\theta = \theta H \cdot HM : XD^2$ by the relation for X, while (b) $XZ : Z\theta = \theta H : HJ$ by similar triangles, so that (c) $XD^2 = HM \cdot HJ$. If now we draw QI through X parallel to HJ, since $QI = HJ$ and $QJ = XD$, it follows that Q lies on a parabola with vertex at H, axis HJ and latus rectum HM (i.e., $QJ^2 = HM \cdot HJ$). Furthermore, from (b) $QI \cdot I\theta (= HJ \cdot XZ) = DH \cdot H\theta$ $(= Z\theta \cdot H\theta)$, so that Q lies on the hyperbola passing through D with asymptotes $H\theta$, θZ. Since the parabola and the hyperbola are given, Q is given as their point of intersection.

This completes the analysis of the problem; the synthesis can be given merely by constructing the conics to determine Q and using this to establish that X does indeed divide ZD as to satisfy the desired relation. Now it is obvious that the original problem of dividing the sphere into segments of a given ratio is always capable of solution, for one or the other of the segments may take on values ranging from arbitrarily small to arbitrarily close to the whole sphere. Thus, to every ratio, however small or great, there will be a corresponding division of the sphere. But the auxiliary problem considered by Archimedes has been generalized, so that it is no longer obvious that a solution will exist for certain choices of E and F, the given area and the given line, respectively. This is apparent from the construction derived in the analysis, for one is not necessarily assured that the conics derived will indeed intersect within the specified range over ZD. Hence, a *diorism*, or determination of the conditions of solvability,

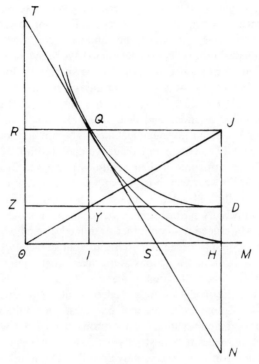

Figure 15

becomes necessary. Archimedes provides only the synthetic proof of the condition to be fulfilled: that the quantity E • F be less than ZY • YD², where 2 ZY = YD. But it is not difficult to perceive the analysis which leads to this condition: the limit of solvability occurs where the hyperbola and the parabola have a common tangent at their intersection Q (Fig. 15). In the case of the parabola, this entails that JH = HN, N taken as the point where the tangent meets the axis JH extended[73]; hence, QJ = 2HS (for S the point where the tangent meets θH), so that DY = 2SI. As for the hyperbola, the point of tangency bisects the segment of the tangent falling between the asymptotes, so that TQ = QS. Thus θI = IS, whence DY = 2YZ.[74] Although Archimedes employs the tangents in the synthesis of this condition, his aim there is to prove that of all divisions X of the line DZ, this one yields a maximum for the value of ZX • XD². It thus follows that if E • F is greater than ZY • YD², the stipulated division of ZD will be impossible.

Archimedes' treatment of this problem is a nice example of how the process of generalizing stimulates the field of geometric studies by raising new questions for solution not posed in the special problems initially. We shall see several instances of this in the work of Apollonius, especially in his studies of *neuses* and plane loci. Moreover, his manner of investigating the normal lines to the conic sections is remarkably close to the Archimedean problem we have been discussing here, including the use of intersecting conics to determine critical points, the properties of tangents to discover the limits of solvability, and even their elaboration in terms of maxima and minima.[75] Nevertheless, certain potentials in Archimedes' results here appear not to have been recognized for further development. Consider that Archimedes has shown how to draw conics to determine X such that $ZX : F = E : XD^2$ for any given line F, area E, and line ZD, where $ZX + XD = ZD$. That is, he has constructed the solution for a general class of cubic relations and determined the condition under which a solution exists for X within a given range. One might formulate other expressions of this type, ultimately to obtain an exhaustive classification of cubic relations together with the construction of their solutions under specified conditions of solvability. This is the very enterprise which Omar Khayyam undertakes in his algebra, and the method of solution he uses depends on intersecting conics.[76] As Eutocius' commentary on the *Sphere and Cylinder* was familiar among the Arabic geometers, we need not doubt that Archimedes' method in this problem could serve as a model for these studies. But if comparable efforts to generalize and classify these cubic constructions were taken up in antiquity, the extant record preserves no direct indication to that effect.

Indeed, without the model of Archimedes' solution to work from, Dionysodorus and Diocles prove quite insensitive to these possibilities of generalization. Their efforts to find solutions to the problem assumed without proof in their copies of Archimedes remain strictly inside the confines of the problem of dividing the sphere. Eutocius inaccurately describes Dionysodorus as unable to cope with Archimedes' form of the auxiliary problem, so that he was forced to derive an alternative reduction starting from the terms of the initial problem.[77] This characterizes the method adopted by Diocles, but Dionysodorus actually follows the line of Archimedes' treatment quite closely. Although Eutocius reports only the synthesis, the analysis underlying Dionysodorus' procedure is not difficult to perceive. Assume that the given line ZD has been divided at X such that $XZ : Z\theta = BD^2 : DX^2$, where $Z\theta$ and BD are also given (Fig. 16).

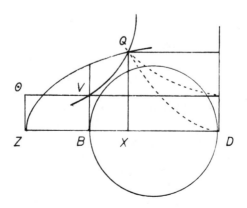

Figure 16

Since XZ : Zθ = XZ • Zθ : Zθ², it we set (a) QX² = XZ • Zθ, it follows that QX² : Zθ² = BD² : DX², that is, (b) QX : Zθ = BD : DX. The locus of Q that answers (a) is a parabola with vertex Z, axis ZD, and latus rectum Zθ; that which answers (b) is a hyperbola with asymptotes ZD and the line perpendicular to it through D, such that the curve passes through the point V for BV = Zθ. Hence, the intersection of these two conics determines that point Q leading to the required division of ZD at X. No diorism is provided; but as we have already noted, none is needed, since the restricted problem of division is always solvable. Although the actual curves derived by Dionysodorus differ from those introduced by Archimedes,[78] the close resemblance of the two methods to one another cannot be missed.

By contrast, Diocles returns to the initial conditions of the sphere division, specifying the points X, L, and P via three proportionalities. Here I present only the construction he derives in the analysis: assume as known the positions of L, X, and P in relation to the diameter DB of the sphere (Fig. 17).[79] Draw Dd at right angles to DB at D and of given length (namely, in the context of the restricted problem, one-half the diameter DB; but here Diocles allows for any value of the given line). Complete the rectangle defg with diagonal dXf, draw QXh parallel to ef, and extend eX to meet dg extended at j. Next determine k and m such that Dk = Bm = Dd, the given length, and draw through them lines perpendicular to DB extended to meet at r and s, respectively, the line

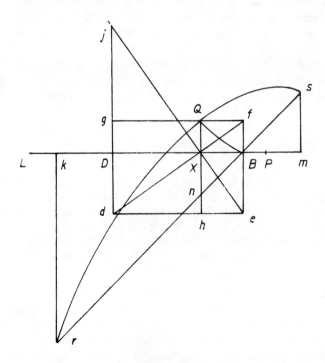

Figure 17

which makes half a right angle with DB at B. Let the line rBs met QXh at n. One finds that (a) $Qn^2 : rn \cdot ns = PX : 2\,XL$, where the ratio $PX : XL$ is given; and (b) $Qh \cdot hd = Be \cdot ed$. The former specifies Qn as the ordinate of an ellipse of diameter rs and whose latus rectum has to rs the ratio $PX : 2\,XL$.[80] The latter specifies Qh as the ordinate of a hyperbola passing through B and having ed, dg as asymptotes. Since these curves are given, Q will be found as their intersection, and from Q one has the required point X dividing DB. Diocles presents only the analysis of the problem; the synthetic demonstration is added by Eutocius.

Diocles, like Dionysodorus, holds the narrow context of the problem of dividing the sphere, and so misses the opportunity for constructing the more general problem with its diorism in the manner of Archimedes. Yet we do well to note the essential similarities uniting these three presumably independent approaches. Doubtless, the procedure of applying intersecting conics for the solution of problems was relatively undeveloped at Archimedes' time, although Menaechmus' cube duplication was of course an established precedent for it. Neither Dionysodorus nor Diocles can refer to Archimedes' solution, while Diocles at least appears to come before Apollonius wrote the *Conics*. Their efforts thus reveal the maturity not only of the theory of the conics at the time near or before Apollonius, but also of the field of problem solving in which the conics were applied.

Another problem of this "solid" class, that is, entailing the use of conic sections for its solution, arises in connection with Archimedes' theorems on the tangents in *Spiral Lines*. As we have seen, the proofs of the determinations of the subtangents in Props. 18–20 assume certain points Q on the curve, and these are found via points P lying on the associated circle. The latter refer back to lemmas (Props. 5–9) in which P is found by inserting a given length PH between the circle and a given line. But how is this *neusis* to be effected? Apparently, Archimedes accepted it as obvious, by virtue of the very terms in which the operation is introduced; that is, it is the abstraction of what one may easily visualize in the form of the manipulation of a sliding ruler. But later geometers were not comfortable with this assumption on his part. Pappus reports a criticism to this effect:

> It seems somehow to be no small error among geometers, whenever one solves a plane problem via conics or [higher] lines, and in general, whenever a problem is resolved [via curves] from a class not akin [to that of the problem]; for instance, the problem of the parabola in the fifth book of the Conics of Apollonius and the neusis of a solid toward a circle which is assumed by Archimedes in the book on the spiral; for without the intervention of any solid it is possible to find the theorem provided by him [sc. *Spiral Lines*, Prop. 18].[81]

One would suppose that the objection relates to the use of solid methods, that is, conics, for solving problems where the more elementary planar methods suffice. That is indeed the case for the Apollonian passage.[82] In the Archimedean instance, however, the alternative constructions of the *neusis* via planar techniques either appeal to nonconstructive assumptions or require major modifications in the structure of Archimedes' proof.[83] Moreover, when Pappus goes on actually to "set out the analysis of the neusis assumed by Archimedes in the book on spirals, so that you might not be at a loss," the method he uses is "solid," for it requires conic sections.[84] While the text is in some disorder, it has been restored to make sense. It is in the form of two lemmas on the construction of the conics, leading to the solution of the *neusis*; in each case only an analysis of the construction, without the synthesis, is given.

The following paraphrase will show its relation both to *Spiral Lines* and to the solving methods used for the solid problem of *Sphere and Cylinder* II. It is required to insert a given length between a given circle and a given line as to pass through a given point on the circle.[85] Here the line is a diameter LN of the given circle of center O and radius OR; let the line PH be inserted between them, where PH has the given length θ and its extension passes through the given point B on the circle (Fig. 18). Since HB and HL are secants to the circle, NH • HL = PH • HB. If we draw KH perpendicular to LH such that (a) KH = HB, it follows that (b) NH • HL = PH • KH; since PH is given, the locus of K is a parabola, for if V is the intercept with the diameter ROR', then OV • θ = NO • OL = ON^2, while KH • θ = NH • (NH + 2ON), so that θ • (KH + OV) = θ • VW = OH^2 = KW^2; that is, K is on the parabola of vertex V and axis VO and latus rectum θ. Furthermore, (a) sets K on a hyperbola of vertex B and diameter BB' (= 2BZ) such that KY^2 (= HZ^2) = BY • YB'. The intersection of these two conics thus produces the point K from which the required position of PH follows.[86]

While this construction does not directly require the "intervention of a solid," as Pappus oddly expresses it, this doubtless is the "solid *neusis*" at the base of his objection to Archimedes' procedure.[87] We must then suppose yet another

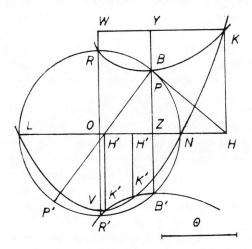

Figure 18

treatment which somehow dispensed with the use of conics. This seems to be indicated in Pappus' remark following upon this construction:

> Now some accuse [Archimedes] of having needlessly employed a solid problem;...they show how it is possible also to find by means of planes the line equal to the circumference of the circle, applying the cited theorems on the spiral.[88]

But now an even graver confusion has set in. The unnamed critics seem to suppose that their alternative to the *neusis* converts the construction into a *planar* solution of the circle quadrature, despite its dependence on the spiral itself, that is, a "linear" method, as well as on the drawing of its tangents. Clearly, the latter cannot be effected without introducing the subtangent, and hence requires a prior solution of the circle quadrature. Thus, neither Pappus nor his source betrays a clear perception of the nature of the problem, that is, under what conditions a proposed circle quadrature will be judged proper.[89]

As for this method of effecting the *neusis* via conics, Archimedes was perfectly able to produce such a construction. But that he did not is clear from Pappus' need to supply it, filling a gap in the Archimedean presentation. Furthermore, Archimedes could hardly have expected his readers to assume such a construction of the *neusis* as obvious. It thus appears that he would not have admitted the force of the objection Pappus raises against the introduction of *neuses*. We shall see additional instances of this difference in attitude, as we turn next to other problem-solving efforts by Archimedes.

PROBLEM SOLVING VIA NEUSES

Arabic sources preserve for us two Archimedean problems which, like the problems in *Spiral Lines*, rely on *neuses* for their solution where alternative procedures making use of conics can be found, and indeed were found by later geometers. One of these is an angle trisection implied by a proposition in the *Book of Lemmas*, the work we consulted earlier for Archimedes' quadratures of the *arbelos* and the *salinon* figures. The second effects the inscription of a regular heptagon in a given circle. In both cases we are indebted to Thābit ibn Qurra for the Arabic translations which preserved knowledge of these results and stimulated a host of related efforts among Arabic geometers.

The writing *On the Inscription of the Regular Heptagon* consists of 17 propositions, most devoted to properties of triangles unrelated to the problem of the title.[90] In its last two propositions one finds first a lemma establishing certain properties of a figure constructed by means of a *neusis*, and then the application of this figure for solving the problem of inscribing the heptagon. The procedure is synthetic throughout, so that the underlying line of thought tends to be obscured. It is of course natural to suppose that the original investigation started off with the analysis of the problem,[91] and on this view Tropfke presented a partial reconstruction. We shall follow his lead as far as he was able to go, and then introduce an auxiliary analysis to complete that part of the argument which he admits baffled him.

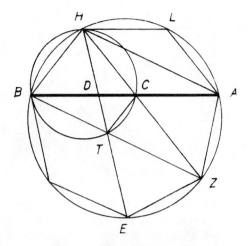

Figure 19

Let it be assumed that the heptagon has been inscribed in the given circle (Fig. 19), so that the side BH is the chord of an arc one-seventh of the circle, the diagonal AH is the chord of the double arc HLA, and the diagonals AB, BZ, HE are chords of triple arcs; let AB, HZ meet in C and HE, ZB meet in T and join CT. In the triangle DHB, the angles at H and B are equal, since they subtend the equal arcs BE, HA; thus lines HD and DB are equal. In the triangles HDC, BDT, the vertical angles at D are equal, while the angles at H and B are equal, since their associated arcs EZ, ZA are equal; thus the triangles are similar, and since HD = DB, they are also congruent. Hence, HC = TB. The quadrilateral HBTC thus has equal sides HC, TB and equal base angles at H and B; it may then be inscribed in a circle, so that angles CTH, CBH are equal (for they are subtended by the same chord CH). Now angle CBH is twice CAH; while DCH equals the sum of CHA, CAH, that is, twice CAH. Thus angle CBH equals each of the angles DCH, CTH. This makes the triangles CTH, DCH similar, so that TH • HD = HC^2. Since TH = CB, HD = DB, and HC = CA, one has (a) CB • BD = AC^2, a relation of the segments of line AB. Furthermore, since CHD and DAH are equal angles, triangles CHD and DAH are similar, so that AD • DC = DH^2. Since HD = DB, it follows that (b) AD • DC = DB^2, a second relation of the segments of AB. Thus the problem of inscribing the heptagon has reduced to that of finding two points C, D dividing the line AB according to the relations (a) and (b).[92]

This derived problem is solved in Archimedes' preliminary lemma. It is there shown that if on the given line CB, one draws the square CBFG with diagonal BG, and the *neusis* of line AF is effected so that the triangles KFG, LCA are equal in area, then the segments of BA will satisfy the relations (a) and (b) (Fig. 20a). Again, only a synthesis is given. But one can construct an analysis which leads from these relations to the condition of the *neusis* in a reasonably straightforward manner.[93] Let it be assumed that the line AB has been divided at C and

180 Ancient Tradition of Geometric Problems

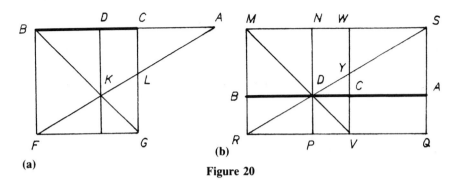

Figure 20

D to satisfy (a) and (b). In accordance with (b), the square on DB will equal the rectangle over AD, where CD = DP (Fig. 20b). We complete the rectangle MRQS whose diagonal RS passes through D (cf. *Elements* I, 43 converse); extend the diagonal of the square MD to meet RQ in V, join VC (= CD) and extend it to meet RS in Y and MS in W. From the similar triangles YWS, DPR, YW : WS = DP : PR. From (a) AC : CB = DB : AC; that is, with reference to triangles YWS, DVR, WS : VR = PR : WS. Thus, compounding the proportions, we have YW : VR = DP : WS. It follows from this that YW • WS = RV • DP, so that the triangles YWS and DRV are equal. If we then take RV as the given line, form its square, and we maneuver RS to cut its diagonal in D, its side VW in Y, and its side MW extended in S such that the triangles YWS and DRV are equal, it will follow via the synthesis that the segments of AB will satisfy (a) and (b). That is, this *neusis* leads to the division of AB required for the inscription of the heptagon.

That Archimedes followed a route of this sort is indicated by the restricted nature of the lemma. As a theorem on the division of a given line segment, it holds not only for the case where the figure on CB is a square, but more generally for any parallelogram on CB. In effect, the geometric aspects of the diagram are not of intrinsic interest; they serve merely to provide a representation for the conditions (a) and (b). That is what our analysis has done: introduction of the square on DB merely initiates the construction of a figure in answer to (a). The investigation of the quantitative relations among the lines in this figure, in particular as they lead to a determination of YW, leads via (b) to the articulation of the condition of the *neusis*. This thus illustrates, I believe, the way in which the ancient geometers used geometric methods for the examination of given "algebraic" relations, albeit in a rather narrow sense of that term.[94]

A difficulty in the logical ordering of these problems remains to be clarified. In the problem of constructing the heptagon, one is required to divide the line AB; but in the auxiliary lemma it is CB which is given. Indeed, in the context of the problem of inscribing the figure, neither of these lines is given, but rather the diameter of the given circle in which the heptagon is to be inscribed. Thus, the formal synthesis must introduce CB as an arbitrary line, effect its division via the *neusis*, and from this construct a heptagon with AB as a diagonal. The problem is then finally solved by inscribing a similar heptagon in the given circle. Thus, Thābit's version of this problem, while entirely correct, falls short of a complete formal synthesis. We have no good cause to impute these lapses to Thābit, for the manuscript at his disposal was already in no little disorder, as he notes.[95] But we have other examples of such informal treatments, like the quadratures of the Archimedean spiral and the spherical spiral reported by Pappus. One might wish to attribute this informality to later editors, presumably less adept at the demanding formal style mastered by the great geometers of antiquity. Yet the commentator Eutocius, for one, had no difficulties over formal matters, and even emphasized formal precision, as we saw in his treatment of Diocles' solution of the division problem in *Sphere and Cylinder* II. I would thus intuit quite the opposite: that informality, when it arises in the later commentators, is not a mark of their intervention, but rather a feature of their sources. It reflects greater concern for actually working through the solution of a problem, than for the precise exposition of a solution already known. This suggests that in such instances writers like Pappus and Thābit have preserved documents nearer the creative phase of the work of men like Archimedes, and less tending to obscure the essential line of thought than the more formal treatments do.

As for the roots of Archimedes' procedure for inscribing the regular heptagon, it may be compared with that underlying Euclid's inscription of the regular pentagon in *Elements* IV, 10, 11 (see Chapter 3). Indeed, the Arabic geometer Abu 'l-Jūd (10th century A.D.) followed this very model in working out an alternative construction of the heptagon. His method is extant only in the form of a synthesis reported by his contemporary, al-Sijzī.[96] On its basis, I will fashion the analysis, the better to display its essential line in relation to the Euclidean construction.

Consider a regular heptagon inscribed in a circle [Fig. 21(a)]. Drawing the two diagonals which join a vertex to the opposite pair, we obtain the isosceles triangle BAD in which each of the base angles is three times the vertex angle.

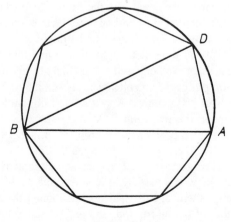

(a)

Figure 21

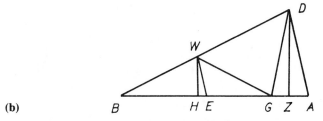

(b)

Thus, just as for Euclid the pentagon construction reduced to the finding of a triangle whose angles were in the ratio 1 : 2 : 2, so now the heptagon construction reduces to finding a triangle of angles a, $3a$, $3a$. Let BAD be such a triangle [Fig. 21(b)]; draw DG such that angle ADG equals ABD ($= a$). Since triangles DAG, BAD are similar, DAG will be isosceles, so that AD = DG. Next draw GW such that angle BGW equals WBG ($= a$). Then angle DWG = $2a$, being external to triangle WBG, while WDG = $2a$, being the difference WDA − GDA; thus triangle WGD is isosceles, so that WG = GD, where also WG = WB and GD = DA.[97] We now seek relations specifying the lengths of AD, AG, AB. From the similarity of triangles we have (i) AG : AD = AD : AB. In the Euclidean figure for the pentagon, where BG = AD, this yielded the "extreme and mean" division by which the problem could then be solved. But here, another relation must be sought. We draw WE parallel to DA and WH, DZ each perpendicular to BA; then from similarity of triangles, EB : BH = AB : BZ.[98] Now, EB = BW = AD, while BH = ½ BG, and BZ = BG + GZ = ½ (BG + BG + GA) = ½ (BG + BA). Thus, from EB : 2BH = AB : 2BZ, it follows that (ii) AD : BG = AB : GB + BA. In Abu'l-Jūd's formulation, one takes AD as given and thus seeks a line AB so divided at G as to satisfy relations (i) and (ii).

Just as for Archimedes, the alternative construction of the heptagon has reduced to the sectioning of a line in accordance with two second-order conditions. As we have seen, Archimedes effected this division via a *neusis* in which two triangular areas are made to be equal. But the Arabic geometers refused to admit this technique, and so attempted to find an acceptable alternative, either via the Euclidean "planar" methods (of compass and straight edge) if possible, or via conic sections. Abu 'l-Jūd himself thought that he had found a solution of the former type; but his error was exposed by al-Sijzī. In his turn, al-Sijzī posed Abu 'l-Jūd's division problem to a colleague, al-ʿAlā ibn Sahl, who worked out a successful analysis of it via conics. A synthesis of this same solution is extant in al-Sijzī's writing on the heptagon.[99] At the same time that he posed this question to al-ʿAlā, al-Sijzī posed another problem, a more general form for the Archimedean *neusis* for the heptagon. It seems that al-ʿAlā, having solved the division problem, nevertheless insisted that the *neusis* problem was unobtainable, on the grounds that one would thereby have solved the Archimedean inscription problem. Apparently, then, al-Sijzī had not mentioned the relation of Abu'l-Jūd's problem to that of the inscription, and al-ʿAlā failed to recognize it.[100]

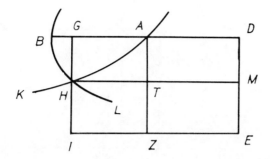

(c)

Following the model of al-Sijzī's synthesis of the problem of division, we may reconstruct al-ʿAlā's analysis thus: let the given line AB be divided at G such that for the line X, where (i) AB : X = X : BG, we have (ii) X : AG = AB: AB + AG[101] [Fig. 21(c)]. If we set GH = X perpendicular to BA, then since GH^2 = AB • BG, H will lie on the parabola of axis BA, vertex B, and parameter AB. If we extend BA to D such that DA = AB, draw perpendicular to BA the line AZ = AB, and complete the rectangle GAZI, it follows from (ii) that GAZI equals rectangle DGHM. The rectangles AM, HZ are thus equal, whence after adding MZ to each, the rectangles ADEZ, HMEI are equal. The point H will thus lie on the hyperbola passing through A and having DE, EZ as asymptotes. In this way, point H will be given as the intersection of a given parabola and a given hyperbola, so that the analysis of the problem has been effected.

This method is remarkably like the one proposed by Dionysodorus for effecting Archimedes' lemma toward the division of the sphere, presented above. In both, a condition of the form (i) entails introduction of a parabola, while another of form (ii) leads to a hyperbola. The coincidence may of course merely follow from the nature of the problems. But the Arabic geometers were well aware of these ancient studies through the translations of Archimedes and Eutocius by Thābit ibn Qurra. Indeed, the geometer al-Shannī, an important source for the Arabic disputes over priority in the solution of the heptagon problem, not only cites Eutocius' commentary on *Sphere and Cylinder* II, 4, where these treatments appear, but also affirms their relevance for solving the lemma to the heptagon construction.[102] It thus seems to me entirely plausible that al-ʿAlā too perceived the nonobvious connection between these two problems, so that a specific knowledge of Dionysodorus' method could assist him in replying to al-Sijzī's challenge.

Arabic geometers continued to search for new solutions to the Archimedean problem, even after the successful effort by al-ʿAlā and al-Sijzī.[103] In each case, the division of the line (e.g., AB in Fig. 20) is effected via the intersection of conic sections, in the manner we have seen in connection with the problems in Archimedes' *Sphere and Cylinder* and *Spirals*. As an example of a solution for the lemma to the heptagon problem, framed directly in the manner adopted by Archimedes in his *neusis*, we may consider the solution given by Alhazen (ibn

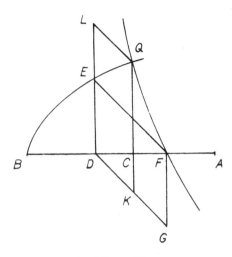

Figure 22

al-Haytham).[104] My paraphrase will be as an analysis patterned after his synthesis. We are to divide the line AB at C and D such that (a) CB • BD = AC² and (b) AD • DC = DB². We assume that the line has been divided in this manner, and treat segment BD as given, rather than BA (Fig. 22). If we draw QC = AC perpendicular to AB at C, then (a) specifies Q as lying on the parabola with vertex B, axis BA, and latus rectum BD. At D the ordinate ED = BD. We set DF = DE and complete the parallelogram DEFG. Extend QC to meet DG in K and complete the parallelogram QKDL. Since DF = FG, DC = CK; also, since AC = CQ, by adding we have AD = QK. From (b) it follows that QK • DC = ED • DF; that is, the parallelograms DEFG and QKDL are equal. Thus, Q lies on a hyperbola passing through F and having DG, DL as its asymptotes. On the assumption that BD is given, then, both the parabola and the hyperbola are given, whence Q is given as their intersection. To effect the synthesis where BA is the given line, say θ, one merely divides θ into segments similar to those found in the case where BD is given.

The resemblance to the solutions of the Archimedean problem in *Sphere and Cylinder* II is of course not accidental; the Arabic geometers could refer to the three versions of its solution preserved by Eutocius as paradigms for their own efforts in related contexts. But as to whether Archimedes himself had in mind such an alternative construction of the *neusis* for the heptagon, the conclusion that he did *not* is even clearer than it was in the case of the *neusis* for the spirals. For in the case of the heptagon the *neusis* is designed to produce a division of the line in answer to (a) and (b). The alternative construction via conics does precisely the same. It is not the *neusis per se* that one requires, but rather the division of the line. The procedure via conics would thus render the *neusis* itself entirely superfluous for the solution. We thus see that Archimedes has *chosen*

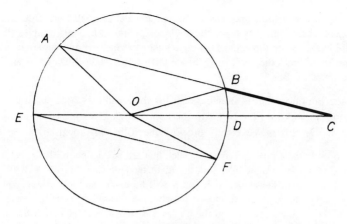

Figure 23

the *neusis* as his way to solve this problem. While this of course does not deny that he could have produced a method via conics had he determined to do so, his adoption of the *neusis* here reveals that this was for him an entirely acceptable means of construction in its own right.

A method via *neusis* also underlies an Archimedean angle trisection preserved in the *Book of Lemmas*. Actually, what appears as its eighth proposition is not a solution of this problem, but rather a lemma appropriate for its synthesis. We give the following paraphrase of the lemma[105]:

> Let arc AE be given in a circle centered at 0.[Fig. 23] The chord AB has been drawn and extended to meet the diameter E0D extended in C such that BC equals the radius of the circle. It is claimed that the arc BD is one-third the arc AE. Draw chord EF parallel to AB and join 0A, 0B, 0F. The angles at E and F are equal, so that angle D0F is twice that at E. By parallels, the angles at E and C are equal; hence angle D0F is twice that at C. Since CB = B0 by construction, angle B0D equals that at C, so that angle D0F is twice angle D0B, or arc DF is twice arc BD. By parallels, the arcs AE and BF are equal. Thus, arc AE is three times arc BD.

In this way, the solution of the angle trisection reduces to the insertion of the given length, the radius of the circle, between the circle and a given line, the extension of its given diameter ED. This is a *neusis* precisely of the class assumed in the constructions of *Spiral Lines*. While we have seen that such a *neusis* can be effected via conics, the Arabic geometers understood by the "ancient method" of angle trisection nothing other than the application of *neusis* to produce the diagram of the *Lemmas*.[106] Our findings relative to the heptagon and the spirals make clear that Archimedes, too, must have intended a construction of this form.

A puzzling feature of the proof in the *Lemmas* is its introduction of the superfluous consideration of the parallel chord EF, for a far more efficient treatment is possible without it:

In the diagram, as before, since CB = BO, the angle at C equals angle BOC. Since ABO equals their sum, it is twice the angle at C. Since OB = OA, angles OBA, BAO are equal. Since the angle EOA equals the sum of OAB and the angle at C, it is thus three times the angle at C, hence three times angle BOD. Thus, arc AE is three times arc BD.[106a]

Then, since the chord EF is unnecessary for the synthesis, one would suppose that it found its way into the proof in the *Lemmas* as a vestige of an analysis of the problem. On this lead we may propose the following form:

> We wish to divide a given arc BF at D such that arc BD is one-third BF. Assume the division has been done. We join the diameter D0E and the radii 0B, 0F. If chord EF is drawn, the angles at E and F will be equal, and angle D0F will be twice the angle at E. Draw through B the chord AB parallel to EF and extend it to meet ED extended at C. Then the angles at C and E are equal. Furthermore, since angle D0F is twice both the angle B0D and the angle at E, the angle B0D will equal the angle at E and hence the angle at C. This makes the triangle B0C isosceles, i.e., CB = B0. Since chords AB, EF are parallel, arcs BF and AE are equal. Thus to trisect the given arc BF, we introduce it in the position of AE, effect the *neusis* of BC, a given length (the radius), between the circle and the given line E0D extended, so as to pass through the given point A. This yields the arc BD as one-third arc BF equal to arc AE.

While it is possible to construct a shorter analysis following the pattern of the abridged synthesis given above, the longer form follows quite naturally from the assumption of the given arc BF divided at D. Within this longer version the chord EF plays a key role. Hence, its presence in the extant synthesis, where it is entirely superfluous, would indicate that the ancient analysis actually had much the form we have given.

Certain modifications of the figure for the Archimedean angle trisection lead to alternative constructions important in later geometric writings. If one draws the line 0H perpendicular to E0D meeting AB in G, then B bisects GC (Fig. 24). Consider the circle which circumscribes the right triangle G0C: GC is a diameter (*Elements* III, 20 and 31), while its center will lie on the perpendicular bisector of 0C (IV, 5). Since B0 = BC, the point B lies on that bisector as well as on the diameter; hence, B is the center, so that GB = BC. From this result

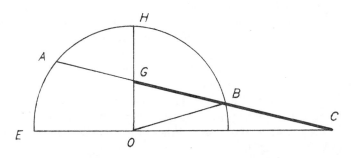

Figure 24

we may conceive two variations of the Archimedean method of trisection: (a) a *neusis* in which the given length (the radius) is to be inserted as BG between the circle and a given line OH (perpendicular to the given line E0) and passing through the given point A; (b) a *neusis* where the given length (the diameter of the circle, or twice 0A) is inserted as GC between two given lines, namely OH and E0 extended.

Variant (a) appears in the writing *On the Measurement of Plane and Solid Figures* by the three Banū Mūsā (or "sons of Moses") in the 9th century A.D.[107] The *neusis* is there effected by means of an actual physical device, in which a sliding length is made to assume the position GB. Although the authors claim this method as their own discovery, its close connection to the method of the *Lemmas* is clear; and as the translator of the *Lemmas*, Thābit ibn Qurra, was their colleague at Baghdad, we may surely suppose they were familiar with the Archimedean approach. Indeed, all of the other materials in this book are based closely on well-known ancient results, like Archimedes' treatments of the measurement of the circle and sphere, the cube duplication of Archytas, the Heronian rule for finding the area of a triangle from its sides, and so on. Therefore, one can admit their responsibility for this particular variant of the construction, yet still insist on its implicit presence in the Archimedean method.

Variant (b) is one of three unattributed methods of trisection presented by Pappus in *Collection* IV. It appears there with two lemmas, one of which shows how to effect the *neusis* through the intersection of a given circle and a given hyperbola, and the other shows how to construct the hyperbola when its asymptotes are given.[108] As we have seen, Archimedes would not have felt the need to eliminate the *neusis* in such a context, and the actual terms of Pappus' text betray the influence of Apollonian terminology. Furthermore, although Pappus gives both the analysis and the synthesis for both of the auxiliary lemmas, the actual solution of the trisection by means of the *neusis* is presented only as a synthesis. This again suggests an origin near the time of Apollonius, late in the 3rd century B.C., when the lemmas might still be viewed as novel, although the trisection would be familiar as a variant of Archimedes' method. Indeed, this variant is the implied basis for an alternative construction using conchoid curves, which one can assign to Nicomedes around the same time.[109]

These instances of the application of *neusis* for the solution of the trisection, the inscription of the heptagon, and the study of the spirals reveal the importance of this method for Archimedes. To be sure, one detects a precedent for such a method in the construction of the third lunule of Hippocrates much before Archimedes; but the potential for its use in the solution of problems was surely never so clearly perceived as through Archimedes' applications of it. The impact of Archimedes' constructions is amply revealed in the diversity of alternative *neuses* devised by later geometers, both in antiquity and in the Middle Ages. But his effort with *neuses* gave rise to another type of inquiry: the search for constructions via conics or other curves which might eliminate them. It thus appears that Archimedes was notable for the ease with which he admitted these methods into his constructions.

AN ANONYMOUS CUBE DUPLICATION

In view of the important contributions Archimedes made to the solution of problems like the angle trisection, as we have just seen, it is surprising that the ancient tradition assigns to him no role in the study of the cube duplication as well. Indeed, it is the need for a construction of two mean proportionals in the course of the first proposition of *Sphere and Cylinder* II which provides Eutocius the context for presenting his series of texts on this problem. Of the eleven, or twelve (if one includes the "Eudoxean" method Eutocius discards as spurious), none is taken from an Archimedean work. But several depend on constructions by *neuses*, a method we have seen was skillfully exploited by Archimedes, and of these, one is of special interest due to the puzzle of its origin.

I refer to the method attributed to Hero. It appears not only in Eutocius' account, but also in Hero's extant *Mechanics* and *Belopoeica*, and Pappus too knows of it from Hero. But another commentator, John Philoponus, assigns it to Apollonius.[110] This indication of a much earlier provenance in the 3rd century B.C. gains support, as we shall see later, from Nicomedes' apparent awareness of the method as a starting point for his own cube duplication.[111] The question of source would arise in any event, since Hero's writings rarely presume to be original; they are intended as compilations of familiar results for the use of practitioners in mechanics and geometry. Now, in the *Mechanics*, Hero cites Archimedean works at least half a dozen times, and in the *Metrica* more than three times that frequently. By contrast, citations of other authorites are extremely rare.[112] Moreover, certain unassigned results, like the rule for triangles and the rule for roots in the *Metrica*, arguably may be associated with Archimedes. The same applies in the case of Hero's rule for estimating the areas of circular segments.[113] Hero also appears to rely on unnamed sources for information of a historical sort, such as that "the ancients" (*archaioi*) applied a certain rule for measuring segments, or that "as some relate" Archimedes measured volumes by immersing the solids into containers of water and taking note of how much water was displaced.[114] Thus, the major portion of Hero's writing on mechanics and metrics is based on Archimedean sources or materials of anonymous origin with some Archimedean associations.

In view of this, when Hero introduces into his *Mechanics* the method for finding two mean proportionals which is "most amenable for practical purposes," we at once suspect the possibility of an Archimedean origin. A consideration of the method itself and the manner of its derivation will assist us here. In all extant versions of Hero's method, the construction is given as a synthesis along the following lines:

> Let the given lines be AB, BG between which the two mean proportionals are to be found [Fig. 25]. Complete the rectangle ABGD and draw its diagonals to meet in H. Conceive now a ruler able to pivot about B, meeting the extensions of DG, DA in E, Z, respectively. Let it be manipulated until it assume the position such that EH = HZ. Then the segments AZ, GE are the required mean proportionals; that is, AB : AZ = AZ : GE = GE : GB.

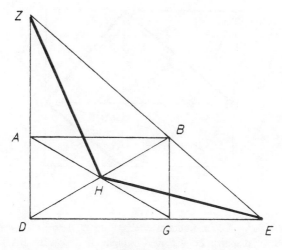

Figure 25

In the proof, from the equality of EH, HZ, one derives GE • ED = AZ • ZD, from which the desired proportions follow through consideration of similar triangles. We defer the full proof until its presentation in relation to other methods of solution.[115] Instead, an analysis reconstructed on the model of that proof will better reveal the manner in which this method was worked out. Let us initiate the analysis with a figure based on the method of Menaechmus in which two parabolas intersect at a point whose coordinates are the required mean lines.

> Let the given lines be AB, BG [Fig. 26]; let them be extended and let two parabolas be drawn on the extensions as axes with vertex at B and with the given lines as their respective parameters. Then the curves intersect in P such that PK^2 ($=LB^2$) = KB • BA and PL^2 ($=KB^2$) = LB • BG. Hence, by adding equals LB^2 + LB • BG = KB^2 + KB • BA, that is, LG • LB = KA • KB. If one draws AG and joins BH at its midpoint H, then BH = HG = AH. Adding BH^2 to each

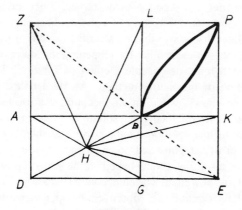

Figure 26

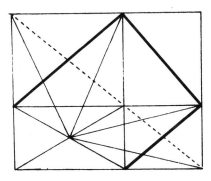

Figure 27

term LG • LB, KA • KB, one obtains $KH^2 = HL^2$.[116] If now the rectangle PZDE is completed, its diagonal ZE will pass through B, since LB : BA = GB : KB (for each equals KB : BL by the hypothesis of the analysis) and HZ = HE (for they are equal, respectively, to LH = HK). Hence the construction will be given via the placement of the line (as ZBE) with respect to the given rectangle ABGD such that ZH = HE.

In this way an analysis starting from the hypothesis of the mean proportionals could lead to the formulation of the *neusis*. While I have introduced that hypothesis in the form of Menaechmus' solution, an alternative based on the figure of the pseudo-Platonic method would make no technical difference (Fig. 27). As far as chronology is concerned, any geometer of the 3rd century could be expected to know of the method of Menaechmus. If, as we have proposed, the pseudo-Platonic method is a form of the approach adopted by Eudoxus, then that too would be familiar in the 3rd century. Moreover, as we shall see, the pseudo-Platonic form is implicit in the method of Diocles.[117] Thus, the basis for an analysis of this type lies in the older methods.

We have already observed that this *neusis* construction was devised around the time of Nicomedes and Apollonius or earlier. Since Eratosthenes fails to mention it in his report on the history of cube duplications, we must suppose the *neusis* was a later discovery, that is, after the middle of the 3rd century.[118] Eratosthenes also sets out the motive for reopening the study of this problem at that time: the search for a practicable method of solution. The same motive underlies the finding of the *neusis* method. As Hero notes, this is the most practical method he knows, while on any account, it arose long after the successful working out of several theoretical solutions. Eratosthenes further specifies contexts in which such a solution would prove valuable: for instance, in the building of ships or of catapults and other engines of war; more generally, in any situation where a solid of given shape is to be scaled up or down according to a set ratio of volume or weight. The latter context occasions Hero's presentation of the *neusis* in the *Mechanica*.[119] Moreover, its appearance in his *Belopoeica* (i.e., *Weapons-Making*) is intended to provide the catapult-builder a ready procedure for carrying out his instructions on devising engines of specified throw weight, namely, given a model of known dimensions and throw weight, the linear dimensions of the model and the engine to be built must have the ratio of the cube root of the ratio of weights.[120] In all these respects, then, the Heronian *neusis* appears to have been made to the order of the demands noted by Eratosthenes. Eratosthenes' own device, for all its ingenuity, did not fill the bill, as Nicomedes' later criticisms reveal.[121] We thus seem to have a context ideally suited in time, motive, and technical preparation for the invention of the *neusis* method of cube duplication.

An Archimedean connection has been noted in the general significance of Archimedean sources for the tradition of mechanics and metrical geometry available to Hero. More specifically, the ancient historians indicate strong associations of Archimedes with the ship building and military engineering centered at Syracuse, while the wondrous performance of the catapults constructed under his supervision receives special comment.[122] In addition, Archimedes maintained a scientific correspondence with Eratosthenes which included not only the sending of the *Cattle Problem*, we may presume, but also of the *Method* and other works on the measurement of volumes and centers of gravity. To propose that Archimedes devised the *neusis* method in response to the efforts by Eratosthenes would go beyond the direct testimony of extant documents, to be sure; but it is hardly far-fetched and serves well for drawing together all the diverse contexts relating to this construction. This view immediately explains why Eratosthenes does not cite the *neusis* in his report on cube duplication. The fact that some later commentators associate this method with Apollonius need only signify that Apollonius used it as the starting point of his own investigation of the construction, just as it appears Nicomedes did, about the same time. As for the eventual anonymity of this method as we meet it in the writings of Hero, we must suppose that, due to its real practical effectiveness, it rapidly found its way into the general body of results of applied geometry and so managed to survive the disappearance of the document in which it was first formally communicated (if indeed there ever was such a document). As we have seen, several discoveries by Archimedes have been preserved via transmission histories of this sort. The paucity of results for which later writers like Hero, Pappus, and Proclus do, or even can, provide express attributions indicates that the route to anonymity was the rule rather than the exception.

A second method of a different sort appears in Hero's *Metrica*: it is a rule for obtaining numerical approximations to cube roots. Characteristically, Hero presents the rule by working through a specific example:[123]

> We shall now say how to take the cubic root [lit.: side] of 100. Take the cube nearest 100 which exceeds it and that which falls short, namely, 125 and 64. The excess of the one is 25, the defect of the other 36. Multiply 5 into 36, result: 180. Add to 100,[124] result: 280. [Divide 180 by 280, result: 9/14.] Add to the smaller side, that is 4, result: 4 9/14. This shall be the cubic side of 100, as nearly as possible.

An obvious difficulty in this style of exposition is that the numbers in the calculation may sometimes be ambiguous for signifying the general rule. For instance, is "5" obtained as $\sqrt{25}$ or as $4 + 1$? Is the second "100" simply the given number, or does it result in some other way, e.g., as 4×25? Adopting the former readings, scholars at first set the rule in a form which in general yields poor approximations, and so treated it merely as an *ad hoc* empirical gesture.[125] But others worked out the derivation of a valid general rule taking the form $x \doteq a + e(a + 1) / [e(a + 1) + da]$, where x is the derived approximation for the cube root of N lying between a^3 and $(a + 1)^3$, for $d = (a + 1)^3 - N$ and $e = N - a^3$. In a case like this, one must of course prefer the interpretation which gives the superior mathematical sense, even though that entails the assumption of a degree of corruption in the text.

How could an ancient geometer have deduced such a rule? In the following analysis I modify the derivation cited above in a more general form which does not demand integral first approximations.[126]

Let N lie between a^3 and $(a + b)^3$, for rational (not necessarily integral) values of a and b. Then, via inspection of the subdivison of the cubes of a and $a + b$ in relation to $N = x^3$, we have $d = (a + b)^3 - N = 3x(a + b)(a + b - x) + (a + b - x)^3$ and $e = N - a^3 = 3xa(x - a) + (x - a)^3$ [Fig. 28]. For each of these, the cubic term at the end is a small fractional part and so may be ignored for the purposes of the approximation. Hence, $d : e \doteq (a + b)(a + b - x) : a(x - a)$, so that $da : e(a + b) \doteq a + b - x : x - a$. By standard manipulations of proportions we obtain $e(a + b) : e(a + b) + da \doteq x - a : b$. We thus obtain the approximation claimed:

$$x \doteq a + b\left(\frac{f}{(f + g)}\right), \text{ or } a + b'$$

for $f = e(a + b)$ and $g = da$.

In Hero's case the rule takes $b = 1$. But in the form given here one sees that the procedure can be applied recursively to produce arbitrarily close approximations; for instance, one may substitute for $a + b$ the derived value $a + b'$ and repeat the procedure to obtain a closer value. Such a refinement was recognized in the context of a rule for square roots, also reported by Hero in the *Metrica*,[127] so that we may well suppose that the ancient geometer who devised the rule for cube roots perceived the possibility of recursion here.

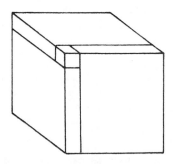

Figure 28

Archimedes: The Perfect Eudoxean Geometer   193

The line of thought underlying the analysis leading to this rule seems to focus on the ratio $a + b - x : x - a$, that is, the ratio into which the true root x divides the interval separating the initial approximations a and $a + b$. Rather than choose the simple arithmetic mean as an estimate on x, the calculator seeks a more refined weighting between the initial values, since in principle one is likely to be a better approximation than the other. Consideration of the ratio $d : e$ then leads him to an estimate of the ratio of the difference of the initial bounds from the root.[128] But one aspect of this analysis strikes me as somehow alien to the ancient view of computation: the casual disregard for the small error terms in the expressions for d and e. To be sure, some approximation will be unavoidable. Is there an alternative form, however, more in the spirit of the ancient methods? I have thus been led to the following version of the analysis:

We seek an estimate of the ratio $a + b - x : x - a$, as before. Now $e = N - a^3 = (x - a)(x^2 + ax + a^2)$. Since ax lies between x^2 and a^2, let us take it as an average value, so that $e \doteq (x - a)(3ax)$. Similarly, $d \doteq (a + b - x) \cdot 3(a + b)x$. Hence, $d : e \doteq (a + b - x)(a + b) : (x - a)a$. The rest of the derivation then proceeds just as given above.

Here the expression for $e = (x - a)(x^2 + ax + a^2)$ was surely familiar in the ancient geometry, for it is implicit in the ancient Egyptian rule for measuring truncated pyramids, while an equivalent form of the same is given by Hero for truncated pyramids and cones.[129] The adoption of the middle term as an average for the extremes in order to simplify the expression seems natural.[130] It also enables us to recognize that the rule obtained here is of the same order of accuracy as the Newton–Raphson rule,[131] and indeed gives a somewhat better estimate. For if instead of ax, we adopt a^2 as the "average" of the three terms a^2, ax, x^2, then we have $e = N - a^3 \doteq (x - a) \cdot 3a^2$, whence $x \doteq a + (N - a^3)/3a^2$, the value derived from the Newton–Raphson rule. Alternatively, using the term $(a + b)^2$ in the expression for d, one obtains the Newton–Raphson value associated with the initial estimate $a + b$. Clearly, the Heronian rule based on the use of ax as the intermediate term leads to a closer result, and indeed its value lies between the two Newton–Raphson values just cited.

This alternative mode of the derivation of the Heronian rule reveals another interesting feature: it is generally applicable for *all* orders of roots. For let us propose to approximate x such that $N = x^n$. As before, we take the initial values a and $a + b$. Now $e = N - a^n = (x - a)(x^{n-1} + x^{n-2}a + x^{n-3}a^2 + \cdots + a^{n-1})$. We may simplify this expression, by analogy with the case for $n = 3$, by taking any one of the terms $x^{n-1-k}a^k$ as an average. Hence, $e \doteq (x - a) \cdot nx^{n-1-k}a^k$, so that $d : e \doteq (a + b - x) \cdot (a + b)^k : (x - a) \cdot a^k$. For $k = 1$ we obtain precisely the same form for the approximation to the nth root that Hero uses for the cube root. Here again, the value will be comparable in accuracy with the Newton–Raphson value which corresponds to $k = n - 1$.[132] However, one sees that a more effective estimate would result by choosing one of the central terms, where $k = [(n-1)/2]$. Furthermore, the rule also applies for $n = 2$, that is, for finding estimates of square roots: when $k = 1$ the form of the rule will be analogous to that of Hero's cube root rule. Furthermore, one can show that if one sets $a + b = N/a$ [that is, $b = (N - a^2)/a$] in this rule, the value computed for x will be equal to that obtained via Hero's rule for square roots, given earlier in the *Metrica*, namely, $\frac{1}{2}(a + N/a)$, for in this case, $d : e = a + b : a$, so that $a + b - x \doteq x - a$, or $x \doteq a + \frac{1}{2}b$.

Clearly, these more general aspects of the rule for cube roots lie far outside the domain of applications relevant to the ancient geometry. But they well reveal the power of the insight which has led that geometer responsible for the rule in making its discovery. Furthermore, although the rule for square roots given by Hero is applied in a cuneiform tablet from the old Babylonian period,[133] so that its knowledge among the Greeks must have resulted from the transmission of older technical materials, it seems quite possible that a Greek geometer seeking a rationale for this rule might come to set it in the form which renders its extension to the cube roots evident. We would thus expect that researches centered on the approximation of square roots provided a suitable context for the discovery of its extension. Now scholars have long perceived that the values Archimedes introduces for the approximation of the square roots of the large numbers in the *Dimension of the Circle* result most naturally from the use of Hero's rule for square roots.[134] One may thus associate the rule for cube roots with the researches implied in Archimedes' work on geometric measurements.

About two-and-a-half centuries separate Hero from Archimedes, the most significant source for his expositions in mechanics and metrical geometry. In the matter of the two methods of cube duplication discussed in this section, considerations of motive and technique have recommended assigning their origin to early in this period, rather than later. Despite their seeming simplicity, both methods are remarkably effective and ingenious. A sign of this has been the somewhat contrived character of the analyses we have proposed for reconstructing their discovery. For no regularized heuristic procedure can naturally duplicate the product of inspired insight. The argument from ingenuity for the defense of attributions is always risky. But this, in conjunction with the other considerations brought forth here, point to Archimedes as the originator of these two methods, or if not him, then some comparably gifted geometer closely associated with his work.

THE IMPACT OF ARCHIMEDES' WORK

The present discussion of Archimedes' study of geometric problems started and ended with sophisticated efforts in the area of computational geometry: the estimates for π in his *Dimension of the Circle*, on the one hand, and the strikingly effective rules for estimating square and cube roots, on the other. The interest in the quantitative measures of geometric figures, for instance, the areas, volumes, and centers of gravity of curvilinear figures, dominates the whole of Archimedes' work. Throughout, his method is theoretical, applying and refining the Eudoxean method of limits. He adapts the older one-sided technique of convergence into the more flexible two-sided technique and applies it to new classes of figures, like those bounded by conics and spirals, and to new types of problems, like the drawing of tangents to the spirals. Furthermore, he investigates the basic principles of mechanics, like the inverse proportionality of weights and distances in the equilibrium of balances, as a field of Eudoxean geometry and in this way comes to introduce and examine its most characteristic concept, the center of gravity.

The more practical tradition of metrical geometry, which among the Greeks already had roots in the far older Egyptian and Mesopotamian traditions, was ready ground for receiving the results newly discovered by Archimedes. Hero's writings are filled with rules for the computational measures of areas and solids, of which about half might be drawn from the *Elements*, but the rest are more advanced, owing their first discovery and proof to Archimedes. For instance, the Archimedean value $3\frac{1}{7} : 1$ for the ratio of the circumference to the diameter of the circle is used in preference to any alternative value, and is applied to the expression of the area of the circle (as $\frac{11}{14}$ the square on the diameter), the volume of the sphere (as $\frac{11}{21}$ the cube on the diameter), and so on. For the derivative tradition which developed from Hero's writing, the exposition of such rules is invariably by means of numerical illustrations; the formal demonstrations of Archimedes find no place here.[135] Even in Hero's *Metrica*, the instances of formal derivations are few. And often when these are given, as for the rule estimating the areas of circular segments, or that for the area of the triangle from its sides, or the procedure for duplicating the cube via *neusis*, it is clear he relies on sources influenced by Archimedes and in all likelihood actually originating with him.

That this practical tradition should not trouble itself over the formal niceties of Eudoxean and Archimedean limiting procedures is hardly surprising. But the virtual abandonment of such interests within the formal tradition of geometry after Archimedes is surely noteworthy. Results on measures, for which the limiting methods are indispensable are absent from the extant work of Apollonius, for instance, and even in his study of tangents and asymptotes, he manages to avoid using its characteristic features, such as successive bisection or the Archimedean axiom.[136] In Pappus' *Collection* only a handful of results fall within this Archimedean class, and most of these—like the quadratures of the plane and spherical spirals, or the alternative demonstration of the surface and volume of the sphere—appear to derive from Archimedes himself, rather than geometers following his lead.[137] Archimedes took great pride in his "mechanical method" for discovering the measures and centers of gravity of figures and recommended its use to Eratosthenes for finding new results, but there survives evidence of only a single result obtained by a later geometer through a comparable method: a rule for the volumes of solids of revolution which we take up in the next chapter.[138] Indeed, the whole field of geometric mechanics, pivotal for the development of Archimedes' work, is but little advanced after him. Hero and Pappus speak of the study of centers of gravity in glowing terms, but what they present merely applies or explicates materials drawn from Archimedes' writings.[139] Thus, the later geometers appear to have contributed nothing of note in the areas most significant to Archimedes. One suspects that his achievement was so impressive that they felt compelled to move in new directions if their own researches were to be fruitful.[140]

By contrast, the field of analytic problem solving actively engaged the efforts of geometers, both before and after Archimedes. Much of Pappus' *Collection* is devoted to examples of analysis or lemmas for their proofs, including the entirety of the enormous Book VII. The latter, a commentary on the corpus of analytic treatises, examines a dozen substantial works, most by Apollonius or Euclid, but not one by Archimedes or even suggestive of his influence. But within the Archimedean corpus one comes upon several fine examples of problem solving. The computational efforts we mentioned above reveal concern over the circle quadrature and the cube duplication. A geometric construction for the former is entailed in Archimedes' theorems on the tangents to the spiral. One of the solutions of the latter by means of *neusis* is arguably of Archimedean origin, for this, like several other known or suspected Archimedean results, appears without attribution in writings by Hero. Archimedes adopts *neuses* for the solution of certain auxiliary problems related to his theorems on the spirals and devises ingenious *neuses* for the construction of the angle trisection and the inscription of the heptagon in a circle. Furthermore, he makes adept use of the properties of conics for the solution of a problem on the division of the sphere in *Sphere and Cylinder* II, and throughout this work, applies the method of analysis effectively.

As we shall see in the next chapter, Archimedes' solutions for these problems attracted considerable interest and gave rise to a variety of alternative efforts by means of conics and special curves. Yet this activity is motivated less by the urge to advance the work initiated by Archimedes than by the desire to set right certain weaknesses perceived in his approach. It is doubtful that even Archimedes himself would accept the construction by means of the tangents to the spiral as an actual solution of the circle quadrature; and when later geometers introduce other curves like the quadratrix for this same purpose, one must infer that they still considered this question very much alive. Most significantly, Archimedes' applications of *neusis* stimulate criticisms rather than imitations. In some cases special curves are introduced as alternative constructions, in others, the intersections of conics. Indeed, the method of *neusis* loses status as a means for constructing problems, for the exercise of effecting *neuses* via planar or solid methods becomes an important research interest.

One thus sees that in his selection of the Eudoxean field as the center of his research, Archimedes set himself off from the general group of geometers in his time and after. For them the analysis of problems, begun by Euclid and consolidated by Apollonius, was of far greater importance. To a certain extent, Archimedes' works continued to be studied and applied, as in the numerical rules of the practical tradition of metrical geometry. But not until the 16th and 17th centuries can one find geometers concerned with taking up and advancing the Archimedean field once again.

NOTES TO CHAPTER 5

[1] The basis of our biographical information on Archimedes is derived from the historian Polybius (2nd century B.C.), whose account of the Second Punic War drew heavily on eyewitness reports. Derivative but generally reliable accounts are given by Livy (1st century B.C.) and Plutarch (2nd century A.D.). The verse chronologist John Tzetzes (12th century A.D.) is representative of the elaborations of the later biographical tradition. For synopses, see the Archimedes surveys by Heath, Dijksterhuis, and Schneider.

[2] Euclid's dates are uncertain; but his activity is generally assigned to around 300 B.C.. Archimedes' year of birth is usually placed at 287 B.C. But this is supported only by the claim by Tzetzes (see previous note) that Archimedes was 75 years old at his death (where the latter is known on good authority to have occurred in 212 B.C.). The earlier and more reliable witnesses, however, say merely that Archimedes was "elderly" (*presbytēs*: Polybius), while Proclus claims that he and Eratosthenes were "of the same age" (*synchronoi*). (For testimonia, see E. Stamatis, *Archimēdous Hapanta*, I, especially pp. 1–3.) As one assigns Eratosthenes' birth to 276 B.C., a comparable forward dating of Archimedes' birth by some ten or more years might well be indicated.

[3] See the argument in my "Archimedes and the Elements" and "Archimedes and the pre-Euclidean Proportion Theory" (abbreviated as "AE" and "APPT", respectively, in the notes which follow).

[4] No ancient testimonia and only a single fragment on the balance extant in Arabic testify to any Euclidean involvement in mechanics. (For a translation and discussion of this fragment, see M. Clagett, *Science of Mechanics in the Middle Ages*, pp. 9 f, 24–30; and my *Ancient Sources of the Medieval Tradition of Mechanics*, Ch. VII.) The fragment is certainly based on a Greek original; but the specific attribution to Euclid is questionable.

[5] See Archimedes' *Sand-Reckoner*, I.9 and my "APPT," pp. 221 f.

[6] For a more detailed discussion of the Eudoxean background to Archimedes' work, see my "AE."

[7] On the "mechanical method," see *ibid.*, and also Dijksterhuis, *Archimedes*, pp. 318–322. On the lost *Equilibria*, see my "Archimedes' Lost Treatise on the Centers of Gravity of Solids."

[8] For general surveys of Archimedes' work, including coverage of the materials to be omitted here, see Heath; Dijksterhuis; and Schneider.

[9] See my "AE" and note 63 below. The *Dimension of the Circle* is discussed by Heath; Dijksterhuis; and Schneider; and its text is reprinted with translation by I. Thomas, *Greek Mathematical Works*, I, pp. 316–333.

[10] Heiberg, *Quaestiones Archimedeae*, 1879, pp. 10–12; Heath, *History of Greek Mathematics*, II, p. 22. (See my "AE," pp. 212 ff).

[11] For a survey of Archimedes' limiting methods, see Dijksterhuis, *op. cit.*, pp. 130–133. Note, however, that he misconstrues the method used in the *Dimension of the Circle* as a form of simultaneous two-sided convergence (see my "AE," pp. 219 f).

[12] Note that in the *Dimension of the Circle* as extant, this rule is expressed in the form of a *right triangle* whose legs equal the circumference and the radius of the circle. Proclus (5th century A.D.) knows the rule in this form, but earlier writers like Hero, Pappus, and Theon cite it in the more general form we have used in the present discussion.

[13] The successive bisection is characteristic of the Eudoxean–Euclidean mode. By contrast, Archimedes resorts to a more general form based on the "Archimedean axiom" in *Sphere and Cylinder* I and other places; cf. my "AE," Sect. III.2.

[14] On alternative definitions of "straight line" see Heath, *Euclid's Elements*, I, pp. 165–169. While I cannot cite a statement of the least-distance aspect of the line before Archimedes, a related property of the circle is asserted by Aristotle in *de Caelo* II, 4: "Of lines starting from a point and returning to the same point the circle is the shortest" (cf. Heath, *Mathematics in Aristotle*, pp. 171 f). Of course, this principle is false; any inscribed polygonal arc will be shorter than the circle, for instance, as is essential for the Archimedean theorem on the circle. But Aristotle might have misconstrued a correct theorem pertaining to lines "containing a given area." When Aristotle elsewhere observes that the physician realizes that round wounds heal slowest, but that the geometer knows why (*Analytica Posteriora* II, 13, 79 a 15–16), he must refer to the same isoperimetric property. Simplicius assumes this sense in his remarks on the passage cited above from *de Caelo* (see *ad loc.* and below, Chapter 6, note 170). We thus have indications of an early interest in the study of isoperimetric figures, whose later proofs by Zenodorus (late 3rd century B.C.) reveal a strong Archimedean influence (see my "AE," Sect. II.6). At any rate, we may admit that certain notions of the relative magnitude of lines and circles were taken as obvious in the pre-Euclidean period, even without their explicit statement as postulates. It is quite wrong, then, to suppose that one *couldn't* prove the theorem in the *Dimension of the Circle* before Archimedes *for want of* the postulates asserted in *Sphere and Cylinder* I.

[15] Cf. the Epicureans cited by Proclus (*In Euclidem*, p. 322) who charge that not even an ass would take the long way around to reach a stack of hay.

[16] Archimedes also posits the analogues for curved surfaces having the same extremities.

[17] We cannot suppose, for instance, that Euclid's Book XII represents the sum total of the Eudoxean geometry, or even that on certain points of technical detail, the extant versions reproduce the earlier methods. Euclid's avoidance of circumscribed figures in the limiting theorems stems not from any formal difficulty here, but rather from the desire to abridge the proofs. It is thus quite possible that he or some prior editor has modified a version in which such figures were used, and which might have served as a precedent for Archimedes' procedure in the *Dimension of the Circle* (see Chapter 3, note 80).

[18] *Collection* V, 12; see also the *Commentary on Ptolemy's Syntaxis* VI, 7 (ed. Rome, pp. 254–260). These are discussed in my "APPT." In the *Collection* these theorems relate to the investigation of isoperimetric segments of circles, a supplement to the results of Zenodorus (see note 14 above).

[19] Cf. Dijksterhuis, *op. cit.*, p. 222 and my "Archimedes and the Measurement of the Circle," p. 133. (The latter abbreviated "AMC" below.)

[20] The Renaissance computers extended Archimedes' method to many decimal places by means of the two-sided method; for references, see Hobson, "Squaring the Circle," pp. 24–27 and my "AMC," p. 118n. One may show that the difference between the circumscribed and inscribed polygons is reduced to just less than its fourth after each doubling of the number of sides. Thus, one need only calculate each circumscribed figure, say, to rather greater than the desired accuracy and then terminate the computation when the difference between the bounding figures is known from successive quadrisection to fall below this same amount. From the computed upper bound, one can thus obtain a lower bound by subtracting this difference.

[21] For a survey of attempts see Heath, *Archimedes*, pp. lxxx–iv, xc–ix; Dijksterhuis, *op. cit.*, pp. 234–238; and Schneider, *op. cit.*, pp. 145 f, 155. I present an alternative method in "AMC," pp. 136–139. Another based on the continued fraction for $\sqrt{27}$ has been proposed by D. T. Whiteside and reported by D. Fowler, "Ratio in Early Greek Mathematics," *Bulletin of the American Mathematical Society*, 1979, p. 844n.

[22] On the role of the continued fraction, in the form of the Euclidean division, within early Greek geometry, see my *Evolution of the Euclidean Elements*, Ch. VIII, Sect. 2. For a more ambitious view of its significance, see the paper by D. Fowler cited in note 21.

[23] The smaller denominator of 66 has resulted from reductions of the terms of fractional values at intermediate steps of the computation.

[24] As $3\ {}^{10}\!/_{71} = 3 + {}^{1}\!/_{7} + {}^{1}\!/_{10})$, in agreement with the first three quotients of the continued fraction for Archimedes' computed lower bound here, one must surely understand his use of a form of this procedure for simplifying fractional values.

[25] One may consult the sections on the measurement of circles and other curvilinear figures in Hero's *Metrica* I, II and the derived metrical collections edited by Heiberg as the *Geometrica* and *Stereometrica* (Hero, *Opera*, IV and V, respectively).

[26] Hero, *Metrica* I, 26; Eutocius' commentary on *Dimension of the Circle*, in Archimedes, *Opera*, ed. Heiberg, III, p. 258. These are discussed in my "AMC."

[27] In particular, the efforts of Tannery, Heiberg, Hoppe, and Bruins are discussed in my "AMC."

[28] More precisely, it exceeds by less than one part in 74,000 of π.

[29] The emended figure falls short by less than one part in 48,000 of π.

[30] This observation undermines Tannery's wish to see 355 : 113 as the approximation implied in the Heronian figures. On this, as well as the Chinese and Renaissance geometers' recognition of this value, see my "AMC," p. 118n.

[31] Here 21,600 is the number of minutes of arc in the circumference; see Toomer, "The Chord Table of Hipparchus and the Early History of Greek Trigonometry," *Centaurus*, 1973, 18, pp. 6–28.

[32] *Syntaxis*, VI, 7.

[33] Note that chord $x = D \sin \frac{1}{2} x$, for $D = 120^p$ (that is, 120 "parts" constitute the whole diameter).

[34] Theon, *Commentaires...sur l'Almageste*, ed. Rome, II, pp. 492–495. An extended table of chords was used by the 11th-century astronomer al-Bīrūnī to recompute π. Taking the value for chord 2°, he derived from the circumscribed and inscribed 180-gons the respective bounds $3^p8'30''59'''10^{iv}$ and $3^p8'29''35'''24^{iv}$, and so took their arithmetic mean $3^p8'30''17'''16^{iv}46^v30^{vi}$ as his value for π (note that the bounds were stated in truncated form). Thus, despite his extra computational effort, al-Bīrūnī has obtained a value less accurate than Ptolemy's $3^p8'30''$. One perceives the difficulties inherent in a computational procedure uncontrolled by careful theoretical concerns. (For al-Bīrūnī's text in translation, see C. Schoy, *Die trigonometrischen Lehren des...al-Bīrūnī*, 1927, Ch. V.)

[35] Cf. Eutocius in *Archimedes*, ed. Heiberg, III, p. 230. A compromise view combining practical and neo-Pythagorean elements has recently been advocated by Schneider, *op. cit.*, pp. 139f.

[36] Cf. O. Neugebauer, *Exact Sciences in Antiquity*, 1957, pp. 46 f.

[37] See Chapter 7.

[38] See Eutocius, *op. cit.*, p. 230. The issue provoked comment in the Middle Ages, as in the Cambridge and Corpus Christi manuscripts of the *Quadratura circuli* and the discussions by ps.-Bradwardine and Albert of Saxony; cf. Clagett, *Archimedes in the Middle Ages*, I, pp. 68, 170, 382 ff, 414 ff.

[39] *Metrica* I, 37; see, also, Pappus, *Commentaires sur l'Almageste*, ed. Rome, I, pp. 254–260. Another rule, directed toward the measurement of circular segments, may also have an Archimedean origin; see below, note 63.

[40] Props. 33 and 42f, respectively.

[41] On the ellipse, see Props. 5, 6; but the statement here is in the form cited by Hero, *Metrica*, I, 34. On the volumes of the solids, see *Conoids*, Props. 21–32.

[42] For the text, see *Archimedes*, ed. Heiberg, II, pp. 510 ff. For discussion, see Heath, *Archimedes*, pp. xxxii f, 301–318; and Dijksterhuis, *op. cit.*, pp. 401–405. The Arabic tradition of this work, including a more accurate rendering of its preface, is surveyed by F. Sezgin, *Geschichte des arabischen Schrifttums*, V, pp. 131–133.

[43] The sense of the name has been debated; see Heiberg, *Archimedes*, II, p. 523n; and Heath, *Archimedes*, p. xxxiiin.

[44] More generally, $CP : PO : OA = a^2 : ab : b^2$, where $AD : DC = b : a$.

[45] Pappus, *Collection* (IV, 14–20), I, pp. 208–232. For discussion, see Heath, *Archimedes*, pp. 305–309; and *History*, II, pp. 371–377.

[46] For a complete exposition of the proof in *Spiral Lines*, see the editions by Heath and Dijksterhuis; and also Schneider, *op. cit.*, Ch. 4.2.

[47] *Collection* (IV, 2), I, p. 234. See my "Archimedes and the Spirals" (abbreviated "AS" below) for discussion of the passage. Scholars typically charge Pappus here for misreading the preface to *Spiral Lines*. To be sure, his testimony is *different* from that in the extant treatise, but not *incompatible* with it. I thus find no cause for dismissing Pappus' report, especially since it provides precious insight into the earlier phases of Archimedes' study of the spirals.

[48] *Collection* (IV, 35), I, pp. 264–268. See also "AS" cited in note 47.

[49] *Sphere and Cylinder* I, 42 f. The proportionality for sectors is proved by Pappus in a different context in *Collection*, V, 12 f, where an Archimedean provenance seems to be indicated (see note 18 above).

[50] *Sphere and Cylinder* I, 33. This is of course a special case of the theorem on segments (see note 49).

[51] In addition to the presentation of the full Archimedean syntheses in their editions of Archimedes, Heath (*History*, II, pp. 556–561) and Dijksterhuis (*Archimedes*, pp. 268–274) provide alternative versions closer to the form of an analysis. My own version is not far different from Dijksterhuis', although less closely wedded to modern differential considerations. Heath, on the other hand, errs in his formulation of the analysis and adopts an invalid definition of tangents; for a synopsis and critique of his attempt and of the similarly flawed analysis proposed by Tannery, see my "Archimedes' *Neusis*-Constructions in *Spiral Lines*" (abbreviated "ANCSL" below). Dijksterhuis shows how Archimedes' results for the tangent leads to a proof of the constancy of the subnormal of the spiral. If from B a line perpendicular to the tangent BE is drawn to meet E0 in K (Fig. 8), then 0B is the mean proportional to K0, 0E for all points B on the spiral.

Archimedes' theorem has established that OE is proportional both to OB and to the angle between the initial and terminal rays of the sector; and as these are proportional, OE is proportional to OB^2. Hence, KO is constant; i.e., $KO \cdot C = R^2$, for C the circumference of the circle whose radius R equals the terminal ray of the spiral. Note that this relation is directly analogous to the property of the quadratrix by which the latter is used for the circle quadrature; see Chapter 6.

[52] The manner of effecting this *neusis* is discussed below.

[53] Cf. Euclid's specification of the tangent of the circle in *Elements* III, 16, namely, that the angle between the tangent and the circle is less than any rectilinear angle, so that any line between the tangent and the circle must meet the circle at another point.

[54] For full formal discussions, see those cited in note 51.

[55] The kinematic conception of the tangent was exploited by G. de Roberval in his *Observations sur la composition des mouvemens et sur le moyen de trouver les touchantes des lignes courbes* (1668); reprinted in the *Mémoires de l'Académie Royale des Sciences*, 6, 1730. Heath expresses the commonly held view when he remarks, "I cannot but think that Archimedes divined the result by an argument corresponding to our use of the differential calculus for determining tangents" (*History*, II, p. 557). But this is surely an unsupported transference of our modern intuitions back to the ancient studies. See "AS," p. 73.

[56] *Collection* IV, 34. See "AS," pp. 63–65.

[57] On the Euclidean background of "surface loci" see Chapter 4.

[58] On Apollonius, see Proclus, *In Euclidem*, ed. Friedlein, p. 105; and Ch. 7 below. The Archimedean invention is discussed by Dijksterhuis, *op. cit.*, pp. 21–23, who doubts the validity of its attribution to Archimedes; and by A. Drachmann, *Mechanical Technology of Greek and Roman Antiquity*, 1963, pp. 150–154, who advocates its authenticity.

[59] For the conception of the cylinder and the cone as deformations of plane figures, see Hero, *Metrica*, I, 36 f.

[60] See the summary by Iamblichus, reported by Simplicius, *In Aristotelis Physica*, ed. Diels, I, p. 60 (reprinted by I. Thomas, *Greek Mathematical Works* I, pp. 334 f).

[61] *Opera*, ed. Heiberg, II, pp. 262–264.

[62] It is so understood by Heiberg (*ibid*, p. 263n) and by Heath (*Archimedes*, p. 233n). For an alternative possibility, see my "AE," p. 231n.

[63] Hero, *Metrica* I, 27–32. On the Archimedean provenance of this rule, see my "AE," Sect. II.3; it is explicitly assigned to Archimedes in an ancient scholium ("AE," pp. 289 f). In the context of the circle measurement, it leads at once to one of the convergence-improving formulas used by Huygens; cf. my "AMC," pp. 133–136. Hero presents a second rule for estimating circular segments: one takes one-half the sum of the base and the height of the segment and multiplies this by the height, and this yields the area (*Metrica*, I, 30). He observes that this assumes the circumference is three times the diameter (i.e., $\pi = 3$), as one infers by applying the rule to the semicircle (where $b = 2h = d$, the diameter). Accordingly, he says, some added $\frac{1}{14} (\frac{1}{2} b)^2$ to make this crude estimate more accurate, in conformity with the Archimedean value of $3 \frac{1}{7}$ (*ibid.*, I, 31). He adds that one should use this rule when b is less than three times h, but the former rule when b is greater. He does not explain this criterion, but one can see how it results from considering where the two rules yield the same result, namely, when $2bh/3 = \frac{1}{2} h (b + h)$, whence $b = 3h$. Now, the crude "ancient" rule given by Hero is

likely to be very old, doubtless of pre-Greek origin; for it is found in demotic (Egyptian) papyri from the Hellenistic period among metrical and computational materials typifying the older Egyptian and Mesopotamian traditions (cf. R. A. Parker, *Demotic Mathematical Papyri*; and my "Techniques of Fractions," Sect. V). The other rule, by virtue of its association with that for the parabolic segment, suggests an Archimedean origin. One suspects that the rather sophisticated effort reported by Hero to assess the relative utility of these two rules for the circular segments is also due to an Archimedean insight.

[64] Hero notes that the analogy between these results for parabolic and circular segments (*Metrica* I, 32), but cites the *Method* for the statement of Archimedes' theorem. One should observe that the *Quadrature of the Parabola* presents two quite different treatments of the theorem on the area of the segment, one relying on mechanical principles (Props. 6–17), the other purely geometric (Props. 18–24) in the manner of the proofs of the area of the circle and its segments. The consequences for dating Archimedes' early work are proposed in my "AE," pp. 248–250.

[65] On the chronology of Archimedes' works, see my "AE." The principal references to Conon appear in the prefaces to *Quadrature of the Parabola* and *Spirals*.

[66] *Opera*, ed. Heiberg, II, p. 428.

[67] *Sphere and Cylinder* I, Props. 42–44.

[68] *Method*, comment after Prop. 2. See my "AE," pp. 220 f on the consequences for the dating of *Dimension of the Circle*.

[69] *Opera*, ed. Heiberg, I, p. 192.

[70] Eutocius, in *Archimedes*, ed. Heiberg, III, pp. 130–132.

[71] See Diocles, *On Burning Mirrors*, ed. Toomer, Prop. 7 f. Toomer compares this text with the version given by Eutocius, *ibid.*, pp. 178-193. Eutocius' text of Diocles' solution appears in *Archimedes*, III, pp. 160-174; of Dionysodorus' solution, pp. 152-160; and of Archimedes' solution, pp. 132-152. These methods are summarized by Heath, *Archimedes*, pp. 66-79; and Dijksterhuis, *Archimedes*, pp. 195-205.

[72] Questions concerning the originality and the textual methods of Eutocius and other late commentators are taken up in the sequel to the present work.

[73] For Apollonius' study of tangents to the conics, see Chapter 7.

[74] In the context of the original problem, this places X at B, so that the smaller segment has vanished.

[75] See Chapter 7.

[76] The *Algebra* of Omar (11th century) has been edited with French translation by F. Woepcke (1851). Omar distinguishes six forms of irreducible cubics in three terms and seven forms in four terms (ed. Woepcke, pp. 32 ff). (For an account, see A. P. Youschkevitch, *Les Mathématiques arabes*, 1976, Ch. 14.) In each instance, the solutions are found via intersecting conics. For the basic form $x^3 = N$, he uses the Menaechmean solution (Woepcke, pp. 28 f).

[77] *Archimedes*, ed. Heiberg, III, p. 130.

[78] The dotted lines in Fig. 16 show the curves used by Archimedes.

[79] In Fig. 17, I continue the lettering introduced in the figures for the solutions by Archimedes and Dionysodorus (designated by capital letters); elements of the diagram which do not appear in those figures are marked with lower-case letters.

[80] Note that this is the relation which defines the conic in the earlier stages of the

theory, e.g. with Euclid and Archimedes (see Chapter 4). Toomer remarks on Diocles' adherence to pre-Apollonian terminology and forms in his *Diocles*, pp. 9–15, 166 f.

[81] *Collection* (IV, 36), I, pp. 270–272. For discussion, see my "ANCSL," especially pp. 89–91.

[82] The manner of drawing the normals to the parabola (*Conics* V, 51) is intended; see Chapter 7.

[83] The versions proposed by Tannery and by Heath appeal to assumptions of continuity; my own utilizes continued bisection, but requires a certain restructuring of Archimedes' proofs; for details, see my "ANCSL."

[84] *Collection* IV, 52; the construction is carried through in IV, 52–54. See, also, Hultsch's appendix (*Collection*, III, pp. 1231–33) for the restoration proposed by R. Baltzer.

[85] The lettering is compatible with that in Fig. 8.

[86] The curves will also intersect at the points marked K'. If we mark the abscissa K'H' and draw BH' to meet the circle at P', then the length P'H' will also equal the given length θ. In this way we effect the *neusis* required in the second part of the theorem on the tangents, relating to the rearward portion of the spiral. Details have been omitted here, but may be found in my "ANCSL."

[87] Alternatively, one might suppose that Pappus' report refers to a different Archimedean construction where actual solids were introduced, for instance, in the manner of surface loci as discussed above. But this would not involve *neuses* as such, nor would the construction be "solid" in the later sense. Thus, this alternative view would still entail a confusion on Pappus' part. On the relation of this construction via conics to later studies of "solid loci," see Chapters 7 and 8.

[88] *Collection* (IV, 54), I, p. 302.

[89] It seems that Pappus' information here may derive from Sporus of Nicaea, a commentator of the 2nd century A.D. For Pappus reproduces Sporus' similar criticisms of the use of the quadratrix for circle quadrature; see Chapter 6, note 65.

[90] For translation and commentary, see C. Schoy, *Die trigonometrischen Lehren des...al-Bīrūnī*, 1927, pp. 74–84; and J. Tropfke, "Die Siebeneckhandlung des Archimedes," *Osiris*, 1936, 1, pp. 636–651. On the construction of the heptagon by Archimedes and several related Arabic constructions, see also Schoy, "Graeco-Arabische Studien," *Isis*, 1926, 8, pp. 21–40. The Arabic text of Archimedes' construction is edited with translation by J. P. Hogendijk, *Greek and Arabic Constructions of the Regular Heptagon*, pp. 122–127.

[91] A comparable analysis has been proposed in Chapter 3 for the problem of inscribing the regular pentagon.

[92] Analogously, the problem of the pentagon reduces to dividing a segment BE at F such that $BF^2 = FE \cdot BE$; an analysis for the latter has been given in Chapter 3.

[93] We surely concur with Tropfke when he remarks that this *neusis* is "a brilliant stroke of insight (*glänzender Einfall*), worthy of admiration (*Bewunderung*)"; but we need not follow him when he further observes that "its origin is unfortunately beyond tracking down" (*op. cit.*, p. 651). Admiration is not incompatible with understanding.

[94] The term "geometric algebra" for describing the type of quantitative identities developed geometrically, as in Euclid's Book II, in the ancient techniques of application of areas, and in the second-order relations for the conics, was introduced, I believe, by

H. Zeuthen (see his *Geschichte der Mathematik im Altertum und Mittelalter*, 1896, Ch. 4 and *Lehre von den Kegelschnitten*, 1886, Ch. 1). It has generally been adopted by historians since then; cf. Heath, *History*, I, pp. 150–154. Misgivings over this view, however, induced Dijksterhuis to institute special notations in his editions of Euclid and Archimedes (e.g., not "AB^2", but "T(AB)" for "the square (*tetragōnon*) on line AB"). Recently, the issue has flared into controversy, centering around the strong restatement of the case against "geometric algebra" by S. Unguru; for a review of the fray, with references to his own papers and the responses to them, see "History of Ancient Mathematics: Some Reflections on the State of the Art," *Isis*, 1979, 70, pp. 555–565. Certainly, the ancient geometry never had access to the special advantages afforded by algebraic notations and conceptions in the full modern sense. On the other hand, one often finds that the ancients introduced geometric diagrams whose purpose was merely to represent given quantitative relations, to provide a vehicle for the investigation of these relations with the intent to apply the results to other contexts. This is precisely what we see in the matter of relations (a) and (b), whose investigation via *neusis* (Figs. 20a and b) is auxiliary to their role within the inscription of the heptagon (Fig. 19). The treatments of the problem relating to *Sphere and Cylinder* II, 4 and *Spiral Lines* are further illustrations. In these instances, the Greek geometers have perceived the structural relations which unite ostensibly unrelated diagrams. A principal virtue of the algebraic manner is its abstraction of such quantitative relations for separate examination, so that results may be transferred from one context to another. Of course, it is in the strict sense fallacious to argue that these ancient techniques *are* a form of algebra because they serve the same function of separating quantitative relations from their special contexts. Nevertheless, the term "geometric algebra" can be useful in alerting us to the fact that, in these instances, diagrams fulfil this function; they are not of intrinsic geometric interest here, but serve only as auxiliaries to other propositions.

[95] See the preface to the *Inscription*, given by Schoy, *al-Biruni*, p. 74; cf. Tropfke, *op. cit.*, pp. 636, 650.

[96] His tract "On the Construction of the Heptagon..." is edited by J. P. Hogendijk in *Greek and Arabic Constructions of the Regular Heptagon*, University of Leiden, Dept. of Math. Preprint No. 236, April 1982. On the life and work of al-Sijzī, one may consult F. Sezgin, *Geschichte des arabischen Schrifttums*, V, pp. 329–334 and W. Thomson, *Pappus' Commentary on Book X...*, pp. 46–51, in addition to Hogendijk, *op. cit.*, Sect. 5.1

[97] Note that the pattern of this figure can be extended to yield an isosceles triangle of angles a, na, na, for any odd n. Figure 29 illustrates Abu 'l-Jūd's probable method for constructing the regular 11-gon (cf. Hogendijk, *op. cit.*, p. 88).

[98] Note that the drawing of WE is actually superfluous; for one has WB : BH = DB : BZ, and this will serve for the derivation of (ii) which follows.

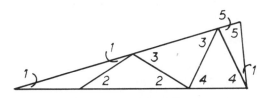

Figure 29

[99] Hogendijk, *op. cit.*, pp. 130–158: Arabic text with translation and notes. Cf. note 104 below.

[100] Cf. Hogendijk, *ibid.*, pp. 82–85.

[101] Cf. *ibid.*, pp. 150 f; note that the lettering is altered from that of Fig. 21(b).

[102] *Ibid.*, p. 24.

[103] Several such constructions are reviewed by Hogendijk, *ibid.*, Sect 4. Versions also appear in Schoy, *al-Bīrūnī*, p. 91; and "Graeco-Arabische Studien," p. 40. Cf also references in Sezgin, *op. cit.*, p. 134.

[104] Alhazen's text is translated by Schoy, *al-Bīrūnī*, pp. 84–91; see also Woepcke, *op. cit.*, pp. 91–93. The latter also presents the version of al-Qūhī (ibid., pp. 103–114). Schoy presents the alternative construction by al-Sijzī in "Graeco-Arabische Studien," pp. 21–27.

[105] I here adopt the lettering of Heath (*Archimedes*, p. 309).

[106] See, for instance, the survey of angle trisections written by al-Sijzī (10th century), summarized by Woepcke, *op. cit.*, pp. 117–125. The "ancient method by means of ruler and mobile geometry" is the eighth of the nine methods given there (*ibid.*, p. 120) and corresponds to the *neusis* implied in the *Lemmas*. The same is basic for the variations on the angle trisection given by al-Bīrūnī; see Schoy, *al-Biruni*, Ch. IV. Other Arabic angle trisections are discussed in Chapter 7(ii) below..

[106a] In view of this reconstruction, it is interesting to note a lemma given by Pappus to Aristarchus' *On the Sizes and Distances of Sun and Moon* (*Collection* (VI), II, pp. 560–562) where the same figure is at issue, save that CB is not of specified length and only the fact that arc AE is greater than arc BD is asserted.

[107] The text of its medieval Latin translation (by Gerard of Cremona, 12th century) has been edited with English translation by M. Clagett, *Archimedes in the Middle Ages*, I, Ch. 4; the angle trisection is Prop. XVIII (*ibid*, pp. 344–348). The Arabic is extant only in the 13th-century edition by al-Ṭūsī; it has been published in the Osmania University series of texts (Hyderabad, 1939–1940), and a translation based on the Arabic has been made by H. Suter (*Bibliotheca Mathematica*, 1902, 3, pp. 259-272).

[108] *Collection* (IV, 36–42), I, pp. 272–280; see Chapter 7.

[109] See Chapter 6.

[110] Eutocius, III, pp. 58-60; Pappus, *Collection* (III), I, pp. 62-64. These and the other texts of the Heronian method will be discussed in the sequel.

[111] One may consult Chapter 6 for references to Eutocius' and Pappus' accounts of Nicomedes' construction.

[112] On the Archimedean sources of Hero's *Mechanics*, see Drachmann, "Fragments from Archimedes in Heron's Mechanics," *Centaurus*, 1963, 8, pp. 91–146; and my *Ancient Sources of the Medieval Tradition of Mechanics*, Part II. Schöne's index to the *Metrica* (*Opera*, III, p. 316) reveals at once the preponderance of citations of Archimedes; Hero includes specific references to *Dimension of the Circle*, *Sphere and Cylinder* I and II, *Conoids* and the *Method*. Otherwise, there is only a single reference to Dionysodorus (*Metrica* II, 13), to an anonymous "Book of Chords" (for the areas of enneagon and hendekagon in I, 22 and 24). The earlier portions appear to rely on unnamed collections of elementary results, while a problem on the division of the circle (III, 18) may be related to a proposition in Euclid's *Division of Figures* (see Heath, *History*, I, pp. 429

f; and II, pp. 339 f). As for other works, Hero cites Eratosthenes for a measure of the circumference of the earth (*Dioptra*, Sect. 35) and Posidonius (the Stoic philosopher) for an incorrect conception of centers of gravity (*Mechanics* I, 24: cf. *Archimedes*, ed. Heiberg, II, pp. 545 ff). Hero's dependence on sources is stated explicitly in the preface to the *Pneumatics*, and has been clarified by Drachmann (*Ktesibios, Philon and Heron*, 1948, pp. 80–84). His special debt to Ktesibius and Philo is evident in the *Belopoeica*; cf. E. W. Marsden, *Greek and Roman Artillery: Technical Treatises*, p. 2 and Chs. II, IV.

[113] On the rules for triangles, see al-Bīrūnī, translated by H. Suter, *Bibliotheca Mathematica*, 1910-11, 11, pp. 39 f. On the rule for square roots, see note 127 below; and on the rule for segments, note 63 above.

[114] *Metrica*, I, 26 and II, 20. The latter bears comparison with the familiar account preserved by Vitruvius of Archimedes' procedure for specific weights and of how he sprang naked from the bath into the streets, crying "*heurēka*" in the excitement of the discovery (*Architectura* IX, Pref. 9–12).

[115] This will be given in the sequel volume.

[116] This lemma is assumed without proof in the demonstration by Hero; the version in Pappus, however, supplies the proof.

[117] See Chapter 6.

[118] On Eratosthenes' account, see Chapter 2.

[119] The construction in *Mechanics* I, 11 supplements the rule for multiplying solids in I, 10.

[120] See Marsden, *op. cit.*, pp. 38 ff. Hero, without doubt following the account by Philo (*ibid.*, pp. 108 ff), introduces the rule $D = 1.1 \sqrt[3]{100W}$ for finding the diameter D of the spring opening of the catapult (to which all other dimensions of the machine have been scaled), where the weight W of the stone is specified; here the units are the *daktyl* ($= \frac{3}{4}$ in., approx.) and the *mina* ($=$ 1 lb., approx.; 100 *drachmas* $=$ 1 *mina*) for D and W, respectively (*ibid.*, p. xvii). For a view of its derivation and significance, see W. Soedel and V. Foley, "Ancient Catapults," *Scientific American*, March 1979, 240, pp. 150–160. The ancient texts give no derivation; but I would suppose that it rests merely on the proportionality of the volume (or weight) of the machine to that of the projectile.

[121] Eutocius reports that Nicomedes "severely ridiculed Eratosthenes' solution"; *op. cit.*, III, p. 98. See Chapter 6.

[122] See Dijksterhuis, *op. cit.*, Ch. 1 and Schneider, *op. cit.*, Ch. 1 for a summary of the reports from Polybius, Livy, Plutarch, and others.

[123] *Metrica* III, 20; reprinted by I. Thomas, *op. cit.*, I, pp. 60–63 and discussed by Heath, *History*, II, pp. 341 f. See also O. Becker, *Das mathematische Denken der Antike*, pp. 69–71; and E. Bruins, *Codex Constantinopolitanus*, III, pp. 336 f.

[124] Lit.: "and 100." Becker (*loc. cit.*) has recognized an omission here and suggests this emendation: "[multiply 4 into 25, result: 100. Add 180] and 100. . ." This is also required for the mathematical sense of the rule, as explained below.

[125] See, e.g., the account by Thomas, *loc. cit.*

[126] For this derivation, see Heath, Becker and Bruins, as cited in note 123.

[127] *Metrica* I, 8: if a is taken as the initial approximation to $\sqrt{N}$, then $\frac{1}{2}(a + N/a)$ is formed as a closer approximation. Hero applies this rule for $N = 720$, $a = 27$, to obtain the approximation 26 ½ ⅓. He notes, moreover, that this value in its turn may be used for obtaining a closer value, if desired.

[128] Note that if one adopts the linear approximation, i.e., $d : e = a + b - x : x - a$, one obtains the computational rule associated with a recursive geometric construction proposed by a contemporary of Pappus and criticized severely by the latter in *Collection* (III), I, pp. 32 ff. For an account of this method, see Heath, *History*, I, pp. 268–270; Steele, "Zirkel und Lineal," p. 356. Its relation to Pappus' views on the classes of problems and solving methods is considered in Chapter 8.

[129] The Egyptian rule is held in the *Moscow Papyrus* from the 2nd millennium B.C. See B. L. van der Waerden, *Science Awakening*, pp. 34 f. The related rules for truncated pyramids and cones appear in Hero's *Metrica* II, 6 and 9; cf. Heath, *History*, II, pp. 332–334.

[130] Indeed, Hero's rule for square roots is comparable in its replacement of two approximate terms by an intermediate term, namely, their arithmetic mean.

[131] On the Newton–Raphson rule, see E. T. Whittaker and G. Robinson, *Calculus of Observations*, 4th. ed., 1944, Sect. 44.

[132] Here we can derive two different bounds, one from d and one from e.

[133] On Tablet VAT 6598 the diagonal of the rectangle of sides 40 and 10 is found to be $40 + \dfrac{10^2}{2 \cdot 40}$, or 41 ¼ (that is 41; 15 in the sexagesimal notation); see O. Neugebauer, *Vorgriechische Mathematik*, pp. 33–38.

[134] A detailed analysis of examples of roots both from Hero and from Archimedes is given by J. E. Hofmann, "Über die Annäherung von Quadratwurzeln bei Archimedes und Heron," *Jahresbericht der Deutschen Mathematiker-Vereinigung*, 1934, 43, pp. 187–210. For a survey, see Dijksterhuis, *op. cit.*, pp. 229–234.

[135] See the collections cited in note 25 above.

[136] Cf. *Conics* I, 33 f (tangents) and II, 14 (asymptotes); the former are discussed in Chapter 7. Apollonius' treatment of the tangents is in the manner of Euclid (cf. note 53 above).

[137] *Collection* IV, 21–25 and 35 (spirals); V, 20–43 (sphere measurement).

[138] Pappus, *Collection* (VII), II, pp. 680–682. On this method and its use, see Archimedes' *Method*.

[139] Hero, *Mechanics* (ed. Nix); Pappus, *Collection* VIII, especially 1–12. For a survey of the Archimedean mechanical tradition, see my *Ancient Sources of the Medieval Tradition of Mechanics*, Ch. VIII.

[140] Cf. also Chapter 8 (introduction). I believe these shifts of interest in the development of the ancient mathematical sciences might provide the basis for parallels with more recent episodes of scientific change. In the aftermath of T. Kuhn's *Structure of Scientific Revolutions* (Chicago, 1962), debates over the nature of scientific change have centered on such revolutionary episodes as "foundations crises," thus losing sight of those less striking shifts which might nevertheless prove more characteristic of the development of the mathematical field.

CHAPTER 6

The Successors of Archimedes in the Third Century

If any period of the ancient geometry can be called a "golden age" of interest in the three classical problems, it would be the latter part of the 3rd century B.C. For these problems form the principal unifying motif in the surviving fragments from the work of a series of geometers from this time: Eratosthenes, Nicomedes, Hippias, Diocles, Dionysodorus, Perseus, and Zenodorus. Doubtless, this impression is to a major extent an artefact of the selective preservation of evidence, for our chief sources are the compilations on cube duplication and other problems in Eutocius and Pappus. There can be little doubt, then, that the expertise displayed in these fragments extended into other fields of geometry and mathematical science as well, and in a few instances our insights in this direction are reasonably good. In particular, we can detect serious efforts in the application of geometry to fields like astronomy, geography, optics, and mechanics, and a corresponding interest in exploring the use of mechanical methods in geometry. This granted, we may still admit the three problems as the hallmark of this period. Erastosthenes and Diocles produced solutions to the cube duplication; Nicomedes studied all three problems; and the geometric methods characteristic of this period find their representative applications in the constructions of these problems. Thus, a survey of the development of the study of these problems must view the work of this group of geometers as a significant phase.

They may be called Archimedes' "successors" only in a special and subtle sense. Only Eratosthenes is definitely known to have had personal contact with Archimedes, and, at that, rather as a colleague than as a disciple. Of the others only Diocles is known for certain to have been active in the 3rd century, although the nature of the work of Nicomedes and Dionysodorus makes a later dating for them quite unlikely. Thus, standard accounts sometimes present them as a 2nd-century footnote to the labors of Archimedes and Apollonius.[1] But as we shall see in this and the next chapter, such a view radically distorts the character of their work. No direct influence by Apollonius is apparent, and where there is overlap in the content and technique of research, one does better to perceive Apollonius as the consolidator of the findings of these geometers than to suppose an influence in the opposite direction. By contrast, the influence from Archimedes is profound. His adoption of mechanical modes in geometry—for instance, the use of motion-generated curves like the spirals, or the application of *neuses*—is prominent in the work of all these geometers. Their interest in the scientific applications of geometry can hardly but be considered an extension of Archimedes' fruitful investigations into geometric mechanics. Even more directly, Archimedes' solutions to specific problems formed the precedents for efforts of these geometers in their search for alternative solutions, that is, in their view, for more satisfactory ones.

This striking affiliation with particular parts of Archimedes' work is the strongest justification for considering these geometers as his successors. As we have already observed, their participation in the field of Archimedes' chief interest, the application of Eudoxean limiting methods, was slight almost to the point of nil. But in developing several of his problem-solving efforts, they prepared the field for the synthesis effected by Apollonius only a few decades later.

ERATOSTHENES

Appointed Librarian at Alexandria some time after the middle of the 3rd century, Eratosthenes of Cyrene was noted for the wide range of his literary, philosophical, and scientific achievements in a career which extended beyond the end of that century.[2] His major work in science lay in the areas of astronomy and geography. From his position at the intellectual center of a thriving Hellenistic kingdom in Egypt, he was able to compile and systematize a wealth of geographical data, a major precedent for the treatises of Strabo and Ptolemy. His best-known contribution to mathematical science arose in connection with this work: the estimate of the circumference of the earth.[3] Assuming that the noon-time position of the sun was directly overhead at Syene (modern Aswan)[4] in Upper Egypt on a day when the sun at noon was inclined one-fiftieth of the full circle from vertical at Alexandria, Eratosthenes deduced from the 5000-stade distance between the two sites that the whole earth was $250,000 (= 50 \times 5000)$ stades around. The figure is alternatively cited by Hero as 252,000 stades; while yet another estimate based on the distance over water between Alexandria and Rhodes works out to $180,000 (= 48 \times 3750)$ stades. The latter values yield estimates of 700 and 500 stades, respectively, for the distance of 1° of arc along a great circle on the earth,[5] and so make evident that Eratosthenes' intent is to derive convenient order-of-magnitude estimates for use in geographical calculations, rather than a precise value. In this, he stands in a tradition seeking measures of the earth and cosmos which dated from at least a century before him and included contributions by Aristarchus and Archimedes.[6]

His competence in mathematics is attested to by Archimedes' favorable notice of his potential as a researcher in the preface to the *Method*. But our knowledge of his work is scant. His efforts in number theory include some study of proportions, the conception of the "sieve" (*koskinon*) for the determination of primes, and perhaps an involvement in the problem Archimedes poses in the *Cattle Problem* addressed to him. We have better information on his geometry. His account of the history of the cube-duplication problem, to which we have already referred, presents a mechanical solution of his own invention; Pappus also cites his analytic treatment of certain locus problems. These efforts thus assign to Eratosthenes a role in the tradition of geometric problems which is our principal concern here.[7]

Eratosthenes proposes to solve the cube duplication through use of an instrument, the *mesolabos* or "taker of means."[8] His presentation of the geometric construction which underlies it takes the form of a synthetic demonstration. But for our purposes, it will be useful to consider the analysis which it implies. Let it be required to find two mean proportional lines BZ, GH between two given lines AE, DΘ [Fig. 1(a)]. We suppose the lines to be constructed and set as parallels in succession between two lines AD, EΘ inclined toward each other and meeting, when extended, at K. Now let the diagonals AZ, BH, GΘ be drawn; it follows that they are parallel to each other. For since AE : BZ = BZ : GH, AK : BK = ZK : HK ; hence, triangles AZK and BHK are similar, so that AZ, BH are parallel. Similarly, BH and GΘ are parallel. This indicates a construction of the means in the following manner [Fig. 1(b)]: take three rectangular plates, each identical to that with diagonal AZ, and let side AE of the first be one of the given lines and the length DΘ marked off the opposite side of the third be the other given line. These plates are now to be slid, the third behind the second and the second behind the first, until the points B and G, where the second and third diagonals meet the occluding edge of the first and second plates, respectively, fall in a line with points A and D. When this is effected, lines BZ and GH will be the desired mean proportionals between AE and DΘ. Eratosthenes' proof is the synthesis on which the above analysis has been framed. In his

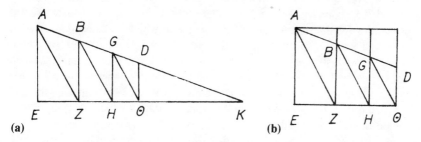

Figure 1

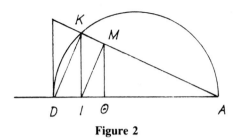 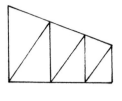

Figure 2   **Figure 3**

account (particularly, in the version reported by Eutocius) some details are provided on the form and materials suitable for the manufacture of the plates, the grooves by which their motion can be controlled, and so on. One presumes a guide rod somehow pivoting on A and D by which the collinearity of A, B, G, and D can be ascertained. In practice, a fair amount of trial and error would be required to achieve the desired terminal position. Eratosthenes observes that the utility of his device is not limited to the finding of two mean proportionals; with the insertion of additional plates, one may find an arbitrary number of means between the two given lines.[9]

From his sketch of the history of the study of this problem, we know that Eratosthenes was familiar with the solutions of Archytas, Eudoxus, and Menaechmus, and quite likely the pseudo-Platonic construction as well.[10] Of these, Archytas' method suggests a natural precedent for the approach adopted by Eratosthenes, in that both depend on the arrangement of a sequence of similar triangles in relation to the given lines. In Archytas' figure it is the lines AK, AI which are found to be means between the givens AD, AM (Fig. 2). But the recursive character of the figure is evident, and if one directs attention to the verticals MΘ, IK, the notion of an alternative figure readily presents itself, in which the only difference from the figure used by Eratosthenes is the reversed orientation of the diagonals (Fig. 3). While geometrically insignificant, this difference seems not to afford any evident mechanical advantage either, so that Eratosthenes' reason for altering the derived figure would be obscure. This suggests that Eratosthenes' design did not arise merely as a variant on Archytas' figure, although familiarity with the earlier method was surely an element in its origin.

The mechanical motif is prominent in this effort of Eratosthenes. Having explained the geometrical rationale of the construction, he goes on to present specifications for the manufacture of the device to which it corresponds.[11] As a model was actually constructed and dedicated in his votive monument, we would suppose that Eratosthenes was serious in his expectation that the invention would be put to practical use. He designates some half-dozen important areas suitable for its application, such as general commercial calculations with weights and volume measures, the production of ships and military engines in scale, and so on. Eratosthenes was chiefly a man of letters, and one suspects that his vision of the practicality of a sensitive special-purpose instrument like his ''mesolabe'' was rather overstated. Still, the ideology behind its invention seems genuine.

We recall that in the alternative version of the cube-duplication legend reported by Theon of Smyrna from Eratosthenes' *Platonicus*, Plato chides the Greeks for their neglect of geometric research.[12] As we have proposed earlier, Eratosthenes may well have used this writing as the vehicle for presenting the mechanical contrivance for cube duplication known to Eutocious under the name of Plato. The strong anti-mechanical position maintained by Plato in Plutarch's versions of the story thus cannot be faithful to Eratosthenes' own view.[13] Doubtless, Plutarch was closer to the spirit of Plato's own thought when he deplores a form of geometric research which spoils its ideal and eternal purity through contamination with the physical and temporal. But Eratosthenes is a man of another century, for whom the commercial and military potential of abstract ideas might matter a great deal. He shares this outlook with Archimedes and others of his generation. Certainly, for them geometry is strictly a theoretical discipline, of intrinsic interest. But one also appreciates its practical power and engages actively in the investigation of such applications.

It is important to realize, however, that this is an ideological position superimposed on the research activities of geometers, for the actual problems and methods which characterize their studies are hardly less theoretical than those representing the most formal part of the ancient geometric tradition, for instance, the work of Euclid and Apollonius. Indeed, among the corpus of works on analysis cited by Pappus in Book VII of the *Collection* there appears a two-volume investigation of *Loci with respect to Means* by Eratosthenes.[14] Its placement among such works as Euclid's *Data* and *Porisms*, Aristaeus' *Solid Loci*, and Apollonius' *Conics*, and other analytic treatises indicates that Eratosthenes' study bore directly on the application of the method of analysis for the solution of problems. Save for the subject matter of the means, however, we have no further direct knowledge of this work, for Pappus provides neither summary nor lemmas relating to it, in contrast to his treatment of the other writings.[15] We may of course surmise that Eratosthenes took up the solution of problems of locus where the condition to be satisfied is expressed in terms of means. At that time the original three means defined by the 5th-century Pythagoreans were familiar, as well as three additional forms introduced by Eudoxus.[16] That is, for two given terms a and b, if an intermediate term x is such that the ratio $a - x : x - b = a : a$, then x is the "arithmetic mean" of a and b (i); if the ratio equals $a : x$ (or $x : b$), x is the "geometric mean" (ii); if it equals $a : b$, then x is the "harmonic mean" (iii); if it equals $b : a$, x is the "harmonic subcontrary" (iv); if it equals $b : x$ (v) or $x : a$ (vi), then x is one of the two "geometric subcontraries." In view of these forms an entirely straightforward hypothesis on Eratosthenes' work is that it sought to describe the loci of points whose distances a, x and b from three given lines satisfied one of these relations.[17] One may observe that Euclid's problem of the three-line locus can be set in such a form, that is, where the distances satisfy relation (ii).[18]

But in this connection Zeuthen proposes an alternative speculation, one of his most striking insights, given its ingenuity and mathematical persuasiveness. In essence, he takes Eratosthenes' work to be an elaboration of the polar properties of the conics. In view of the incredible slimness of the documentary evidence here, any attempt at reconstruction must face the risk of appearing arbitrary. It is unfortunate, then, that Zeuthen omits mention of certain extant treatments of related properties of the circle which might provide a natural context for the more advanced effort he assigns to Eratosthenes. Let us then consider these materials first, after which we can present a brief sketch of Zeuthen's account.

Although ancient discussions of the means usually arise within the context of arithmetic theory, at least one of the means, the geometric (ii), owes its natural formulation to geometry. It is thus appropriate to seek geometric constructions for the other means as well. In the case of the harmonic mean, three ancient methods can be cited. One is reported by Pappus from an unnamed contemporary.[19] Given segments AD, DG, if their sum ADG is bisected at E, AE will be their arithmetic mean [Fig. 4(a)]. If the circle of diameter AG is drawn and the line perpendicular to AG at D is drawn to meet the circle at B, then BD will be the geometric mean of AD, DG (cf. *Elements* VI, 13). If a line is now drawn from D perpendicular to EB, meeting EB at Z, it follows that ZB is the harmonic mean of AD, DG. Pappus does not understand this last conclusion and criticizes his informant for having merely said that ZB "is the third proportional of EB, BD." Ironically, this is all one need have said to display BZ as the desired harmonic mean, and Pappus' failure to recognize this reflects poorly on his own expertise in geometry. For the ancient arithmetic writers well knew the relation $m : h = a : n$, linking the arithmetic (a) and harmonic (h) means between given terms (m, n); indeed, they dubbed this the "perfect proportion."[20] Since, also, $m : g = g : n$, for the geometric mean (g), it follows at once that $h : g = g : a$, as required by Pappus' source. Pappus himself establishes the relation in a far more complicated manner, and then introduces an alternative construction of the harmonic mean, as the line EK in relation to EB, EZ, where $\theta B = BH$ [Fig. 4(b)].[21]

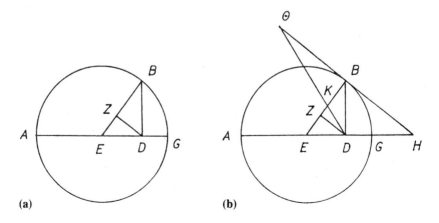

Figure 4

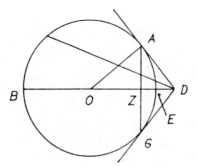

(c)

Yet a third construction, and for our present purposes the most interesting one, relates to a lemma (28) given by Pappus toward Euclid's *Porisms*.[22] As we have already noted in connection with the Euclidean work, this lemma establishes that any secant DEB to a circle will meet in Z the chord joining the points of tangency A, G of lines from D such that BD : DE = BZ : ZE [Fig. 4(c)]. That is, in view of the definition in (iii) above, DZ is the harmonic mean of ED, DB. In this way, AG is seen to be the locus of the harmonic division of secants drawn from D to the given circle: in the modern theory, it is called the "polar" of D with respect to the circle, and the construction can be generalized for any conic (cf. Apollonius, *Conics* III, 37).[23] In the special case of the secant passing through the center of the circle, we thus derive a construction of the harmonic mean of given lines ED, DB. Draw the circle on BE as diameter, draw DA tangent to this circle, and draw AZ perpendicular to BED; then DZ is the required mean. A proof is easily seen by drawing A0, for 0 the center of the circle. Since AD is tangent, $AD^2 = ED \cdot DB$; since AZ is the altitude of the right triangle 0AD, $AD^2 = ZD \cdot D0$. Thus, $ED \cdot DB = ZD \cdot D0$. Since D0 is the arithmetic mean of ED, DB, it follows from the "perfect proportionality" cited above that ZD is their harmonic mean. The proof of Pappus' lemma leads to a different version of the proof. In this case, one would note that $AZ^2 + ZD^2 = AD^2$, while $AD^2 = ED \cdot DB$ and $AZ^2 = BZ \cdot ZE$; thus, by substituting we have $BZ \cdot ZE + ZD^2 = ED \cdot DB$. Pappus concludes at once that BD : DE = BZ : ZE. He has surely elided several steps to secure this result for the general case of secant[24]; but in the particular case of interest to us, we need only subtract from each side the term $ED \cdot DZ$ to obtain $ZE (BZ + ZD) = DE \cdot BZ$, from which the desired proportion follows.

The association of such lemmas with Euclid's *Porisms* would indicate that some version of these materials was familiar at the time of Euclid and could thus serve as background for Eratosthenes' study of means. Indeed, the geometric study of the means is likely to have been underway relatively early in the 4th century. For instance, Theaetetus' theory of the irrational lines distinguished three types: those formed as the geometric, arithmetic, or harmonic mean of two given lines commensurable in square only.[25] Within such a theory, a geometric construction of the harmonic mean would be indispensable. Furthermore, one

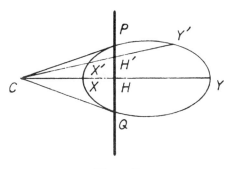

Figure 5

may note that the definition we gave for the harmonic mean in (iii) above is, despite its rather opaque formulation, the traditional form, as we see in its adoption by Archytas.[26] Since this form is especially well suited to the configuration for its construction via the tangents to the circle, we may have in this a sign of the early recognition of this construction.[27]

We may now turn to Zeuthen's proposal relating to Eratosthenes' study of the means.[28] Let us first consider a given central conic, let a point P be taken on it with abscissa XH and ordinate PH, and let the tangent at P be drawn to meet the diameter in C (Fig. 5). It follows that YH : HX = YC : CX (cf. Apollonius' *Conics* I, 34); since YH = YC − CH and HX = HC − CX, this means that CH is the harmonic mean of YC and CX according to the definition given in (iii) above. If now we extend PH to meet the conic in Q and draw the tangent at Q, it too will meet the diameter in C. Furthermore, if we draw any other line through C to meet the conic in X' and Y' and the line PQ in H', it follows that H' cuts the line CX'Y' harmonically; that is, Y'H' : H'X' = Y'C : CX' , or CH' is the harmonic mean of CY' and CX' (*Conics* III, 37). We may recast this construction as a locus: given a conic and a point C, the locus of points H which effect a harmonic division of the secant lines CXY is a given line, namely the line PQ which joins the points P and Q where the tangents from C meet the conic.[29]

This much merely sets familiar propositions from Apollonius in an alternative form. But Zeuthen continues. Let A bisect the line XY, so that CA is the arithmetic mean of CX, CY (Fig. 6). The locus of points A' which divide the secants from C such that CA' is the arithmetic mean of CX', CY' is a second conic section, similar to that given, also passing through P and Q and having C0 as diameter, for 0 the center of the given conic. Furthermore, let G divide CXY such that CG is the geometric mean of CX, CY (or, equivalently, of CH, CA); then the locus of such points G' is a third conic section, similar to the other two, passing through P and Q and having C as its center.[30] Zeuthen's derivations require but a few manipulations of proportions based on the standard form of the sections used by Euclid and Archimedes. They thus would not overreach the technical basis available to Eratosthenes in the production of his analytic treatise *On Means*.

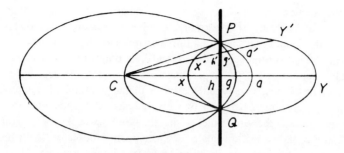

Figure 6

Zeuthen now develops a hint implied in one of Pappus' allusions to this work. Pappus regularly distinguishes loci and problems into three classes: the planar, the solid, and the linear. But he also recognizes a fourth class, the "loci with respect to means" which fall under the former classification, but somehow differ from the other classes. The following comment is attached to his remarks on Apollonius' *Neuses*:

> Now these things in the *Topos analyomenos* are planar, which also are proved first, in separation from the means of Eratosthenes; for those are last [or: later]. Then following after the planar [problems] the order demands investigation of the solids; but they call "solid" not those problems which involve solid configurations, but rather those which cannot be proved by means of planes, but are proved by means of the three conic sections; thus it is necessary first to write about these. Now the earlier edition of the conic elements were the five books of Aristaeus the elder, written rather succinctly, however, as if for those already competent in these matters.[31]

Needless to say, this passage is packed with obscurities. Zeuthen proposes that it has suffered through the confusion of a late editor without direct knowledge of Eratosthenes' work and its relation to the others. For instance, there appears to be a mistaken reference to Aristaeus' *Conic Elements*, where the source text would have intended his *Solid Loci*. But Zeuthen calls attention to the context of studies of *neusis* which has occasioned this passage.[32] The problems associated with Eratosthenes' loci with respect to means fall under several headings, but some are planar; yet they cannot be treated at the same time as the planar *neuses* of Apollonius, because they still require a background in the conic sections. Of course, one can construe the passage in other ways. But this sense relates nicely to a set of problems directly related to Zeuthen's earlier observations on Eratosthenes' loci.

Consider the following problem of *neusis*: to insert a given length as a chord in a given conic section so as to pass through a given point. Zeuthen introduces the conic and the line answering to A and H, respectively, in the prior constructions of the means and then reduces the problem to one whose solution entails the drawing of a hyperbola. This, in turn, can be replaced by a circle given under the conditions of the problem. Thus, although the problem relates to a given conic section, its solution requires only an auxiliary circle, so that it might be classified as "planar." Furthermore, Zeuthen shows that methods familiar in the ancient theory of conics permit a full description of the diorism, or conditions under which the insertion is possible. Finally, he points to an unusual difficulty: one might suppose that if the hyperbola were viewed in the degenerate case of two intersecting lines, one would derive a planar solution of the problem of inserting a given length between two given lines. Unfortunately, the solving circle degenerates at the same time, so that the point one seeks for the completion of the figure remains indeterminate. As it happens, the *neusis* with respect to two lines is a solid problem; its solution via the intersection of a circle and a hyperbola is given by Pappus and appears to have originated around the time of Apollonius.[33]

This outline of Zeuthen's view makes clear the range of problems he proposes as comprehended within Eratosthenes' study of the means. Mathematical considerations afford a natural place for such a study within the 3rd-century theory of conics. Unfortunately, the textual evidence is so meager that one cannot adduce any firm support on its behalf. On the negative side, one might wonder whether Eratosthenes had sufficient competence in this advanced theory to carry through the sometimes elaborate derivations presented by Zeuthen. Here again, our information on Eratosthenes' mathematics in general is too scant to say yes or no. Certainly the construction of the cube-duplicating instrument entails nothing of comparable complexity to the propositions on loci discussed here; but we can hardly take the one extant document as a full measure of Eratosthenes' geometric competence.

A more explicit difficulty concerns the relation between Eratosthenes' work as reconstructed and Apollonius' study of *neuses*. The latter analyzed only those instances of *neuses* solvable by planar methods, and among them appears the *neusis* of a given length as a chord in a given circle.[34] This solves for the circle, then, what Zeuthen proposes Eratosthenes solved for the conics. But if the latter did indeed effect the planar form of the solution, as Zeuthen shows, then there would be no justification for a further treatment by Apollonius. To sustain Zeuthen's thesis one might perhaps confine Eratosthenes' solutions to their solid form, that is, where the auxiliary solving curve is a conic; then Apollonius' work would supply the alternative planar constructions for the circle.[35] Alternatively, one might note that Eratosthenes' career stretched over several decades, overlapping the period of Apollonius' activity by twenty or thirty years. In this event, it is easily possible that a work by Apollonius, say the treatise on planar *neuses*, prepared the way for the investigation of *neuses* drawn to conics as Zeuthen suggests for the work of Eratosthenes. Indeed, the topicality of a work of this description at precisely this time, as well as its dependence on just those theorems on conics one finds in a section of Apollonius' Book III, would seem to heighten the credibility of Zeuthen's conjecture. If the absence of textual supports compels scepticism, the mathematical ingenuity and historical appropriateness of his view render it virtually irresistible.

NICOMEDES

Reports of the work of Nicomedes concern only his contributions to the solutions of the three classical problems. Pappus and Eutocius reproduce extended extracts from Nicomedes' book *On Conchoid Lines*, where various forms of the conchoid curve are generated and then applied to the cube duplication.[36] Presumably, the same work showed how these might be used for the angle trisection as well, for Pappus informs us that Nicomedes did in fact apply these curves to this effect.[37] Furthermore, Nicomedes introduced the quadratrix for the purposes of the circle quadrature; indeed, one source claims, in effect, that Nicomedes was the one responsible for giving the curve its name, on the basis of this application.[38]

It is not difficult to detect Eutocius' disapproval as he records Nicomedes' treatment of the cube duplication:

> Nicomedes too in the composition ascribed to him *On Conchoid Lines* describes the construction of a device serving the same function [sc. of cube duplication]. The man seems greatly taken with his accomplishment and heaps ridicule on the inventions of Eratosthenes as being impracticable and at the same time devoid of geometric interest. So that none of those who have labored over the problem should lack anything for the sake of comparison with Eratosthenes we add him [sc. Nicomedes] to those already written.[39]

Thus, Eutocius appears to include the method of Nicomedes, last in a series of a dozen, more for the sake of completeness than for the intrinsic merit of the method. The specific opposition of Nicomedes to Eratosthenes is noteworthy. For Nicomedes' method actually has no special connection with Eratosthenes', while several rather effective mechanical methods became familiar through treatments by Diocles and Apollonius, for instance, before the close of the 3rd century.[40] This would place Nicomedes as a contemporary of Eratosthenes and the writing *On Conchoids* as a direct response to the latter's treatment of the cube duplication. The polemical tone is quite in keeping with the intellectual climate of this period,[41] yet it is hard to imagine how the practical deficiencies of Eratosthenes' method could remain controversial for any great length of time. Indeed, the search for such mechanical constructions, amounting in Nicomedes' case to the use of auxiliary curves of higher order, would be criticized on formal grounds soon after the time of Apollonius.[42] Thus, Nicomedes' effort is most closely associated with approaches adopted by geometers early in the last third of the 3rd century. As we shall now see, this is further supported by the Archimedean influence evident in Nicomedes' construction and applications of these curves.

In its basic form the conchoid is produced as the locus of points a fixed distance from a given line as measured in the direction inclined toward a fixed point. The instrument for tracing it consists of a fixed length EK (the "distance") to be moved such that a pin fixed at E can slide along a groove in a given line AB (the *kanōn*), while the extension of EK is free to slide past a fixed pivot set at the point D (the "pole") [Fig. 7(a)]. Thus, K moves along a curve such that

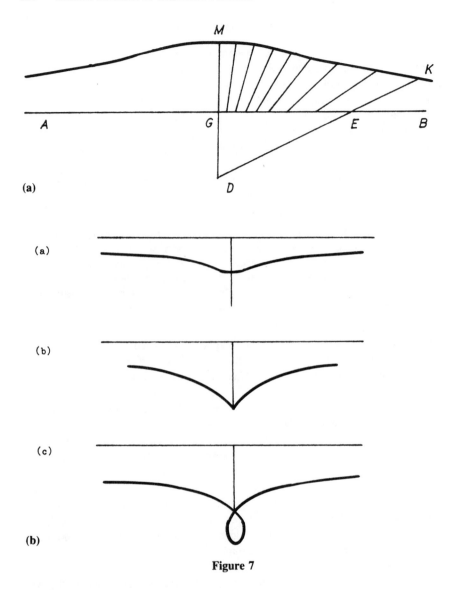

Figure 7

KE, its distance from the given line AB, is constant, while KE extended always inclines toward D. This curve is called the "*first* conchoid." Pappus states that three other forms of conchoids may be constructed,[43] and one can readily infer what these must be. The "first conchoid" has been obtained by marking off the fixed "distance" on the side of the "rule" (*kanōn*) *away* from the "pole." But if we mark it off *toward* the "pole," we will obtain three different shapes of curve, according to whether the "distance" is less than, equal to, or greater than the distance of the "pole" from the "rule" [Fig. 7(b)]. The first of these (a) resembles the primary conchoid in rising smoothly toward the "pole" without reaching it, and then dropping back asymptotically toward the "rule." The second form (b) has a cusp at the "pole," while the third (c) enters into a loop through the "pole." Like the other forms of conchoid, these tend asymptotically toward the "rule." Since Pappus does not say which form is the "second," the "third," or the "fourth" conchoid, let us designate the above cases as "secondary conchoids of form (a), (b), or (c)."

It is evident that the conchoids are merely an elaboration of the procedure of *neusis*, so that any problem solvable by means of the insertion of a given length between a given line and a second given curve, so as to pass through a given point, can be solved alternatively by means of the appropriate conchoid. For instance, the method of angle trisection reported by Pappus (introduced in Chapter 5 (iii) as variant (b) of the Archimedean *neusis*) requires the insertion of a given length between two given lines; this may now be effected by finding where the given line E0 extended meets the primary conchoid of "pole" A, "rule" 0H, and "distance" equal to twice 0A (Fig. 8). Alternatively, one may use the similarly situated conchoid of "distance" equal to 0A, by finding its intersection with the given circle of center 0 and radius 0A (cf. variant (a) of the Archimedean *neusis*). Since both Pappus and Proclus assert that Nicomedes solved the angle trisection via conchoids, we can hardly but conclude that the form reported by Pappus, and doubtless also the alternative form (now extant only in Arabic), were worked out in the course of Nicomedes' study of this curve.[44]

One observes that the Archimedean *neusis* itself, as well as its variant version (b), can be effected by means of a "secondary conchoid of form (c)" (Fig. 9).[45] The fact that Proclus claims that Nicomedes trisected the angle "by means of conchoids" would seem to indicate that one or more of these alternative methods was used.[46] Note, furthermore, that the *neuses* required by Archimedes in *Spiral Lines* may also be effected by secondary conchoids whose form will depend on the length to be inserted between the given line and the given circle (Fig. 10).

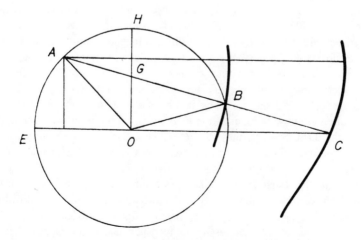

Figure 8

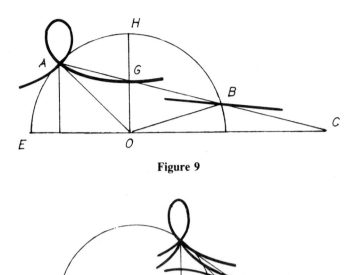

Figure 9

Figure 10

In the instance of such problems as these, involving a circle as one of the given figures, a further variation on the form of the conchoid adapted for their solution is immediately suggested. Taking as "rule" the given circle of center 0 and radius 0A, we can conceive of the curve whose points lie at a given "distance" from the circle, as measured in the direction of a given point. For instance, when the "distance" equals the radius of the circle and the given point A lies on it, the curve will consist of two parts: an outer arc (where the distance is measured away from A) and an inner loop (where it is measured toward A) joined by a double point at A (Fig. 11). Then the Archimedean *neusis* for the angle trisection may be effected by finding the intersection of the outer arc with the extension of E0, that is, the point C. Similarly, variant (a) of the *neusis* is effected via the intersection G of the inner loop with the radius 0H. Clearly, the *neuses* for the spirals follow via curves of the same sort, when the given length is assigned as the "distance." The family of curves resulting from this mode of generation are now familiar as the *limaçons*, first proposed by Étienne Pascal early in the 17th century; they are a subclass of the "conchoids of circular base."[47] The natural association of these curves with the Archimedean *neuses*, however, has led some scholars to recommend assigning their study within the ancient geometry. Indeed, under one view their invention preceded that of the Nicomedean conchoids and their original name was "cochloids," until, through a confusion, Pappus came to apply that name to the Nicomedean curves.[48] These reflections on the names of curves are not fully persuasive; for Pappus' discrepant terms can easily be accounted for on the basis of simple scribal errors.[49] But the essential intuition that the ancients were familiar with the circle-based variations

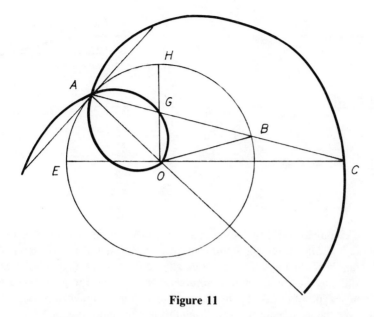

Figure 11

on the conchoids is, I believe, a sound one. Further aspects of this question will be considered below.[50]

The extracts which Pappus and Eutocius present from Nicomedes' book are principally concerned with the application of the conchoid toward the cube duplication. Although the curve (in its "primary" form) is first described with reference to the instrument by which it is traced, the exposition thereafter is entirely geometric in the standard formal manner. There follow two propositions on the curve and the construction of a preliminary problem: (a) that the curve continually approaches the "rule"; (b) that any line drawn in the space between the curve and the "rule" will intersect the curve; (c) the problem to draw a line from a given point across a given angle so that the segment intercepted by the sides of the angle shall have a given length. The proofs, omitted by Pappus, are set out in full by Eutocius.[51] For (a) one considers two rays GZ and Gθ, where angle MLG is greater than NKG (Fig. 12); it follows that the angle at Z is greater than that at θ, and since θK = ZL, inspection of triangles θKN and ZLM shows that θN is greater than ZM. One observes that in so establishing the *"continual approach"* of the curve toward the line, Nicomedes has not yet shown that they become *arbitrarily* close to each other. That is implied in the proof of (b). Here Nicomedes introduces an arbitrary parallel ZHθ and constructs its point of intersection with the conchoid (Fig. 13); namely, if one defines the length K via the proportion DH : HG = DE : K and draws about the center G a circle of radius K to meet ZHθ in Z, it follows that ZL = ED, so that Z must lie on the conchoid. Nicomedes asserts as obvious that any other line passing through H must also meet the curve.[52]

The asymptotic approach of the curve toward the "rule" is not explicitly stated here, although it would follow at once since we may take HD arbitrarily

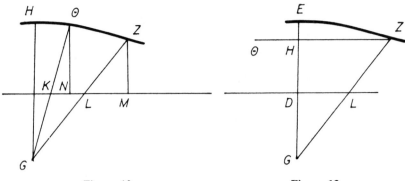

Figure 12   **Figure 13**

small.[53] A comparison with Apollonius' treatment of the asymptotes of the hyperbola in *Conics* II reveals significant differences. Analogous to Nicomedes, Apollonius shows that any line drawn parallel to the asymptote meets the curve (II, 13), although he uses an indirect argument. Next, by an argument comparable to Nicomedes' proof of (a), Apollonius establishes the continual approach of the hyperbola to the asymptote (II, 14). From this and the preceding he then obtains that the curve and the line will come closer than any preassigned distance. Although some of the gaps in Nicomedes' treatment may be due to the omissions by the later commentators,[54] the differences in both order and precision between his method and that of Apollonius conform well with our view that Nicomedes' work came before Apollonius. It is striking that neither geometer attempts to frame the treatment of asymptotes in the manner of Eudoxus and Archimedes, for instance, by means of a limiting procedure based on successive bisection.[55]

The construction of (c) is straightforward. Given the angle BAH and the external point G, one constructs the conchoid of "pole" G and "rule" AB and "distance" the given length (Fig. 14). By (b) the ray AH meets the conchoid; let it be at H. Then the ray GKH effects the construction, for the intercept HK will be the given length in accordance with the generation of the curve. It is

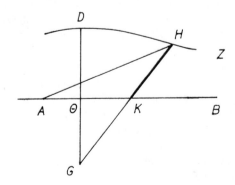

Figure 14

interesting that Nicomedes chooses to treat this general case of *neusis* as a separate application of the conchoid. In this way, further problems will be solved as special cases of this *neusis*, rather than through direct introduction of conchoids.

This has prepared the ground for a new construction of the two mean proportionals between two given lines. The method employed is rather elaborate and rarely fails to evoke awe among those who discuss it, with the result that the manner of its origin remains a mystery.[56] In the hope of producing a credible and naturally motivated account, I will adopt the procedure, followed several times above, of seeking an analysis modelled on the extant synthetic treatment. The key lies in perceiving that Nicomedes' method is an extension of the "Heronian" solution we have already discussed in connection with the work of Archimedes.

Let us first consider the diagram produced via the *neusis* of the "Heronian" method. By situating the line MLK so that MP = PK, one finds that GK, AM are the two mean proportionals between the given lines LG, AL [Fig. 15(a)]. Now the tentative nature of the procedure can be remedied by seeking a regular geometric construction determining the position of K. Since MP = PK, one has $MD^2 + DP^2 = PE^2 + EK^2$. Thus, if we determine EZ via the right triangle GEZ whose hypotenuse GZ equals PE and whose leg GE equals DP, then since $EZ^2 = PE^2 - DP^2$, one has that the hypotenuse KZ of triangle KEZ is equal to MD [Fig. 15(b)]. We may divide KZ at θ such that Kθ = DA and θZ = AM. Since DA = ½ GL and by similar triangles, KG : GL = LA : AM, it follows that Kθ : θZ = KG : 2 GB. If we then set H such that HB = BG and draw HZ and Gθ, the latter lines will be parallel to each other. Now, from the initial conditions of the construction, the point Z is given, as is the length Kθ (= ½ AB) and also the position of H; hence also the line HZ and from it the line Gθ are given. Thus, the determination of K reduces to a *neusis*: the insertion of the given length as θK across the given angle KGθ such that it inclines toward the given point Z.

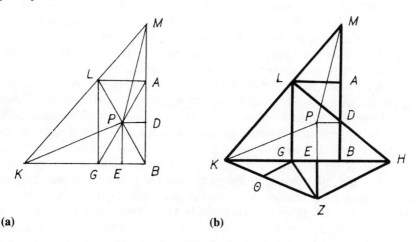

(a)

(b)

Figure 15

It is via this derived *neusis* that Nicomedes solves the problem of finding the two mean proportionals. To effect the *neusis* he introduces the primary conchoid, as shown in the preliminary construction (c) above, where GΘ will be the "rule" and Z the "pole." Through the argument prefigured in the analysis just given, augmented by what amounts to a modified form of the proof of the "Heronian" construction, Nicomedes shows that GK • KB = BM • MA, whence GK, AM are the required mean proportionals. Thus, the diagram, method, and proof of the "Heronian" method are implicated within the construction used by Nicomedes, and this interdependence is further revealed through some striking *literal* correspondences linking his text of the proof with Eutocius' version of the "Heronian" method.[57] Thus, we must surely infer that Nicomedes had access to that method and produced his own as a modification of it. The converse view—that the "Heronian" form emerged later as a simplification of Nicomedes—is of course conceivable, but rather improbable. For the motivation underlying the former is reasonably apparent, while the heuristic mode behind the more elaborate Nicomedean method remains obscure until viewed in the context of the other method. This would then assign the invention of the "Heronian" *neusis* to some time after the middle of the 3rd century, that is, before Nicomedes, but after Eratosthenes' treatment of the cube duplication. As the later tradition never associates the name of Nicomedes with this construction, he is unlikely to have been responsible for inventing it or for producing the textual description by which it was known to later writers like Hero and Eutocius. Furthermore, since Nicomedes' studies of the conchoids reveal familiarity with Archimedes' *neusis* constructions for problems like the angle trisection, we may well suppose that Nicomedes also learned of the "Heronian" method of cube duplication through Archimedean sources. We may thus view this as important support for the earlier argument assigning an Archimedean origin to that method.

The Archimedean influence on Nicomedes' work is further shown in the use Nicomedes makes of the quadratrix for effecting the circle quadrature; for this curve, like the Archimedean spiral, is generated via the composition of two motions, one linear, the other circular.[58] As the ray AB rotates uniformly on A toward the terminal position AD, the point K on it moving uniformly from B to A in the same time will trace the arc of the spiral BKA (Fig. 16); if, simultaneously, BG moves uniformly toward AD, its intersection Z with the rotating ray will trace the quadratrix BZH. It follows that the ordinate ZΘ of the quadratrix equals the radius vector AK of the spiral, where A, K, and Z lie in a straight line. From this Pappus obtains an alternative construction of the quadratrix, based on the projective correspondences linking the spiral with the conical and cylindrical helices.[59] He also shows how both the spiral and the quadratrix are useful for solving not only the problem of trisecting the angle, but more generally that of dividing the given angle into parts according to any given ratio. Let ΘAZ be the given angle marking off Z on the quadratrix and K on the spiral; set AL = ZΘ (= AK) (Fig. 17). Then divide AL at N such that AN : NL equals the given ratio. By setting PM = QA = AN, we thus find points P and Q on the quadratrix and spiral, respectively, and from their defining properties it

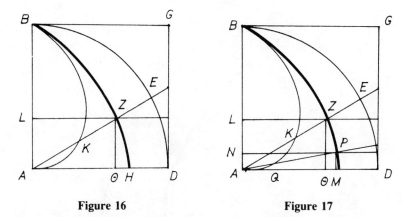

Figure 16   **Figure 17**

follows that the radius vector AQP divides angle θAZ into two angles whose ratio equals AN : NL, as required.

Furthermore, we may see how Archimedes' theorems on the tangents to the spiral reveal a way to use the quadratrix for the circle quadrature. If we draw the tangent to the spiral at B, it has been shown that its intercept AZ with the line AD extended equals the arc of the circle BD (Fig. 18). But at B the motions which give rise to the spiral are the same as those which generate the quadratrix, namely the uniform linear motion of B toward A and, perpendicular to this, the uniform radial motion of B on ray AB rotating toward AD. Hence, the tangent to the quadratrix at B is the same line BZ, so that its intercept has the same effect for rectifying the arc BD.[60] One would suppose the ancients perceived this property of the quadratrix. For Nicomedes doubtless went beyond the special applications of the curve in his study of its properties, just as he did for the conchoids; while Proclus' citations of a treatment of the "quadratrices" by the geometer Hippias seem to indicate a work devoted entirely to this curve.[61] Nevertheless, it was through quite a different property from the above that the ancients solved the circle quadrature.

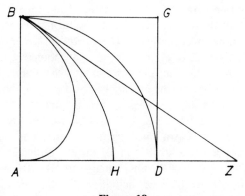

Figure 18

228 Ancient Tradition of Geometric Problems

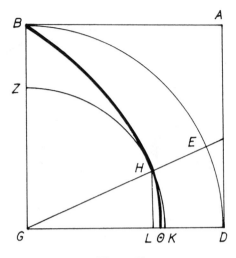

Figure 19

We introduce the quadratrix BHθ and, from the motions which generate it, derive the "property" (*symptōma*) that HL : BG = arc ED : arc BD; that is, that its ordinates are proportional to the corresponding arcs (Fig. 19) It is claimed that the terminal position of the curve θ is such that θG : GB = GB : arc BD. Since θG and GB are presented as lines, one may construct the line which is their third proportional, say C. Then, C = arc BD, so that we have solved the problem of rectifying the arc BD. In Pappus' report on the quadratrix, one cites Archimedes (sc. *Dimension of the Circle*, Prop. 1) to the effect that the product of the circumference of the circle, or $4\,C$, times its radius equals twice the area of the circle; whence the quadrature of the circle is effected.[62]

Since, at the limiting position θ, the curve is not well-defined, in that its generating rays coincide here, an indirect argument is used for proving the claim that θG, GB, arc BD are continuously proportional. In its dependence on inscribed and circumscribed arcs and its assumption that a circular arc is bounded below by its chord and above by the segment of its tangent, we meet further instances of Archimedean influence; for the same results are critical for the proofs and calculations presented in the *Dimension of the Circle*. While Pappus does not identify the source of his version, its application of a Eudoxean method relying on specific Archimedean results sets it after the time of Archimedes. We will suppose that it represents the essential line adopted by Nicomedes.

Following an indirect argument, let us first assume that θG is not the third proportional, but that arc BD : BG = BG : GK, for GK greater than θG (see Fig. 19). The arc ZK is drawn, meeting the quadratrix in H, the line GH is joined and extended to meet arc BD in E, and the ordinate HL is drawn. By assumption, arc BD : BG = BG : GK; but also arc BD : BG = arc ZK : GK[63]; hence, BG = arc ZK. From the definition of the curve, arc BD : arc ED = GB : HL, while arc BD : arc ED = arc ZK : arc HK. Since arc ZK = BG,

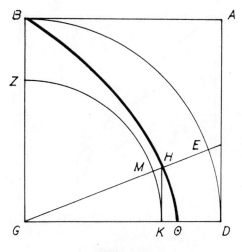

Figure 20

it follows that arc HK = HL, which is impossible. If it is then claimed that GK is less than Gθ, one again draws the arc ZK, determines H where the perpendicular to GK at K meets the quadratrix, and draws GH meeting arc ZK in M and arc BD in E when extended (Fig. 20). By an argument similar to that just given, it follows that arc ZK equals BG and arc MK equals HK, the latter being impossible.[64] Thus, arc BD : BG equals the ratio of BG to a line which is neither greater nor less than θG, that is, to θG itself.

Implied by this proof is that BG equals the arc of a quadrant of the circle of radius θG. This is not established as such, but is effectively entailed by the step appearing in both parts of the proof, that BG equals arc ZK, the quadrant of the circle of radius GK. However prominent this step appears, it is not actually necessary for the proof. For instance, in the first part, where GK is assumed greater than θG, we have, by hypothesis, GK : BG = BG : arc BD; by similar arcs, GK : BG = arc HK : arc ED; by definition of the curve HL : BG = arc ED : arc BD, or HL : arc ED = BG : arc BD. Comparing, we find arc HK : arc ED = HL : arc ED, or arc HK = HL, an impossibility. In a similar manner, one proves the second part without referring to the fact that BG = arc ZK. We may thus suspect that its presence in Pappus' version is a vestige of the preliminary planning of the theorem.

Adopting this view, we can propose the form of an analysis leading to the discovery of this theorem. Let us set out to determine θG, the intercept of the quadratrix with its terminal ray GD (Fig. 21). We draw the quadrant Zθ of radius θG; we mark off any point H on the curve, then draw GH meeting arc Zθ in M, and extend GH to meet arc BD in E and the perpendicular to Gθ at θ in N. By definition of the curve, HK : BG = arc Mθ : arc θZ. If H is taken very close to the terminal position θ, then HK and arc Mθ will be nearly equal; for both lie between the half-chord ML and the half-tangent Nθ, and the latter may

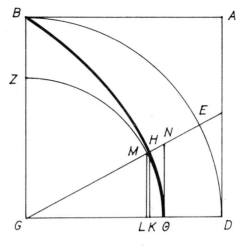

Figure 21

be made as close as desired by an appropriate choice of angle HGθ (cf. Archimedes' *Sphere and Cylinder* I, 3). Assuming their equality in the limiting position, we have BG = arc θZ, that is, BG equals the quadrant of the circle of radius θG. But BG, in turn, is the radius of the quadrant BD. Since similar arcs have the ratio of their radii, it follows that θG : BG = BG (= arc Zθ) : arc BD. This is the form of the theorem stated by Pappus, and his indirect proof follows readily as the formal synthesis answering to this analysis.

Pappus reproduces certain criticisms raised by the ancient commentator Sporus against the conception of the quadratrix[65]: (i) that its generation is mechanical and entails the difficulty of synchronizing a linear with a circular motion; (ii) that the terminal position is not specifiable without prior access to a construction of the lines in the ratio of the circumference and diameter of the circle. As we have seen, Pappus presents an alternative construction, deemed not to be mechanical, where the quadratrix is the planar projection of a curve ("surface locus") formed as an oblique planar section of a spiral surface.[66] Since this construction addresses objection (i), but not (ii), it may have been introduced originally when the terminal position of the curve was not yet considered of interest, that is, when the curve was known for its use toward the angle trisection, but not toward the circle quadrature. We have also noted a second construction given by Pappus, relating the quadratrix projectively to the Archimedean spiral. This form reveals a property, not noted by Pappus, that the two curves have the same tangent at their initial position, so that for the quadratrix as well as the spiral, the subtangent AZ equals the arc BD (Fig. 22).[67] One thus sees that the subnormal EA equals the intercept AH, for both lines are third proportionals to ZA, AB.

The extent to which results of this sort were discovered by the ancients is not specified in the extant sources. But when Proclus assigns to the geometer Hippias

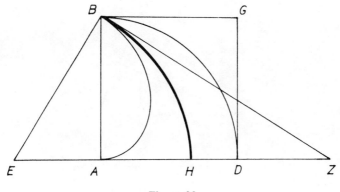

Figure 22

a treatment of the properties of the *quadratrices*,[68] the use of the plural term would surely indicate that Nicomedes' construction was somehow generalized to produce a class of related curves. One obvious way in which this might be done is suggested by Archimedes' treatments of spherical and spiral figures, where theorems established for a principal case (e.g., in correspondence with the full circumference of a given circle) are next extended to the cases of segments. One would thus be led to adapt the quadratrix for the quadrature of sectors. For instance, if we mark off the arc ED from the quadrant BD, we may define a quadratrix-related curve by the relation KL : EZ = arc MD : arc ED (Fig. 23). The limiting position θ of this curve will satisfy the relation θG : EZ = DG : arc ED, so that a line constructed as the fourth proportional of θG, EZ, DG will equal arc ED. The quadrature of the sector then follows, since its area is one-half the product of DG times arc ED.

Becker makes an interesting application of the quadratrix for approximation of π.[69] Consider the triangle GHL, where GH = GL [Fig. 24(a)]; if we bisect HL at H' and draw GH', then GH' is perpendicular to HL. Now draw perpendicular to GL the lines HK, H'K'; since H'H = ½ HL, H'K' = ½ HK. Fur-

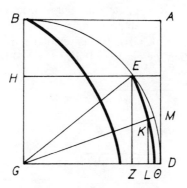

Figure 23

232 Ancient Tradition of Geometric Problems

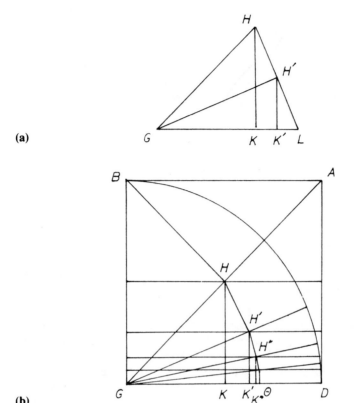

Figure 24

thermore, angle H′GL is one-half angle HGL. Thus, if H lies on the quadratrix of origin G and terminal ray GL, then H′ also lies on that curve. The construction is recursive, producing a sequence of points H, H′, H″..., tending toward the point θ on GD [Fig. 24(b)]. To evaluate Gθ one may consider either the sequence of abscissas GK, GK′, GK″,..., approaching it from below, or the sequence of rays GH, GH′, GH″,..., approaching it from above. Now, GK′/GH′ = GH′/GH = $\sqrt{½ + ½ (GK/GH)}$.[70] This relation is recursive, so that if one begins with HG bisecting the quadrant BGD, then HG/BG = $\sqrt{½}$;

$$GH'/GH = \sqrt{½ + ½\,(GH/BG)}, \text{ or}$$

$$GH'/BG = \sqrt{½} \cdot \sqrt{½ + ½\sqrt{½}}\,;$$

$$GH''/GH' = \sqrt{½ + ½\,(GH'/GH)}, \text{ or}$$

$$GH''/BG = \sqrt{½} \cdot \sqrt{½ + ½\sqrt{½}} \cdot \sqrt{½ + ½\sqrt{½ + ½\sqrt{½}}}\,.$$

Hence, in the limit, $G\theta/BG$ will be expressed as an infinite product of such radical terms. Now, from the theorem on the quadratrix, one knows that $G\theta/GB = BG/\text{arc } BD = 2/\pi$. One thus obtains an alternative derivation of Vieta's rule for the circle quadrature.[71] Vieta himself found it through considering the sequence of regular polygons of 4, 8, 16,...sides, inscribed in the same circle. It is clear then that this rule is equivalent to the Archimedean procedure, as far as computation is concerned. But the connection with the standard bisection method may indicate an avenue by which the manner of applying the quadratrix toward the circle quadrature was at first recognized.

Our sources do not specify the context of the original invention of this curve, the way its use for the circle quadrature was discovered, or the full extent of the ancient study of its properties. But the striking correspondences between the quadratrix and the Archimedean spiral, together with the strong dependence on Eudoxean methods and Archimedean results displayed in the proof of the limiting theorem for this curve, set its study firmly in the tradition of Archimedes' work on the measurement of the circle. Although one earlier geometer, Dinostratus, studied the curve, he was apparently interested in its use for the angle trisection; for Nicomedes is the one associated with the actual naming of the curve *tetragōnizousa*, for its use for the circle quadrature.[72] The Archimedean element underlying the study of the quadratrix is thus surely a mark of Nicomedes' treatment, just as Archimedes' applications of *neuses* served as the stimulus for Nicomedes' investigation of the conchoid curves.

DIOCLES

From a work by Diocles *On Burning Mirrors*, the commentator Eutocius draws two extended fragments, neither on burning mirrors, the one dealing with a method of cube duplication, the other providing an alternative solution to Archimedes' problem of dividing the sphere.[73] The work as a whole does not now exist in Greek, but an Arabic translation is extant and has recently been edited by G. J. Toomer.[74] In its present form the work is a compilation of loosely related geometric materials, brought together perhaps through their shared association with constructions employing conic sections. Doubtless, the Arabic version represents the work as known to Eutocius; for wherever it is possible to check, their agreement with each other is quite satisfactory, once a certain allowance is made for Eutocius' editorial revisions of details.

The principal value of the newly edited work is that it establishes a dating for Diocles rather earlier than that formerly assumed, and thus provides materials toward a better, if still extremely sketchy, understanding of the geometric field between Archimedes and Apollonius. In the preface, Diocles alludes to prior studies of the problem of burning mirrors: that Pythion and Conon posed the problem of finding a surface which will reflect the sun's rays toward a given circle; that Zenodorus posed to Diocles and his colleagues the related problem of finding the surface reflecting the rays to a given point; and that Dositheus had effected some form of construction for the latter. Now, Conon's activity falls around the middle of the 3rd century B.C., that of Dositheus only slightly later; both engaged in correspondence with Archimedes. Toomer assigns Diocles to early in the 2nd century, on the strength of his dating Zenodorus to that same time, and thus sets them both as close contemporaries of Apollonius.[75] But one cannot but be disturbed by the implication that Dositheus' contribution to the problem of burning mirrors pended for several decades before its reexamination and successful resolution by Zenodorus and Diocles. Toomer has observed that Diocles treats the conics in the form that one finds in the work of Archimedes, rather than in that of Apollonius.[76] Furthermore, our principal testimony to the work of Zenodorus, his study of isoperimetric figures, stands directly in the line of results on the circle and the sphere established by Archimedes. It would thus appear more natural to associate Diocles with work in progress in the closing third of the 3rd century, that is, slightly earlier, but surely not incompatible with the dating advocated by Toomer.[77]

On Burning Mirrors contains several efforts of interest for our survey of the research on geometric problems. It falls into three main sections. The first part on burning mirrors presents a proof that the parabolic contour has the property of reflecting rays to a given point, and from this, constructions answering the requirements of both of the problems posed on burning mirrors are described (Prop. 1). Diocles next takes up spherical contours, showing that reflection is not to a given point, but to a given segment of the axis of the surface (Props. 2, 3). He then gives a construction of the parabola defined as the locus of points whose distances from a given point and a given line are equal (Props. 4, 5). In the second part Diocles presents his own solution to the problem assumed by Archimedes for the division of the sphere into segments of given ratio (Props. 7, 8). We have already described his method, using the intersection of a hyperbola and an ellipse.[78] The fact that Diocles, even in the absence of Archimedes' solution, produces one comparable to it points to the familiarity at that time of the use of conic techniques for the solution of problems. In the third section Diocles presents two methods for the duplication of the cube: the first merely reproduces in an alternative form the two-parabola method of Menaechmus (Prop. 10), while the second utilizes a special curve, known in modern times by the name *cissoid* (Props. 11–16).[79]

Although Diocles has introduced his study of burning mirrors in the form of two problems of construction, what he presents in his opening propositions is a set of theorems on the reflective properties of specified surfaces, namely the paraboloidal surface and the spherical segment. The treatment is entirely synthetic. But certain incongruities in the order and relations of his separate theorems appear to suggest the lines of the underlying analysis.

First consider his theorem on the parabola, to the effect that any incident ray entering parallel to the axis will be reflected to a given point. Let the parabola have vertex B, axis BZ, and latus rectum L; hence, for any point θ on the curve, $\theta G^2 = GB \cdot L$ for coordinates GB, θG (Fig. 25). Set BE = ½ L, draw the tangent at θ to meet the axis in A, draw Sθ parallel to the axis, and join θD, where D bisects BE.[80] It is claimed that Sθ, θD make equal angles with the tangent; that is, angles T and P + Q are equal.[81] In his proof Diocles introduces

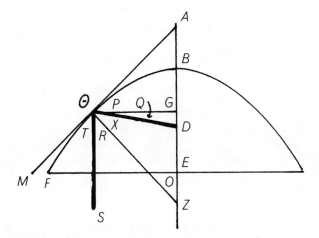

Figure 25

the normal line θZ perpendicular to the tangent, so that GZ = BE (hence, he assumes as known that the subnormal of the parabola is constant).[82] Thus, BG = EZ. Since D bisects BE and AB = BG (for θA is tangent),[83] ZD = DA. Hence, θD is the median drawn to the hypotenuse of right triangle AθZ, so that θD = DZ. Thus, angles O and X equal each other, while angles O and R are equal (by parallels), so that X = R. Hence, angles T and P + Q are equal.

On the strength of this result Diocles provides constructions answering the demands posed in the original problems. The equality of the angles entails that a ray entering along the line Sθ will be reflected to D, as will any other ray parallel to the axis BZ. If the parabola is rotated about its axis to produce a paraboloidal segment, this surface will thus reflect all rays parallel to BZ to the same point D. If, alternatively, the figure is rotated about the base of the segment FE, so as to generate a surface, while simultaneously the point D traces part of the circumference of a circle, Diocles appears to assert that this contour solves the other problem, in that rays entering parallel to the axis will be reflected to that circle. As Neugebauer and Toomer have shown, however, this assertion is actually incorrect, for the sections of the surface by planes parallel to the base plane will not be parabolas, so that the reflections in those other planes will not be toward the associated position of D.[84] Perhaps Diocles merely slipped into a wrong, but superficially straightforward construction. In the light of another construction he goes on to describe, however, I suggest that he here intended to form a surface via the parallel displacement of the parabola above and below the plane, such that D moves along a given circumference. This surface, quite obviously, would solve the problem at hand. If this is what Diocles had in mind, the extant text would thus include incorrect revisions made by the later editors and translators.[85] On the other hand, it is conceivable that Diocles himself initiated the error by supposing that his surface of revolution amounted to the same thing as this parabolic sheet.

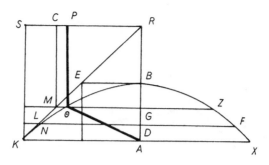

Figure 26

Diocles' proof can be streamlined somewhat, as is shown in an ancient mathematical fragment on parabolic and spherical mirrors.[86] But one encounters real puzzles when he turns, in Prop. 4, to the actual construction of the contour of the burning mirror corresponding to the result in Prop. 1 just discussed. Diocles sets AB as the axis of the mirror, B as the vertex, and A as the focal point (Fig. 26). Then setting AK = 2 AB and erecting the square on AK with diagonal KR, he divides AB into arbitrarily many equal parts (say, by the points of division G, D) and draws line GM perpendicular to the axis to meet the diagonal in M. Using length GM as radius, he draws the circle centered on A, meeting GM at θ. The process is repeated to determine N on DL such that AN = DL, and so on. To the points θ, N,...thus found, Diocles marks off Z, F,... in symmetrical relation to them with respect to the axis AB, observing that they all "lie on a parabola, *as we shall prove subsequently.*"[87] To draw the curve he suggests the use of a flexible ruler joining these points. Then:

> if we make a template from that curve and with that template arrange the curvature of a surface, the burning from that surface will occur at A, *as was proved in the first proposition.*[88]

The promise to prove that this curve is indeed a parabola is fulfilled in the proposition which follows (Prop. 5). Diocles employs the method one would anticipate, relying on the definition of the parabola via the familiar relation between ordinate and abscissa, that $\theta G^2 = BG \cdot T$, where the constant length T equals 4 AB.

The fact that Diocles goes through this construction and proof with such care—indeed, with a certain clumsy overelaboration—would appear to support Toomer's contention that Diocles has played a role in the discovery and proof of the focus-directrix property of the parabola, that is, its specification as the locus of points equidistant from a given point and a given line.[89] We have proposed, however, that some form of this locus problem was taken up within Euclid's study of surface loci.[90] In that event, Diocles' contribution would lie in setting the proof more directly in terms of the coordinate relation, for instance, rather than through the consideration of cones and their sections, as one might do via the Euclidean conception. One notes, moreover, that Diocles presents this result not in the *form* of a locus problem or of a theorem on the focus-directrix property, but rather as the pointwise construction of a certain curve. Indeed, the role of line SR as directrix is almost entirely submerged within the construction, so that one strains to view Diocles as engaged in the solution of a locus problem as such.

More striking still is that, contrary to Diocles' assertion at the end of Prop. 4, the curve determined via the pointwise procedure does *not* satisfy the property of the burning mirror established in Prop. 1. For the latter defines the parabola with reference to the ordinate-abscissa relation, not the focus-directrix property, while it is only in Prop. 5 that the curve satisfying the latter is shown also to satisfy the former. In fact, from the viewpoint of the ray-focussing property at issue in Prop. 1, the focus-directrix property is entirely irrelevant. Why, then, has Diocles introduced the focus-directrix at all, and given the impression that it is somehow an essential aspect of the problem of the burning mirror? One might suppose, as one possible answer, that he wished to highlight his own contribution to the locus problem and so was led to single it out artificially as the way to specify the burning curve, perhaps viewing this as an especially efficient mode for its construction.[91] Alternatively, the focus-directrix construction may have been interpolated into Diocles' text by a later editor.[92] In either case, however, one must assume a defect in the logical relationship of Prop. 1 to Prop. 4, whether due to Diocles himself or to the post-Dioclean editor.

We have seen, however, that the synthetic proof of a problem might sometimes find dispensable an aspect which was essential, or at least natural, in the context of the preliminary analysis; hence, that the retention of irrelevant features in a synthesis might be employed as a key for retrieving that analysis. In the present instance, we seek a way of reducing the ray-focussing problem of the burning mirror to the focus-directrix property directly, without reference to the relations which define the parabola. For having done this, a geometer could then go on to solve the problem of the locus with respect to given focus and directrix, or perhaps already know that this locus is a parabola. With this at hand, he could fashion the synthesis of the burning-mirror property directly from the definition of the parabola, in the style of Diocles' proof, without having to consider the focus-directrix property. This would thus account for Diocles' treatment of the latter in Props. 4, 5, as if essentially linked to the proof of the burning-mirror property, when it actually has no ostensible bearing on his Prop. 1.

The principal difficulty in attempting to reconstruct such an analysis of the burning-mirror property is that the curve to be found is specified via the known orientation of its tangent at each point; in effect, we require the geometric solution of a differential equation. But the ancient methods for finding tangents, as to the circle, the conic sections, and the spiral, are not general, and they entail no technique comparable to the use of the differential triangle developed by 17th-century geometers by which we might model such a reconstruction. Fortunately, however, in a fragment from the *Paradoxical Mechanisms* written by Anthemius of Tralles, we possess a treatment which answers our needs precisely. Anthemius, a younger contemporary and correspondent of Eutocius early in the 6th century, is most noted for his services as architect in the restoration of Hagia Sophia under Justinian.[93] His interest in higher geometry is revealed in Eutocius' directing to him his commentaries on Apollonius' *Conics*,[94] while the extant fragments display competence both in geometric and mechanical areas. What survives of his writing is a discussion of several forms of reflecting surfaces designed to satisfy specific conditions of reflection. In particular, Anthemius takes up the issue of the reputed construction of vast burning mirrors by Archimedes to set ablaze the Roman ships besieging Syracuse. Not quite convinced of the truth of this legend, he proposes an assembly of independently directed plane mirrors by which such a feat might be performed. He next turns to the problem of finding the surface which has this property of focussing the sun's rays to a given point.

To construct this surface, Anthemius adopts a pointwise procedure.[95] Let DG be the axis of the mirror, D the burning point, and let the aperture AB be bisected at right angles by DG extended at E (Fig. 27). Divide BE into as many equal parts as desired, and draw lines parallel to the axis through the points of division. Let ZB = BD, and join ZG, equal and parallel to BE. We now introduce the line BX bisecting angle ZBD, where X lies midway between the first two dividing parallels; that is, ZX = XM. We join XD and observe that it equals ZX, since the triangles ZBX, XBD are congruent. Next draw line XTY to bisect angle MXD, where Y lies midway between the next two parallels (i.e., MY = YN). Since triangles MXT, TXD are congruent, MT = TD. At this point the fragment breaks off; but Anthemius' construction can be completed without difficulty. Each repetition of the procedure yields an additional point on the contour, arriving at last at θ, the midpoint of DG. If one conceives of a plane mirror set in the line of BX, another along XTY, and so on, then from the bisection of the associated angles it follows that rays PB, RT, and so on, drawn parallel to the

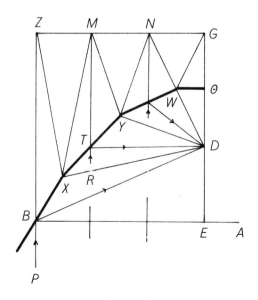

Figure 27

axis will all be reflected through D. Anthemius shows this, and thus has effected to within any desired degree of approximation the construction of the required reflector.

The fragment as extant preserves neither the statement nor the proof that the curve corresponding to this construction is the parabola. But as he criticizes the ancient writers for "saying that these [curves] are conic sections of some sort, but not of which sort and how produced,"[96] and then introduces the construction we have just considered with the stated aim to "set out certain constructions of contours of this sort, and these not without proof but confirmed via geometric methods,"[97] it is clear that he knew this fact and provided some proof for it. Indeed, in an earlier section of the fragment he effects the construction of another surface by the same type of pointwise procedure and then observes that the curve determined is an ellipse.[98] Heiberg has suggested that another portion of Anthemius' writing is extant in the Bobbio mathematical fragment. If he is correct, as I have argued he is, then the latter preserves most of Anthemius' geometric demonstration of the focal property of the parabola.[99]

A critical feature of the pointwise construction given by Anthemius is that $DB = BZ$, $DT = TM$, *etc.*, that is, that all the points determined on the curve $(B, T, \ldots, \theta)$ are equally distant from the given point D and a given line ZG. In effect, then, Anthemius presents just the form of analysis which we have sought as the basis of Diocles' study of the parabolic burning mirror. Furthermore, one notes that neither treats the focus-directrix property in the form of a problem of locus *per se*; indeed, neither views the directrix as a given of the construction, but rather only as an auxiliary line introduced in the course of the generation of the contour of the burning mirror. Thus, one is led to consider whether Anthemius has somehow preserved the manner of a treatment of burning mirrors before Diocles. Clearly, Anthemius did not have direct access to Diocles' book—a surprising fact in view of Eutocius' extensive borrowings from Diocles in his commentaries on Archimedes. But Anthemius could hardly have charged the ancients with not having specified which conic section the burning mirror was, if he had consulted Diocles; for this very fact is proved in Diocles' first theorem. Furthermore, when Anthemius describes those ancient studies as "rather mechanical and lacking any geometric proof," he seems to echo the remark Diocles makes relative to the construction of Dositheus. For if, as Toomer supposes, Diocles is contrasting a form of practical construction with his own geometric proof,[100] then not only does he express the same opinion of the earlier studies as Anthemius, but also he could well have in mind the type of approximate construction which Anthemius actually gives. This confirms what we have perceived through consideration of the technical method followed by Anthemius in the construction of the ray-focussing surface: that he is working from a source which represents a phase of the study of this problem from the time before Diocles, and perhaps specifically that form of the solution due to Dositheus.[101]

The rest of Diocles' writing presents three further efforts in the solution of geometric problems. We have already discussed his solution of Archimedes' problem on the division of the sphere, where Diocles, like Archimedes, reduces the given problem to the intersection of two conics.[102] In Prop. 10 Diocles turns to the cube duplication, showing how it may be solved by means of two intersecting parabolas. Although this is precisely the same two-parabola method Eutocius presents as an alternative construction following the parabola–hyperbola method assigned to Menaechmus, we cannot infer that Diocles' inclusion of the two-parabola form must deny its origination with Menaechmus.[103] It is conceivable, if unlikely, that Diocles happened not to know of Menaechmus' effort. But one may observe that Diocles effects the construction of the cube duplication by means of curves constructed to satisfy the focus-directrix property in the pointwise manner of Prop. 4. Only as an afterthought, at the very end of the construction, is the comment made that these curves are in fact parabolas.[104] It would thus appear that Diocles' intent here is not to offer a new solution to the problem, but rather to provide another example of his method of drawing parabolas. His approach becomes the more understandable under the view we proposed earlier, that Menaechmus himself had employed a form of pointwise procedure for these curves.[105]

The derivative character of Diocles' constructions is evident not only in this first form of the cube duplication, modelled on that of Menaechmus, but also in the case of the second, developed in his Props. 11–16; for although the latter is ostensibly an original contribution, it too can be viewed as the elaboration of an earlier method, namely, the one Eutocius ascribes to Plato. Let us consider the figure sought in the latter, three similar right triangles such that the sides θH, HG are given lines (Fig. 28). A device is to be manipulated where θ and G are fixed points, D is held on the line GH extended, and the angles at D and at M are right angles; when M comes to cut the line θH extended, the construction will be obtained. Now the conditions on the two right angles may be secured alternatively by keeping D and M on the circumference of the circle of diameter GD, where GD is taken as a fixed line. Then θ will be the intersection of two lines, the perpendicular to MD at D and the perpendicular to GD passing through

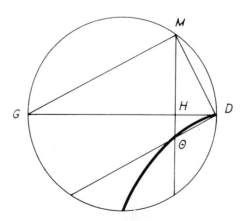

Figure 28

M. As M varies, θ will trace a curve, and the ratio θH : HG will continually decrease, assuming each value from 1 : 1 down to arbitrarily small values, as M passes from the position bisecting arc GD along the arc toward D. At every position, DH, HM will be the two mean proportionals between θH, HG. Hence, to find the means X, Y between given lines S, T, we need only find the point θ on this curve such that θH : HG = S : T (assuming T to be the greater line) and then construct the two lines X, Y via the proportions X : DH = Y : HM = T : HG (Fig. 29).

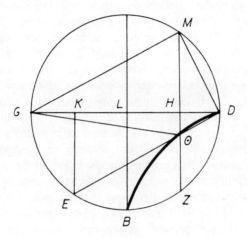

Figure 29

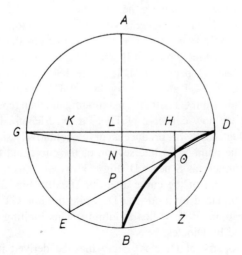

Figure 30

The curve used here is identical to that introduced by Diocles. He constructs it by dividing the given circle ADBG into quadrants, then marking off pairs of equal arcs, like DZ, GE, from the ends of diameter GD in the lower semicircle (Fig. 30). Then the intersection θ of the vertical ZH with the chord DE defines a point on the curve. Diocles establishes that for each such point, the lines DH, HZ are the two mean proportionals between θH, HG. Then from the given lines S, T, one sets NL : LG = S : T and extends GN to meet the curve in θ. The two mean proportionals can then be found via the continued proportionality of θH, HD, ZH, HG, as before.

Variant forms of Diocles' construction appear in the late commentators Sporus and Pappus. Both frame the procedure as a *neusis*, in which a chord from D is to cut the given line GN extended at θ such that the intercept θE is bisected by LB. In this way, the curve itself is eliminated. From details in the figures one may detect that Sporus is working directly from Diocles (as in Fig. 30), while Pappus seems to mount an independent reworking of the pseudo-Platonic method in the manner of Fig. 29. Indeed, Pappus alone of these three treatments presents his method in a form which manifests the underlying role played by the pseudo-Platonic method, although, as we have maintained, the latter is surely implicit in the approaches adopted by Diocles and, through him, of Sporus as well.[106]

An interesting relation between the curves produced through the pseudo-Platonic and Dioclean methods suggests an alternative route by which Diocles could have hit upon his approach. Returning to the pseudo-Platonic method, we consider the points D, R as fixed and line DR the hypotenuse of the right triangle DSR [Fig. 31(a)]. If we introduce an arbitrary line through R meeting DS extended in V, draw through D the line parallel to RV, and mark as E the point where this line meets the line perpendicular to RV at V, then E traces a curve as the initial ray is made to rotate on R. The name "ophiuride" was coined for these curves in the 19th century, but we have proposed that this very curve was so generated by Eudoxus for solving the cube duplication,[107] for its intersection with RS extended yields the two lengths VS, SE as mean proportionals between the given lines RS, SD. If we continue the construction, the curve will enter into a loop through the double point at D, emerging to tend asymptotically from below toward the line GT parallel to DS, where TS = SR. On the other side of D, the curve will meet GT along the ray through D perpendicular to DR, will increase its separation from GT, attain a maximum, and then tend asymptotically back toward it. If we now conceive of TR as a fixed length, but allow SD to diminish, each value of DS will produce a curve qualitatively similar to that just described, where the point W of intersection of the curve and the line GT will tend outward indefinitely as S nears D [Figs. 31(b) and 31(c)]. In the limiting case, however, the form of the curve alters drastically [Fig. 31(d)]. The loop disappears, being replaced by a cusp at D, while the line GT becomes a true asymptote, for the point W now lies at infinity. This limiting curve is in fact the same curve used by Diocles.

The defining property of Diocles' curve may be derived in the following manner [Figs. 31(e) and 31(f)]: any point θ on it is found as the intersection of

The Successors of Archimedes in the Third Century 243

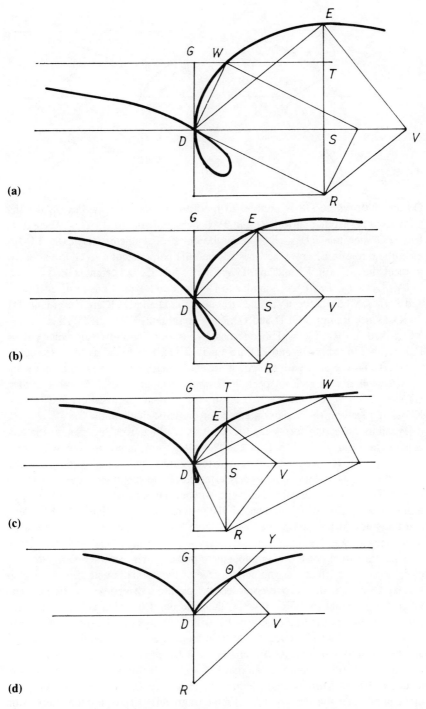

(a)

(b)

(c)

(d)

Figure 31

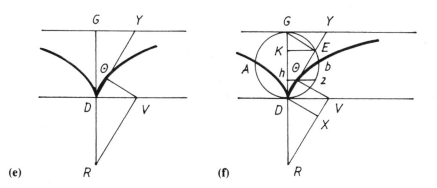

(e)   (f)

Dθ and θV, where Dθ is parallel to the arbitrary ray RV, V lies on the line through D perpendicular to DR, and θV is perpendicular to RV at V. We extend RD to G such that RD = DG, draw the circle of diameter GD, extend Dθ to meet the circle in E, draw the lines EK and θH perpendicular to GD, extend Hθ to meet the circle in Z, and join GE. Since angle GED is right, GE is parallel to θV. Draw the line perpendicular to DG at G and extend DE to meet it in Y, and draw DX parallel to θV. Since triangles GYD and DVR are congruent, DY = RV; since triangles GED and DXR are congruent, DE = RX. Thus, XV = YE. Since further XV = Dθ, one has YE = θD. It follows also that GK = HD, so that the arcs GE and ZD are equal. Thus, θ is found as the intersection of DE, HZ associated with the equal arcs GE, ZD, as Diocles demands for the construction of his curve. Inspection of similar triangles establishes that DH (= GK) and HZ (= KE) are the two mean proportionals between θH and HG (= KD), thus leading to Diocles' method for the cube duplication.

We thus have two forms for the analysis leading from the pseudo-Platonic construction to that used by Diocles. The latter form, although somewhat more complicated, gives rise more naturally to the figure actually employed by Diocles, when the corresponding synthesis is effected. It presupposes a detailed investigation of the properties of the curve; but under our view of the Eudoxean origin of the technique on which it is based, the generation of the related "ophiurides" and the study of their properties would have been a prominent part of the work underlying Diocles' effort.

The alternative construction happens to make evident an interesting property of Diocles' curve: that it is the pedal curve of the parabola with reference to its vertex.[108] Specifically, if one considers the parabola of vertex D, axis DR, and latus rectum equal to 2GR = 4RD, then the point θ which is the foot of the perpendicular from D to the tangent PT will lie on this pedal curve [Fig. 32(a)]. To see this is Diocles' curve, one introduces the perpendiculars to the axis at D and G and draws the circle of diameter DG. Since PT is tangent, the abscissa SD equals the subtangent DT, so that PV = VT. It follows that RV is perpendicular to PT, as is proved in the lemma to the demonstration of the focal property of the parabola in the version of the Bobbio fragment.[109] Thus, θ is the intersection of the perpendicular lines Dθ, θV, where V lies on the given line and Dθ, RV

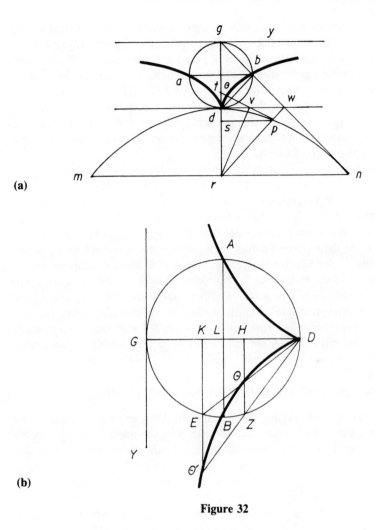

Figure 32

are parallels emanating from the given points D, R; that is, θ lies on Diocles' curve, as constructed in the alternative manner.

Although our sources do not indicate to what extent, if at all, Diocles examined the properties of his curve beyond its role for the cube duplication, the precedent of Nicomedes' study of the conchoids and Archimedes' study of the spirals would suggest that he did pursue such a line of research. Of course we cannot know on the basis of extant evidence what other properties Diocles discovered; but one important feature of the curve could hardly have been missed: that its extension beyond the quadrant DB continues indefinitely toward the line GY as asymptote. For one need only continue the sequence of chords and verticals in the obvious way, so that points θ' are found through the intersection of KE and DZ extended [Fig. 32(b)]. The same definition produces a symmetrical branch in the upper semicircle. The curve's asymptotic extension was evident to the geometers of the 17th century. Roberval also introduced as alternative the mechanical conception of the curve as the locus of points at the intersection of a first line rotating uniformly on D from the initial position DG and a second line HZ, Z moving along the circle uniformly from D toward G.[110] This conception proved especially fruitful for finding the tangents via the composition of motions, a method apparently absent from the ancient geometry. But other methods, in particular the remarkable mechanical construction devised by Newton,[111] are very much in the spirit of the ancients.

In the modern theory of algebraic curves, one defines Diocles' curve via the condition $D\theta = EY$ [as in Fig. 31(f)], a property which arose during the analysis based on the pseudo-Platonic construction. The modern form is readily extended to a family of curves where each is specified by the condition that the radius vector drawn from a given fixed point (pole) to any point on the curve equals the length along the extended radius vector intercepted between two given reference curves.[112] For Diocles' curve, the pole is at D and the reference curves are the circle and the tangent at G. But other choices of lines or circles will yield new curves, from which, for instance, alternative constructions of the Archimedean *neuses* might be given. Both Diocles' curve and the family defined through its generalization are now called "cissoids." But the ancient references to curves having this name cannot be to Diocles' curve. This is the question we take up next.

ON THE CURVE CALLED "CISSOID"

In the writings of Roberval and other geometers from about the middle of the 17th century, one finds a consensus on assigning the name "cissoid" to the curve defined by Diocles in connection with the cube duplication.[113] Although no ancient text known then or now makes this identification explicitly, its status as an inference seems already to have disappeared from the mathematical literature by this time, so that the original grounds for maintaining it are no longer clear. Presumably, the question arose in discussion and correspondence, and a comparison of references from Proclus to a curve called "cissoid" but not there defined, with Eutocious' account of Diocles, where a curve is introduced but not named, somehow led some to suppose that the two contexts referred to the same curve. As Toomer notes in his edition of the full text of Diocles, however, the absence of the name from Diocles' version is established, so that doubts on this matter have been confirmed: the identification is not only inferential, but quite surely false.[114]

To appreciate this, we need only take note of the several references Proclus makes to the "cissoid." Although he never actually defines the curve, he indicates several of its characteristics in a qualitative manner: (i) it is a finite curve which encloses a figure, as do the circle and ellipse, for instance; in this it is specifically to be distinguished from the hyperbola and parabola and other such curves which extend to infinity; (ii) it comes to a point, as do the leaves of ivy,

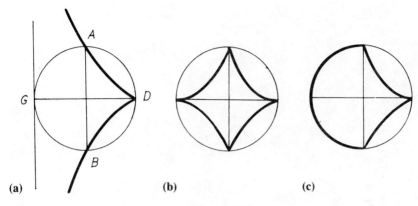

Figure 33

whence the name "cissoid" (i.e., "ivy-shaped"); (iii) it makes an angle with itself, as do the hippopede, the conchoid, and the spiric curves—that is, it bears comparison with curves which enter into loops, like a figure-8; (iv) it is a "mixed" curve, like the spirals, for instance, but distinct from "irregular mixed" figures bounded by two different curves, e.g., segments of circles.[115] Since Diocles defines his curve only with respect to a single quadrant of its reference circle, his form of it has none of the properties indicated by Proclus. Yet we have noted that his procedure makes evident the manner of its completion, and this produces a curve with a cusp at D, extending symmetrically on both sides through the points A and B toward the asymptotic line GY [Fig. 33(a)]. It thus has property (ii), but is incompatible with (i). In fact, Roberval and other 17th-century geometers were well aware of the infinite extension of the branches of their "cissoid," and so must have assumed that the ancient conception of the curve somehow differed from their own.[116] In particular, property (i) requires that the curve be made to enclose a figure. Tannery proposes that a double reflection of Diocles' single-quadrant curve was envisioned, yielding a four-cusped shape [Fig. 33(b)].[117] Loria suggests another form, where Diocles' segment is reflected and then filled out by inclusion of the opposite half of the circumference of the reference circle [Fig. 33(c)].[118] The latter might be especially convincing as suggestive of an ivy shape, and modern treatments invariably refer to this figure to account for the name of the "cissoid;" but the very fact that Tannery and Loria are in the position of having to propose such figures reveals that neither had direct knowledge of the rationale adopted by Roberval and his colleagues, not to mention that of the ancients. As it happens, both figures are composed of different curves, and so are of Proclus' "irregular mixed" class of curves; property (iv) of Proclus' cissoid thus directly excludes both suggestions from being the form of curve intended by the ancients.[119]

Which other curve, then, could the ancient "cissoid" have been? A most promising clue to the answer, I believe, lies in Pappus' linking this curve with those of Nicomedes in both of the passages where he uses the term:

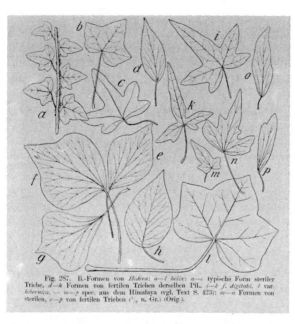

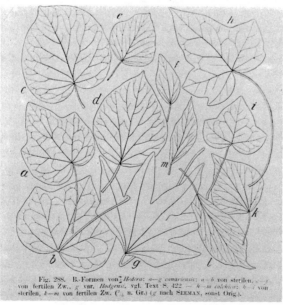

Plate I. Forms of ivy: *hedera helix* (above); *hedera canariensis* and *hedera colchica* (below). *H. helix* is common to central and western Europe and North America; *h. canariensis* now flourishes in northwest Africa; *h. colchica* is the predominant form in the Caucasus. Note that the familiar multilobed leaves typify the sterile phases of growth, while lobeless obovate and cordate forms mark the fertile phase. (Reproduced from C. K. Schneider, *Illustriertes Handbuch der Laubholzkunde*, Jena, 1912, II, pp. 421, 423.)

The specimens in Pl. II-VIII display several different uses of the ivy motif, always with the cordate form of leaf, made by the Attic vase painters: on wands and streamers in depictions of the Dionysian rituals, on wreaths, or simply as border decoration. To appreciate how common these uses were, one may survey the following list of specimens, each displaying ivy among its painted figures, as illustrated in a highly regarded reference account (P. E. Arias and M. Hirmer, *A History of Greek Vase Painting*, rev. B. B. Shefton, London: Thames and Hudson, 1962): the numbers refer to full-page photographic plates, the Roman numerals to color plates; items in parentheses are details or alternative views of the preceding item. The artisans are named in roughly chronological order, spanning the period from the mid-6th century to the late 5th century B.C.:

Amasis painter XV (57); Exekias 65; Lysippides painter (or manner of) XXI (69); Tarquinia mask-painter XXII; Lysippides painter 89; Phintias 92 (94, 95); Oltos 100 (104); Kleophrades painter 122 (−124, XXX, XXXI); Makron 133; Brygos painter XXXIV (143); Berlin painter 150(−153); Pan painter 160(−163); Phiale painter XLIV; Kleophon painter 196(−199); Dinos painter 206(−211); Prononomos painter 219; Pompe painter 224; Karneia painter 230(−233).

We may also cite two items where ivy appears in a rounded, dot-like form: Rycroft painter 70; Nikoxenos painter (near) XXIII. Modified cordate forms are evident in an item from the 4th century: Asteas 240.

This list indicates that many of the major painters from this period produced noteworthy specimens in which the ivy motif was prominent. The list cannot be taken as exhaustive, of course, since any number of the excluded artists may have figured ivy in other works not shown. But considering this to be a random sample of fine vases (for one can hardly suppose that Arias and Hirmer selected these vases on the basis of their depictions of ivy), we count 18 specimens bearing the cordate motif among the drawn figures and 3 others with modified ivy patterns; among the same sampling 71 do not bear the motif (although some of these might bear it in portions not reproduced by Arias and Hirmer). As a rough estimate, then, about one-fifth of the specimens include ivy among the figures. One may observe that the frequency of ivy patterns owes considerably to the strong association of ivy with the Dionysian rites, a favorite subject for the vase-painters, and fully appropriate in view of the contents of many of the vessels so decorated.

Many other vases display ivy as a border pattern, framing the scene depicted on the body of the vessel, for instance, or quite commonly, tracing along the curved line of the handles of amphoras. This is the case for the following items reproduced by Arias and Hirmer (as above, a number in parentheses refers to an alternative view of the preceding item): 62 (63), 66 (68), XXI, 71, 72, 80, 82 (87), 88 (89), 90 (91), 92, 112, 113, 116 (117), 130 (131), 150 (151), 165, 168 (169), 170, 176 (177, 181), 224, 230, 236 (237) — all of which employ the cordate form; 192 (dot-form); L (three-sided, concavo-convex). In most of these cases the ivy motif appears on amphora handles (as in Pl. V, VI and VII), often on the handles only; where the motif also appears among the figures on the body of the vase, it need not be precisely the same as on the handles (cf. Pl. VI). In view of the invariance of the ivy motif of the handles, it seems to me pointless to associate its appearance with any one or another artist; presumably, the execution of the border ivy could be delegated to less skilled members of the workshop. Common in Attic work after the mid-6th century, the ivy motif is rare before then. A Rhodian vase from the 7th century (shown by E. Buschor, *Griechische Vasen*, new ed., Munich: Piper, 1969, no. 58) bears a horizontal band of ivy-like cordate forms, somewhat in the manner of the Amasis painter (cf. Plate III). But the handles of amphoras, if decorated at all, sometimes bear simple triangles (as in the 7th-century Attic vase, *ibid.*, no. 50), and conceivably, this evolved into the later ivy motif (cf. also the grape-leaf border in *ibid.*, no. 118, and Arias, no. 40).

The cordate forms of leaf, as noted, characterize the entire group of vases from the 6th and 5th centuries; if we add to our previous count those specimens which bear ivy in border decoration, we find that it has intruded into almost two-fifths of the sampling. But the accuracy of the statistic does not matter. What these data make clear is that the ivy, in its cordate form, is an extremely popular motif in the classical art, and that the appearance of any other shape for the leaf is merely a rare exception to the stereotype.

Plate II. Dionysos and attendants: detail from a vase by the Amasis painter (after the middle of the 6th century B.C.). Long ivy branches bear leaves of cordate form, blunt at the tip. Leaves of similar shape are visible in the wreath on Dionysos' head. (From the Staatliche Antikensammlungen, Munich, no. 1383 #5; reproduced with the permission of the authorities of the Antikensammlungen.)

Plate III. Maenads presenting a sacrifice to Dionysos: detail from a vase by the Amasis painter. Note, as in Plate II, the branches of ivy and the ivy wreaths with leaves of the cordate form. This amphora is in the Cabinet des médailles of the Bibliothèque nationale, Paris (no. 222) and is reproduced here with the permission of the Bibliothèque nationale.

Plate IV. Details of figures of Dionysos (above) and a maenad (below) from a vase by the Kleophrades painter (early 5th century). Ivy leaves are conspicuous in the wreaths worn by each, as well as at the ends of the thyrsos-wands carried by the god's attendants (top left above; top left below). Note how the cordate shapes of the leaves are considerably rounded in the maenad's wreath. (From the Staatliche Antikensammlungen, Munich, no. 2344; reproduced with permission.)

Plate V. Dionysos with maenad and satyrs from a vase by the Lysippides painter (late 6th century). As with the Amasis painter (Plate III), ivy branches stream behind the god, but here (as with Kleophrades, Plate IV), they are balanced by grape vines to his front. The cordate ivy motif is evident also in the wreaths worn by all figures, as well as in the border decoration on the handles. (From the Musée du Louvre, no. F 204; reproduced with permission.)

Plate VI. Dionysos and satyrs: detail from a vase in the manner of Exekias (mid-6th century). As with the Amasis painter (Plate II), a branch of cordate ivy leaves trails behind the god, while an ivy wreath circles his head. The virtual explosion of swirling grape vines together with the ivy motif traced along the handles of the amphora fixes a style continued by later vase painters (cf. Plate V). (From the Museum of Fine Arts, Boston: Henry Lillie Pierce Residuary Fund and Bartlett Donation, no. 63.952; reproduced with the permission of the Museum of Fine Arts.)

Plate VII. Hermes and satyr: amphora and detail by the Berlin painter (early 5th century). Cordate ivy leaves appear in a wreath worn by the satyr; a branch of sharply-pointed ivy leaves forms a band about the neck of the vase. As in the Exekias vase (Pl. VI) branches of rounded ivy leaves decorate the handles. (From the Staatliche Museen (Antikenmuseum) Preussischer Kulturbesitz, Berlin, no. F 2160, reproduced with the permission of the Staatliche Museen.)

256   Ancient Tradition of Geometric Problems

Plate VIII. A cup of the so-called "Cassel group" (and detail) displaying an ivy pattern of cordate leaves with exaggerated lobes and blunt tips. Several dozen cups have been associated with this group, among which those marked by this ivy design appear to form a principal subgenre. (From the Stanford University Museum of Art, Leland Stanford Junior Collection, no. 17437, reproduced with permission.)

[In addition to the "planar" and "solid" classes of problems] there is left yet a third class, which is called "linear." For lines other than those just mentioned [sc. the straight line, the circle, and the conics] are taken for their construction; they have a more diverse and complicated generation, such as happen to be the . . . and quadratrices and conchoids and cissoids, having many paradoxical properties about themselves.[120]

A second passage, also explaining the three-class division of problems, illustrates the "linear" class by "helices, quadratrices, as well as conchoids and cissoids."[121] The coupling of "conchoids and cissoids" might of course be merely fortuitous. But it invites us to consider whether Nicomedes' study of the former might lead to a related class of curves having the properties Proclus assigns to the "cissoids." We have seen how Nicomedes produced the conchoids through the elaboration of certain *neuses* employed by Archimedes. Since the *neuses* both for the angle trisection and for the problems relating to the spirals require that the given length be inserted between a given line and a given circle, we at once perceive an alternative to Nicomedes' construction: instead of generating the auxiliary curve with reference to the line as base, so as to obtain the conchoid, we may generate another curve in the same fashion but with the circle as base. That is, we conceive of the curve each of whose points is a given distance from the given circle, where this distance is to be measured in the line toward a given point. The curves so formed are now familiar under the name of the "conchoids with circular base," and some scholars have already proposed that they were introduced by the ancient geometers in connection with these Archimedean constructions.[122]

Is it possible that these forms of the conchoid are the ancient "cissoids"? Let us first set out the various forms it can assume and then see how these relate to Proclus' descriptions. The shape depends on the relative sizes of the "distance" (d) and the radius (r) of the base circle and the distance between that circle and the "pole" (p). Just as for the linear conchoid, one obtains different forms corresponding to the measurement of the "distance" from the base either toward the pole or away from it. Furthermore, the forms will differ according to the pole's location on, outside or inside the circle.

The *neuses* for Archimedes' problems relate to cases of the curve where the pole lies on the circle. Three forms may be distinguished [Figs. 34(a)–34(c)]: (a_1) when d is less than $2r$, the curve has a double point at the pole, leading into an interior loop; (a_2) when $d = 2r$, the curve has an inward-directed cusp at the pole; (a_3) when d is greater than $2r$, it has a smooth indentation towards (but not reaching) the pole. These curves were known as the *limaçons* by 17th-century geometers, and Roberval assigns their invention to "M. Pascal," that is, to Étienne Pascal, father of the noted philosopher-mathematician.[123] The cusped form (a_2) is often selected for special study under the name "cardioid."[124]

By analogy with the Nicomedean conchoids, four forms of the circle-based conchoids result when the pole is taken outside the circle [Fig. 35(a)–35(d)]: (b_1) when the distance is marked off away from the pole, one obtains a reniform figure like a distorted ellipse; this corresponds to the "first conchoid." When

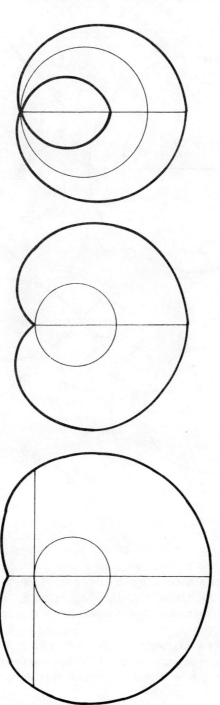

Figure 34

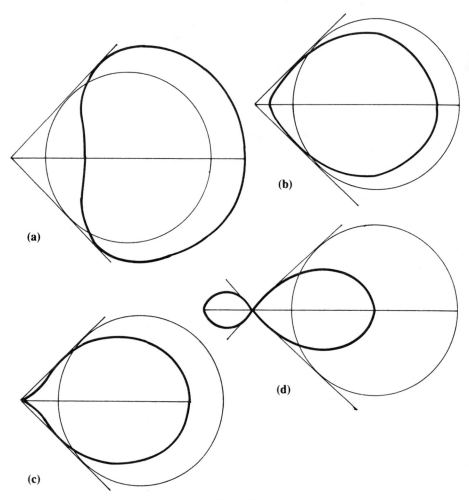

Figure 35

the distance is marked off toward the pole, three cases corresponding to the three forms of the "secondary conchoid" are produced: (b_2) when d is less than p, the curve encloses a rounded figure rising toward (but not reaching) the pole; (b_3) when $d = p$, it has an outer-directed cusp at the pole; the appearance is like that of a leaf, rounded at its attachment by the stem, pointed at its end; (b_4) when d is greater than p, the curve has a double point at the pole leading into a loop; it resembles an asymmetrical hippopede or figure-8.[125] Two further forms result when d equals or is greater than $p + 2r$; these resemble forms (b_3), (b_2), respectively, save that their orientation relative to the pole is in the opposite direction from the base circle.

When the pole lies within the circle, we obtain a family of nested convex curves (c), and when the pole is at the center of the base circle, these are concentric circles.

Since this curve admits a diversity of forms, we should be prepared to find that Proclus sometimes assigns to the "cissoid" a property appropriate to one or a few forms, but not to all. The situation in the case of the "conchoid" is analogous. When Proclus includes the conchoid among curves which converge indefinitely toward an asymptotic line, he could have in mind any of its four basic forms. But when he speaks of the conchoid now as extending to infinity *without* making a figure, later as making a figure, and then extending to infinity, he appears to refer to just one form of the secondary conchoid in the latter case, and to some other form of the conchoid in the former.[126] As for Proclus' observations on the "cissoid," his property (i) applies to all the forms we have distinguished for the circle-based conchoids, since they all are finite and enclose figures. They are all "mixed" curves, as required in property (iv), for their generation entails the combination of linear and rotational components.[127] For a form which crosses itself in the manner of the hippopede, as cited in property (iii), form (b_4) is suggested, although form (a_1), with its inner-directed loop, might also be possible. The key passage, however, relates to property (ii):

> whenever the cissoid lines converge to one point, as the leaves of ivy—for it is from this they have their name—they make an angle.[128]

This seems best suited to the cusped forms (a_2) and (b_3); for although the looped forms (a_1) and (b_4) also come to a point by virtue of crossing themselves, it is difficult to think of them as resembling leaves. If no single form of the circle-based conchoids satisfies all four of the properties mentioned by Proclus, each property applies quite naturally to some one or more of the forms. Under this view, then, one of the two cusped forms suggested the association with the ivy leaf and from this the name "cissoid" was applied to the whole family of curves.

Beyond Proclus' testimony we have a few other kinds of evidence which might help us learn what shape the Greeks thought of in association with ivy. The most straightforward approach would be inductive: actually to gather ivy leaves and examine their shapes, or to consult reference works in botany. One will come up with a welter of different forms: the leaf is typically described as broadly ovate, narrowing to a tip at the base (cf. form b_3), but often widening into a cordate, or heart-shaped, leaf (cf. forms a_2, a_3, b_1). These characterize the fertile stage of growth; in the sterile stage the familiar three-lobed and five-lobed leaves predominate.[129] We are thus no closer than before to a determination of the essential ivy shape. Even worse, can we be certain that the Greeks' *kissos* was indeed one of the many varieties we now signify by the name "ivy"?

In this matter, ancient writings on botany provide valuable assistance. In particular, Theophrastus (c. 300 B.C.), Aristotle's most distinguished disciple and successor as director of the Lyceum, compiled in his *Enquiry into Plants* a reference treatise on the plant species then known, parallel to Aristotle's zoological works.[130] Theophrastus takes up the discussion of ivy at several places, and devotes an entire section to it in Book III, Ch. 18. Among its principal characteristics, he notes (i) that ivy is distinguished into three types, called "black," "white," and "spiral," of which the last has the greatest variety of forms. Specifically, the spiral form of ivy has leaves which are smaller, more angle-shaped (*gōnoeidē*) and better balanced (*eurhythmotera*), while other forms of ivy are generally more rounded and simpler. (ii) There is a change in the shape of the leaf when the branch passes from its vertical climbing stage to its horizontal flowering stage; the leaves now tend to lose the angular and divided form and to become more rounded and simpler. This transformation, he observes, had prompted some researchers of his time to suppose that the spiral-formed ivy naturally turned into the other varieties. (iii) The leaf of the lime is likened to the ivy, save that the lime leaf "rounds more gradually, being most curved at the part next the stalk, but in the middle contracting to a sharper and longer apex."[131] From these comments one may be confident that Theophrastus' spiral ivy is the same plant as Linnaeus' *hedera helix*, whose varieties are familiar in Europe and America today. Confirmation is obtained through the *Materia Medica* compiled by Dioscorides in the 1st century A.D. This work includes a section on the description and medical uses of some 500 species of plants, one of these being ivy. An elaborate illustrated manuscript was commissioned for the Emperor early in the 6th century, and from this one can distinguish the form of ivy common today.[132]

Although these ancient writers might seem merely to confirm our own experience of the diversity of shapes of ivy leaves, Theophrastus clearly treats the simple leaf of the fertile phase as a paradigm, in preference to the multilobed shapes of the sterile phase. This serves to weaken the suggestions made by Loria and Tannery on the identification of the "cissoid" and to support our view that this curve should be taken as one of the circle-based conchoids. The cusped forms b_3 and a_2 are of particular interest, especially the latter in view of the association with the cordate lime leaves.

Yet another line of inquiry is open to us: to consider how ancient artists drew the ivy leaf. Indeed, this is surely the most appropriate line to pursue. For the geometer seeking a suitable name for a curve is not likely to consult explicit, starkly realistic representations of natural forms, but rather to refer to the general impressions of these forms which he shares with others in his culture. The situation of the artist can be quite the same, for often the purposes of communication are as well or better served by introducing the generally accepted images of things than by a slavishly realistic picture. A survey of ancient vase painting, for instance, reveals that the ivy motif was a popular one throughout the classical period. It is prominent in the iconography of Dionysos, god of intoxicated inspiration: his *thyrsos* wand is entwined by ivy, while he and his frenzied followers are crowned with ivy wreaths. The borders and handles of vases commonly bear chains of ivy.[133] Here again, one encounters a diversity of forms, including both the simple and the multilobed leaves. But on balance, the prevalent form appears to be the cordate or heart-shaped leaf—with two cusps, one pointed inward, the other opposite and outward. As often as not, the cusps are rounded, leaving a smooth V-shape, somewhat like an exaggeration of forms a_3 and b_1 of the circle-based conchoid.[134]

These considerations show that for the ancients, as for us, there is no single shape which can be set out as "*the* ivy shape." Yet among these forms, that of the cordate leaf seems to be recognized as a paradigm for the others. In similar fashion, Proclus' remarks on the curves called "cissoids" refer to a class of related forms, rather than to a single curve. This in itself lends support to our identification of the "cissoids" with the circle-based conchoids; for we may distinguish at least eight basic forms for the latter, many of which have a convincing resemblance to one or another of the ivy forms. There would thus be a certain poetic justice in assigning these curves this name, since the diversity of the mathematical forms so neatly matches that of their natural eponyms. One also perceives the parallel with the line-based conchoids of Nicomedes. After all, what is *"the* shell shape"? What aspect of the *konchē*, i.e., the mussel or cockle, suggested the name "conchoid," and in connection with which of the four forms of Nicomedes' curve? Conceivably, the mutual diversity of forms was a factor encouraging their association.

Admittedly, none of the forms of the circle-based conchoid precisely matches the cordate shape which we have found to be the paradigm for ivy. Closest to it is form a_2, having the sharp indentation, but lacking the pointed base. As we have seen, this curve together with the related looped form a_1 has a firm geometric context growing out of Archimedes' *neusis* constructions, just as do Nicomedes' conchoids. Does it resemble the cordate ivy leaf of the vases closely enough to have attracted this name? We noted earlier that Étienne Pascal assigned the name *limaçons* (or "snails") to the family of circle-based conchoids. But since Castillon, near the middle of the 18th century, the cusped case a_2 has been known under the special name of the "cardioid" for its resemblance to the heart shape. But as one learns from artistic representations of hearts, as in playing cards, jewelry, and symbolic pictures of the period, the general image of the shape was the same then as now: two rounded lobes separated by a sharp indentation at the top, tapering to a sharp point at the base. This is of course the same shape as the cordate leaf which we found to be the ancients' favored image for ivy. Under the present view, then, one and the same curve received the name "cissoid" among the ancients and "cardioid" among the moderns by virtue of its resemblance to one and the same shape. The names may be different, but the psychological rationale underlying the process of naming turns out to be the same.

DIONYSODORUS, PERSEUS, AND ZENODORUS

Three geometers of the late 3rd or early 2nd century B.C. further reveal the interest in elaborating upon the problem-solving efforts of Archimedes. Dionysodorus, whose alternative solution via conics of Archimedes' problem of dividing the sphere we have already taken up, employed a method based on centers of gravity to determine the measures of solids of revolution, in particular, of the torus, or anchor ring. Perseus took up the study of this solid, to work out the properties of the curves formed as its planar sections. Zenodorus contributed to astronomy and to the study of burning mirrors, as we noted in connection with Diocles; his principal effort in geometry deals with the relative sizes of isoperimetric figures. Although these geometers surely had more to their credit than is contained in this brief description, it is the whole of their work as we learn

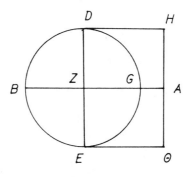

Figure 36

of it through the extant ancient sources. It is enough, however, for appreciating the strong influence of Archimedes, both through his general methods, as in geometric mechanics, and through specific results he discovered, as on the measurement of plane and solid figures.

In *Metrica* II, 13 Hero sets out the measurement of the volume of the torus as follows:

> It has been proved by Dionysodorus in his writing *On the Torus* that the ratio which the circle BGDE has to half of the parallelogram DEHθ is the same as that which the torus generated by circle BGDE has to the cylinder of which the axis is Hθ and the radius of the base is Eθ.

Here the torus is produced by rotating the given circle of diameter BG about the fixed axis Hθ such that the center Z of the circle is kept at the fixed distance AZ from the axis (Fig. 36). Hero illustrates the rule by computing the volume of the torus for which AB is 20 and BG is 12. But the calculation is effected a second time, where a different rule is obtained "after the torus has been rolled out to become a cylinder"; that is, it equals the cylinder whose base is the rotated circle BGDE and whose height equals the circumference traversed by its center Z. For AZ = 14 and BG = 12, the volume is worked out as $(22/7) \times 28 \times (1\!1/14) \times 12^2$, or 9956 $4/7$. This alternative form is familiar today as an instance of the general rule for the measurement of solids of revolution presented by Paul Guldin in his *De Centro Gravitatis* (1640): if a solid is generated through the revolution of a given area, its volume equals that of the cylinder whose base is the given area and whose height is the distance traveled by its center of gravity.[135] Guldin's originality in the independent discovery and proof of this rule has been persuasively argued by P. Ver Eecke.[136] Yet its appearance in Book VII of the *Collection* has moved some to assign priority in its discovery to Pappus. We read there as follows:

> the ratio of complete figures of revolution equals the ratio compounded of the ratio of the figures rotated and the ratio of the lines drawn similarly from their centers of gravity; the ratio of incomplete figures of revolution [sc. their sectors] is that compounded of the rotated figures and of the arcs which their centers of gravity

make, where the latter ratio is that compounded of the ratio of the lines drawn [from the centers of gravity to the axes] and the ratio of the angles which their extremities make at the axes of the figures of revolution.[137]

This is of course equivalent to the rule of Guldin. One notes that Pappus' rule covers cases not only of solids of revolution, but also of areas so formed. Moreover, his version of the rule assigns a prominent role to the technique of compounding ratios.[138] In this it echoes the first form of the measurement of the torus which Hero takes from Dionysodorus. There is another coincidence: when Pappus goes on to signal the generality of the rule, he observes that it permits in a single demonstration the proof of the measures of many figures, "some not up until now proved, others already proved, such as those in Book XII of these very *Elements* [sc. of Euclid]."[139] In the same way, in the *Definitions* based on a tradition from Hero, the account of the torus generalizes the conception to include not only circular cross sections, but also squares and others. The latter produce cutouts of cylinders and many other sorts of figures.[140] It thus appears that both Hero and Pappus refer to a discussion of solids of revolution which uses the torus as a paradigm instance and which specifically indicates the figures treated in Euclid's Book XII (i.e., circles, cones, cylinders, spheres) as special cases. In view of the striking conformity of the expression of the measuring procedures by Dionysodorus and by Pappus, we can hardly doubt that the general theorem, with its proof and applications, was given by Dionysodorus in his writing on the torus.

The Archimedean background to such a study is clear, for to Archimedes is due both the introduction of the concept of center of gravity as the basis of the formal geometric study of mechanics and also its application toward the discovery of theorems on the measurement of geometric figures via the "mechanical method." A route toward Dionysodorus' rule is readily seen with the assistance of the concept of "moment," the product of mass and distance. If we divide the given area A into indivisible elements of width w, for each of which r is the distance of its center of gravity from the axis of rotation, then the sum of all the moments $r \cdot w$ equals $A \cdot x$, where A is the area of the given figure and x is the distance of its center of gravity from the axis (Fig. 37). If we next consider the solid generated by revolving A about the axis, where the distance x remains constant, then each element w corresponds to an annulus of area $2\pi r \cdot w$, and

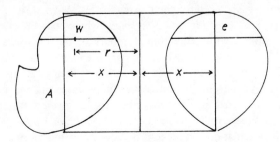

Figure 37

the sum of all of these elements is the volume V of the solid. Comparing sums, one sees that $V = 2\pi x \cdot A$, so that the solid equals the cylinder of base A and height $2\pi x$, the circumference traced by the center of gravity of A. It was by a method of indivisible moments of this sort that Cavalieri derived this theorem,[141] while the indivisibles can be replaced by finite elements of area and volume, as in the formal "exhaustion" proof adopted by Guldin.[142]

One should not deny that Dionysodorus might possibly have depended on an intuitive use of moments for the discovery of his theorem. But the ancient style of geometric mechanics as instituted by Archimedes did not admit the concept of "moment" as such.[143] Within a geometric theory framed about the ratios of homogeneous magnitudes, the notion of a product of different kinds of magnitude, like distance and weight, must be viewed as formally incoherent, even if they might enter into informal treatments. Now, both Hero and Pappus appear to know of a regular proof of Dionysodorus' theorem, and its expression in terms of compound ratios and proportions indicates that Dionysodorus adhered to the usual forms in geometric mechanics. We must thus seek an alternative manner for deriving his theorem.

An appropriate model may be found in Archimedes' mechanical manipulation of indivisibles in *The Method*, together with the technique of proportions used in Pappus' versions of the measurement of the Archimedean and spherical spirals.[144] Consider, as before, an indivisible element w of the given area A. By the principle of equilibrium, a second element of width $(r/x)\,w$ set at the distance x will exactly balance w set at the distance r. Hence, the sum of all such elements $(r/x)\,w$ set at x will balance the entire figure A set in position. But the latter will balance a figure of the same area set at its center of gravity, that is, at the distance x. Thus, the sum of all the elements $(r/x)\,w$ equals A. In the solid of revolution generated from A, as above, the element w corresponds to the area $2\pi\,r \cdot w$. Let us set $e = (r/x)\,w$ and $f = 2\pi r \cdot w$ and introduce in a second area B the corresponding terms $e' = (r'/x')\,w'$ and $f' = 2\pi r' \cdot w'$. Then, $(e : e')(x : x') = f : f'$. Hence, it follows that the ratio compounded of $x : x'$ and the ratio of all elements e to all elements e' equals the ratio of all elements f to all elements f'. That is, $(x : x')(A : B) = V_A : V_B$. This is precisely the form of the theorem stated by Pappus. In the special case of the torus and the cylinder, $x : x' = 2:1$, thus yielding the rule Hero draws from Dionysodorus. In this derivation, the step deducing the proportionality of the sums of elements merits special note. The same step is essential throughout Archimedes' *Method*; its formal proof is given as Prop. 1 of *Conoids and Spheroids*, where it is taken only for finite sums and is applied within the formal proofs of the measures of conics of revolution.[145] We have seen it also in the measurement of the spirals as reported by Pappus. The formalization of this derivation of Dionysodorus' theorem presents no difficulty; the elimination of the indivisibles via the introduction of finite elements inscribed and circumscribed to the given areas follows on the pattern of the indirect proofs given in *Conoids and Spheroids*.[146]

It is unfortunate that no further information on Dionysodorus' writing on the torus has been passed down. His able treatment of Archimedes' problem of dividing the sphere makes clear his competence in the formal geometric tradition, so that his study could well have been an impressive contribution in the Archimedean manner. The precedent of Archimedes' *Method*, moreover, reveals how Dionysodorus' theorem is prolific with results not only on the measures of areas and volumes, but also on the determination of centers of gravity, for instance, of circular arcs and segments. There survive a few other traces of an ancient conception of figures generated through the motion of indivisibles,[147] but no evidence of its development to any degree comparable to that after Kepler and Cavalieri in the 17th century.

We do know, however, that the special properties of the torus attracted further attention. For Proclus cites two lines from an epigram:

> Upon finding three spiral-like lines on five sections
>   Perseus thanked the spirits on their account.[148]

Elsewhere Proclus more closely specifies that there are three cases of torus, according to whether the rotated circle is of radius equal to, greater than, or less than the distance of its center from the axis of revolution. These are called the "continuous," "open," and "interlaced" (or "interchanged") cases, respectively, and curves are formed as sections by planes parallel to the axis of revolution.[149] Proclus' statement of the number of such sections raises some difficulties for which Tannery's solution is usually adopted. Setting r as the radius of the generating circle and d as the distance of its center from the axis, we have the open case when $r < d$ (Fig. 38). (a) If the plane cuts the torus at a distance k between d and $d + r$, an oval is formed. (b) When k lies between $d - r$ and d, a barbell shape results. (c) For $k = d - r$, the section is shaped like a figure-8. Proclus observes, quite reasonably, that it "*resembles* a hippopede"[150]; but since the latter is a space curve, this cannot be a formal identification. (d) For smaller values of k, the section consists of two symmetrically disposed ovals; for $k = 0$, these become circles. In the continuous case, where $d = r$, one obtains forms of curves (a') and (b'), analogous to the first two cases for the open torus. The one remaining case, for $k = 0$, consists of two mutually tangent circles. For the interlaced torus, one class of sections (a'') have the form of ovals and correspond to the first class of sections of the open torus. A second class (b'') consists of barbell shapes, each with an oval about its center. For $k = 0$, the section consists of two intersecting circles. To these Tannery adds the three particular cases, where $k = d$, and works out a division into "three curves together with five sections" by treating the continuous torus as a case of the open torus.[151] But perhaps the phrase means only that three different forms of curve (i.e., simple, self-crossing, and disjoint) arise from five different dispositions of the cutting plane. For instance, if we group together the oval and barbell sections as a single class of closed curves, then the "five sections" might be taken from (a/b), (c), (d), (a'/b'), and (a''/b''), to yield representatives of each form of curve from each kind of torus. But, of course, the epigram is hardly a medium for precision, and its interpretation cannot affect one's view of the curves Perseus studied.

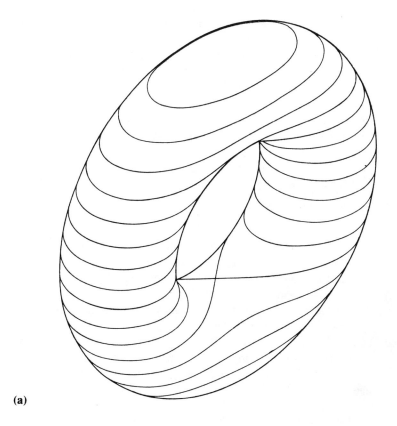

(a)

Figure 38

Proclus notes that Perseus worked out the *symptomata*, or principal properties of the sections of the torus, just as Apollonius had done for the conics, Nicomedes for the conchoids, and Hippias for the quadratrices.[152] An appropriate form for the quantitative relation expressing these sections is not difficult to derive. Let the generating circle of radius r ($=$ AB) be revolved about the axis OP such that the plane of the circle always passes through OP and the distance of its center from 0 is the constant length d ($=$ OA) (Fig. 39). We section the torus so formed by a plane parallel to OP and perpendicular to OAB at the distance k ($=$ OG) from the origin. Consider a circular cross section HZE of the torus, intersecting that plane in the line ZD, so that Z is a point on the section of the torus. Then, DGO is a right triangle and ZD is the mean proportional of ED, DH, where OE $=$ OB $=$ OA + AB and HE $=$ 2AB. If we set ZD $= y$ and DG $= x$, it follows that $y^2 = (r + d - \sqrt{x^2 + k^2}) \cdot (r - d + \sqrt{x^2 + k^2})$, whence $2d\sqrt{x^2 + k^2} = x^2 + y^2 + d^2 + k^2 - r^2$, or $2 \, \text{OI} \cdot \text{OD} = \text{OZ}^2 + \text{O}\theta^2$, for $\text{O}\theta$ tangent to the generating circle and I its center.[153]

The Successors of Archimedes in the Third Century   269

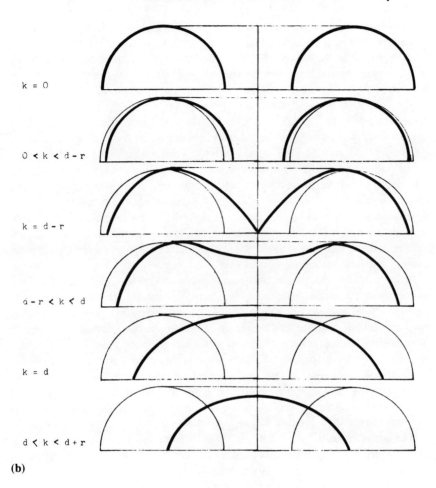

k = 0

0 < k < d - r

k = d - r

d - r < k < d

k = d

d < k < d + r

(b)

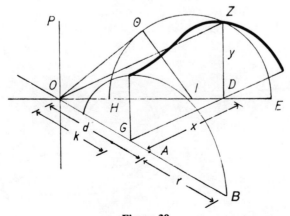

Figure 39

What further properties of these curves was Perseus capable of proposing? If one adopts the view of Heath, that Perseus wrote near the time Euclid,[154] hence only shortly after the first synthesis of the theory of the conics, then the range of techniques available to him for studying the sections of the torus could not have been great. On the other hand, we have maintained that the other geometers mentioned by Proclus in this context are to be dated rather later than this: Nicomedes after the middle of the 3rd century, Hippias somewhat later, and Apollonius toward the end of the 3rd and beginning of the 2nd century. Moreover, the studies of Dionysodorus would provide a natural stimulus for the further investigation of the torus. To be sure, Archytas' solution of the cube duplication introduces the curve formed by the intersection of a torus and a cylinder, and there is a certain affinity between Perseus' sections and the hippopedes of Eudoxus (constructible as intersections of the sphere and a cylinder).[155] But these sections are not in fact the ones taken up by Perseus, as far as we can tell, and neither context sets out for study the specific field of figures of revolution in as explicit a manner as Dionysodorus does.

A later dating for Perseus thus seems preferable, setting him near the time of the more advanced problem-solving activities toward the close of the 3rd century. Now, our survey of the studies of curves by the geometers in this environment has revealed that even if in principle one can generate any number of solids and then examine the curves formed as their sections,[156] those curves actually singled out for special attention—conics, spirals, conchoids, cissoids, quadratrices, the curve of Diocles—were of interest primarily for their applications toward the solution of geometric problems. We should thus expect that Perseus, too, discovered such applications of his curves. Among their properties, two are notable for a relation to locus problems.

The ancients well knew that the ellipse solved the locus of points whose distances from two given points have a constant *sum*.[157] But consider the complementary problem, where it is the *product* of the distances which is constant. Let Z be the locus of points such that $KZ \cdot ZL = N$, for given points K, L and given area N (Fig. 40). The curve assumes a variety of shapes—oval, barbell,

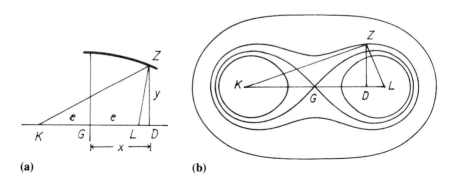

Figure 40

figure-8, disjoint ovals—according to whether N is greater than, equal to, or less than GK^2, for G the midpoint of LK. If we set $ZD = y$, $DG = x$, and $GK = e$, we require that $[(x - e)^2 + y^2] \cdot [(x + e)^2 + y^2] = N^2$, whence $(x^2 + y^2 + e^2)^2 = N^2 + 4x^2e^2$, or $x^2 + y^2 + e^2 = 2e \sqrt{(N/2e)^2 + x^2}$. Comparing with the primary *symptoma* of the sections of the torus, we see that by setting $e = d$, $k = r$, and $N/2\ e = k$, there results a section which satisfies the property of the locus. That is, the section of the torus for $k = r$ is the locus of points whose distances from points L, K (where $LK = 2\ d$) have the constant product $2\ dr$.[158]

An ostensible obstacle for the ancient examination of this locus problem lies in the need to consider fourth-order relations. This difficulty seems unavoidable in the analysis of the problem, as just given. But the ancient geometers were prepared to admit fourth-dimensional terms within informal derivations, as we see in Hero's account of the rule for the area of the triangle of given sides. He obtains the rule in the form $A^2 = s(s - a)(s - b)(s - c)$, for s the semiperimeter of the triangle of sides a, b, c and area A, and his method is entirely geometrical, even if the wider context is metrical-arithmetical.[159] Since Hero's rule is arguably of Archimedean provenance,[160] we may infer that geometers long knew of this ploy and could exploit it for the analysis of problems. As for the synthesis, the formal treatment can be regularized by introducing a unit line u and the auxiliary lines v, w such that $uv = 2\ xd$ and $uw = 2\ rd$. This device is utilized throughout the theory of irrationals in Euclid's Book X to circumvent the need for terms of dimension greater than 3.[161] Thus, in this artificial manner Perseus might produce a formally correct synthesis of the locus problem of constant product.

A second application of the sections of the torus is reminiscent of the role of the conics in the solution of the three-line locus. Given an equilateral hyperbola and two lines parallel to its ordinate axis, if a line parallel to the abscissa axis cuts them at points G, F, and H, respectively, then the locus of points I such that $HI^2 = FG \cdot GH$ is a section of the torus (Fig. 41). This theorem first appears in the modern literature under the name of R. de Sluse, in correspondence with Huygens in 1657.[162] But it is surely the type of problem which geometers around

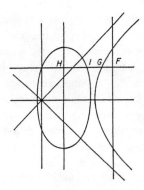

Figure 41

the time of Apollonius were investigating, and one might suppose that a skillful student of the sections of the torus could then have worked out its solution. It is of course impossible, on the basis of the extant documentation, to claim more than that results of this sort were possibly included in the work of Perseus.

The study of isoperimetric figures is another important area extending the work of Archimedes. In his commentary on Ptolemy, Theon reproduces an ample "epitome" based on the tract *On Isoperimetric Figures* by Zenodorus. Theon uses this to explicate Ptolemy's remark:

> that of the different figures having equal perimeter, the more polygonal ones are greater, so that the circle is greater among the plane figures, while the sphere is greater among the solid ones.[163]

In Theon's version, the theorem is established first for regular polygons, then for the circle in relation to these; two lemmas are included, one proving an inequality holding for the ratios of tangents and of angles,[164] the other proving in full the theorem on the area of the circle from Archimedes' *Dimension of the Circle*.[165] The theorem on isoperimetric plane figures is next made more general through a series of propositions building from irregular polygons toward the regular ones. Relative to the solid case of the theorem, the sphere is shown to be greater than any of the regular polyhedra, and then it is shown to maximize the composite conical solids which Archimedes introduces for proving his measurements of the sphere in *Sphere and Cylinder* I. Clearly, despite the general terms in which the isoperimetric theorem has been posed, the planar case has been effected only for the circle in relation to rectilinear figures, while the solid case deals with the sphere only in relation to a very small set of solids. This lack of generality is noted in a remark tacked onto the end of an alternative version of this material, a tract compiled on the basis of Theon's version by an anonymous editor, it would appear, in the 5th century, and included with several other short writings serving as an introduction to Ptolemy's Book I.[166]

A third version of the theorems on isoperimetric figures, given by Pappus in *Collection* Book V, is generally in good agreement with Theon.[167] To the planar section, Pappus adds a theorem on the segments of circles, to the effect that of segments having equal arcs the one which is a semicircle bounds the greatest area. This includes a number of lemmas on the area of segments and sectors which, as noted earlier, can be associated with Archimedes' study of the circle.[168] Pappus includes Theon's case comparing the sphere to the regular solids, but he omits the comparison with the Archimedean conical solids. Instead, he adds a theorem comparing the sphere to a certain cone and a certain cylinder (but not the general case of either), and uses this as a preface to a set of theorems on the regular polyhedra, showing, through a consideration of the appropriate special cases, that for these five figures having equal surface area, any one will have a greater volume than those of fewer faces. Within this section Pappus inserts an account of the thirteen semiregular solids of Archimedes and provides a full exposition of the theorems on the measurement of the sphere in a form alternative to Archimedes' *Sphere and Cylinder* I.[169]

One sees from these versions, especially Pappus' version, that the study of isoperimetric figures depended on Archimedes, not only for the proofs of essential theorems on the measurement of the circle and the sphere, but also for the context giving rise to an interest in this subject. In fact, Archimedes himself proves a theorem of this type in Prop. 9 of *Sphere and Cylinder* II: that of all spherical segments having equal spherical surface, the one which is a hemisphere bounds the greatest volume. This is the solid analogue of Pappus' case for circular segments and indicates that the study of isoperimetric figures had already received considerable attention around the time of Archimedes, well before Zenodorus composed his treatment of it. Indeed, the basic theorem was familiar even earlier, for Aristotle observes in *De Caelo* II, 4 that:

> the swiftest of all motions is that of the heaven. But of those proceeding from the same point to the same point, the circle is least. Furthermore, the swiftest motion is that corresponding to the least path. Hence, if the heaven is carried along a circle and moves most swiftly, it must be spherical in shape.[170]

In connection with this passage, the 6th-century commentator Simplicius notes:

> it has been proved even before Aristotle, in any event if he does indeed use it as having been proved, and by Archimedes and by Zenodorus more generally, that of isoperimetric figures the more spacious one among the planes is the circle, among the solids the sphere.[171]

Despite the clear importance of Archimedes' measurements of the circle and the sphere for the general background of this study, it would seem odd for Simplicius to speak in such terms of Archimedes, unless he had contributed notably to the specific area of isoperimetric figures. Further suggestions of Archimedes' explicit involvement with this study are to be found in his *Sand Reckoner*. Not only does he show himself to be familiar with the same inequality of ratios of tangents and angles essential for the comparison of isoperimetric regular polygons,[172] but also he invokes a theorem of a closely related kind:

> it has been proved that of every circle the diameter is less than its third part, or the third part of the perimeter of any polygon which is equilateral and more polygonal than the hexagon inscribed in the circle.[173]

This points to a prior proof that of regular polygons inscribed in the same circle, the ones with the greater number of sides have the greater perimeter. The proof would closely parallel that of the isoperimetric polygons.[174] Its dependence on the theorem on the area of the circle from *Dimension of the Circle* would assign to Archimedes himself responsibility for its proof, even though he does not expressly make this claim here.

It would thus appear to be no small task to extricate Zenodorus' own contribution to the study of isoperimetric figures from results earlier worked out by Archimedes. An unusual feature of the order of presentation in the various versions of this material may provide a clue. The anonymous version culminates in the proof that the circle maximizes the regular polygons after a series of proofs considering the relative sizes of regular and irregular polygons. This is the order one would expect, since the theorem on the circle is the principal result of the work. But by contrast, both Theon and Pappus prove the theorem on the circle *before* those on the regular and irregular polygons. This suggests that Zenodorus effected the latter results by way of generalizing what Archimedes or others had shown earlier. In this event, the Archimedean element in this effort by Zenodorus, already quite evident, amounts to a preliminary treatment of its chief results.

IN THE SHADOW OF ARCHIMEDES

Throughout the fragmentary evidence of the work of the geometers we have been considering in this chapter, the name of Archimedes appears in only two contexts: as the source of the problem of dividing the sphere taken up by Diocles and Dionysodorus, and as the source of the theorems on the area of the circle underlying the studies of the quadratrix and of the isoperimetric figures. But the importance of Archimedes' work to these efforts has been clear at every turn:

—Archimedes' studies of the spirals are an important factor behind Nicomedes' application of the quadratrix for solving the circle quadrature, while the Eudoxean limiting methods, so characteristic of Archimedes' work, prove indispensable for examining the properties of this curve.

—Archimedes' theorems on the measurement of the circle and the sphere are essential to Zenodorus' study of isoperimetric figures, a field to which Archimedes himself made significant contributions.

—Archimedes' concept of center of gravity and his elaboration of it into a "mechanical method" for finding the measures of curvilinear figures are the basis of Dionysodorus' theorem expressing the measure of figures of revolution, like the torus.

—Archimedean problem-solving techniques, like *neusis* constructions and the use of special curves, are a clear precedent for Nicomedes' use of the conchoids and may be perceived as a factor in Diocles' use of his special curve for the cube duplication.

—An Archimedean context arguably gave rise to the "Heronian" *neusis* for the cube duplication, and this in turn inspired Nicomedes' method of solution.

Since Archimedes' methods and results so strongly guided the research efforts of these geometers, we would naturally be inclined to view them as being nearly contemporary with him. Eratosthenes, of course, is known to be almost exactly contemporaneous with Archimedes, and served as correspondent in the receipt of the *Method* and other Archimedean writings. Since Nicomedes presents his own study of the conchoids as a biting response to Eratosthenes' manner of solving the cube duplication, we surely do well to place him about the same time; for the *ad hominem* tone would hardly suit a writing composed long after the event. As for the others, the argument for dating must be rather less direct. The central figure in this matter is one Philonides, a native of Laodicea in Syria, who early in his career became seriously interested in geometry, and in his maturity earned note as an Epicurean philosopher.[175] Apollonius calls him "the geometer" in the preface to *Conics* II and encourages the correspondent Eudemus to show Philonides the treatise at the latter's next visit to Pergamon. Extensive details of Philonides' life are recorded in a biography preserved among the remains unearthed at Herculaneum: he studied geometry both with Eudemus and with Dionysodorus of Kaunos (in Caria, Asia Minor); he tutored Demetrius, the heir-apparent to the Seleucid throne, around 175 B.C. or earlier, and was an influential teacher in Syria in the days of the elder Seleucid, Antiochus Epiphanes (reigned 175–164 B.C.) and his son, Demetrius Soter (reigned 162–150 B.C.); these scholarly activities brought him twice to Athens, site of the principal Epicurean school, and thus he came to meet one Zenodorus. Assuming that these individuals Dionysodorus and Zenodorus are the same mathematicians we have been discussing here, their period of activity would seem to fall around the early and middle parts of the 2nd century B.C. Since, furthermore, Diocles cites Zenodorus "the astronomer" as having posed the problem of the burning mirror, he too would be placed around this time.

These considerations conform well enough with the present view of the Archimedean background to the work of these geometers. But certain details present difficulties and suggest that this dating is too late by several decades. The data on Philonides have been taken to set his birth to approximately 200 B.C., so that he might have encountered Apollonius at about age eighteen (c. 182 B.C.) and tutored Demetrius at twenty-five or younger (that is, before 175 B.C.). Now admittedly, Apollonius was a mature man when he recommended "Philonides the geometer" in *Conics* II, for his own son was then already fully grown.[176] But surely it would be remarkable to refer thus to Philonides, if indeed he was barely twenty. Moreover, as Apollonius commenced his studies at the time of the third Ptolemy (reigned 246–221 B.C.) and is said to have studied under students of Euclid,[177] even the very conservative estimate of his birth as approximately 240 B.C. would assign the composition of *Conics* II to quite late in his life. We have already mentioned in passing another difficulty. Diocles mentions two problems on the contours of burning mirrors: the earlier form requiring reflection to the arc of a given circle was posed by Pythion and Conon; the later form seeking reflection to a given point was posed by Zenodorus and investigated first by Dositheus, then by Diocles himself.[178] Now, Conon died around the middle of Archimedes' career (c. 240 B.C.), after which Dositheus became Archimedes' regular correspondent. On the view that Zenodorus was a near contemporary of Philonides, his work on this problem could hardly have begun before approximately 180 B.C., so that any contribution by Dositheus to the same problem must have been entirely independent of Zenodorus' proposition and prior to it by around half a century. This extraordinary gap between the earlier and later stages of this study must cause uneasiness, as it surely violates the plain impression made in Diocles' preface. The late dating of Diocles also makes puzzling the absence of any clear sign of dependence on Apollonius' study of conics, either through explicit reference or implicitly through the use of Apollonian terminology or techniques.

The resolution of these difficulties lies outside the scope of the present work. But their implication seems fairly clear: that Diocles and Zenodorus should be set much nearer to the time of Dositheus. Even if the latter can be viewed as active as late as 220 B.C., we would still have to assign the other two a date about fifty years earlier than estimated previously. This would surely rule out that the same Zenodorus was Philonides' acquaintance at Athens in the 160s. Toomer himself notes several gaps in his painstaking effort in favor of this identification: specifically, that the name "Zenodorus" was common among Hellenized Semites, Philonides' compatriots, while the Athenian clan among whom Toomer seeks to locate Zenodorus boasts at least four others of the same name, and he finds yet others from the region around Athens.[179] We may observe that nothing in the fragments of Philonides' biography suggests that his Zenodorus had anything to do with geometry. By contrast, his Dionysodorus was indeed a teacher of geometry, so that we do well to maintain the identification with the mathematician we have discussed here. But we have seen that the late dating of Philonides raises difficulties. A date of approximately 220 B.C. is compatible with the biographical data for Philonides' birth and would conform nicely with a date of approximately 240 B.C. for Dionysodorus, setting him squarely in Apollonius' generation. Assuming, then, that *Conics* II refers to Philonides at about age thirty, its composition would fall at approximately 190 B.C., near Apollonius' fiftieth year. The fit is comfortable. As for Diocles and Zenodorus, they must now be dated independently of our testimonies on Philonides. The association with Dositheus thus becomes our principal datum and assigns their period of activity to about 230 B.C. Even if we cannot presume to have demonstrated these proposals, they surely avoid the real difficulties entailed by the current chronological views.

This matter of chronology is significant, because the Archimedean element in the work of these geometers is stronger than one would expect, were they to have conducted their studies after the time of Apollonius. As we have noted, Archimedes' emphasis on inquiries relating to the Eudoxean method of limits contrasts with the field of research most prominent in the work of Euclid and Apollonius, typified by the investigation of problems of locus, and especially those occasioning the use of conics. Diocles and Dionysodorus participate in this latter tradition when they solve the Archimedean problem of dividing the sphere by means of a method of intersecting conics. Although Archimedes himself solved the problem in this manner, his solution was not available to Diocles, and doubtless not to Dionysodorus either, to serve as a model. The problems of the burning mirrors have no Archimedean association, but rather fall into the problem-solving tradition.[180] The reconstructions of Eratosthenes' study of loci with respect to means and of Perseus' applications of the sections of the torus are extremely tentative. But as suggested here, they too would have a strong affinity with the problem-solving field. Furthermore, the quest for alternative solutions to problems like the circle quadrature and the angle trisection might well betoken a dissatisfaction with the solutions actually proposed by Archimedes, even if the new methods owe much to his efforts.

A metaphor drawn from mechanics is surely appropriate: we appear to see in these geometers a generation striving to shift its center of gravity away from the region dominated by Archimedes' achievement. The Euclidean field of problems provided a possible outlet, but its exploration demanded techniques different from those most strongly advanced by Archimedes. The challenge overreached the competence of this generation, but would stimulate the best efforts of the next.

NOTES TO CHAPTER 6

[1] Heath, *History of Greek Mathematics*, II, Ch. xv. The chronological issues are taken up below, especially in the closing section.

[2] For general surveys of Eratosthenes' life and work, see the articles by G. Knaack in *Pauly Wissowa Real-Encyclopädie*, Vol. 6, Pt. 1, col. 358–389; D. R. Dicks in the *Dictionary of Scientific Biography*, Vol. IV, pp. 388–393; and E. H. Warmington and J. F. Lockwood in the *Oxford Classical Dictionary*, p. 405.

[3] See the summaries by O. Neugebauer, *History of Ancient Mathematical Astronomy*, p. 653; and Heath, *Aristarchus of Samos*, pp. 339 f. Our information on Eratosthenes' measurements derives from later authors, specifically, Strabo, Hero, and Cleomedes.

[4] Syene is near the Tropic of Cancer, so that the sighting would be made at noon at the summer solstice. This doubtless points to the use of astronomical observations for setting up terrestrial coordinates. On Eratosthenes' possible involvement in such work, see Neugebauer, *op. cit.*, pp. 333 ff.

[5] This might imply a subdivision of the circumference of the circle into 360 equal parts. Precisely when the Babylonian computational mode entered the Greek astronomical system is in question. But clearly it was not used by Aristarchus or Archimedes in the first half of the 3rd century B.C. See Neugebauer, *op. cit.*, p. 590. Around Eratosthenes' time, the Greeks appear to have adopted a 60-division of the circle. Presumably, by estimating the shadow angle to be 6/5 of one of these parts, Eratosthenes could obtain his value of 1/50. Altering the final figure of 250,000 stades into 252,000 could result from desiring an even value (4200 stades) for each 60th of the circumference. In this case, it would be only coincidence that the figure produces an even value for 360ths as well.

[6] See Heath, *Aristarchus* and Neugebauer, *op. cit.*, for discussions of Archimedes' *Sand Reckoner* and Aristarchus' *Sizes and Distances of Sun and Moon*.

[7] For references, see Heath, *History*, II, pp. 104–109. The text on cube duplication appears in Eutocius' commentary on Archimedes' *Sphere and Cylinder* II, 1 (Archimedes, *Opera*, III, pp. 88–97); on its use for the general history of this problem, see Chapter 2. For Pappus' references to the loci with respect to means, see notes 14, 15, and 31 below.

[8] Pappus, *Collection* (III), I, pp. 56–58; cf., also, Eutocius in Archimedes, III, pp. 94–96.

[9] Eutocius, *op. cit.*, p. 96. This aspect of the construction is not mentioned in the version given by Pappus.

[10] *Ibid.*, pp. 90, 96; cf. Chapter 3.

[11] Eutocius, *op. cit.*, pp. 92–94. It is this part of Eutocius' version to which the one in Pappus is in closest agreement (see note 8 above).

[12] Theon of Smyrna, *Expositio rerum mathematicarum*, ed. Hiller, p. 2; cf., also, Chapters 1 and 2.

[13] Plutarch, *Vita Marcelli*, 14.5–6; *Quaest. Conv.* VIII, 2 (718 e–f). The epithet of the "new Plato" which Eratosthenes acquired in antiquity doubtless reflects the impact of his *Platonicus* and a prominent Platonizing element in his thought; it is consistent with the view proposed here and in Chapter 2 (cf. also Chapter 3, note 22), to the effect that Eratosthenes, by his emphasis on the mechanical aspect of geometry, sought the readjustment of an unsympathetic Platonic attitude.

[14] *Collection* (VII), II, p. 636.

[15] Pappus' scattered references to "loci with respect to means" (*ibid.*, pp. 652, 662, 672) suggest contexts of their study which Zeuthen has exploited in his search for a reconstruction; see the discussion below.

[16] Extended treatments of means are given by Pappus in *Collection* III and by Nicomachus in his *Introduction to Arithmetic*, II, Ch. 26–28. For a survey, see Heath, *History*, I, pp. 84–90; and I. Thomas, *Greek Mathematical Works*, I, Ch. III (g). I present some suggestions on Eudoxus' contributions to the theory in my *Evolution of the Euclidean Elements*, Ch. VIII, Sect. IV.

[17] In essence, this is the suggestion of Tannery; see Heath, *History*, II, p. 105.

[18] On the 3-line locus, see Chapter 4.

[19] *Collection* (III), I, pp. 68–70. This passage is discussed by M. Brown in "Pappus, Plato and the Harmonic Mean," *Phronesis*, 1975, 20, pp. 173–184.

[20] Nicomachus, *Introduction to Arithmetic*, II, 25 and 29; cf. Iamblichus, *In Nicomachum*, ed. H. Pistelli, p. 111. A proof of this relation is given by Proclus, *In Platonis Timaeum*, ed. E. Diehl, II, pp. 173 f.

[21] *Collection* (III), I, pp. 80–82.

[22] *Ibid.*, (VII), II, p. 904. Cf. also Chapter 4.

[23] For a derivation of this general result via a method of solid projections, see Chapter 4, note 103.

[24] See the emendation proposed by Hultsch, *Collection*, II, p. 905n.

[25] These correspond to the classes of medial, binomial, and apotome irrational lines in the form of the theory given by Euclid in *Elements* X. Our authority for Theaetetus' alternative use of the means is Eudemus, as reported by Pappus; see my *Evolution*, Ch. VII/IV.

[26] Fragment B 2, from Porphyry's commentary on Ptolemy's *Harmonics*; cf. H. Diels and W. Kranz, *Fragmente der Vorsokratiker*, 6. ed., I, 47 B 2, pp. 435 f. The text is given by Thomas, *Greek Mathematical Works*, I, pp. 112–115.

[27] One would suppose that the first geometric construction for the harmonic mean was in this manner. This provides the context for the following discovery: if any line is drawn through Q, and AB extended meets it in M, while PB extended meets it in K, and if PM meets BQ in N, then the line KN passes through H [Fig. 42]. Furthermore, one may find that B need not lie on the circle, as originally posited; that is, one may assume an arbitrary point as the center of projection. We thus obtain the general configuration for the harmonic division, as given by Pappus in Lemma 5 to Euclid's *Porisms*

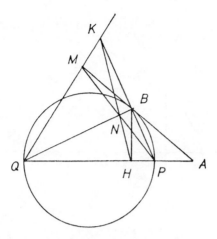

Figure 42

(*Collection* (VII), II, pp. 874–876). The fact that he goes on to consider, in Lemma 6, the case where QH = HP (whence MB is parallel to QP) may suggest the larger context of conics; for this case corresponds to the drawing of the tangent to the parabola (cf. Chapter 4). On any account the discovery of these constructions of the harmonic division and related projective configurations must be considered a remarkable insight. To the extent these were not yet worked out in Euclid's *Porisms*, one may look to Eratosthenes as a contributor in his study of "loci with respect to means."

[28] Zeuthen, *Lehre von den Kegelschnitten im Altertum*, 1886, Ch. 14, especially pp. 318–342. A brief, but rather unappreciative summary is given by Heath in his *History*, II, pp. 105 f.

[29] In the modern theory of conics, PQ is called the "polar" of C with respect to the given conic; see C. Zwikker, *Advanced Plane Geometry*, pp. 78 f.

[30] The configuration for circles can be worked out from propositions given by Pappus in his discussion of Euclid's *Porisms*; the extension to ellipses is straightforward via parallel projection (cf. Chapter 4, Fig. 26 and note 103). But to obtain the cases for parabola and hyperbola, planimetric methods are required. For the parabola PXQ, the locus of the harmonic mean is the line PQ; that of the arithmetic mean is the parabola PCQ; that of the geometric mean is the pair of lines parallel to the axis CXY and passing through P and Q ([Fig. 43(a)]. For the hyperbola PXQ, the harmonic locus is again the line PQ; the arithmetic locus is the hyperbola PCQ; the geometric locus is a hyperbola whose axis is the perpendicular through C and whose opposing branches each pass through one of the points P, Q [Fig. 43(b)]. Note that the asymptotes of each of these hyperbolas will be parallel to those of the other two. We have already seen that the equivalent of the harmonic locus for any conic is given by Apollonius (III, 37). The arithmetic locus for the hyperbola may be given thus: to the given hyperbola PXQ we draw a secant CSU from the assumed pole C [Fig. 43(c)]. We wish to describe the locus of point A, the midpoint of chord SU (and thus the arithmetic mean of SC, CU). Draw the ordinates ST, UV, and AW parallel to them. Then, $ST^2 : XT \cdot TY = UV^2 : XV \cdot VY = L : XY$, for L the latus rectum and XY the diameter. Thus, $UV^2 - ST^2 : XV \cdot VY - XT \cdot TY = (UV + ST) \cdot (UV - ST) : VT \cdot (TX + XV + XY) = L : XY$. But AW : WC =

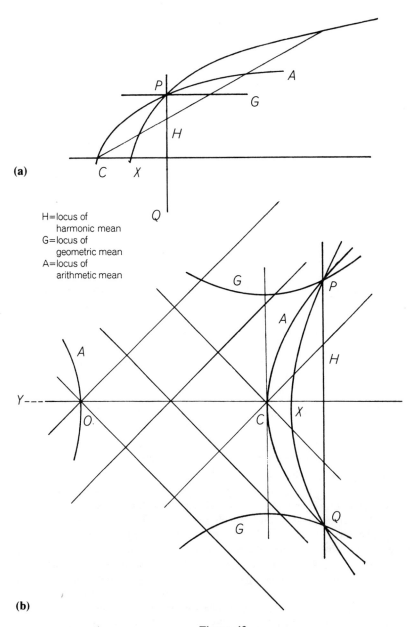

Figure 43

UV : VC = ST : TC = UV − ST : VT; and 2AW = ST + UV, while 2 WC = VC + CT = TX + XV + 2CX. Thus, substituting, 2AW • AW : WC • (2WC − 2CX + XY) = L : XY; that is, A lies on the hyperbola with vertex C, diameter ½ XY − CX and latus rectum L' (for L' : L = ½ XY − CX : XY). Note that the second branch of this hyperbola passes through the midpoint of XY. When C is the midpoint of XY, the

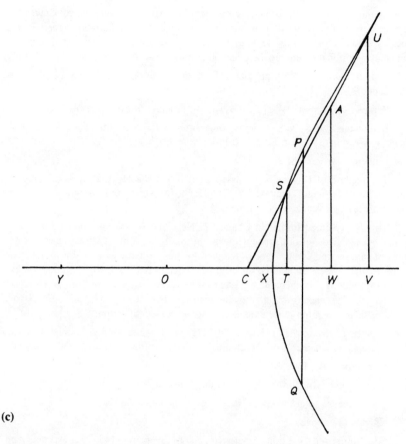

(c)

asymptotes to the given hyperbola are its "tangents" from C; by symmetry, C itself will be the arithmetic locus. When CX is greater than ½ XY, C will be the vertex of the second branch, while 0 (the midpoint of XY) will be the vertex of the first. The arithmetic loci for parabola and ellipse can be derived analogously from the ratios of areas which define them. One may consult Zeuthen for the derivation of the loci of the geometric means (*Kegelschnitte*, pp. 323–325). As for the loci with respect to the subcontrary means, these yield curves of higher order; cf. *ibid.*, pp. 325 f.

[31] *Collection* (VII), II, p. 672.

[32] Zeuthen, *op. cit.*, pp. 326–329.

[33] *Collection* (IV), I, pp. 272–274. See Chapter 7.

[34] Apollonius' work *On Neuses* is discussed in Chapter 7.

[35] Given the scantiness of our information on Eratosthenes' work, the range of possibilities is quite large. One might suppose, for instance, that his investigation of the means effected the locus theorems relative to the circle (as in Pappus' lemmas to the *Porisms*) and those cases of conics amenable to a method of solid projections (as outlined in Chapter 4; Zeuthen mentions the method of parallel projections as a convenient heuristic for the elliptical case; cf. *Kegelschnitte*, p. 323 and note 30 above). Their further extension, their treatment via the planimetric methods, and their elaboration for the solution of solid

problems might then be left for later researchers like Apollonius and his colleagues (see Chapter 7). Even this reduced effort would be no small feat, and would suggest Eratosthenes as a source of Pappus' lemmas to the *Porisms*. Of course, the extant evidence permits no determination among these and other possible accounts of Eratosthenes' study.

[36] Pappus, *Collection* (IV), pp. 242–250; cf. the excerpt in (III), pp. 58–62. Eutocius, *op. cit.*, pp. 98–106.

[37] Pappus, *Collection* (IV), I, p. 56; so also Proclus, *In Euclidem*, p. 272.

[38] Iamblichus, cited by Simplicius, *In Aristotelis Physica*, ed. Diels, I, p. 60; cf. Pappus, *Collection* (IV), I, p. 252; and Proclus, *In Euclidem*, p. 272.

[39] Eutocius, *op. cit.*, p. 98.

[40] The method of Diocles is discussed later in this chapter; that of Apollonius is a variant of the "Heronian" method (cf. Chapter 7), for which we have already proposed an Archimedean provenance (cf. Chapter 5).

[41] This is explained further in Chapter 7. Note that Apollonius (preface to *Conics* IV), names one Nicoteles of Cyrene who was especially harsh in his criticisms of Conon's studies of intersecting conics. One is tempted to suggest that a scribal slip in the transmission of the *Conics* has altered an original "Nicomedes." Clearly, in both contexts an abrasive individual is encountered, engaged in advanced geometric studies associated with the work of geometers around the middle of the 3rd century or a bit later. But beyond this, there seems to be no other evidence to argue the identification.

[42] Pappus, *Collection* (IV), I, pp. 270–272; see Chapter 5 on Archimedes' *neuses* for the spirals and Chapter 7 on Apollonius' problems of the normals to the parabola.

[43] *Collection* (IV), I, p. 244.

[44] See Chapter 5 for further details on these *neuses*.

[45] The former is found via point B in Fig. 9; the latter via G.

[46] Cf. note 37 above.

[47] For details, see G. Loria, *Ebene Kurven*, I, Ch. 6, pp. 143–152, especially pp. 146 ff.

[48] Cf. R. Böker, "*Neusis*," in *Pauly Wissowa*, Suppl. IX, 1962, col. 418–423. Hultsch supposes that Pappus follows the original terminology in speaking of "cochloids," so that later writers, like Proclus and Eutocius, would be adopting an altered term in their "conchoids"; cf. *Collection*, III, Index, p. 62.

[49] One may observe that all appearances of the term "cochloid" in the *Collection* are associated with a variant or emendation to "conchoid" or "conchloid" in one or another manuscript (e.g., *Collection*, pp. 54, 56, 60, 244, 246, 248, 270). It thus seems far more likely that in the preparation of the manuscript which served as the prototype of those now extant, the copyist wrote κοχλοειδης instead of κογχοειδης (perhaps through confusion with the name κοχλιας, *cochlias*, for the cylindrical spiral), and that this was detected and emended by later scribes.

[50] See the discussion of the "cissoids" in section (iv).

[51] Note that the lemma in this form is reminiscent of the terms of the *neusis* preceding Pappus' angle trisection (*Collection* (IV), p. 272), especially in the version preserved in Thābit's Arabic rendering. The texts will be presented in the sequel study.

[52] This implies a nonconstructive assumption that a magnitude which varies continuously from being less than a given value to being greater than the same value will,

sometime within that interval, be equal to it. Becker discusses the ancient appearances of this assumption in his "Eudoxos-Studien II" (*Quellen und Studien*, B : 2, 1933, pp. 369–387). It is implied in the Eudoxean appeals to the existence of the fourth proportional of three given magnitudes, for instance; see my "Archimedes and the pre-Euclidean Proportion Theory," especially Sect. IV. Interestingly, Nicomedes might have dispensed with the constructive part of his proof of (b) by invoking this same principle.

[53] Note an ambiguity in the term "asymptote." Its literal meaning is "line which does not meet," and hence need not be the line toward which a curve approaches arbitrarily closely. Thus, *any* line lying parallel to the "rule" of the conchoid, but outside the interval ED, is properly called "asymptotic" to it under the ancient definition.

[54] Nicomedes' proof of (b) is started as a direct argument, but concludes as if indirect (see Heiberg's remark, *Archimedes*, III, p. 103n). There is no profound difficulty in this, but it does indicate a certain degree of tampering with the text on the part either of Eutocius or of other editors.

[55] This might be done via the postulate discussed in note 52 above.

[56] Cf. B. L. van der Waerden, *Science Awakening*, p. 237; and A. Seidenberg, "Some Remarks on Nicomedes' Duplication," *Archive for History of Exact Sciences*, 1966, 3, pp. 97–101.

[57] The textual comparisons are presented in the sequel study.

[58] The principal discussion is by Pappus in *Collection* (IV), I, pp. 250–258.

[59] *Collection* (IV), I, pp. 260–262. This projective correspondence is discussed in Chapter 5 in connection with Archimedes' studies of the spiral. An alternative construction of the quadratrix via a "surface locus" (*ibid.*, pp. 258–260) is cited in Chapter 4.

[60] Note that at points other than B, the components of the motions have different magnitudes and directions, so that the curves will have different tangents there.

[61] Proclus, *In Euclidem*, p. 356; cf. p. 272. We have argued in Chapter 3 that, contrary to the general view, this is not the 5th-century Sophist, Hippias of Elis. According to Iamblichus, Nicomedes used the curve which received its name *tetragōnizousa* in special recognition of its role in the circle quadrature, while Apollonius called the same curve the "sister of the cochlioid," the latter being the cylindrical spiral, or *cochlias* (Simplicius, *In Physica*, I, p. 60). See my "Archimedes and the Spirals," especially pp. 66, 74. The "sisterhood" of these curves is doubtless due to the projective correspondences cited in note 59.

[62] *Collection* (IV), I, pp. 256–258.

[63] Note that this step assumes the proportionality of circular arcs of equal angle and their radii. Pappus provides two proofs of this lemma in other contexts, from which an Archimedean origin is arguable; see my "Archimedes and the pre-Euclidean Proportion Theory," especially Sect. III and Appendix II.

[64] This assumes Archimedes' axiom on arcs from *Sphere and Cylinder* I, Post. 2, or its implicit form in *Dimension of the Circle*, Prop. 1; cf. the discussion in Chapter 5 above.

[65] *Collection* (IV), I, pp. 252–254.

[66] See Chapter 4.

[67] See Chapter 5, and especially note 51.

[68] *In Euclidem*, pp. 272, 356.

⁶⁹ *Mathematisches Denken in der Antike*, pp. 97 f. The same construction was advanced by G. Fontana in 1784; see Loria's summary (*op. cit.*, II, pp. 27 ff).

⁷⁰ This is equivalent to the rule for the cosine of the half-angle. Becker (*loc. cit.*) works out a proof framed within the standard geometric manner of the ancients.

⁷¹ Oddly, Loria omits noting the connection with Vieta.

⁷² See note 61 above.

⁷³ Eutocius on Archimedes' *Sphere and Cylinder* II, 1 and 4 (Archimedes, *Opera*, III, pp. 66–70 and 160–176).

⁷⁴ G. Toomer, *Diocles: On Burning Mirrors*, 1976.

⁷⁵ *Ibid.*, pp. 1–2; cf. also Toomer, "The Mathematician Zenodorus," *Greek, Roman and Byzantine Studies*, 1972, 13, pp. 177–192.

⁷⁶ Toomer, *Diocles*, pp. 9–15.

⁷⁷ This question of dating is considered further below (see note 175).

⁷⁸ See Chapter 5.

⁷⁹ That this name was *not* applied by the ancients to designate Diocles' curve has been persuasively argued by Toomer (*Diocles*, pp. 24 f); see the discussion below.

⁸⁰ That is, D is the "focus" of the parabola. The term, however, is modern. Diocles employs no special term for it. Nor does Apollonius, who discusses the analogous points for central conics in *Conics* III, 45–52, but there adopts for them a construction which covers the parabola only implicitly as the limiting case, where the diameter is infinite.

⁸¹ Note here the notation for angles which corresponds to that used by Diocles. It is not used by the classical geometric authors, but is virtually standard in the ancient optical literature, in particular, the pseudo-Euclidean *Catoptrics*, Hero's *Catoptrics*, and Theon's recension of Euclid's *Optics*. Its use in the pre-Euclidean geometry is known through a mathematical passage in Aristotle (*Prior Analytics* I, 24). For comments, see Toomer, *Diocles*, p. 151.

⁸² This is the converse of Apollonius' *Conics* V, 27; see Toomer, p. 151.

⁸³ Cf. *Conics* I, 33 and Chapter 7 below.

⁸⁴ Toomer, pp. 153 f and Appendix D.

⁸⁵ That the work has suffered some disruption by careless editing is evident from its Props. 6 and 9, both of which are lemmas unrelated to the main sections and are quite ineptly executed. See Toomer's commentary, *Diocles*, pp. 161 f, 168 f.

⁸⁶ See note 99 below on the Bobbio mathematical fragment.

⁸⁷ Toomer, *Diocles*, pp. 66 f (line 108). Emphasis mine.

⁸⁸ *Ibid.*, lines 110 f; emphasis mine. This use of the flexible ruler provides one manner of obtaining a curve from a pointwise construction. As one has no cause for supposing Diocles to have originated this technique, it doubtless relates to a method known much earlier. Toomer cites a passage from Aristotle which speaks of a leaden rule used by architects (*Diocles*, pp. 159 f). We have suggested that Menaechmus himself resorted to some such pointwise construction (see Chapter 3). Interestingly, Eutocius eliminates reference to the flexible ruler, but instead recommends that the constructed points be joined by a succession of straight lines. For his own pointwise construction of conics, see note 91.

[89] Toomer, *Diocles*, pp. 16 f.

[90] See Chapter 4 (iv).

[91] Note, however, that when Eutocius presents what he takes to be a practical mode for the construction of the conics, it is a pointwise method based on the ratio property, its principal *symptoma* in the older theory; see his commentary on *Conics* I, 20–21 in Apollonius, *Opera*, II, pp. 230–234.

[92] On other such interpolations in the extant text of Diocles, see note 85.

[93] On Anthemius, see G. Huxley, *Anthemius of Tralles*, 1957. The single extant fragment of Anthemius' book is presented there (in the text as established by Heiberg, *Mathematici Graeci Minores*, 1927, pp. 78–87) with translation and commentary.

[94] See Eutocius in Apollonius, *Opera*, II, pp. 168, 290, 312, 354.

[95] Anthemius, in Heiberg, *Mathematici Graeci Minores*, pp. 85–87.

[96] *Ibid.*, p. 85.

[97] *Ibid.*, p. 85.

[98] *Ibid.*, pp. 78–81. Anthemius accepts as known that the ellipse is such that the distances of each of its points from two given points have constant sum. Presumably, then, he has access to Apollonius, *Conics* III, 52 or some comparable treatment. It is noteworthy, however, that Apollonius establishes this result on the basis of the prior proof that any two lines drawn from the foci to a point on the conic make equal angles with the tangent at that point (III, 48). This oddly circuitous manner of proof has led some to suppose that the former property was surely discovered in a simpler way (cf. Coolidge, *op. cit.*, p. 20). Certainly, if Anthemius knew *Conics* III, 48, there would be little sense in his working through his construction and so deduce the constant-sum property as a more characteristic identifying feature of the ellipse. This seems to indicate that his account relies ultimately on a treatment of the conics different from Apollonius'. The latter could well have adopted this order of proof in view of a construction of the elliptical reflecting surface in the form preserved by Anthemius.

[99] Heiberg, "Zum Fragmentum mathematicum Bobiense," *Zeitschrift für Mathematik und Physik*, hist.-litt. Abth., 1883, 28, pp. 121–129, especially pp. 127 ff. For the text of the fragment, see Heiberg, *Mathematici Graeci Minores*, pp. 87–92 (text reproduced in Huxley; see note 93 above). Heiberg's view identifying Anthemius as author of this fragment has been accepted by Huxley, but is contested by Heath (*History*, II, p. 203) and more recently by Toomer (*Diocles*, pp. 19–21). I provide a reformulation and defense of Heiberg's position in my "Geometry of Burning Mirrors in Antiquity." One may note that the Bobbio fragment presents the geometric proof of the ray-focussing property of the parabolic contour, and so provides a natural supplement to the construction extant in the fragment from Anthemius' book. The Bobbio fragment then goes on to the properties of spherical reflecting surfaces, much in the style of Diocles' Prop. 2. In the above-cited article I elaborate the view that Anthemius in effect served as the medium for the transmission of Diocles' results into the Arabic optical tradition, as with al-Kindī and ibn al-Haytham.

[100] *Diocles*, p. 140.

[101] I would suppose that Anthemius' treatment of the elliptical reflector likewise derives from pre-Dioclean sources. As for the spherical mirrors, I have argued that Anthemius was the vehicle for the transmission of Diocles' theorems to the Arabic tradition repre-

sented by al-Kindī and ibn al-Haytham (cf. note 99). It is thus possible that the configurations of annular mirrors discussed by the Arabic writers might also have their origins in Dioclean or pre-Dioclean efforts. For the discussion of these views, see my "Geometry of Burning-Mirrors in Antiquity."

[102] See Chapter 5.

[103] See Toomer (*Diocles*, pp. 169 f). The question is taken up in the sequel, in the context of Eutocius' textual methods.

[104] *Diocles*, p. 96 (line 207).

[105] See Chapter 3.

[106] For further details on the comparison of these three methods, see the discussion in the sequel. The method of Pappus is given in *Collection* (III), I, pp. 64-68; Eutocius describes the methods of Pappus and Sporus, noting their essential identity with that of Diocles (Archimedes, *Opera*, III, pp. 70-78).

[107] See Chapter 3. On the modern studies of the "ophiurides" (from Greek *ophis + oura + eidos*, "snake-tail-like") see Loria, *Ebene Kurven*, I, p. 50; and Wieleitner, *Spezielle Ebene Kurven*, p. 37. Loria has noted that the "Heronian" construction leads to a curve of this same class; see his "Sopra una relazione che passa fra due antiche soluzioni del problema di Delo," *Bibliotheca Mathematica*, 1910–1911, 11$_3$, pp. 97-99. Such a connection need not surprise us, in view of the relation of the "Heronian" and pseudo-Platonic methods noted above (see Chapter 5, especially Fig. 27).

[108] Cf. Zwikker, Ch. XI, especially p. 155.

[109] On the Bobbio fragment, see note 99.

[110] G. de Roberval, "Observations sur la composition des mouvemens...," *Mém. de l'Acad. des Sciences*, VI (1666–99; repr. 1730), Sect. 10, pp. 53-58. For remarks, see Loria, *Ebene Kurven*, I, pp. 37 f.

[111] In Newton's construction, one sets ND = DL and conceives the broken line PMQ with right angle at M and PM = 2 BL to move such that P remains on LB extended, while MQ passes through N. The path of θ, the midpoint of PM, will be the curve of Diocles. For from the congruence of triangles PNL, PNM one finds NM = LP; since Pθ = ND, lines NP and Dθ are parallel. Furthermore, triangles LDE and PθY are found to be isosceles and congruent. Thus, DE = θY, or Dθ = EY; that is, θ lies on Diocles'

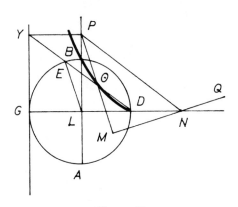

Figure 44

curve in accordance with the standard modern form of its construction (see references in note 112). (Fig. 44) For further details, see Loria, pp. 39 f; and Wieleitner, p. 79.

[112] Cf. Wieleitner, Sect. I, Ch. 1; and Loria, I, Sect. I, Ch. 4–5.

[113] See note 110.

[114] Diocles, pp. 24 f. Toomer notes that this identification was already taken for granted by Fermat and Roberval, approximately 1638–40, and possibly even earlier by Descartes; but he and others cite no earlier appearances.

[115] Proclus, *In Euclidem*, (i) pp. 111, 152, 177, 187; (ii) p. 126; (iii) p. 128; (iv) pp. 126, 164.

[116] Cf. Loria, pp. 37 f.

[117] Tannery, "Lignes et surfaces courbes," *Bull. des Sci. Math.*, 1884, 19, pp. 109–112. But he notes the inferential character of this identification and hazards that some other curve, a hypocycloid or epicycloid, say, might be an equally probable candidate for the ancients' "cissoid."

[118] Loria, *Le Scienze esatte nell' antica Grecia*, Milan, 1914, pp. 413 f; *Ebene Kurven*, II, p. 37.

[119] One notes that the ancients clearly did admit *figures* delimited by two or more intersecting curves. Examples where the curves are arcs of circles include the lunules of Hippocrates (see Chapter 2), the *arbelos* and *salinon* of Archimedes (cf. Chapter 5), the "arch" and "hatchet" of Hero (*Opera*, IV, pp. 34, 38), and the biconvex, biconcave, and concavo-convex figures cited by Proclus (*In Euclidem*, pp. 127, 163). But none of these would be viewed as bounded by a "regular" curve. This is the force of Toomer's principal objection to this identification of the "cissoid" (*Diocles*, p. 24).

[120] *Collection* (III), I, p. 54. Hultsch conjecturally fills the lacuna by "spirals" on the basis of the passage cited in note 121.

[121] *Ibid.*, p. 270. Problems in the word order, however, suggest that "spirals" may have been interpolated (cf. Hultsch's note, *ibid.*, p. 270n). If so, the emendation mentioned in note 120 is called into question. One may observe that in a comparable context Proclus speaks of "spirics, conchoids and cissoids" (*In Euclidem*, p. 113), where the "spirics" are sections of the torus (see below).

[122] See note 48 above.

[123] Cf. Loria, I, p. 147.

[124] The name was introduced by Castillon in 1741; see Loria, *ibid.*, p. 153. Further details on the history and properties of the curve are given by R. C. Archibald, *The Cardioide...*, Strassburg, 1900.

[125] The tangents from the pole to the base circle are also tangent to these conchoids. In Fig. 35(d) the "distance" happens to equal the tangent to the circle, so that the point of tangency to the conchoid falls at the pole itself.

[126] Contrast *In Euclidem*, pp. 111 f, 177.

[127] Among other "mixed" curves are the conchoids, quadratrices, and spirals (*ibid.*, p. 272). Clearly, the circle-based conchoids fall into this class.

[128] *Ibid.*, p. 126.

[129] Consider the variety of forms illustrated in Plate I, reproduced from C. K. Schneider, *Illustriertes Handbuch der Laubholzkunde*, Jena, 1912, pp. 421, 423.

[130] Cf. the edition by A. Hort in the Loeb Classical Library, 2 vols., 1916.

[131] Theophrastus, ed. Hort, (i) III, 18, 6–10; (ii) III, 18, 7; cf. I, 10, 1; (iii) III, 10, 5. The lime leaf resembles a jagged-edged form of shape (c) in Plate I(b).

[132] See the frontispiece; from the facsimile edition (Graz, 1966) of this manuscript of Dioscorides.

[133] See the illustrations in the plates.

[134] Examining the dozens of vases represented in the survey by E. Buschor (*Griechische Vasen*, new ed., Munich, 1969), one finds that a sizeable portion (on the order of one-fourth) introduce ivy leaves either among the portrait figures or as border decoration. The cordate leaves predominate.

[135] Guldin adheres to a formal "exhaustion" procedure, and indeed criticizes contemporaries like Kepler for their use of informal methods of indivisibles (cf. C. Boyer, *History of the Calculus*, repr. 1959, p. 138). Note that the informal procedure of "rolling out" used by Kepler (1615) for the measurement of solids, including the torus, is reminiscent of Hero's alternative form. See Boyer, *ibid.*, pp. 107–109; and M. Baron, *Origins of the Infinitesimal Calculus*, Oxford, 1969, pp. 108–116.

[136] Ver Eecke, *Pappus: Collection Mathématique*, I, pp. xciv–vii.

[137] *Collection* (VII), II, pp. 680–682.

[138] Indeed, the use of compound ratios effectively explains the appearance of Pappus' passage on this theorem. For it interrupts his account of Apollonius' *Conics*, specifically, his elaboration of the problem of the 3- and 4-line locus. Pappus notes that the problem can be expressed for 5- and 6-line locus, but for more lines than this, one encounters a difficulty with dimensions, since a product of four lines (a term of the proportion defining the 7- or 8-line locus), for instance, has no geometric sense. But the problem can be recast in terms of compound ratios and thus generalized for any number of lines. (A special case of this, one notes, is the reduction of the cube duplication to the interpolation of two mean proportionals. The problem of interpolating means can be expressed for any number of terms, as Eratosthenes observes, although it has no direct interpretation in terms of geometric measures.) One regrets that neither this passage nor others suggests that the ancients made any headway in the study of these higher-order problems.

[139] *Ibid.*, p. 682.

[140] Hero, *Opera*, IV, ed. Heiberg, p. 62.

[141] Cavalieri's procedure is presented in the *Geometria Indivisibilibus*, Bologna, 1635 and 1653. See his *Geometria degli Indivisibili*, ed. L. Lombardi-Radice, Turin, 1966, pp. 847 ff.

[142] See the references in notes 135 and 136. For excerpts from Guldin's controversies with Cavalieri, see the Appendix to L. Lombardi-Radice's edition of the latter (cited in the previous note).

[143] The concept of *rhopē* is important within the ancient science of weights, and might be termed a sort of effective weight, that is, the strength of a weight when applied at a given distance from a fulcrum. But theorems on equilibrium (*isorrhopia*, or "equal effective weight") are managed via proportions, and so never require the direct quantitative measure (sc. weight times distance) answering to this concept. An extensive review

of ancient and medieval writings relating to the mathematical study of equilibrium is provided in my *Ancient Sources of the Medieval Tradition of Mechanics*.

[144] See Chapter 5.

[145] In Archimedes' form of the lemma on sums, the ratio $x : x'$ would be $1 : 1$. In Dionysodorus' case, it will assume a constant value. This poses no formal difficulty, however, for we need merely sum over the elements g, where $g : e = x : x'$.

[146] For details of such proofs, see my "Archimedes' Lost Treatise on the Centers of Gravity of Solids."

[147] For instance, Hero conceives of the measurement of cylindrical solids via consideration of parallel plane sections (*Metrica*, II, preface), and Theon effects the measurement of the cylinder in a similar fashion via the proportionality of the volumes and the areas of their cross sections (*Commentaires...sur l'Almageste*, ed. Rome, II, pp. 398 f). See also note 135 above.

[148] *In Euclidem*, p. 112; cf. pp. 119, 356. From Proclus' description of the forms of the sections, it is evident that they are produced by a cutting *plane*.

[149] *Ibid.*, p. 119. The illustration of the "open" torus and its sections in Fig. 38(a) is taken from E. Brieskorn and H. Knörrer, *Ebene algebraische Kurven*, p. 20.

[150] *Ibid.*, p. 112; but contrast p. 127 where Proclus seems to call the hippopede "one of the spiric lines," that is, a section of the torus.

[151] Tannery, "Lignes et surfaces courbes," pp. 24–27. For further discussion of his view, see Heath, *History*, II, pp. 203–206; I. Thomas, *Greek Mathematical Works*, II, pp. 362–366; and Loria, *Ebene Kurven*, I, pp. 124 ff.

[152] *In Euclidem*, p. 356.

[153] Note that this alternative relation can be derived directly from consideration of the triangle OIZ in the light of the generalization of the "Pythagorean theorem" to obtuse triangles (cf. *Elements* II, 12), and noting that $ZI = I\theta$ and that triangle $I\theta 0$ is a right triangle. When $d = r$ (the continuous case), θ and 0 coincide. When the torus is closed (i.e., $d < r$), one takes the triangle $OI\theta$ with right angle at 0 and changes the sign of $O\theta^2$.

[154] *History*, II, pp. 203 ff; cf. Tannery, "Lignes...courbes," pp. 27 f.

[155] See Chapter 3.

[156] Compare the remarks on the "unbounded" (*aperanton*) character of the mixed curves, formed as such sections, in Proclus, *In Euclidem*, p. 112.

[157] See above note 98.

[158] For further details, see Loria, *Ebene Kurven*, Sect. 61; and Wieleitner, *Spezielle Ebene Kurven*, pp. 32 f. The special curves constructed in answer to the locus condition are called the "curves (ovals) of Cassini," for G. D. Cassini (1625–1712). Cassini, noted for his work at the Paris Observatory, introduced these curves with the intent of describing planetary orbits better than via Kepler's ellipses; an account is given by his son, J. Cassini, in his *Éléments d'astronomie* (1749, p. 149, as cited by Loria, *op. cit.*, Sect. 90). These curves are now treated as a special class of the curves of Darboux (1873); cf. Zwikker, pp. 291 f. The diagram in Fig. 40(b) is modified from Brieskorn, *op. cit.*, p. 21.

[159] *Metrica* I, 8; cf. *Dioptra*, 30. One notes that Hero could easily have dispensed with this anomalous aspect of dimension by viewing A as the mean proportional between the areas $s(s-a)$ and $(s-b) \cdot (s-c)$, for instance. In the medieval Arabic treatment of this rule by the Banū Mūsā, something of this sort is attempted, in that A is worked out as the mean proportional of s and the volume $(s-a)(s-b)(s-c)$. Nevertheless, they persist in expressing the area as the square root of a four-dimensional term. See their *Verba filiorum*, Prop. vii, in M. Clagett, *Archimedes in the Middle Ages*, pp. 278–289.

[160] The Arabic astronomer al-Bīrūnī knew of a tradition assigning this rule to Archimedes; see H. Suter, "Buch der Auffindung der Sehnen im Kreise von...el-Bīrūnī," *Bibliotheca Mathematica*, 1910–11, 11₃, pp. 39 f.

[161] B. L. van der Waerden, *Science Awakening*, pp. 168 ff.

[162] Loria, *Ebene Kurven*, I, pp. 127 f. Fig. 41 is taken from the same work, Plate III, Fig. 26.

[163] Ptolemy, *Syntaxis*, I, 3. For Theon's commentary, see *Commentaires...sur l'Almageste*, ed. Rome, II, pp. 355–379. This material is surveyed by Heath in his *History*, II, pp. 206–213.

[164] Theon, *op. cit.*, p. 358. Specifically, $\tan a : \tan b > a : b$, for $a > b$. This is proved in Euclid's *Optics*, Prop. 8, and is applied by Aristarchus (*Sizes and Distances*, Props. 5–7, 10) and by Archimedes (*Sand Reckoner*, in *Opera*, II, p. 232). The companion relation for sines, also assumed by Aristarchus, is proved by Ptolemy in *Syntaxis*, I, 10.

[165] Theon, *op. cit.*, pp. 360–364.

[166] For the Greek text, see Hultsch's edition of Pappus' *Collection*, III, pp. 1138–65, especially p. 1164. The concluding note is absent from the medieval Latin recension of this work; see H. L. L. Busard, "Der Traktat *de Isoperimetris*," *Mediaeval Studies*, 1980, 42, pp. 61–88. J. Mogenet has argued that the series of writings forming this introduction to Ptolemy is derived from Eutocius (see his *Introduction à l'Almageste*, Brussels, 1956, Ch. II, especially pp. 22 ff, 33 ff). This may be. But I suspect that the section on isoperimetric figures, at any rate, is somewhat older. It is surely later than Theon, in the 4th century. But as its special term *koilogōnion* (for rectilinear figure with a concave indentation) is already known to Proclus in the 5th century, I would assign this writing to a time before this; for the term is absent from the alternative versions on isoperimetric figures in Pappus and Theon (cf. Theon, ed. Rome, II, p. 371n).

[167] *Collection* (V), I, pp. 308–334.

[168] See Chapter 5.

[169] Specifically, *Collection* V, 1–10 cover Zenodorus' plane case; 11–18 cover the circular segments; 19–20 survey the Archimedean semiregular solids and then work out the isometric theorem for the sphere in relation to regular solids; 20–43 treat of Archimedes' *Sphere and Cylinder* I; and 44–65 compare the regular solids to each other.

[170] *De Caelo*, 287 a 27. As we have observed above, Aristotle's statement of the "least path" is technically incorrect unless one adds the condition of "enclosing equal areas" (see Chapter 5). This is, of course, Simplicius' understanding of the passage.

[171] *In Aristotelis De Caelo*, ed. Heiberg, (*Commentaria in Aristotelem Graeca*, VII), p. 412. Note that Simplicius' term "more spacious" (*polychōrētoteros*) betrays his dependence on the anonymous version of this work (see note 166); for Ptolemy, Pappus, and Theon merely speak of the circle's being "greater" than the other figures.

[172] See note 164 above.

[173] *Sand Reckoner*, in *Opera*, II, p. 234; see my discussion in "Archimedes and the Elements," Sect. II.6. One notes that the term "more polygonal" (*polygōnoteron*) follows Heiberg's very reasonable emendation of a senseless reading (*ibid.*, p. 235n). Accepting it, we observe that this term used throughout the development of the study of isoperimetric figures, as cited in Ptolemy, Pappus, and Theon, would have its precedent in Archimedes' writing.

[174] One notes that the analogous theorem for isometric regular polyhedra does *not* generalize in the same manner to the polyhedra inscribed in the same sphere. Among the regular solids, the dodecahedron will have greatest volume, followed by the icosahedron, cube, octahedron, and tetrahedron. This progression can be inferred from the theorems on the construction and comparison of these solids in Pappus, *Collection* III (e.g., 51–52) and V (e.g., 55, 57). These have their provenance with Aristaeus and Apollonius in the 3rd century B.C. and Hypsicles in the 2nd century B.C.; cf. Chapter 7.

[175] For a survey of the evidence on Philonides, see R. Philippson, "Philonides (5)" *Pauly Wissowa*, 1941, 39, col. 63–73. Toomer reviews the material with special reference to the dating of Zenodorus in "The Mathematician Zenodorus," *Greek, Roman and Byzantine Studies*, 1972, 13, pp. 177–192. In particular, Toomer proposes that our Zenodorus was Philonides' associate in the 160s and can be identified as the dedicator of a certain Attic inscription from 183/2 B.C.

[176] On the chronology of Apollonius, see Toomer, "Apollonius," *Dictionary of Scientific Biography*, I, pp. 179–193.

[177] Eutocius says of Apollonius that "he was born (*or* he flourished) in the days of Euergetes." Reading in the former sense, Toomer assigns his birth to approximately 240 B.C.; but Heath must have in mind the latter sense in his estimate of 262 B.C. (*History*, II, p. 126). Heiberg renders as *vixit* the Greek verb *gegone*, and so reads in the sense of Heath (Apollonius, *Opera*, II, p. 171). Pappus preserves a note that Apollonius spent much time at Alexandria in study with students of Euclid (*Collection* (VII), II, p. 678), which would support the earlier dating. The latter passage is bracketed by Hultsch and dismissed by Toomer, for reasons not specified.

[178] Diocles, preface to *On Burning Mirrors*, ed. Toomer, pp. 34 f.

[179] See Toomer's "Zenodorus" (cited in note 175), pp. 186n, 187–190. Note that at the time of the inscription (183/2 B.C.), Zenodorus had at least five younger brothers, three with sons of their own. This would appear to indicate an individual of considerable years, and certainly quite older than a Philonides supposed to have been born in 200 B.C. A further complicating factor in this matter is that the name "Zenodorus" in the manuscript of the Arabic translation of Diocles has been greatly corrupted, so that the identification may be questioned. The name appears twice, as "'byūdām-s" and as "'ynūdām-s" (Toomer, *Diocles*, p. 139; cf. "Zenodorus," p. 191). Toomer emends to "zīnūdūrus,"

defending the changes through considerations of the orthography of the manuscript. His view has the decided advantage of conforming with the name of a mathematician already known from our ancient sources. Unfortunately, it is not necessarily the case that such a figure named by Diocles would be known to us from elsewhere (cf. his citation of one "Pythion," otherwise unattested). The Arabic texts seem in better conformity with a name like "Hippodamos," for instance. In this event, Diocles' preface would no longer be witness to the work of Zenodorus, and the dating argument would have to rest on the other testimonia, like the possible association with Philonides and the strong Archimedean element in Zenodorus' isoperimetric study.

[180] Diocles' failure to mention Archimedes among the prior researchers on burning mirrors is especially significant in this regard. For a review of the ancient testimonia to Archimedes' construction of such devices, see Dijksterhuis, *Archimedes*, pp. 28 f and my "Geometry of Burning-Mirrors in Antiquity."

CHAPTER 7

Apollonius:
The Culmination
of the
Tradition

The corpus of analysis which Pappus describes in *Collection* VII under the title *topos analyomenos* (lit.: "analyzed locus") is dominated by works of Apollonius. These include two extant works, the *Conics* and the *Section of a Ratio*, as well as five others now lost. Since Pappus represents this entire corpus as compiled in the interests of abetting the efforts of geometric problem solving, a purpose expressed by Apollonius himself in prefaces to the several books of the *Conics*, Apollonius' impact on this field could hardly have been other than immense.[1] Thus, although the *Conics* (specifically, the first four of its eight books) constitute a systematic and rigorous exposition of the theory of the conics in the synthetic mode, comparable to the Euclidean *Elements*, one ought to seek the measure of Apollonius' achievement in geometry through these problem-solving researches. In this respect, Apollonius and Euclid pose analogous problems for historical investigation: for in both cases, their original efforts came within the area of analytic studies of problems, while the extant documents consist only of synthetic treatises composed as ancillary to these studies.

Consciously following the model of Euclid in his advancement of the analytic tradition, Apollonius seems equally keen to renounce the field of researches explored by Archimedes. The Eudoxean method of limits and its applications to such problems as the circle quadrature receive but the barest minimum of attention from Apollonius. By contrast, the field of problems for which the method of analysis is especially effective, and of which the cube duplication is a representative example, appears to have limited significance in the work of Archimedes, while it dominates the work of Apollonius. If this was the result of deliberate choice on Apollonius' part, it may indicate the concern of a highly talented mind to define an area of inquiry where his original findings would not be overshadowed by the accomplishments of his superbly gifted precursor. Indeed, this ambition to obtain the fullest possible recognition for his work emerges in another aspect of Apollonius' career, the element of controversy in his relations with the disciples of Euclid and Archimedes in his generation, whose work he was elaborating and extending.

APOLLONIUS, ARCHIMEDES, AND HERACLIDES

The purpose of Archimedes' *Dimension of the Circle*, according to Eutocius, was to provide approximations useful for the needs of everyday life. He takes the assertion from one "Heraclides, in the *Life* of Archimedes,"[2] from whom he seems also to have drawn the following remark, given later in his commentary:

> One must know that Apollonius of Perga also proved this in the *Easy Delivery* [*Ōkytokion*] through other numbers [sc. other than 3 1/7 and 3 10/71] having reckoned out rather more closely. Now this seems to be more accurate, but is not useful for the purpose of Archimedes. For we have said that he had the purpose in this book to find it closely enough for the uses in life.[3]

Of Apollonius' refined computation of values for π, we know nothing further. But at least part of his motivation, it appears, was to engage in a bit of one-upmanship at Archimedes' expense, so prompting the defense from Archimedes' biographer. Heraclides' claim of the mere practical intent behind Archimedes' effort is far from convincing, as we have seen, and Apollonius was certainly operating within the spirit of that effort in seeking closer values. But the implicit polemical tone is noteworthy.

Modern scholars have speculated on further episodes in this controversy.[4] Supposing that the scheme for expressing and rapidly computing with large numbers, which Pappus assigns to Apollonius in Book II of the *Collection*, originally appeared in the same *Easy Delivery*,[5] it is proposed that Archimedes retaliated with the *Cattle Problem*. This is a word problem seeking integer solutions for eight unknowns satisfying seven linear relations. As if tacked onto the problem, its closing lines impose an additional restriction, that a certain term be square and a certain other be triangular in form, from which would arise the demand to solve the Pellian equation $u^2 - 4729494 v^2 = 1$ in integers. Thus, however skillful one must be to solve the initial form of the problem (where the solutions run to seven or eight decimal digits), earning praise as being "neither ignorant nor unversed in numbers," greater skill by far is needed for the augmented version, whose solver will indeed "carry away victory in the assurance of being deemed among the wise." That is a phenomenal understatement. For the solution gives rise to a continued fraction with a period of 91 terms resulting in answers whose digits number in the hundreds of thousands. Although this analysis of the problem was effected by Amthor a century ago, the complete solution has only recently been feasible through the assistance of electronic computers.[6] Thus, Archimedes has devised a labor of truly Herculean proportions for the would-be master calculator.

Unfortunately, our sources give no encouragement for supposing Apollonius to have been the implied target of this problem. Archimedes addresses the text of the problem to Eratosthenes. It has the form of a verse epigram, a form nowhere else to be found in the extant Archimedean corpus, although it was exploited on occasion by the artier Eratosthenes, as we recall from his verses on the cube duplication. One might thus more naturally suppose that Eratosthenes had composed the verses presenting the initial form of the problem, itself a rather complicated example of a traditional type of arithmetical puzzle,[7] and that Archimedes had redirected the text to him with the addition of the lines converting it into a truly difficult exercise in number theory. In any event, Archimedes' influence on Apollonius' arithmetic researches is evident: the computations with large numbers reported by Pappus link Apollonius with the arithmetic manipulations in the *Sand Reckoner*,[8] while the *Easy Delivery* is an extension of the calculations in *Dimension of the Circle*. With or without the hypothesis of the Archimedean counter-rebuttal in the *Cattle Problem*, the competitive aspect of Apollonius' efforts is evident here.

Heiberg has suggested, not unreasonably, that one might identify this Heraclides with the associate (presumably young) named in Archimedes' preface to *Spiral Lines* as the bearer of this writing to its addressee Dositheus.[9] If so, Heraclides would be a contemporary witness of the work of Archimedes and its influence on Apollonius. We shall recognize further support for this view below. But it has an interesting implication in the present context. While in the *Dimension of the Circle* as extant Archimedes establishes $3 \frac{1}{7}$ and $3 \frac{10}{71}$ as bounds on π, we have seen that Hero reports an alternative set of bounds indicating a rather closer degree of accuracy, which he cites from another Archimedean writing *On Plinthides and Cylinders*.[10] There thus appears to be a discrepancy between the reports of Hero and Heraclides. For the latter would surely have known of such an alternative computation by Archimedes and so framed his rebuttal to Apollonius by invoking this, rather than by insisting on the restricted objective of Archimedes' treatment to practical utility. One cannot hope to resolve such a problem with certainty. But a simple possibility is that Hero has merely misconstrued his source as being a genuine Archimedean writing. If, for instance, the *Plinthides* was actually a collection of geometric materials by Archimedes and others, its compiler could easily have introduced the closer computation from another source. In that event, it seems the most likely origin would be Apollonius' *Okytokion*; that is, Apollonius' effort to outdo Archimedes produced the alternative values which ultimately would reappear in Hero's *Metrica*. There is a certain poetic justice in the idea that Apollonius' effort, conceived with the aim of diminishing Archimedes, should end up magnifying him instead, through scribal inadvertence. Whether this captures the true relation of the efforts of Apollonius and Archimedes, however, must remain a subject for speculation.

Another sign of Archimedean influence relates to Apollonius' study of the curve called *cochlias*, the cylindrical helix. As we have seen, tradition assigns

296 Ancient Tradition of Geometric Problems

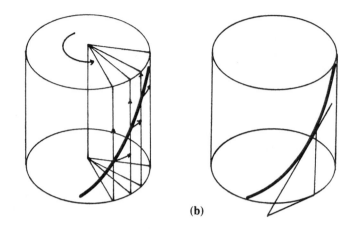

(a)   (b)

Figure 1

to Archimedes the invention of the water-raising device which takes this form, while its projective relation to the Archimedean plane spiral can assist the investigation of the tangents of the spiral.[11] As Pappus shows, the cylindrical helix has a comparable projective relation to the quadratrix of Nicomedes, and since Apollonius is said to have named the latter the "sister of the cochlioid" (where "cochlioid" must surely refer to the *cochlias*), relations of this sort may well have figured in his study.[12] Proclus specifically informs us that Apollonius prepared a writing on the *cochlias* in which he established that this curve, like the straight line and the circle, is homoeomeric, that is, that any part of it can be superimposed over any other part uniformly.[13] This property is evident from the manner of the curve's formation given in the Heronian *Definitions* (no. 7). There, the plane spiral is defined, in accordance with Archimedes' way, as the trace of a point moving uniformly along a line as that line rotates about its fixed endpoint; analogously, the cylindrical spiral (here called simply the "spiral," sc. *helix*) is the trace of a point moving along one side of a rectangle which in turn rotates about its opposite side as axis [Fig. 1(a)].[14] Since it is possible to translate any point of the curve to any other point such that the generating motions are identical, the homoeomeric property of the curve is manifest. The same is clear through an alternative conception of the curve, obtained by bending a rectangle around to form a cylinder (cf. the "wrapping" procedure for cones and cylinders in Hero, *Metrica*, I, 36, 37); the diagonal of this rectangle becomes the cylindrical helix [Fig. 1(b)]. The latter mode also makes evident that the tangent to the helix inclines to the perpendicular cross section of the generating cylinder at the same angle for all points. This, in turn, can enter into the proof of another property of the curve which Proclus cites from Geminus, doubtless also from Apollonius' book: that two equal lines drawn from a point to two points of the helix will make equal angles with the curve. (*In Euclidem*, p. 251) Although we have no direct information on the proof, a method using superposition would be reasonably straightforward. Consider the helix between points A and B, and

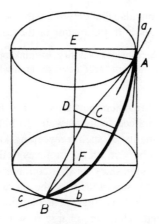

Figure 2

let lines be drawn from C such that AC = BC (Fig. 2). If one next takes the perpendicular sections at A and B, that is, the circles of radius EA and FB, respectively, and draws the tangents to these circles, they will make equal angles *a* and *b*, respectively, with the tangents to the helix at A and B, by the property of the constant inclination of the helix. Furthermore, the vertical angles *b* and *c* formed at B are equal. Hence, if one conceives of a transposition of the curve, such that B is set over A, BF will then fall over EA and CB over CA, while angle *c* will fall over *a*. Hence, the tangent at B now comes to lie over the tangent at A, so that the angle at which CB meets the former tangent must equal that at which CA meets the latter, as we wished to prove. One cannot now say whether Apollonius attempted to frame a formal proof of this theorem along the lines just proposed, or whether he followed the model of Euclid's treatment of the analogous property of the base angles in plane isosceles triangles *(Elements* I, 5) and so exploited considerations of congruence rather than superposition. Yet it is clear that Apollonius' study of this curve was an extension of prior work by Nicomedes and Archimedes. For the latter, the primary interest lay in its use for solving the circle quadrature in connection with the spiral and the quadratrix. For Apollonius, by contrast, topological, rather than metrical, properties of the curve were the principal concern.

Further hints of the relation of Apollonius to Archimedes emerge from a consideration of the figure of Heraclides. As we have seen, he composed a biographical account of Archimedes which included an apologia for Archimedes' circle measurement. Now, in his opening remarks in the commentary on Apollonius' *Conics*, Eutocius draws from an Archimedes biography a statement of the relative contributions of Archimedes and Apollonius to the foundation of the theory of conics. In our manuscripts the biographer's name appears as "Heraclius" (Greek: *Hērakleios*), but one immediately suspects a scribal slip for "Heraclides" (*Hērakleidēs*); for not only is it unlikely that two different individuals with such closely resembling names should both have composed biographies of Archimedes, but also that they share the polemical attitude with respect to the work of Apollonius.[15] In the matter of the conics Eutocius reports:

> [Heraclius] says that Archimedes was first to think out the conic theorems, but Apollonius arrogated to himself those he discovered which were not given out by Archimedes.[16]

Eutocius understandably registers puzzlement. For the sources available to the later commentators, as indeed to us as well, do not leave the impression that Apollonius claims originality of such a sweeping kind for the elementary materials on the conics. To be sure, in the prefaces to the first and fourth books of the *Conics*, Apollonius emphasizes the shortcomings of earlier treatments, which he deems were effectively superseded by the more systematic and exhaustive treatment he provides.[17] But the precedent of prior work, specifically that of Euclid, is explicitly cited. Thus, the charge of unfairness here levelled against Apollonius is not one that a later writer would think to make. Adopting the view that Eutocius' source, the biographer Heraclides, is a younger contemporary of Archimedes, we can account for these discrepancies. For such an individual could have gained through direct involvement with the principals a distinctive perspective on Archimedes' contribution, influenced by oral communications and subtly at variance with a view based only on the published record; he could also have obtained a sense of Apollonius' claims to originality different from the view presented in those writings known to us. Apollonius mentions with some regret that an earlier edition of his *Conics* had begun to circulate, but that its defects had recommended the preparation of an improved treatment.[18] While it is this latter edition which is the basis of the extant manuscript tradition of the *Conics* (allowing, of course, for later editorial alterations, by no means always of a trivial sort), it is certainly possible that the earlier version had occasioned Heraclides' objections. In this event, the acknowledgments Apollonius makes to his precursors in the extant edition may have been introduced to mitigate such misunderstandings.

Pappus inserts into the series of lemmas to Apollonius' *Neuseis* the complete solution via analysis and synthesis of a related problem by a geometer named "Heraclitus" (Greek: *Hērakleitos*).[19] The problem is of the sort we have already seen in connection with efforts of Archimedes and Nicomedes: to extend a line from a given point such that its intercept between two given lines has a given length. In the present instance, the given point is one vertex of a square, while the given lines are one of its opposite sides and the extension of the other. The problem has an obvious association with one of Apollonius' theorems, solving the same problem for a given rhombus instead of a square. Although the solution methods of the two treatments differ, they agree in adopting planar means exclusively, that is, admitting only such lines and circles as can be constructed in accordance with the Euclidean postulates of plane geometry. Now, the motive behind working out the special case of the square in detail would be difficult to understand if the more general Apollonian solution were already known. On the other hand, Apollonius' demonstration of the extension of a problem solved for the square to the more general case of the rhombus would be perfectly in keeping

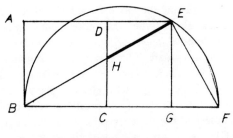

Figure 3

with his efforts in other fields, as we have seen in the instances of the conics and the circle measurements. In this case, the solution by Heraclitus would precede his own effort, so that we may suspect yet another scribal variant for "Heraclides" in the name now held in the Pappus manuscripts.[20] Furthermore, this suggests that a consideration of these two methods of solution will assist our understanding of Apollonius.

We may sketch the analysis of Heraclides (Heraclitus) as follows[21]: given the square ABCD with AD extended, let us suppose that line BHE has been drawn such that the intercept HE between CD and AD extended has a given length (Fig. 3). Draw from E the perpendicular to EB, meeting BC extended at F, and draw EG perpendicular to BF. Now, $CF^2 = CB^2 + HE^2$, as is known from a lemma just proved (and which we shall consider below); since CB and HE are given lengths, CF is thus also given. Since, furthermore, the angle BEF is a right angle, E lies on the circumference of the semicircle of diameter BF, and hence is the intersection of a given circle with the extension of AD, a given line. This solves the problem and makes the synthesis of the construction and proof obvious.

The analysis given here is unusual in that it fails to motivate the construction. In fact, the construction is inherited from the preliminary lemma, so that the heuristic line must somehow be retrieved from this. Now, the lemma is a straightforward manipulation of identities involving right triangles and depends on two features of the diagram: that the triangles BHC and EGF are congruent; and that the points E, H, C, F lie on the same circle (for the angles at E and at C are right angles)[Fig. 4(a)]. It is possible to imagine that Heraclides effected the

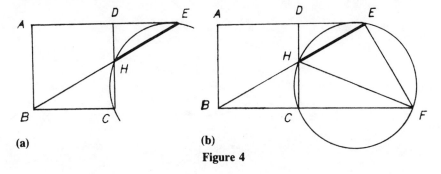

Figure 4

analysis at first by introducing the circle through C, H, E and then working out the resultant identities. But it seems far more natural to suppose that he had earlier examined problems of *neuses* with respect to circles and was led by these to take up the case of the square. For instance, one might seek to draw from a given point B to a given circle CHEF the line BHE such that the intercepted chord HE has a given length [Fig. 4(b)]. In the special case where EF can be arranged to be equal to BH (entailing a specific quadratic relation between the diameter of the circle and the distance of B from the circle), one produces a figure with ABCD as a square and with BCF as a chord such that $CF^2 = BC^2 + HE^2$. The converse problem is suggested at once: given the square ABCD, to insert the given length HE so as to incline toward B; and its solution is virtually a corollary to the case of the circle just considered. Although it is only this latter problem which Pappus presents under the name of Heraclitus, we would expect that this geometer had investigated a range of related problems, among which *neuses* with respect to circles surely appeared. Significantly, *neuses* of the last-named type constituted a major portion of the material in Apollonius' treatment *On Neuses*.

One of the problems in Apollonius' book solves the *neusis* with respect to a rhombus, and like Heraclitus' problem for the square, suggests an associated problem of *neusis* with respect to a circle. The solution for the rhombus is not extant, but Pappus proves in full an auxiliary lemma from which it may be recovered. Heath's attempt to reconstruct the analysis suffers, however, from not revealing any natural line in the development of the construction.[22] The version I offer here turns on a fact not noted by Heath, that embedded in the diagram to Apollonius' lemma is his construction for another problem, the locus of points whose distances from two given points are in a given ratio. Consider the figure of the analysis for the rhombus: ABDC is the given rhombus and the line BKH has been extended such that the intercept KH between CD and AC extended has a given length (Fig. 5). We seek to display the point K as a given.

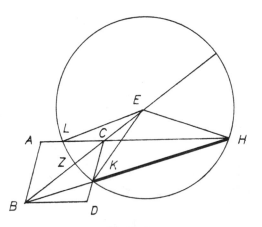

Figure 5

Now, one approach to this might be to display the ratio CK : KB as a given; since the locus of such points is a given circle, K would then be the intersection of that circle with CD. As we have seen, this locus is effected thus[23]: draw the diagonal BC, then draw from K the line KE meeting BC extended at E such that angle CKE equals CBK. Then the circle of center E and radius EK is the locus of points whose distances from C, B have a constant ratio CK : KB. Let this circle meet AC at L; since CB is the diagonal of a rhombus, it bisects arc KL at Z. The angles ECH and BCK are equal, since each equals angle CBD (the former via the parallel lines AC, BD; the latter via the isosceles triangle BCD). Adding to each the common angle HCK, we have angles HCB and ECK equal to each other; also, angles CEK and CBK are equal (by construction), so that the triangles ECK and HCB are similar. Hence, angles CHK and CEK are equal; as angle CEK is one-half LEK, while angle LHK subtends the arc LK, it follows that H lies on the circle of center E. Now consider the isosceles triangle EKH: the base angle EHK is the sum of angles CHK and CHE, where (a) angle CHK = KBD (by parallels CH and BD); and (b) angle CHE = ELC (since EH = EL) = CKE (since triangles ECK, ECL are congruent) = angle KBC (by construction). Adding equals, angles EHK and CBD are equal. Hence, triangles EHK and DBC are isosceles triangles with the same base angles; thus they are similar, so that EK : KH = DB : CB. Since KH, DB, and CB are given, so also is EK. Furthermore, since EK is the mean proportional between EB and EC (in accordance with the construction of the locus problem) and BC is given, EC may be determined via the application of areas (via Euclid's *Data*, Prop. 59). Thus, the point E is given in position and the line EK is given in magnitude, so that the circle of center E and radius EK is given, as is its intersection K with the given line CD. Thus, the analysis of the problem is complete.

As conceived here, the analysis intended to specify the ratio CK : KB as given, but eventuated in an alternative manner for determining the auxiliary circle, via the similar triangles EKH, DBC. The synthesis is evident: it entails specifying a length m (cf. EK) via the proportion $m : k$ = DB : CB (where k is the given length to be inserted and DB, CB are the side and diagonal, respectively, of the given rhombus) and the point E on BC extended via the proportionality of the locus construction (sc. EC : m = m : EB). The circle of center E and radius m meets CD at K; if BK is joined and extended to meet AC extended, the length KH will equal the given length k, as required. But the proof critically depends on the fact that H lies on the circle. It is this fact which forms the content of Pappus' lemma. Apparently, Apollonius did not provide the proof in the *Neuses*, doubtless considering it evident from the construction in the analysis. The omission is especially fortunate for us, however, since it provided Pappus the context for producing the lemma and thus preserving the manner of Apollonius' solution of the *neusis* problem.

If we view Heraclides as Apollonius' precursor in these studies of *neuses*, the problem of the square worked out by the former and reproduced by Pappus was not a mere alternative proof for a case of Apollonius' problem of the rhombus, but, to the contrary, served as a precedent for the Apollonian study. Indeed, one may suppose that Apollonius initially cast his method of solution as an alternative treatment of the case of the square and then recognized that it applied to the more general case of the rhombus. Most significantly, the Heraclidean study of *neusis* marks the first time in the extant mathematical literature where the objective is evident to effect a *neusis* via planar methods, that is, admitting into the construction only such lines or circles as are given via the Euclidean postulates. One commonly supposes that this objective characterized the Greek geometric tradition from its earliest stages. In persistently emphasizing the inappropriateness of this supposition for interpreting geometric research through the time of Archimedes and his successors, we face the issue of determining when and for what reasons this objective did in fact enter into the tradition. Our present findings hardly sustain the claim that Heraclides was the first to articulate this objective; indeed, the mere act of providing planar constructions for several cases of *neuses* need not assign to him the recognition that this formed the starting point for a scheme of classification extending to the whole field of geometric constructions. Nevertheless, the precedent did not fail to impress Apollonius; for in addition to his compilation of planar constructions in the *Neuses*, he produced another work *On Plane Loci* and a third *On Tangencies*, both of which are restricted to planar methods.[24] The second amounts to a systematic inquiry into the various ways a line or circle is specified as a locus with respect to given points or lines; we have already seen from this Apollonius' solution of the locus determined by the ratio of distances from two given points, where the method here was derived from much earlier sources. The Cartesian expressions for lines in the form $y = ax + b$ and for circles in the form $x^2 + y^2 = r^2$ have their analogues to special cases treated in the Apollonian work. As for the *Tangencies*, it showed how a circle may be constructed to pass through or be tangent to any combination of three given points, lines, or circles. The case of drawing the circle tangent to three given circles has attracted an enormous interest among mathematicians since the 16th century in search of alternative proofs and extensions; its Apollonian solution can be reconstructed in view of a lemma provided by Pappus.[25]

The present account indicates Heraclides to have been a geometer of no small impact. An associate of Archimedes, his biographer and defender against priority claims by Apollonius, he investigated the operation of *neusis*, so important for several constructions by Archimedes, to determine cases which can be effected by planar methods. Heraclides' effort set a precedent for the extensive researches into planar problems by Apollonius and so was an important factor in the emergence of the research program to regularize the methods of geometric construction. History has been less than generous to him, however. His works have been lost, but for the fragment on the *neusis* problem for the square given by Pappus, while his achievements were assimilated and superseded by the efforts of Apollonius. Unkindest of all, the later tradition appears even to have failed to get right the spelling of his name.

APOLLONIUS AND NICOMEDES

These efforts of Heraclides and Apollonius reveal a transition in thought concerning geometric constructions to the effect, in particular, that *neuses* might be viewed no longer as a fundamental operation, but rather as subordinate to other means of construction. We thus expect to find attempts to recast other problems where *neuses* had been employed, for instance, the cube duplication and the angle trisection. This turns out to be the case. Where the Archimedean solutions used sliding-ruler conceptions and where these were elaborated by Nicomedes in the generation of the conchoid curves, alternative methods were found which replaced the *neuses* with conic sections. In each instance, the extant sources betray convincing signs of Apollonius' responsibility.

Such a use of conics characterizes the first of several methods of angle trisection presented by Pappus (*Collection* IV, 36–40).[26] It relates to that variant of the Archimedean *neusis* we have already considered, in which a given length is inserted between two given lines.[27] The angle to be trisected is figured as that made by the diagonal AB and the side BG of a given rectangle (Fig. 6); if line BDE is extended from B such that its intercept DE between side AG and side ZA extended equals twice the diagonal of the rectangle, then angle DBG will be one-third the given angle ABG. The *neusis* is of the general sort considered by Heraclides and Apollonius, save that instead of a square or a rhombus, the initial figure is a rectangle. Pappus prefaces the angle trisection by a lemma even more general than this, showing how to effect the insertion of any given length with respect to any given parallelogram. He provides both analysis and synthesis in full; for our purposes, a sketch of the analysis ought to suffice.[28] Let the given rectangular[29] parallelogram be ABGD, and let the line AEZ be drawn such that EZ has a given length (Fig. 7). If DH, HZ are drawn parallel to EZ, ED, respectively, then DH will equal the given length EZ, and thus H will lie on a given circle, sc. that of center D and radius equal to the given length. Furthermore, since the product of BG, GD is given and equals the product of BZ, ZH,[30] the latter is thus also given and so specifies the locus of H as a hyperbola of asymptotes BG, AB passing through the given point D. Thus, H is given as the intersection of a given circle and a given hyperbola.

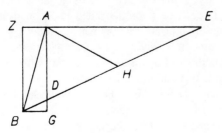

Figure 6

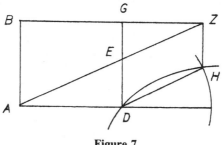

Figure 7

The synthesis takes the expected form where a critical step involves the construction of the hyperbola with given lines as asymptotes and passing through a given point.[31] The property at issue is that at each point of this curve the product of the ordinate and the abscissa has a constant value. As one sees in such early studies using conics, such as the solution of the solid-sectioning problem by Archimedes and by Diocles, or indeed the solution of the cube duplication by Menaechmus, the introduction of the curve satisfying this same property occasioned no special consideration.[32] But here Pappus takes pains to provide complete details of the construction of the hyperbola, via analysis and synthesis in full. The construction introduces the tangent to the conic at the given point and from this works out the Apollonian parameters: its diameter (*diametros*), latus rectum (*orthia*), figure (*eidos*), and orientation of ordinates (*katagomenai tetagmenōs*). One should have supposed this to be superfluous, for a synthesis of this problem in just this manner appears in Apollonius' *Conics* II, 4. But further inspection reveals the latter proposition to be an interpolation by Eutocius or some subsequent editor in the 6th century A.D.[33] Nevertheless, the Apollonian provenance of this construction as presented in the context of the angle trisection is indicated not only by the strength of its Apollonian terminology, but also by the fact that the activity of investigating problems for which conic sections were necessary could not long dispense with it. One should note the structure of the present problem: the angle trisection depends on the lemma effecting the *neusis* (in a more general form) via conics; the latter, in turn, depends on the lemma constructing the hyperbola to given asymptotes (also more general than necessary, in that the case of orthogonal asymptotes would suffice); furthermore, this depends on a lemma (cited in the analysis) showing how to find a cone and its appropriate section which produce a hyperbola of given Apollonian parameters (sc. *Conics* I, 54, 55). Note, moreover, that the construction of the hyperbola of given asymptotes has wide applicability, entering as auxiliary wherever two variable lines are to contain a given area. In addition to its appearances already mentioned here, it is needed for finding the normal lines to given conics, as Apollonius treats this in *Conics* V.[34] These considerations thus indicate that Pappus' text of the angle trisection and its accompanying lemmas are based closely on a treatment by Apollonius or a contemporary adhering to his version of the theory of conics.

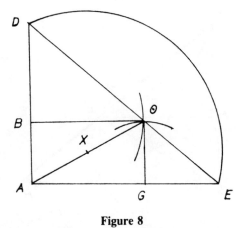

Figure 8

That Apollonius did indeed contribute to the study of the angle trisection remains a matter of inference, for our ancient sources do not cite him explicitly in this connection. The case of the cube duplication is different. Three closely related versions of the solution have come down under Apollonius' name, and there is a passing allusion to his responsibility for a fourth. The one reproduced by Eutocius is equivalent to the method of Hero, discussed above in the context of Archimedes and Nicomedes,[35] save that instead of a sliding ruler DθE, there is conceived a circle centered at X such that the vertex θ and the two intercepts D, E lie in a straight line (Fig. 8). Two variants are given by Joannes Philoponus, Eutocius' contemporary, citing as authority one Parmenion[36]; if the latter is the same "Parmenio" mentioned in conjunction with Apollonius and others by Vitruvius for their inventions of sundials, then Philoponus' source here may well derive from quite close to the time of Apollonius.[37] Of these two solutions, one is precisely the method of Hero, while the other is the same as that given by Eutocius under the name of Philo of Byzantium, a mechanical writer of the late 3rd century B.C., in agreement with the version extant in the latter's *Belopoeica*.[38] Compounding the confusion which appears to afflict our sources is Pappus' mention of yet another solution by Apollonius, this one by means of conic sections.[39] Although Pappus does not give the actual construction or any further information about it, his remark will assist greatly in accounting for the discrepancies in the other sources.

For clarity's sake, I shall speak of the "Heronian" and "Philonian" methods, although the above has shown that somehow both figure in Apollonius' treatment of this problem. The Philonian form may be derived from the Heronian in the following way: we set AB, BG equal to the two given lines between which the two mean proportionals are to be found and from them form the rectangle ABGD, with diagonals meeting in E and sides DA and DG extended (Fig. 9). Then by conceiving a line to pass through B and to meet the extended sides in H and Z such that EH = EZ, one may show that AZ and GH are the required means. Now, let us draw the circle about ABGD, meeting line ZH at a point K as well

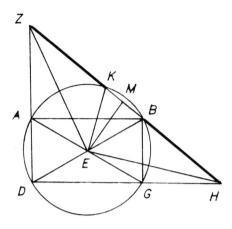

Figure 9

as at B. If we draw EM perpendicular to ZH, it follows that the triangles EMH, EMZ are congruent, as are the triangles EBM, EKM; hence, ZM = MH and KM = MB, so that ZK = BH. Thus, we have an alternative mode for effecting the *neusis*: to manipulate the line ZBH until the segments ZK, BH intercepted between the circle and the extended sides are equal. This is the method of Apollonius in its "Philonian" form.

This form of the construction has certain advantages over the Heronian form: the *neusis* is somewhat more easily executed in practice, as Eutocius observes; while consideration of the secant lines drawn from H and Z helps to simplify the proof.[40] But merely inferring the one form from the other seems not to signify any major insight. One is led to consider whether Apollonius hit upon the new method via an alternative route, without recourse to the Heronian method. Just as the latter can be derived from the Menaechmean analysis of the problem, so also can the Apollonian.[41] To see this, let K be the intersection of the parabola and the hyperbola, such that its coordinates KL, KN satisfy the relations $KL^2 = KN \cdot GD$ and $KL \cdot KN = AB \cdot BG$, where the given lines are AB, BG (Fig. 10). This makes KL, KN the required mean proportionals. Now, draw BK and extend it to meet DG and DA extended at H and Z, respectively. It follows that GH = KN (= DL) and AZ = LK (= DN).[42] We draw KD. Since K is on the parabola, $KL^2 = GD \cdot DL = HL \cdot DL$ (for GD = HL), so HKD is a right angle. Thus, $KD^2 = HK \cdot KZ$. But as K and B lie on the hyperbola, HK = BZ (an important property of the secants to hyperbolas, established in *Conics* II, 8); thus, $HK \cdot HB = KD^2 = KL^2 + LD^2 = HL \cdot LD + LD^2 = HD \cdot HG$. We thus have $KD^2 = HD \cdot HG = HK \cdot HB$. Hence, if a circle is drawn around the rectangle ABGD, the point K will lie on it, since HGD and HKB are secants. Thus, the point K may be determined as the intersection of this circle with either of the conics. An analogous derivation can be given via the assumption of the two-parabola form of the solution. In each instance, one should note the central role played by the equality of the segments HK, BZ.

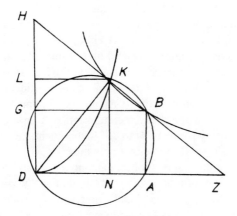

Figure 10

The proof in the synthesis for all these forms will depend on this equality, HK = BZ, applied to the secants drawn from H and Z. That is, DH • HG = BH • HK = KZ • ZB = DZ • ZA, so that HD : DZ = AZ : HG. By similar triangles, HD : DZ = HG : GB = BA : AZ. Hence, BA, AZ, HG, GB are in continued proportion, establishing AZ, HG as the required means between BA, GB. This is, in fact, the manner of proof adopted in all our texts of the Apollonian–Philonian method. But we may observe that the conception of the construction presents several options: we might seek the intersection of the circle with one or the other of the parabolas, or that of the circle and the hyperbola. Alternatively, we might seek that position of the moving line HBZ such that HK = BZ. This last is the form preserved in our sources. In the texts received under the name of Philo, the notion of "a ruler turning about B" is explicit. But in Philoponus' text under the name of Apollonius, it is, at most, implicit; for the construction requires only that "ZH be joined such that ZD equals EH" (here points D, E correspond to B, K, respectively, in our lettering); the commentator notes that "this is assumed as a postulate without proof." The same applies to the Apollonian solutions in their "Heronian" form, as given by Philoponus and Eutocius: one demands certain features of the diagram (e.g., equality of lines or collinearity of points) without specifying how these are to be obtained.[43]

This suggests that Apollonius' intent was not to produce a different form of *neusis*, but rather to replace the *neusis* by an alternative construction. This is entirely in keeping with the project of his other studies of *neusis* discussed earlier. In the present case, a method arises entailing the intersection of the circle with a conic, and this conforms with Pappus' remark that Apollonius used the conics for solving this problem. The variant using the circle and the hyperbola is especially interesting in this regard; for although no such treatment is extant in Greek, it occurs frequently in the Arabic tradition of classical geometry. For instance, Naṣīr al-Dīn al-Ṭūsī writes this construction and its proof in the margin of his manuscript of *Conics* V, 52 (on the normals to the hyperbola and the ellipse); he provides a more ample treatment along the same lines in his remarks on Archimedes' *Sphere and Cylinder* II, 1, thus setting aside the dozen "instrumental" methods of solution which Eutocius provides at the corresponding place in his commentary.[44] An earlier version, dating from the 9th or 10th century, derives from one Abū Bakr al-Harawī.[45] Textual considerations reveal that it was drafted with one or another of the Apollonian–Philonian texts in view, and one may better view it as produced as a translation from a Greek source rather than as an original composition in Arabic.[46] In this event, the ultimate provenance of this method would surely lie with Apollonius.

The precise chronological relationship among the several Apollonian versions remains elusive. In the form of *neuses*, the "Philonian" version can be obtained immediately from the "Heronian," and conversely. Considerations of practical efficiency and elegance of proof recommend viewing the former's introduction of the circle as a refinement of the latter; and the assignment of priority in time to the "Heronian" version finds some support from its implication in the method used by Nicomedes, as we have seen. I find it most plausible to view Apollonius' presentation of the "Philonian" version, where the explicit mechanical conception of the sliding ruler is lacking (as in the text given by Philoponus), as a step toward the elimination of the *neusis* via the use of conics, specifically, through the intersection of the circle with a hyperbola. The alternative view, that Apollonius hit upon the "Philonian" form of the *neusis* through direct investigation of the Menaechmean procedure, without reference to the "Heronian" *neusis* as intermediary, seems to me less likely; for there would seem to be little point in replacing one construction by conics (a hyperbola and two parabolas) with another of the same type (the same hyperbola and a circle), while the *neusis* would be purely superfluous.

Apollonius followed Eudoxus and Aristarchus as one of the major formulators of Greek geometric astronomy in the period before Hipparchus. His studies of the geometry of circular motions form another link with the earlier studies of special motion-generated curves by Archimedes and Nicomedes. Ptolemy (*Syntaxis* XII, 1) assigns to Apollonius the determination of the stationary points of a planet.[47] In the epicyclic model, where the observer on earth is at Z and the planet at H lies on an epicycle whose center E lies on the deferent circle centered on Z, then the planet will appear stationary at H as viewed from Z when ½ BH : ZH = $w_e : w_h$, where w_h, w_e are the angular velocities of H and E, respectively, with reference to the centers of their circles (Fig. 11). This property is readily derived from considering the motion of H as compounded of two circular motions, so that when the projections of the latter perpendicular to the line of sight are equal but opposite, H will appear to be stationary. But just as Archimedes had employed no such kinematic notions in his investigation of the tangents to the spirals, so, also, Apollonius reduces this theorem to static considerations, showing that this position of H separates a prior range of apparent direct motion from a subsequent range of apparent retrograde motion. Ptolemy views the same theorem in the context of the alternative eccentric model for

Apollonius: The Culmination of the Tradition   309

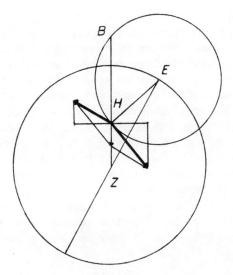

Figure 11

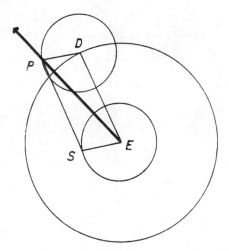

Figure 12

planetary motion, so that one infers that Apollonius well knew the equivalence of the two modes: that is, the orientation of the diagonal EP of the parallelogram EDPS is the same whether one determines it via the rotations of ED, DP (the epicyclic model of the motion of planet P with respect to the earth E) or via those of ES, SP (the eccentric model for the outer planets Mars, Jupiter, or Saturn at P and the sun at S) (Fig. 12). Historians of astronomy often note the implicit irony in Apollonius' recognition of the equivalence of these two models, for it assigns to him a recasting of the prior heliocentric scheme of Aristarchus via eccentrics much as Tycho Brahe modified the Copernican hypothesis in the 16th century; yet for Apollonius this was a step en route to the epicyclic model, which through its later refinement by Hipparchus and Ptolemy established the geocentric scheme which the Copernican astronomers challenged.[48]

Our sources do not reveal what led Apollonius to the notion of the epicycle. Was it a consideration of the kinematic equivalences entailed in the Aristarchean system? Or was it an investigation of the orthogonal projections of the Eudoxean hippopedes onto the plane of the ecliptic? At any rate, it is clear that Apollonius' interest lay in the more abstract geometric study of this configuration. The more practical effort of modifying the models and determining the parameters to agree with data on planetary motion and of developing the computational techniques of trigonometry to convert these models into an effective predicting system engaged Hipparchus and later geometers in the period after Apollonius. Their work profited from access to Mesopotamian data and techniques to an extent far greater than seems to have been possible for Greek geometers before the 2nd century B.C.[49]

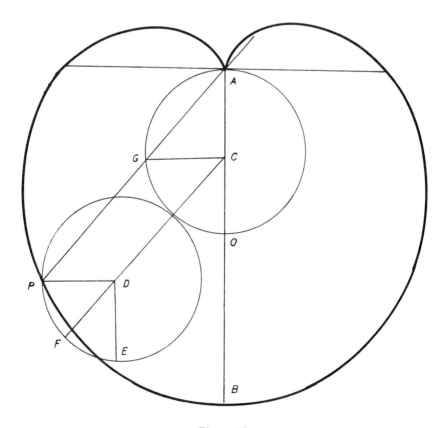

Figure 13

For one such as Apollonius, then, we would expect that geometric interests in epicyclic motions led him to examine the properties of the curves so traced.[50] For instance, the circle-based conchoids with pole lying on the base circle can be generated as epicycles. Consider two equal circles centered at C and D such that CD always equals twice the radius of either (Fig. 13). Let the point P rotate about D with twice the angular velocity at which D rotates about C, so that P traces an epicycle with respect to the deferent circle of radius CD. If we draw in a reference axis AB, so that P coincides initially with B, at a later position, angle EDP will be twice DCB, for ED parallel to AB. Let A be the outer point of the diameter of the second circle, centered at C, such that AC falls on the axis initiating the epicyclic path of P. Set angle CAG equal to BCD and join CG. Since angle 0CG is twice CAG, it equals EDP, so that CG and DP are parallel and equal. Hence, CD and GP are also parallel and equal, and PG lies in a straight line with GA. Thus, P also traces a conchoid with respect to the "pole" A, having the circle centered on C as its base and the constant distance measured off from that base equal to CD, or twice the radius of the base circle. As we have seen, this curve is the modern "cardioid" which I have argued to be the primary form of the curve called "cissoid" by the ancients.[51] More generally, any epicycloid for which the angular velocity of the epicycle is twice that of the deferent can be expressed alternatively as a conchoid whose circular base has the same radius as the epicycle, whose "distance" equals the radius of the deferent, and whose "pole" lies on the circle (Figs. 14 and 15).[52] We can thus perceive a natural link between the kinematic interests of Apollonius and the special curves studied by Nicomedes.

This section has considered the evidence for Apollonius' contributions to the area explored by Nicomedes and other geometers following Archimedes, in particular, to the problems of angle trisection and cube duplication and to the study of motion-generated curves. The precise manner and extent of Apollonius' work here has not been preserved in our sources. But we learn enough to sense that his involvement was significant. His constructions for the two problems raise instances where *neuses* may be effected via conics, and so extend his researches on *neuses* constructible via planar methods. Further, it is entirely plausible that Apollonius' astronomical studies of eccentric and epicyclic motions were based on geometric studies of the kinematic configurations giving rise to the various forms of epicycloids.[52a] It seems impossible to determine whether these were originally suggested as plane projections of Eudoxus' hippopedes, or as geocentric equivalents of Aristarchus' orbits for the planets, or in some other way. But the richness of their geometric properties is clear, especially when viewed against the background of Nicomedes' conchoids. Indeed, the theorem on the equivalence of eccentric and epicyclic models, the heart of Apollonius' contribution to planetary theory, is hardly other than a generalization of the geometric theorem linking the formation of the *limaçons* as conchoids and as epicycloids. These astronomical studies thus testify indirectly to Apollonius' extensive interest in these curves.

312 Ancient Tradition of Geometric Problems

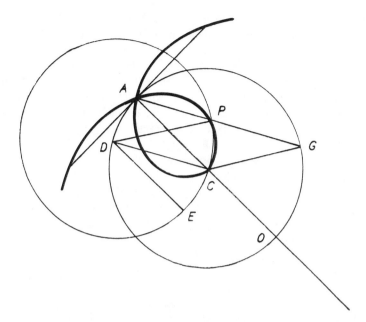

Figure 14

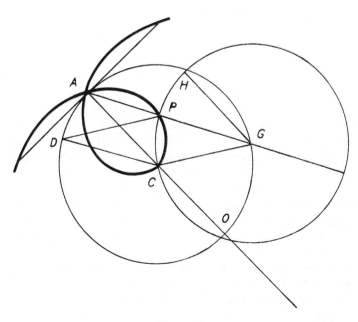

Figure 15

APOLLONIUS AND EUCLID

Apollonius is most famous for his contributions to the theory of the conics. His massive treatise collecting the elements of this theory and applying it toward the solution of problems in advanced geometry was composed in eight books. Of these only the first four are now extant in Greek; but the first seven and a fragment of the eighth survive in their Arabic translation, while commentaries by Pappus (*Collection*, Book VII) and by Eutocius further augment our knowledge of his achievement in this area.[53] The *Conics* systematized a major portion of the field as already known, introducing new materials where prior treatments were incomplete and applying these toward the investigation of new areas. An effort of this sort was bound to blur the line between Apollonius' original work and those results he could derive from his predecessors. But if this makes understandable the controversies which arose in connection with this work, Apollonius seems to have provoked sharp responses through his own aggressiveness. To be sure, he is magnanimous toward Conon of Samos, whose studies on the intersection of conics, while flawed, still had utility in his view, as expressed in the preface to Book IV.[54] On the other hand, Apollonius somehow gave the impression of arrogating to himself the discoveries of others, notably Archimedes, thus earning the rebuke of disciples like Heraclides.[55] The reaction among Euclid's followers was especially strong. Apollonius criticizes Euclid for having worked out the solutions of conic problems, like that of the locus with respect to three and four lines, poorly and haphazardly. Euclid's apologists object that their master's resources were still limited, that he had made no presumption to comprehensiveness, and that Apollonius could hardly have made the advances he did without the firm precedent of Euclid's achievements.[56] Clearly, this was a geometric field of major competitive interest, where reputations stood to be made or lost.

Apollonius adheres to the synthetic mode throughout the *Conics*, with but a few exceptions in Book II where analyses are also given. His conscious imitation of the formal style of Euclid is often in evidence, even to the details of the wording of certain propositions.[57] Nevertheless, Apollonius, like Euclid, intended compilations of this sort to serve the interests of the analytic solution of problems; he makes this claim explicitly in his prefaces and is confirmed by Pappus who assigns to the entire *topos analyomenos* the same role.[58] In the *Conics* the problem-solving aspect is especially prominent in the later books, treating the more advanced portions of the theory. Book V, for instance, develops toward the solution of problems on the drawing of normal lines to given conics from given points. Book VI examines problems of the construction of given conic sections in given cones; an Archimedean influence suggests itself through comparison with the problems on segments of spheres and cones presented in *Sphere and Cylinder* II, as well as through the theorems on similar conics which are based on ideas introduced in *Conoids and Spheroids*.[59] Book VII of the *Conics* considers properties of the conjugate diameters of central conics; although

Book VIII is lost, we learn that it consisted of problems applying the results from Book VII, and from this Halley was able to propose a full reconstruction.[60]

In the more elementary books (I–IV), the role of problems is less apparent. The first book concludes with the syntheses of problems to produce a cone and a section of it resulting in a conic whose parameters are given (I, 52–60); these are in effect the converses of the theorems (I, 11–14) in which the curves are defined as sections of cones and the parametric representations are derived. The second book includes a set of problems on the drawing of tangents to given conics from given points (II, 49–53) and on the finding of the axis or center of a given conic (II, 44–48); for these, but nowhere else in the *Conics*, Apollonius provides the analyses along with the syntheses. These apply the basic properties of the curves established in Book I: e.g., for the parabola, the tangent at A meets the axis at E such that EB = BD (I, 33, 35), where B is the vertex, BD the abscissa, and DA the ordinate corresponding to A [Fig. 16(a)]; similarly, for the hyperbola and ellipse, the tangent at A meets the axis at H such that GH : HB = GD : DB (I, 34, 36), where GB is the diameter and BD, DA the coordinates of A, as before [Figs. 16(b) and 16(c)]. From these one may then find tangents, first for points on the curve itself, then for points on the axis, and then for other exterior points (II, 49). The placement of these problems is somewhat puzzling. Only a few of the theorems in Book II deal with the properties at issue

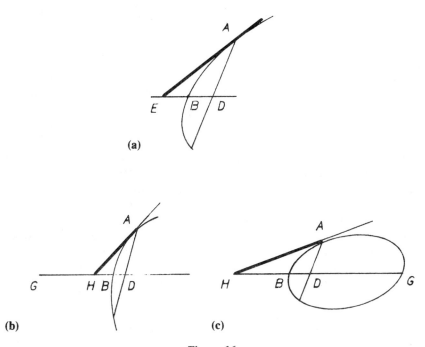

Figure 16

here (specifically, tangents to the hyperbola are considered in II, 1–3, 9–11), for its principal topic is the investigation of the interrelations of the two branches of the hyperbola.[61] But the problems develop around certain properties well established in the pre-Apollonian literature; the intercept of the tangent to the parabola, for instance, is assumed without proof both by Diocles and by Archimedes.[62] These considerations, together with the unusual adherence to the double method of analysis and synthesis, suggest that Apollonius has attached these problems as an appendix to Book II. Presumably, they were the outgrowth of an earlier stage of his study of the conics, for they make no reference to the conception of the hyperbola as a double-branched curve; that is, they work exclusively within the domain of the earlier theory, applying familiar results, working principally from the "symptom" of the conics in the form of a ratio of areas (as in I, 20, 21) which was basic to Archimedes, and showing no applications of notions one usually views as Apollonius' original innovations.[63]

The problems on normals in Book V, in addition to providing an illustration of the problems in the *Conics*, raise certain interpretive questions. Apollonius defines the normal as the minimal line drawn from a given point to the given curve.[64] In V, 27–29, Apollonius establishes that the minimal line drawn to a point on a conic meets the tangent to that point at right angles. For the parabola (V, 27) this refers back to the property that the subnormal always equals one-half the latus rectum (V, 13; converse of 8); that is, if BO is the minimal line drawn from O on the axis to B, and BC, AC are the coordinates of B on the parabola (i.e., $BC^2 = AC \cdot L$, where L is the latus rectum), then $CO = \frac{1}{2} L$

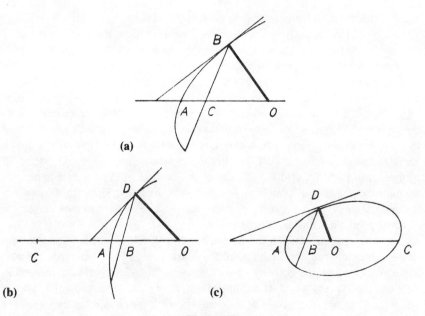

Figure 17

316 Ancient Tradition of Geometric Problems

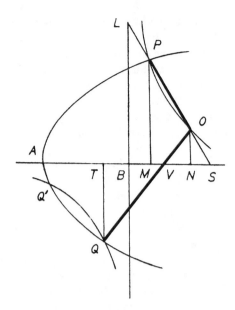

Figure 18

[Fig. 17(a)].[65] For the hyperbola and ellipse (V, 28), one refers to the analogous property (V, 14, 15, converses of 9, 10) that for point D on the curve, having coordinates DB, BA (that is, $DB^2 : AB \cdot BC = L : AC$, for L the latus rectum and AC the diameter), the normal meets the axis at 0 such that CB : B0 = AC: ½ L [Figs. 17(b) and 17(c)]. Let us then consider one of the cases of drawing the normals to the parabola, namely when the given point 0 lies within the curve. Although Apollonius gives only the synthesis, we may sketch a form of the analysis to reveal the basic idea. Let 0 be a given point within the given parabola of latus rectum L, and let it be supposed that P0 is normal to the curve (Fig. 18). Then, if P0 is extended to cut the axis in S, and PM is the ordinate at P, the segment MS equals ½ L. Now if 0N is drawn parallel to the ordinate direction to meet the axis in N, then N is given; if we set NB = ½ L, B will be given and BM = NS. Now draw BL parallel to the ordinate direction and extend 0P to meet it in L. Since BM = NS, LP = 0S, so that we may pass a hyperbola through the points 0, P as to have BS, BL as asymptotes. Since the asymptotes and point 0 are given, so also is the hyperbola; hence, P is given as the intersection of this hyperbola with the given parabola.

This analysis, modelled after the synthesis in V, 62, makes key use not only of the subnormal property, but also of the construction of the hyperbola of given asymptotes and of the equality of the segments of the secant line intercepted between the hyperbola and its asymptotes. Both of these latter properties appear in Pappus' lemmas to this book of the *Conics*[66]; while the equality of the intercepts was critical for the solution of the cube duplication via the hyperbola, as we saw above.[67] Indeed, the introduction of two mean proportionals between two given lines is required for the investigation of the normals to the hyperbola and the ellipse appearing within this same set of problems. It was for this reason that al-Ṭūsī interpolated his solution of the problem of the means via the hyperbola.[68] Since all these related problems introduce the auxiliary hyperbola by virtue of the same property, the equality of the intercepts, one appears to obtain further support for the view that Apollonius was responsible for this manner of solving the cube duplication. Furthermore, one notes that the need for a construction of the hyperbola here is not supplied within the body of the *Conics*[69]; it would appear then that Apollonius can assume such results from an alternative context, like the treatment of conic problems we proposed in connection with the angle trisection.

We may consider the problem of the normals further: is it possible that a normal be drawn to the other portion of the parabola from 0?[70] Let us suppose that 0Q is such a line, so that if QT is its ordinate, and V is the point where 0Q meets the axis, TV = ½ L (see Fig. 18). Hence, since NB = ½ L also, NV = BT. By similar triangles 0VN, QVT, 0N : VN = QT : TV, that is, 0N : BT = QT : NB. Hence, QT • BT = 0N • NB. It is thus apparent that Q lies on the same hyperbola which passes through 0, for the product of its ordinate and abscissa with respect to the same asymptotes LB, SB (extended) equals that for 0. Not just P, then, but all the points of intersection of this hyperbola with the parabola will give rise to normals from 0, where the hyperbola is conceived as a single curve with two branches. Now it is possible that the second branch does not meet the parabola; or they might meet at a single point of mutual tangency; or they might meet at two distinct points, but never more than that. These relations are the content of another theorem (V, 51) where Apollonius provides the condition discriminating among these cases for a normal drawn from a given interior point so as to cut the axis. It is that the ordinate 0N of the given point satisfies the relation 0N : TQ = TB : BN, where BN = ½ L, as before, and T is the specific point such that BT = 2TA. The auxiliary hyperbola is introduced only late in his synthesis, but it helps one to see the reason behind these conditions. The proportionality assures that 0 and Q lie on the same hyperbola with respect to the asymptotes BS, BL, thus assuring that 0Q is indeed a normal (i.e., that TV = ½ L) (Fig. 19). The condition that BT = 2TA entails that the hyperbola and the parabola have the same tangent at Q; for if CQ is tangent to the parabola, CA = AT; while if it is tangent to the hyperbola, CQ = QD, whence CT = TB.[71] Thus, BT = CT = 2AT. In the actual synthesis, the uniqueness of Q for this value of 0N follows from inequalities showing that for any other Q, the corresponding line TV will be less than ½ L, so that 0Q will not be a normal. Apollonius determines analogous conditions for normals cutting the axis internally for ellipses and hyperbolas (V, 52). One may note that the converse application of these conditions for each of the conics leads to the determination for each point Q on the curve of a point 0 on the normal to that point such that 0Q is the unique normal from that point which crosses the

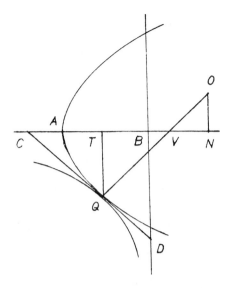

Figure 19

axis. In the modern theory of conics (e.g., since the 17th century), this point has been identified as the center of curvature of the curve for that point, while the locus of all centers of curvature gives rise to the "evolute" of the curve. The defining conditions for the evolutes of the conics are immediate corollaries to the condition for the uniqueness of the normal in V, 51, 52. For the parabola, the coordinates ON, NA of 0 on this curve satisfy the relation $\frac{1}{4} L \cdot ON^2 = (\frac{4}{27})$ $(NA - \frac{1}{2} L)^3$, that is, a semicubic parabola with cusp at the focus T (where $AT = \frac{1}{2} L$) (Fig. 20).[72] Extant materials, however, provide no direct support for the view that Apollonius developed his results on normals in this way, or that any ancient geometer defined the center of curvature for the points of a curve or the evolute of any curve.

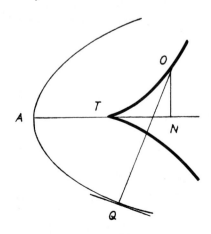

Figure 20

In view of the analysis just given, the treatment of normals Apollonius presents differs notably from what we would have expected, in that he never considers both branches of the auxiliary hyperbola simultaneously. His principal theorem (V, 51, 52) establishes the condition determining whether from a given point there may be drawn two, one, or no normal lines to a given conic, so as to cut its axis. This is followed by theorems to the effect that from a given point lying below the axis of an ellipse, exactly one normal can be drawn to the opposite quadrant (V, 55–57); that from any point outside a conic (but not on its axis) it is possible to draw a normal (V, 58–61); that from any point inside a conic it is possible to draw a normal (V, 62, 63).[73] In these latter cases, one does not apply the result of V, 51, 52, but rather effects independent constructions of the sort we used in the first part of our analysis above. Thus, one loses sight of the fact that the auxiliary hyperbolas in V, 62, 63 are the same as those in V, 51, 52. Viewing these as double-branched curves, Apollonius would have been in the position of transferring results on the manner of intersections of conics worked out in IV, 24–55. He could thus have more comprehensively set out the position of the given point with respect to the conic such that merely one normal can be drawn, or several up to as many as four. I find it difficult to accept that Apollonius would have omitted mention of these aspects of the auxiliary hyperbola had he perceived them, and so am inclined to view these problems on the construction of normals as the product of research completed before his introduction of the two-branched conception. Under this view, an early treatment of the constructions, where the normals were presented as the lines drawn perpendicular to the tangents,[74] was later revised through the alternative notion of the normal as minimal line. In this way, the latter version, extant as *Conics* V, replaced one of the former sort based on certain pre-Apollonian results and conceptions about normals.

Pappus reports an objection against the treatment of "the problem of the parabola in the fifth book of Apollonius' *Conics*," to the effect that it employs a "solid construction" (that is, one involving auxiliary conics) when a planar method is possible. With good reason this has been read as a reference to the problems on the normals to the parabola we have been considering.[75] Apparently, some geometer in antiquity recognized that the auxiliary hyperbola used in these constructions could be replaced by a circle. Pappus does not inform us of the actual manner of this alternative solution; but one approach might take the following form[76]: let us consider the figure used for determining the normal OP and ask whether P can be positioned on a given circle, rather than, as here, on a given hyperbola. We suppose C to be the center of such a circle, and since this still leaves us a latitude of choice, let us also suppose that this circle passes through the vertex A (Fig. 21). If we draw the perpendiculars CE, CD to PM, MA, respectively, since CP = CA, we have $PE^2 + EC^2 = CD^2 + DA^2$, whence PM · (PM − 2 ME) = AM · (AM − 2 MD).[77] Now, ON : PM = NS : SM = MB : ½ L, so ON · PM : PM^2 = MB : ½ L. Since P is on the parabola, PM^2 = L · MA; thus, ½ ON · PM = MB · MA. Subtracting, PM · (PM − ½ ON) = AM · (L − MB). Comparing with our earlier result, we recognize that a solution results from PM − 2 ME = PM − ½ ON and AM − 2

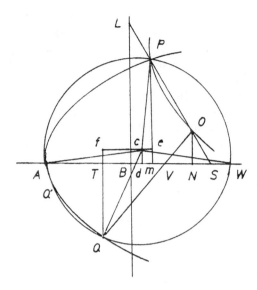

Figure 21

MD = L − MB. The former yields ME = ¼ ON, so that ME is given. In the latter, setting ½ L = NW = NB, we have AM − 2 MD = NB + NW − MB, or AM − MD = MD + BN − MB + NW, so that AD = DW. Thus, given 0, we construct the circle whose center is at C, for CD = ¼ ON and AD = ½ AW (= ½[AN + ½ L]), and passes through A. If one performs the same analysis on Q, one finds that the very same circle passes through it, so that this circle on the center C, as above, and with the radius AC determines *all* the normals (whether one, two, or three) that can possibly be drawn from 0 to the parabola.

Since Pappus knows this result, the above analysis or some comparable alternative is implied in his source, and its originator was surely a significant geometer. But the fact that Pappus raises the issue as a criticism of Apollonius reveals that Apollonius was not that geometer. This does not imply that he was incapable of such an analysis, but only that it did not occur to him to pursue this matter in the present context. Heath's observation seems appropriate: that it would be mere pedantry for Apollonius to have introduced the circle for the normals to the parabola, when the entire set of problems lends itself so naturally to the method using auxiliary hyperbolas.[78] Nevertheless, this *is* a pedantry which competent geometers after him engaged in, if he did not. Thus, the categorical insistence on planar procedures, where possible, in preference to solid procedures is a formal dictum which Apollonius seems not yet to have subscribed to. To be sure, his efforts on the construction of *neuses* by planar methods, as we saw above, reflect an interest in regularizing the treatment of an important class of geometric problems. But it need not follow that he went beyond that, to assign a comparable formal status to the ''solid'' problems, in distinction from the planar ones. His massive contribution to the consolidation and extension of the techniques of conics must certainly have induced others to make that move, perhaps in his own lifetime. Indeed, the fact that the criterion was used to fault Apollonius' work suggests that it arose within that contentious environment centering on Apollonius' theory of conics. If this is the case, its author could be the same from whom Pappus derived his account of the rebuttal by Euclid's disciples in defense of his theory of conics against the charges levelled by Apollonius.[79]

APOLLONIUS AND ARISTAEUS

The third book of Apollonius' *Conics* is rich with implications for the field of geometric problems. Yet it contains not a single proposition presented and solved in the form of a problem. Consider the following items and the questions to which they give rise:

Apollonius' treatment of the focal properties of central conics entails, at once, the solutions of such problems as determining the locus of points whose distances from two given points have a given sum or difference (cf. III, 51, 52). When the late commentator Anthemius of Tralles derives the former as a property of a reflecting surface he has constructed, he immediately concludes that the contour is an ellipse.[80] Yet the contour he has sought is initially specified by the condition that all rays emanating from one of the given points shall be reflected through the other. That this obtains for any central conic having the given points as foci is proved by Apollonius in III, 48. Thus, a knowledge of this result would have rendered superfluous Anthemius' whole construction. Moreover, he goes on to a separate investigation of the reflective properties of the parabolic mirror, specifically, that all incoming rays parallel to its axis will be reflected through a single given point. While this is precisely the analogue for the parabola of the property just cited for central conics (that is, where the second focus is conceived at infinity), it finds no place in the *Conics* of Apollonius.[81] It is thus clear that Anthemius' source is not based directly on Apollonius' treatment of the theory of conics. Then, what treatment *was* it based on?

Apollonius takes up a wide variety of theorems on the ratios of products of segments of lines drawn as chords or tangents to conics. The results on intersecting tangents given in III, 41–43 might form the basis for constructing the conics of which given lines are tangents. It has been recognized that the problem for the parabola gives rise to a series of special cases requiring the determination of their own auxiliary constructions. Precisely these constructions comprise the lemmas of Apollonius' analytic work, *On the Cutting of a Ratio* (preserved in its Arabic translation).[82] Furthermore, the analogous lemmas for the ellipse and hyperbola were compiled in the lost writing *On the Cutting of an Area*, as we may gather from the lemmas which Pappus provides for it in *Collection* VII.[83] These two works, however, do not take up the actual solution of these problems on the construction of the conic sections. What can be the reason for Apollonius' omission? If he did indeed present the solutions of these problems in yet another analytic work, why has Pappus failed to include any mention of it in the *Collection*?

The problem of the three- and four-line locus engaged the efforts of Euclid, as we have already discussed, but Apollonius himself observes that the complete solution only became possible on the basis of theorems presented in *Conics* III.[84] In fact, the basis of a solution for this locus can be derived from III, 16–23.[85] But neither here nor anywhere else known to us does Apollonius take up the investigation of these problems as such. What can explain this omission?

From the four-line locus one can develop solutions to the problems of constructing conics passing through five given points. The manner of an Apollonian procedure can be inferred from *Conics* IV on the intersections of conics.[86] By a method of the suggested sort, Pappus presents a solution constructing the ellipse passing through five given points (*Collection* VIII, Props. 13–14).[87] In this there is a natural affiliation with the problem of finding the intersections of a line with a conic specified as a four-line locus. That Apollonius addressed this problem in exhaustive detail is revealed in Pappus' account of another lost analytic work by Apollonius, *On the Determinate Section*.[88] As before, this work appears to be a repository of ancillary results useful for the solution of the motivating problem. A characteristic case, for instance, is to find a point P on a given line whose distances from four given points A, B, C, D are such that the ratio PA • PB : PC • PD has a given value. This may be related to the case of the problem of finding the intersection of a conic with this line, where the conic is a four-line locus whose reference lines meet the given line in the four given points. In projective geometry, one introduces the "involution" which projects A, B, C onto B, A, D, for instance. The associated configuration of the "complete quadrangle" is shown to satisfy the corresponding cross-ratio equality in Pappus' Lemmas 1–4 to the *Porisms* of Euclid.[89] In the context of the conics, this result gives rise to the remarkable theorem of Desargues: all the conics which pass through four given points and meet a given line will meet that line in point pairs or double points of the involution determined via the quadrangle of the given points.[90] On the basis of the *Determinate Section*, Zeuthen wishes to assign to Apollonius effective knowledge of this theorem, even though Pappus' lemmas to it never introduce the notion of point pairs associated with different choices for the given ratio.[91] But we surely do better to view Apollonius' effort as the systematic working out of the lemmas toward a certain set of conic problems, rather than as an inquiry into projective involutions as such.[92] Even with this, however, we still do not know where the solutions to these problems were effected. In the case of the five-point ellipse, for instance, what was Pappus' source for his treatment?

In summary, these parts of Apollonius' contribution to the analytic tradition suggest an intense interest in the effort to solve problems dealing with conics.[93] The class of problems seeking the construction of conics under the conditions of given points and tangent lines forms a natural complement to the project of Apollonius' *On Tangencies*, where planar methods were worked out for the determination of circles passing through given points or tangent to given lines or circles.[94] But for the conic problems, the pertinent works in the *topos analyomenos* discussed by Pappus seem not to have addressed their solution in the form of problems; they are devoted, instead, to the elaboration of subordinate lemmas. We are thus at a loss to know what contexts Apollonius exploited for the exposition of these problems.

Among the works in the analytic corpus discussed by Pappus in *Collection* VII, there is only one whose title would mark it as pertinent to this field, the treatise in five books, *On Solid Loci*, by one "Aristaeus the Elder." At first glance, however, this is an unpromising candidate to have served as a compendium of advanced problems of the solid class. For although Aristaeus was known to have produced an elementary presentation of the theory of conics, Pappus set him in the generation of Euclid—it was Aristaeus' theory which formed the basis of Euclid's researches into the three- and four-line locus, for instance.[95] But it is scarcely credible that at this early time the fledgling theory of conics could sustain such a massive treatment as would seem to be indicated in the multivolume treatise on solid locus. Moreover, the term "solid locus" would not then have signified the restricted class analyzed by means of conics, as it did for writers after the time of Apollonius; for we have seen that even Apollonius did not view the distinction separating planar and solid problems in precisely the way that Pappus did.

Thus, either the Aristaean treatise covered a wider field than that of conic problems alone, or else there has been some confusion in the chronology. That the latter may be the case is suggested in Pappus' remark that during his own time the *Solid Loci* was still extant,[96] for major textbooks like Euclid's *Elements* and Ptolemy's *Syntaxis* invariably contributed to the extinction of the precursors in their genre, and Apollonius' *Conics* would certainly have done the same for a treatment of the theory of conics antedating it by almost a century. Note that the *Conics* had precisely that effect on the transmission of Euclid's efforts in this field. We would thus suppose that the treatise on conics still available to Pappus derived from a time appreciably later than Euclid. Now, we learn from the geometer Hypsicles, writing around the middle of the 2nd century B.C., that one Aristaeus had composed a work *On the Comparison of the Five Figures* (sc. the five regular solids) bearing an interesting relation to a writing by Apollonius *On the Comparison of the Dodecahedron and the Icosahedron*: both provided proofs of the theorem that if these two figures are inscribed in the same sphere, their volumes will have the same ratio as their surfaces, for the line drawn from the center perpendicular to any face will have the same length in both cases.[97] Hypsicles notes that Apollonius' work circulated in two versions. The earlier of these held a mistake in the proof which was rectified in the later edition. He adds that his own father had been spurred by that error to investigate the theorem, while he himself had examined Apollonius' later version and deemed it in good order. On the view that this Aristaeus was the same Aristaeus who wrote on conics at the time of Euclid, we would be compelled to accept that Apollonius again took up the study of the regular solids almost a century after its analogous treatment by Aristaeus, yet despite the precedent managed the demonstration incorrectly. The implausibility of this view has already led historians to suppose that two different geometers are at issue, so that the Aristaeus mentioned by

Hypsicles would be a later figure, perhaps a contemporary of Apollonius.[98] In the light of this, we would surely assign the treatise *On Solid Loci* to the latter Aristaeus whereby it becomes precisely that repertory of solved conic problems filling the gap in the ancient literature of this field as known to us. By assigning this work to "the Elder," Pappus would then have made the mistake of conflating two geometers of the same name and similar interests, but a mistake fully understandable in view of the half-millennium separating him from the original researches.

One expects that when Pappus follows Apollonius or a geometer working directly in his tradition, this dependence would be reflected in certain details of the proofs, for instance, in the choice of theorems cited or assumed, or in terminology. In the case of the angle trisection discussed above,[99] an Apollonian provenance seemed to be indicated in the adoption of Apollonian usage within the affiliated hyperbola construction. The same holds for an alternative solution, in which one seeks to divide a given arc in the ratio 2 : 1. The analysis leads to the relation $3AD \cdot DH = BD^2$ (Fig. 22), from which one immediately observes that B lies on a hyperbola "of which the latus transversum (*plagia*) of the figure (*eidos*) toward the *axis* is AH and the latus rectum (*orthia*) is three times AH."[100] The inference follows from the property of conics proved in *Conics* I, 21; the terminology unmistakably reveals the author's reliance on Apollonius. Pappus presents another version of the solution in which the same hyperbola is specified in different terms. This is the form we have already discussed in connection with Euclid's *Surface Loci*.[101] The analysis specifies the locus of B via the relation $BZ^2 + ZG^2 = 4 EZ^2$ (Fig. 23); the author concludes at once that since $EZ^2 : BZ^2 + ZG^2$ is a given ratio, B lies on a hyperbola. Implied by this is dependence on a solution of the locus problem in which the distances of each point from a given point G and a given line DE are to have a given ratio (here, 2 : 1). Pappus presents the solution among his lemmas to the *Surface Loci*, showing that the locus is a parabola, hyperbola, or ellipse, according to whether the ratio is equal to, greater than, or less than unity, respectively.[102] The connection is all the more clearly seen in that the latter is solved after recasting the condition in terms of the ratio $AD^2 : BD^2 + DG^2$ instead of the ratio AD : BG, as done initially. While this is convenient in the present context, it was an entirely superfluous step in the angle trisection. Thus,

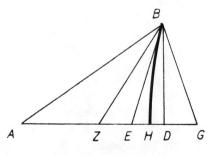

Figure 22

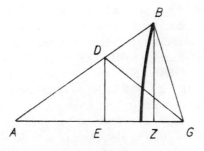

Figure 23

the latter has evidently been framed with this particular treatment of the point-line locus in mind. Now, in the analysis of the locus problem, where the ratio is unity, one derives that $GD^2 = 2\ AB \cdot DZ$, whence it is seen at once that G lies on a given parabola (Fig. 24); where the ratio is different from unity, the analysis establishes that $GD^2 : \theta D \cdot DH$ has a given value, so that the locus is an ellipse or a hyperbola (Fig. 25). Correspondingly, in the synthesis the parabola is defined "over the axis ZB such that if any point G be taken and the perpen-

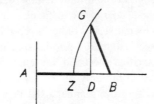

Figure 24

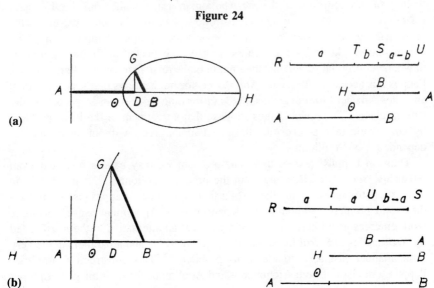

Figure 25

dicular GD be drawn, $R \cdot ZD = DG^2$," for $R = 2\,AB$ and Z the midpoint of AB; it is then proved that $AD^2 = GD^2 + DB^2$. In the other case, the ellipse and the hyperbola are to be drawn "such that if one takes any point G and draws the perpendicular GD, then $\theta D \cdot DH : DG^2$ equals the ratio compounded of TS : SU and TS : SR and $RT^2 : ST^2$, the given ratio"; in both instances $AB : BH = SU : ST$ and $AB : B\theta = RS : ST$ to determine the diameter θH. Thus, if we express the given ratio as $a^2 : b^2$, when a is the greater, the compounded ratio will be $(b : a - b)(b : a + b)(a^2 : b^2)$ to yield an ellipse; when a is the lesser, the same ratio, with $b - a$ in place of $a - b$, will yield a hyperbola. It is then shown that for the points G on these curves, $AD^2 : GD^2 + DB^2 = RT^2 : TS^2$, the given ratio, thus fulfilling the required condition of the locus.

In this locus problem, as in the second of the three angle trisections, the hyperbola is conceived as satisfying the condition that the ratio $y^2 : x(x + k)$ has a given value, for y, x the coordinates at each point and k a given line as the diameter. In the angle trisection, however, the given ratio $h : k$ is understood in Apollonian terms, h being the latus rectum and k the diameter (latus transversum). By contrast, in the locus problem no such interpretation is placed on this ratio; the curve is defined by actually stating the second-order relation satisfied at each point. Now Pappus is of course perfectly familiar with Apollonian usage, as the angle trisection here and innumerable other passages make plain. Thus, his deviation from this usage in the instance of the locus problem reveals that he is not producing his own text, but rather is reproducing one from a source, and that this source is conceived apart from the specifically Apollonian framework. It is surely significant that the condition chosen here for specifying the conics is just that one most commonly adopted by Archimedes as, in effect, their defining relation.[103] Thus, the manner used here in the treatment of the locus problem conforms to the pre-Apollonian usage of the conics. Of course, a later writer might merely choose to follow the older mode; however, in the present instances, it offers no advantage over the Apollonian mode, while later commentators like Pappus and Eutocius can be expected to follow Apollonius in whatever original statements they might make relative to conics. Furthermore, Pappus has access to Aristaeus' *Solid Loci*, for he mentions that it is still extant and cites Aristaeus for the archaic mode of forming the sections by perpendicular cutting planes.[104] We would thus suppose that a treatment of the locus problem by him would take a form like the one Pappus gives, without overt signs of dependence on Apollonius.

Thus, in Pappus' text of the locus problem we may have a fragment from Aristaeus' *Solid Loci*, deriving from the time of Apollonius, but still based on the earlier structure of conic theory. Furthermore, where Pappus elsewhere handles solid problems without clear dependence on Apollonian usage, the same work emerges as a likely source. For instance, an alternative to the *neusis* used by Archimedes in *Spiral Lines* introduces a hyperbola and a parabola in the very same manner just seen in the locus problem.[105] This construction amplifies Pappus' criticism of both Archimedes and Apollonius for their alleged improper use of solid constructions, the latter being the hyperbola-assisted method for finding normals.[106] Such an interest in articulating the distinction between planar and solid methods is precisely what we would expect in a work devoted to the examination of solid loci, as Aristaeus' treatise was. Moreover, the implied give and take between the efforts of Aristaeus and Apollonius conforms with what we have discerned from Hypsicles' remarks on their studies of the regular solids. Note also that both appear to have applied conics toward the solution of the cube duplication,[107] while our discussion here assigns versions of the angle trisection to both. If we view the first of Pappus' methods of trisection as derived from Apollonius in refinement of the earlier *neusis*-method, then the third would derive from Aristaeus, being at once an amplification of a Euclidean effort as well as a response to Apollonius. The second method is so closely modelled after the technique of the third, despite its use of Apollonian terms, that I would view it as a later editor's adaptation of Aristaeus' method. Such alternatives could arise as annotations written into copies of the *Solid Loci*, and in this way come into the hands of Pappus.[108] Yet another method of angle trisection turns up in Arabic texts. Its basic configuration and its introduction of a hyperbola expressed in terms diverging from the canonical Apollonian manner encourage comparison with the second of Pappus' methods and suggest that one consider seriously the likelihood of its ancient provenance.[109]

In the light of these suggestions relating to the *Solid Loci*, we may briefly reconsider three examples of solid problems. First, the theorems on conic reflecting surfaces presented by Anthemius assume a fair range of the theory of conics, yet clearly not in the form set down by Apollonius. Aspects of the constructions, in particular the interest in setting out a practical procedure for approximating the contours, conform with what Diocles tells us of the earlier studies of these devices.[110] Second, in his construction of the ellipse passing through five given points, Pappus only once employs the Apollonian terms "latus rectum" and "latus transversum," although, just as in Pappus' second angle trisection, these relate to the specification of the conic in terms of the ratio of areas and so may have been superimposed as revisions of an earlier treatment. It has been noted, furthermore, that the method of solution differs from what is implied in the lemmas to the *Determinate Section*.[111] Now, the proof depends on results found in the *Conics*, for instance, a relation of the segments of intersecting chords (cf. III, 17) and properties of conjugate diameters (I, 15, 16). But we have seen that the former relation must already have been available for use in the early studies of the four-line locus by Euclid; and presumably some form of the study of conjugate diameters is likely to have existed in the pre-Apollonian theory. Third, Pappus employs no specifically Apollonian terms of relations in his explication of Euclid's observation in the *Optics* that obliquely viewed circles appear "pressed in."[112] Here, Pappus works out the center and axes of the apparent figure which he asserts to be an ellipse. It is remarkable that he does not actually prove it to be elliptical; but perhaps that might be taken as obvious from the configuration of the cone generated by the lines of sight and the manner of its meeting the plane of the object seen.

In each of these instances the source is unknown. If one supposes dependence on works by Apollonius or followers in his tradition, the low level of the commitment to specifically Apollonian usage would be puzzling. We would be compelled, in any event, to hypothesize the availability of writings the existence of which is never expressly evidenced in any of our sources. On the other hand, the *Solid Loci* of Aristaeus was known and available to commentators like Pappus in later antiquity. Its placement in relation to the efforts of Euclid and his disciples and of Apollonius would lead us to expect precisely the blend of archaic and advanced features which characterizes these particular examples. It thus becomes plausible to view this work as one of the important sources for the solutions of conic problems used by the ancient geometric writers.

ORIGINS AND MOTIVES OF THE APOLLONIAN GEOMETRY

The previous section opened with a puzzle and closed with the suggestion for part of its answer. Apollonius insists that the theory of conics has value principally for its use in the investigation of geometric problems. For instance:

> the third book [of the *Conics*] contains many striking [*paradoxa*] theorems useful for the *syntheses* of solid loci and diorisms...without them it is not possible to complete the *synthesis* [of Euclid's problem of the locus with respect to three and four lines].[113]

Moreover, he does indeed provide in his theorems a foundation suitable for the formal syntheses of such problems. Yet he rarely actually engages in the activity of solving problems as such, either in the *Conics* or in the other analytic works, as known to us. If we then accept the second Aristaeus as a contemporary of Apollonius, his work *On Solid Loci* would serve admirably for filling this gap: for a treatise in five books on this field, as Pappus describes Aristaeus' work, must surely have presented a large portion of the solutions of problems like those mentioned by Apollonius. In effect, we would then view Apollonius to be concerned primarily with the formalization of the results attained within the analytic tradition up to his time.

Two such massive compendia as the *Conics* and the *Solid Loci* must have consolidated the researches extending over a long period in the careers of their respective authors. One thus expects a certain dialectical relationship to have marked the evolution of these efforts, and a number of passages support this intuition. For instance, criticism of Apollonius' failure to use planar methods where these are possible for finding normals seems to arise from a context such as would be provided in a treatise on solid loci. On the other hand, the *Solid Loci* drew from the older tradition of conic studies, as developed by Euclid, the elder Aristaeus, Archimedes, and their followers. Thus, a major concern of Apollonius was to recast, extend, and rigorize the results of that work. We find in the theorems of *Conics* III, for instance, as well as in associated analytic writings like those *On the Cutting of a Ratio* and *On the Cutting of an Area*, a systematic effort to establish the auxiliary theorems required for the solution of problems of the sort appearing in the *Solid Loci*. Apollonius presents himself as providing the same service of revision and extension by means of the theorems in *Conics* III and IV in relation to the earlier efforts of Euclid and Conon; for he maintains that only through these can one secure the results which their incomplete and even defective treatments had brought forth.

A similar dialectical relation characterizes Apollonius' work in other areas. His study of circle quadrature attempts to outdo Archimedes. He draws upon results long familiar in the field of plane loci: a method for solving the two-point locus, known to us through the Aristotelian *Meteorologica*, is worked into a systematic survey of the specification of loci via planar conditions in the tract *On Plane Loci*. That same locus construction is modified into the solution of the *neusis* with respect to the rhombus, so generalizing a known result relating to the square. It would appear that Apollonius' *Neuses* extended a study of planar constructions by Heraclides. Similarly, his solutions via conics for the cube duplication and the angle trisection showed how to eliminate the *neuses* and special curves used by Archimedes and Nicomedes.

Working so intensively in fields so actively researched by others, Apollonius was bound to clash with them. Heraclides comes to the defense of Archimedes on the significance of the Archimedean treatments of circle measurement and of conics, while a disciple in the Euclidean tradition (perhaps Aristaeus) does the same for Euclid's work on conics. Indeed, Apollonius is notably ungenerous in his comments on the achievements of such precursors as Euclid, Conon, and Nicoteles. As for contemporaries, he omits referring to any of them by name, save for his Pergamese friends. His grudging acknowledgment of others' merit must surely have been an enormous irritant to many and so led them to seize opportunities for rebuttal, such as those afforded by Apollonius' own errors in his early versions of the study of the regular solids and the conics. His pattern of silence on the work of contemporaries also explains why there is no mention of Aristaeus in spite of the great importance we have assigned to his *Solid Loci* within the current field.

Thus, the mathematical field at the close of the 3rd century takes on the appearance of an active and contentious environment. One might look on the needless personal frictions with disapproval and regret, and doubtless many of the ancient participants did so too. Yet the outgrowth of this competition was a phenomenal advancement of geometry. The *Conics*, the *Solid Loci*, and the other analytic works reported by Pappus represented the high point of the ancient classical tradition. Their later editions, even in fragmentary form, continued to challenge geometers until well into modern times.

NOTES TO CHAPTER 7

[1] On the role of the analytic corpus within the field of problem solving, see Pappus, *Collection* (VII), II, p. 634, and Apollonius, *Conics*, I, p. 4; II, p. 4 (Prefaces to Books I and IV). For general surveys of this corpus, the accounts by Zeuthen (*Kegelschnitte*)

and Heath (*Apollonius of Perga; History of Greek Mathematics*, II, Ch. XIV and pp. 399–427) are extremely useful and will be cited frequently here. See also the summary in Ver Eecke, *Pappus d'Alexandrie*, I, pp. liv–ci.

[2] Eutocius in *Archimedes*, ed. Heiberg, III, p. 228. On the Archimedean work, see Chapter 5.

[3] *Ibid.*, p. 258.

[4] See the references in Dijksterhuis, *Archimedes*, p. 398; he reports them with no little scepticism.

[5] *Apollonius*, ed. Heiberg, II, pp. 124–132; and Heath, *History*, I, pp. 54–57, II, p. 194. To illustrate the Apollonian theorems on calculating, Pappus says he had quoted a line of verse ("Call forth, ye nine maidens, the eminent power of Artemis") and proposed to multiply together all its letters, exploiting the fact that the Hellenistic notational scheme used the letters of the alphabet for numerals. G. Huxley affirms Heiberg's view that the materials derive from Apollonius' *Easy Delivery*, pointing out that Artemis was traditionally invoked as the guardian at childbirth ("Okytokion," *Greek, Roman and Byzantine Studies*, 1967, 8, pp. 203 f).

[6] Dijksterhuis, *op. cit.*, pp. 398–401; Heath, *Archimedes*, pp. 319–326 and *History*, II, pp. 97 f. A complete solution has recently been worked out and published by a computing group at the Lawrence Livermore Laboratory of the University of California (Berkeley) under the direction of H. L. Nelson; a brief report appears in *Scientific American*, June 1981, p. 84. See also D. Fowler, "Archimedes' Cattle Problem and the Pocket Calculating Machine" (1981, in preprint from the University of Warwick Mathematics Institute).

[7] For an example of such a verse problem, see I. Thomas, *Greek Mathematical Works*, II, pp. 512 ff.

[8] For a synopsis of the *Sand Reckoner*, see Heath, *History*, II, pp. 81–85.

[9] *Archimedes*, III, p. 447; cf. II, pp. 2, 4.

[10] See Chapter 5.

[11] Dijksterhuis, *Archimedes*, pp. 21–23. On the relation to the spirals, see Chapter 5.

[12] Pappus, *Collection* (IV), I, Props. 28, 29. On Apollonius' name for the quadratrix, see Iamblichus as cited by Simplicius, *In Aristotelis Physica*, I, p. 60; reproduced by I. Thomas, *op. cit.*, I, pp. 334 f.

[13] *In Euclidem*, p. 105.

[14] *Heronis Opera*, ed. Heiberg, IV, p. 20.

[15] Note that the likeness of the names, in and of itself, would not be particularly significant, for both variants are quite common. For instance, *Pauly Wissowa* lists 17 entries under "Herakleios" and 64 under "Herakleides."

[16] *Apollonius*, II, p. 168.

[17] *Ibid.*, I, p. 4; cf. the prefaces to Books V and VI, as in Heath, *Apollonius*, pp. lxxiv f.

[18] *Ibid.*, I, p. 2.

[19] *Collection* (VII, Props. 71–72), II, pp. 780–784.

[20] Here again, however, the variant "Herakleitos" is not uncommon; 16 are listed in *Pauly Wissowa* (cf. note 15 above). Thus, Böker's wish to assign this method to the Presocratic philosopher Heraclitus is unconvincing, since the mere coincidence of names here cannot override doubts that geometric technique around the close of the 6th century B.C. had not yet advanced to this level (cf. his *Neusis* article in *Pauly Wissowa*, Suppl. IX, col. 440 ff, especially 444). Nevertheless, he displays some interesting parallels between this *neusis* and the planar divisions of figures familiar in the ancient "geometric algebra."

[21] See Heath, *History*, II, pp. 412 f for a more detailed paraphrase.

[22] Pappus, *Collection* (VII, Prop. 70), II, pp. 778–780; paraphrased by Heath, *History*, II, pp. 189 ff.

[23] See Chapter 4.

[24] For accounts of these three works, see Heath, *History*, II, pp. 181–192.

[25] Heath, *ibid.*, pp. 182–185. A review of the modern efforts to provide alternative proofs and extensions of this problem is given by Ver Eecke, *Pappus*, I, pp. lxvi–lxxii.

[26] *Collection* (IV), I, pp. 272–280. Two other methods are given by him in subsequent propositions and will be discussed below and also in the sequel. The third of these methods has already been introduced above in connection with Euclid's *Surface Loci*; see Chapter 4.

[27] See Chapter 5.

[28] A translation of the synthesis appears in the sequel in company with an Arabic version attributed to Aḥmad ibn Mūsā.

[29] The restriction to a rectangle, as stated by Pappus, is unnecessary, since all the conic theorems required in the proof obtain for both orthogonal and oblique coordinates; cf. Apollonius, *Conics* II, Prop. 12.

[30] Pappus omits the justification for this step, but it is easily provided from elementary geometric theorems: since triangles BZA, AED are similar, BZ : AB = AD : ED. But AB = GD, AD = BG, and ED = ZH, so that BZ : GD = BG : ZH, or BZ • ZH = BG • GD, as claimed. This method is followed in the Arabic version of Thābit (to be presented in the sequel); it may be contrasted with the cumbersome reconstruction offered by Hultsch (*Collection*, I, p. 273n), following Commandinus.

[31] *Collection* (IV), I, pp. 276–280. This lemma is of concern below; cf. note 69.

[32] See Chapters 3 and 5.

[33] See below, note 66, and Eutocius, III, p. 176. Full details of the argument assigning this proposition to interpolation by Eutocius or a later editor appear in my article "The Hyperbola Construction in the *Conics*, Book II."

[34] These constructions will be given below.

[35] Eutocius in *Archimedes*, III, pp. 64–66. See Heath, *Apollonius*, pp. cxxv–vii; and *History*, I, pp. 262–264; and my discussions in Chapter 5 and in the sequel.

[36] *In Aristotelis Posteriora Analytica*, ed. Wallies, pp. 104 f.

[37] *De Architectura* IX, 8, 1.

[38] See E. W. Marsden, *Greek and Roman Artillery: Technical Treatises*, p. 110; and Eutocius, in *Archimedes*, III, pp. 60–62. This method is discussed in the sequel.

[39] *Collection* (III), I, p. 56: "They agree this problem [sc. of cube duplication] is a solid problem, and made the construction of it only mechanically, in accordance with Apollonius of Perga who also effected its analysis via the conic sections, and others via the solid loci of Aristaeus, but none via planes [sc. planar methods] in the strict sense [*idiōs*]."

[40] In the sequel the full proof will be given in parallel with the solution in Arabic by Abū Bakr al-Harawī.

[41] See our derivation of the Heronian form in Chapter 5.

[42] We have proved this step in connection with the *Meno* passage in Chapter 3, note 59. One may note the corollary, that HK = BZ, which arises later in the present *neusis* construction.

[43] For references, see notes 35, 36, and 38 above. As I will argue in the sequel study, Eutocius' version of the Apollonian method seems to carry an interpolation, doubtless by Eutocius himself, introducing a *neusis* of the Philonian type into a text lacking any such explicit mechanical element.

[44] Texts will be presented in the sequel. Note that this version of the solution, using the equal intercepts of the secant drawn to the hyperbola, is an essential auxiliary to Apollonius' constructions of the normals (see discussion below).

[45] See note 40 above.

[46] See the sequel for details. Thaer has made a similar proposal, on the basis of constructions by al-Ṭūsī; see his "Würfelverdoppelung des Apollonios," *Deutsche Mathematik*, 1940, 5, pp. 241–243.

[47] Heath, *History*, II, pp. 195 f; van der Waerden, *Science Awakening*, pp. 238–240. Fuller discussions of Apollonius' contributions to geometric astronomy are given by O. Neugebauer in "The Equivalence of Eccentric and Epicyclic Motion According to Apollonius," *Scripta Mathematica*, 1959, 24, pp. 5–21; and *History of Ancient Mathematical Astronomy*, pp. 263–265, 267–270.

[48] Heath, *History*, II, p. 196 and *Aristarchus of Samos*, pp. 268 f (there citing Tannery for this insight).

[49] Neugebauer views as open the question of whether Apollonius had access to any form of the Babylonian data (*History*, pp. 270–273), although it is clear that Hipparchus had such access (*ibid.*, pp. 274 ff).

[50] A related conception of the ellipse as a motion-generated curve is reported by

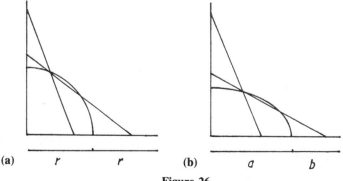

Figure 26

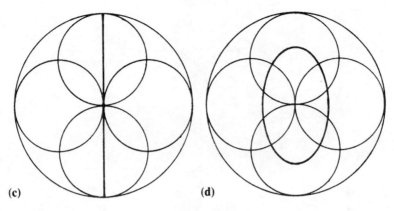

Figure 26

Proclus (*In Euclidem*, p. 106): if the extremities of a line segment of given length slide along the two sides of a right angle, the midpoint of the segment will trace a circle, but any other point on it will generate an ellipse. One can apply this result toward a case of hypocycloidal motion: if a circle revolves without slipping along the inner circumference of a circle twice its diameter, then any point on the circumference of the revolving circle will trace a diameter of the larger circle, while any other point attached to the former will trace an ellipse. Coolidge (*Conic Sections*, pp. 151 f) presents this result of La Hire in the context of citations from Proclus and al-Ṭūsī.

[51] See Chapter 6. We have seen that a case of these curves results from the plane projection of the Eudoxean-related space curves introduced by Riddell in his reconstruction of Eudoxus' cube-duplication; cf. Chapter 3. For further details of this construction, see Loria, *Ebene Kurven*, III.6; Wieleitner, *Spezielle Ebene Kurven*, II.13, III.18; and Zwikker, *Advanced Plane Geometry*, Ch. XX. Note that in this special case of the cardioid, since the two circles maintain rolling contact throughout the motion, the curve generated is also a hypercycloid. The more general class of epicyclic curves can be traced as trochoids; see Zwikker, pp. 235, 276.

[52] The epicyclic generation of this class of conchoids (the *limaçons* of E. Pascal) is illustrated in Figs. 14 and 27(a). The epicycle is centered on D which, in turn, revolves about C. We arrange that P rotates with twice the angular rate about D that D does about C; that is, angle PDE is twice angle ACD for DE parallel to AC. Viewing A as "pole" of the conchoid and the circle with radius CG as its base, it follows that the distance PG maintains the constant value CD throughout the motion and is the "distance" of the conchoid. Consideration of the parallelogram CDPG enables passage from the epicycle to the eccentric mode of generation (see Fig. 12), illustrated in Figs. 15 and 27(b). Here P turns on the eccenter G with half the angular rate that G turns about C; that is, angle OCG (= CGH) is twice AGH for GH parallel to CA. It follows immediately that the curve traced by P is the conchoid, for GP becomes the constant distance measured toward the pole A from the base circle of radius CG. For GP (= CD) = 2 CG, the curve becomes the cardioid with cusp at A (Fig. 13). For yet larger values of GP, it will be depressed toward A without passing through that point.

[52a] This view is parallel to that recently proposed by B. R. Goldstein and A. C. Bowen relative to the astronomical work of Eudoxus: that the hippopedes could have originated not through the problem of saving the phenomena of planetary motion, but

334   Ancient Tradition of Geometric Problems

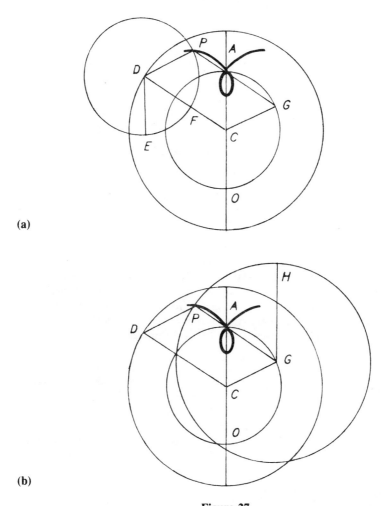

Figure 27

rather through an investigation of the geometric properties of spherical configurations. Cf. "A New View of Early Greek Astronomy," *Isis*, 1983, 74, pp. 330–340.

[53] For surveys of the *Conics*, see Heath, *History*, II, Ch. XIV; and van der Waerden, *op. cit.*, pp. 240–261. More detailed accounts of the background and contents of this work are provided by Zeuthen (*Kegelschnitte*), O. Neugebauer ("Apollonios-Studien," *Quellen und Studien*, 1933, B 2, pp. 215–254); and in the editions of the *Conics*, by Heiberg, Heath, and Ver Eecke.

[54] Heath, *Apollonius*, pp. lxxii–iv. A full synopsis of Apollonius' relations to his predecessors appears in the same work, Part II, Ch. I.

[55] See the discussion of Heraclides earlier in this chapter.

[56] Pappus, *Collection* (VII), II, pp. 676–678.

[57] See Heath, *Apollonius*, Part II, Ch. II for some striking examples.

[58] See note 1 above.

[59] Archimedes asserts the theorem that all parabolas are similar in the preface to *Conoids and Spheroids* (*Opera*, I, p. 250) and uses this property in *Plane Equilibria* II, 3 and 7, and *Floating Bodies* II, 10. It is proved by Apollonius in *Conics* VI, 11, who devotes the whole of Book VI to the investigation of similar conics.

[60] On Book VIII, see Heath, *History*, II, p. 175. Halley's reconstruction is given in the German edition of the *Conics* (Balsam), but is omitted by Heath and Ver Eecke in their editions.

[61] See specifically the series of theorems II, 15–23 and 31–43.

[62] See Diocles, *On Burning Mirrors*, Prop. 1. Archimedes states the result as Prop. 2 of *Quadrature of the Parabola*, omitting the proof as having been given "in the *Conic Elements*." The latter, one must suppose, refers to the treatments of conics by Euclid or Aristaeus the Elder.

[63] On the early expressions for the conics, see Chapter 4 and Heath, *Apollonius*, Part I, Ch. II, III, and pp. lxxix ff. A useful account is given by Dijksterhuis, *Archimedes*, pp. 55–69. That the two-branched conception of the hyperbola is due to Apollonius is maintained by Zeuthen and Heath, for instance; see Heath, *Apollonius*, pp. lxxxiv f, cxl f. But van der Waerden notes that in the preface to *Conics* IV Apollonius seems to assign recognition of this conception to an earlier geometer, Nicoteles (*Science Awakening*, pp. 248 f).

[64] See the preface to Book V. There is a suggestion here that the notion of the minimal lines had been introduced in earlier studies, although, at best, in a superficial way.

[65] Note that this property is assumed without proof by Archimedes (*Floating Bodies* II, 5) and Diocles (*Burning Mirrors*, Prop. 1; cf. Toomer's comments, pp. 150 f). In both instances, the normal line is conceived as the perpendicular to the tangent, not as a minimal line in the Apollonian manner.

[66] Pappus, *Collection* (VII), II, pp. 952–960.

[67] See notes 42 and 44.

[68] See notes 44 and 46.

[69] Although the hyperbola construction appears in *Conics* II, 4, this is revealed as an interpolation by a later editor, as one recognizes from the fact that Pappus provides the identical construction among his lemmas to *Conics* V. The same is supported by a passage in Eutocius. See note 33 above.

[70] For an alternative treatment of this case, cast entirely in the modern style of Cartesian coordinates, see Heath, *Apollonius*, pp. cxxvii–ix.

[71] This relation of the tangent to the hyperbola is a special case of the equality of the segments of the secant, used earlier; see note 44. It is also critical for the hyperbola construction accompanying the first of Pappus' angle trisections; see note 31.

[72] See the development in terms of Cartesian coordinates by Heath, *Apollonius*, pp. 171, 177 f.

[73] An interesting feature of Apollonius' formal style throughout this section on the constructions of the normals to the conics is that although these are clearly a set of *problems* of construction, they are presented in the form of *theorems* of *possibility* of construction, sc. "it is possible to draw lines...''; hence, as *diorisms*. In standard usage among ancient writers, a diorism is that section of a theorem or problem establishing the

conditions under which solution is possible (cf. Pappus, *Collection* (VII), II, p. 636; and Proclus, *In Euclidem*, pp. 202 ff). A Euclidean example is *Elements* XI, 22, giving the condition under which three triangles give rise to a solid angle; the problem of constructing a solid angle from three given triangles is then worked out in XI, 23. Diverging from this usage, Apollonius appears to cast the entire problem of construction in the form of a diorism when, as here in V, 51, 52, it depends on such a discriminating condition. Thus, when he says that his theorems are useful for "the synthesis of *diorisms*" (prefaces to Books I and IV), he would seem to refer to *problems* of this sort.

[74] See note 65.

[75] *Collection* (IV), I, pp. 270–272. In the same passage this same criticism is directed against Archimedes' *neuses* for the spirals (see Chapter 5 and note 106). Zeuthen and Heath (*Apollonius*, pp. cxxviii f) are clear on the identification of this "problem of the parabola" with that of constructing the normals in V, 51. But Hultsch, misled by the very anomaly of formalism we noted in note 73 above, maintained that since there are no *problems* in Book V, Pappus must refer to another place, namely, the problem of constructing the parabola in I, 52 (*Collection*, I, p. 273n). But Hultsch's alternative will hardly do, for the construction in I, 52 requires producing the cone whose section is a parabola of given parameter; it would not make sense for Pappus to fault the introduction of a solid (i.e., the cone) in such a context.

[76] A derivation of the circle is given by Heath, *Apollonius*, pp. cxxviii f. It is managed entirely with reference to Cartesian coordinate expressions, but Heath insists "there is nothing in the operations leading to this result which could not have been expressed in the geometrical language which the Greeks used." The version I give here is an attempt to show precisely how that might be done.

[77] Note that this is exactly the step employed in the proof of the cube duplication in the Heronian form; see Chapter 5, note 115.

[78] *Apollonius*, p. cxxix.

[79] See note 56 above.

[80] Anthemius, "On Paradoxical Devices," ed. Heiberg, *Mathematici Graeci Minores*, pp. 77–92, especially pp. 78–81 (reproduced with translation by G. Huxley, *Anthemius of Tralles*). For a synopsis, see Heath, *History*, II, pp. 541–543. On his treatment of the parabolic mirror and its relation to that by Diocles, see Chapter 6 and my article, "The Geometry of Burning-Mirrors in Antiquity."

[81] This curious omission by Apollonius has been noted by Heath (*Apollonius*, p. 114; *History*, II, p. 119) and by Toomer (*Diocles*, pp. 15–17).

[82] Heath, *Apollonius*, Part II, Ch. IV; and *History*, II, pp. 175–179; see, also, Zeuthen, *Kegelschnitte*, Ch. 15. The work survives through its Arabic translation, and this was used by Halley for his Latin edition.

[83] *Collection* (VII), II, pp. 640–642, 684–704. See Heath, *History*, pp. 179 f, 404f. The close relation of these two Apollonian works is indicated by the fact that Pappus merges his commentary on them into a single set of lemmas.

[84] See the preface to *Conics* I.

[85] See Chapter 4.

[86] For the detailed treatment of what follows, see Zeuthen, *Kegelschnitte*, Ch. 9; summarized by Heath, *Apollonius*, Ch. vi.

[87] The proof is sketched by Zeuthen, *op. cit.*, pp. 192 f and Heath, *op. cit.*, pp. cliii f. One may note that although there is some Apollonian terminology (i.e., "transverse" and "right diameters" in *Coll.* III, 1078, 16; 1080, 1), there is no particular Apollonian cast to the treatment over all. The results assumed here on the segments of intersecting chords were known to Archimedes and were necessary for Euclid's study of the four-line locus (see Chapter 4); the property taken as determining the ellipse is not the Apollonian parametric form, but the ratio form characteristic of the older theory. In these respects, the treatment is comparable to that of the focus-directrix problem given by Pappus; see the discussion below.

[88] *Collection* (VII), II, pp. 642–644, 704–770.

[89] *Ibid.*, II, pp. 866–874. Cf. van der Waerden, *op. cit.*, pp. 288 f.

[90] For Desargues' statement of this theorem, see Taton, *Desargues*, p. 143. It is of major interest in all treatments of projective geometry.

[91] Zeuthen, *op. cit.*, pp. 200, 495 f. In a similar manner, Zeuthen maintains that the ancients knew the effective equivalent of Pascal's theorem on the hexagon inscribed in a conic (*ibid.*, pp. 168–170, 496). This is of course the generalization of the hexagon inscribed between two lines which Pappus establishes as a lemma to the *Porisms* (see Chapter 4). Zeuthen speculates that another of the porisms (the same one used in connection with extending the solution of the four-line locus from the trapezium to the general quadrilateral), when it is proven with respect to a conic (Pappus' statement only applies to given lines), leads to the construction of the conic as the locus of the vertex of a variable triangle defined with reference to five given points (cf. Coxeter, *op. cit.*, pp. 85 f for the construction). Pascal's theorem is a ready inference from this result. But this view cannot draw convincing support from the extant documentation. The porisms cited by Zeuthen are never asserted for the case of conics in the ancient sources; when Pappus actually solves the five-point ellipse, the method adopted is unrelated to a construction via a variable triangle; more generally, the essential notions of projective geometry are never introduced, even in our later sources, like Pappus. We must thus look toward the field of problems of conics as the essential interest of the ancients. If the configurations which are thereby introduced became the basis for the later development of the projective field, this can hardly justify the view that the ancients were somehow engaged in studies of this latter sort.

[92] Cf. Heath, *Apollonius*, p. clvi, summarizing Zeuthen's view of the *Determinate Section*: "the treatise contained what amounts to a complete *theory of Involution.*" (Emphasis and capitalization his.) Heath recognizes the view to be an overstatement, yet goes on to conclude, "There is therefore nothing violent in the supposition that Apollonius had already set up many landmarks in the field explored eighteen centuries later by Desargues" (*ibid.*). He has in mind precisely those claims made by Zeuthen which are discussed in the preceding note.

[93] Heath judges that on balance these inquiries by Apollonius must have substantially prefigured the developments of conic theory by Desargues and Pascal in the 17th century (*Apollonius*, p. clvi). One must beware, however, not to assign to the ancients any such notions as those of the projective and perspective relations which arise in the latter theories. We have seen, for instance, that the lemmas to Euclid's *Porisms*, however closely related to the *content* of projective geometry, received from Pappus a strictly metrical, rather than projective, treatment in *form*. Thus, neither Euclid nor Apollonius nor any other geometer in Greek antiquity appears to have perceived such projective conceptions. The

correspondence in content linking the ancient and modern theories of conics is due in no small part to the intense diligence with which 16th- and 17th-century geometers studied the works of Apollonius and Pappus.

[94] See above, notes 24 and 25.

[95] See Chapter 4. I have collected and discussed the testimonia on Aristaeus in my article, "On the Early History of the Conics."

[96] *Collection* (VII), II, p. 672.

[97] Hypsicles in the preface to Book XIV of the *Elements*. See also Heath, *Euclid's Elements*, III, p. 512; and *History*, I, pp. 419 f. One may note that of the traditional fifteen books of the *Elements*, only the first thirteen are actually by Euclid. The fourteenth is by Hypsicles; the fifteenth is of uncertain authorship, but parts of it are by a 6th-century disciple of Isidore of Miletus.

[98] See K. Vogel, "Aristaeus" in the *Dictionary of Scientific Biography*. In the article cited in note 95 I suggest that this second Aristaeus may have been the father of Hypsicles, for their relation to the effort by Apollonius on the regular solids would appear to be of the same sort.

[99] See notes 26 and 33.

[100] Pappus, *Collection* (IV), I, pp. 282–284. The construction is paraphrased by Heath, who mistakenly makes the latus rectum $\sqrt{3}$ AH instead of its triple (*History*, I, pp. 241–243). The correct reading is adopted by Hultsch and Ver Eecke.

[101] See Chapter 4.

[102] *Collection* (VII), II, pp. 1004–1014.

[103] On the early expressions for the conics, see Chapter 4 and note 63.

[104] *Collection* (VII), II, p. 672.

[105] *Ibid.*, (IV), I, pp. 298–302. See also my article, "Archimedes' *Neusis*-Constructions in *Spiral Lines*," and Chapter 5(ii) above.

[106] See note 75 above.

[107] See note 39 above.

[108] Dozens of times throughout his commentary on the *Conics*, Eutocius explicitly cites the "scholia in the copies" on which he depends for alternative proofs and other explicatory material. One thus has little doubt that Pappus, too, depended on such manuscripts with annotations for the lemmas he provides to the *Conics* and other analytic works in *Collection* VII. Furthermore, one would expect that annotations in a work like the *Solid Loci* would reflect the significance of Apollonius' form of the theory of conics among later geometers. Thus, a source of this nature could well account for the major portion of the solid problems presented by Pappus.

[109] I present in the sequel the texts, with discussion, of the methods of al-Sijzī and al-Qūhī.

[110] I elaborate the view that Anthemius' study is somehow based on pre-Dioclean sources in the article cited in note 80 above.

[111] See note 91.

[112] *Collection* (VI), II, Props. 53–54. We have mentioned this section in connection with Euclid's *Porisms* in Chapter 4.

[113] *Conics*, preface to Book I (ed. Heiberg, I, p. 4).

CHAPTER 8

Appraisal of the Analytic Field in Antiquity

It is generally supposed that after the time of Apollonius the study of geometry entered a period of decline, interrupted only briefly by the work of Pappus in the 4th century A.D. This view agrees with the single most obvious and significant fact that the later tradition of geometry never again witnessed the rise of a first-rate mathematical mind comparable to Archimedes or Apollonius. On other counts, however, the view is questionable. The assumption, for instance, that "Pappus stands out as an accomplished and versatile mathematician" has never been subjected to scrutiny and is not likely to survive it.[1] More importantly, the view implicitly assumes that the later figures sought to make original contributions, but merely failed through technical incompetence. Until we obtain a better sense of the methods and motives of these figures, we shall thus be barred from an adequate assessment of their achievement in relation to that of their precursors.

We may usefully distinguish three aspects of their efforts: technical, textual, and evaluative. The technical competence of later writers like Hero, Theon, and Eutocius to expound mathematical texts, add minor lemmas or appropriate citations to the standard textbooks, and so on is manifest throughout their works; equally clear is the absence of original findings on their part, or even the intent to produce such. By contrast, the work of Pappus, in particular his *Collection*, contains such a rich diversity of more advanced materials that one naturally tends to assume his originality, save where he explicitly acknowledges his dependence on a prior source. The rigorous testing of this hypothesis would entail a detailed examination of all parts of the *Collection*, and this evidently overreaches the scope of the present context. A start is made in the sequel with the discussion of his treatments of the geometric problems, and this reveals that his effort is not essentially different from that of any other commentator in this later tradition. If Pappus might on occasion submit a result as a finding of his own, as in the case of his method of cube duplication, we can perceive the clear dependence on earlier methods which effectively undercuts his claim. His extensive review of the works on analysis in Book VII only rarely rises above the level of what one might expect to find in the margins of annotated copies of these treatises, precisely the source exploited by Eutocius for the production of his commentary on Apollonius.[2] The more interesting parts of Book VII usually bear signs of a provenance in sources; for instance, the enunciation of the rule for measuring solids via centers of gravity is related to the theorems "in the 12th book of these *Elements*," and so appears to betray its placement within a commentary on Euclid and, as a comparison with a passage from Hero has indicated, its origination with Dionysodorus late in the 3rd century B.C.[3] It is especially significant that the entire analytic corpus surveyed in Book VII derives from 3rd-century geometers, primarily Apollonius and Euclid. Surely one must infer that the problem-solving activity embodied in those works received no appreciable advancement at the hands of later geometers.

The decreased attention to advanced geometry beginning in the 2nd century B.C. should not be taken as a decline of the mathematical field in general, but rather a shift of interest from the special field of problem solving toward others, especially spherics, trigonometry, and numerical methods, which bore directly on the development of geometric and computational astronomy. Such shifts in the fashion of research, as it were, are a recurrent phenomenon in the history of science and mathematics, and indeed, changes of this sort had happened at least twice before within the ancient geometry. The study of irrational lines, for instance, was a prominent concern among geometers in the pre-Euclidean period; it engaged the efforts of superior geometers like Theaetetus and Eudoxus, while the formal synthesis of the classification of the irrationals constitutes the largest and most tightly structured part of Euclid's *Elements*.[4] Yet we learn of no work in this field among later geometers beyond a study of "unordered irrationals" by Apollonius and a scattering of isolated results in Pappus which lie entirely within the Euclidean framework.[5] Similarly, Eudoxus' limiting methods were the chief inspiration for the work of Archimedes and of necessity remain an element in the study of the circle quadrature and related problems of curvilinear measurement. Nevertheless, their significance to geometers other than Archimedes was minimal, as one appreciates most strikingly in their utter absence from the works of Apollonius.[6] Doubtless, in the cases both of the study of irrationals and of the application of the Eudoxean techniques, the prospect of new and interesting discoveries was discouraging, so that later geometers largely abandoned their study for other fields. In much the same way, the field of analysis must have made the impression of having reached saturation through the accomplishments of Apollonius and his colleagues.

Thus, the later writers on geometry are not likely to be of great interest in connection with the development of technical methods or the discovery of new results. They are extremely important, however, with respect to the issue of the textual transmission of these technical materials. The editorial tradition of collecting and commenting on interesting geometric results extends at least as far back as Hero, Geminus, and Menelaus in the 1st century A.D. and continues almost without interruption through late antiquity into the circle of Arabic scholars in the 9th and 10th centuries. Theon and Eutocius contributed critically to the preparation of editions of the works of Euclid, Archimedes, and Apollonius.[7] It is thus through these commentators that we receive the overwhelming majority of our data relating to the ancient geometry. Understanding their editorial objectives and methods is thus indispensable not only for a view of geometric activity in later antiquity, but for one of the earlier researches as well. The detailed comparison of alternative texts which such an inquiry demands will be undertaken in the sequel to the present volume.

A third aspect of potential interest in these later writings lies in their efforts to evaluate the geometric achievement of their own and earlier times. The commentators typically volunteer observations on general matters dealing with the structure and methods of the field. In effect, they preserve a shadow of ancient metamathematical thinking, touching on such issues as the classification of problems, the relative status of the different constructing methods, the nature of the distinction between theorems and problems, and the role of the method of analysis. From their remarks we may hope to determine their positions on two questions bearing on the special problems of construction which have been our continuing interest in the present study: What precisely were the conditions to be imposed on a proper solution of these problems? Did the ancient search for solutions attain its goal? As we turn now to a survey of these matters, we may anticipate that the ancients' assessments differed in significant ways from those one would expect in a modern account of geometric constructions.

THE ANCIENT CLASSIFICATIONS OF PROBLEMS

In two passages preliminary to separate discussions of the special problems, the one (*A*) devoted to the cube duplication, the other (*B*) to the angle trisection, Pappus sets out in virtually identical terms the classification of geometric problems which one now generally accepts as the standard view among the ancients:[8]

> The ancients (*B*: We) say that there are three kinds of problems in geometry: some of them are called "planar," others "solid," and others "linear." Now those which are capable of being solved (*lyein*) by means of straight line and circular arc would quite reasonably be called "planar"; for the lines by means of which such problems are found (*heuriskein*) have their genesis in a plane. But those problems which are solved when there is assumed toward their discovery (*heuresis*) one or several of the sections of the cone are called "solid"; for their construction (*kataskeuē*) necessarily employs surfaces of solid figures, namely the conic surfaces. Yet a third kind of problems is left, the one called "linear"; for there are taken for their construction lines different from those just mentioned, having a more diverse and rather contrived genesis (*B* adds: being generated from less regular surfaces and intertwined motions. Such are the lines found in the so-called "loci

with respect to surfaces"[9] ... and those found from the intertwining of plectoid surfaces and all manner of other surfaces and having many wondrous (*thaumasta*) properties about them). Such happen to be (*B*: Of the same kind are) the spirals and quadratrices and conchoids and cissoids (*A* adds: having many marvelous (*paradoxa*) properties about them).[10] Since there is such a differentiation of the problems, the ancient (*B*: former) geometers were not able to construct (*B*: find) the above-mentioned problem of the two mean lines (*B*: of the angle) when they adhered to the geometric manner (*logos*) (*B*: when they sought it by means of the planes), for it is by nature solid.[11]

Pappus recognizes an exactly analogous division of loci, as he explains in a remark prefacing his commentary on Apollonius' *Plane Loci*:

> Those loci which concern us here are called "planar" and in general are such as are straight lines and curves [!] or circles. But solid loci are such as are sections of cones—parabolas or ellipses or hyperbolas. Loci are called "linear" such as are lines other than straight lines or circles or any of the cited conic sections.[12]

This classification scheme is of course not original to Pappus. He himself ascribes it to "the ancients," while his designation of the problem of cube duplication as "solid" is confirmed much earlier by Hero.[13] As for the loci, Apollonius composed a work *On Planar Loci* which does indeed examine locus problems solvable via straight lines or circles. If the Aristaeus who compiled the treatise *On Solid Loci* was in fact a contemporary of Apollonius, as I have argued in Chapter 7, then this work would surely have dealt with those loci solvable as conics; for Apollonius describes the third book of his own *Conics* as containing "many remarkable theorems useful for the syntheses of the solid loci."[14]

Commenting on the last-mentioned passage, however, Eutocius presents a somewhat different view of the division of the loci:

> It was customary for the ancient geometers to speak of "planar loci" whenever in the case of problems the problem comes about not only from a single point, but from several.[15] ... Such as these are called "planar loci." But the loci called "solid" take their name from the fact that the lines by means of which the problems corresponding to these loci are drawn have their genesis from the section of solids, in the manner of the sections of the cone and many others. There are also others, called "loci with respect to a surface," which have their name from the special property relating to them.[16]

In this instance, Eutocius is clearly drawing from a different body of information and perhaps inferior to that accessible to Pappus two centuries earlier. Yet he appears to maintain a more primitive view than Pappus, one bringing together all cases generated from solids, rather than differentiating the "solid" from the "linear" type. Both writers call attention to the "loci with respect to surfaces;" for Pappus these fall within the "linear" category, while for Eutocius they would appear to constitute a class of their own. Now, Euclid wrote a work *On Surface Loci*, and it was seen in Chapter 4 to be connected not only with certain locus problems involving conics, but also with certain space curves, including forms of spirals, and the plectoidal curve which projects onto the quadratrix in the plane. One would thus suppose that the wide variety of cases introduced in this Euclidean work came to be viewed as a class of loci additional to those loci connected with the constructions in the *Elements*, and only later was this new class reconsidered within the tripartite system known to Pappus. The same thing appears to have happened to the "loci with respect to means" arising from the work of Eratosthenes discussed in Chapter 6; for Pappus himself treats these loci as a class separate from his basic three, yet case by case assignable to one or other of them.[17]

Surprisingly, Proclus more resembles Eutocius than Pappus in his view of the division of loci:

> I designate as "theorems of locus" those in which the same property obtains over some entire locus, and as "locus" (*topos*) the placement (*thesis*) of a line or a surface which makes one and the same property. For of the locus theorems some are framed with respect to lines, others with respect to surfaces. And since some lines are planar, others solid—planar are those whose conception in a plane is simple, such as that of the straight line; while solid are those whose genesis arises from some section of a solid figure, such as of the cylindrical spiral and the conic sections—I would then say that of the locus theorems with respect to lines, some have a planar locus, others a solid locus.[18]

In admitting both conics and spirals into his "solid" category, Proclus agrees with Eutocius rather than with Pappus. Yet in speaking of these as forms of "line locus" he differs from both, since they would term these "surface loci." Proclus goes on to report, on the authority of Geminus, a view of Chrysippus: that he likened the class of locus theorems to the "ideas" by virtue of their bringing together an infinity of cases answering to the specified terms.[19] It is reasonable to infer that the more flexible classification of loci shared by Proclus and Eutocius also has standing in the older sources, and that the tighter view proposed by Pappus need not express the consensus of the ancient geometric field.

What of the provenance of Pappus' view? Clearly it would be impossible before the 3rd century B.C., since distinguishing the "solid" class must indicate a substantial development of the theory of the conic sections. One may detect from Pappus' passage on the loci that he is not reporting a very ancient view there, for he inserts it to elucidate a distinction of quite a different sort drawn from Apollonius' *Planar Loci*.[20] Apollonius is said to denote loci as "ephectic," "diexodic" and "anastrophic," according to whether the locus of a point, line, etc. is respectively of its same dimension or one or two dimensions higher.[21] Pappus then indicates that planar, solid, and linear loci in his own sense can have one or another of these designations. It would appear that in his work restricted to planar loci, Apollonius did not have to remark on the special nature of other kinds of locus, like the solid class. This suggests that Pappus' long accounts of the division of the problems (passages A and B above) might include his own editorial elaborations of notions only partly prefigured in his sources.[22] Indeed, one may well suspect that his rationale for their naming lacks any real authority from the older tradition, for it is essentially incoherent. How can one presume to account for the "planar" class through its dependence on lines which are formed in a plane, when the same is no less true of the conic lines, the plane spiral, the quadratrix, the conchoid, and so on? Similarly, the "solid" class employs only the sections of the cone, indeed only its planar sections, rather than the full range of planar and space curves formed by the sectioning of the cylinder, cone, sphere, torus, and other solids.[23] Pappus himself elsewhere insists that the solid class is *not* to be understood through such a reference to solids, but only to the conic sections.[24] Discrepancies like this indicate that neither Pappus nor his sources had a clear rationalization for these designations.

If Pappus' view is thus a faithful reflection of the original sense of this classification, we would have to suppose that at the time when the field of problems had been sufficiently advanced so that geometers wished to distinguish between those solvable by means of straight lines and circles and those solved by means of conic sections, they seized upon the salient difference in the manner of generating these curves, the former lying in the plane, the latter arising through the sectioning of solids, and so named the respective classes "planar" and "solid," without considering the inconsistencies entailed. Extant evidence cannot deny the possibility of this view. But one can hardly be blamed for seeking a more satisfying alternative. Now, the terms "planar locus" and "solid locus," in the sense used by Pappus, are attested with Apollonius, but not earlier.[25] If they were coined by him, or very near his time, it is surely possible that his reasons for the choice were more profound than those just given. Now, in the case of problems, whenever a line segment is to be found in answer to a second-order relation, the construction can be effected by means of circles and straight lines, as one knows from the older techniques of the application of areas. Furthermore, when the segment is specified via a third-order relation, its construction may be effected via the intersection of conics; we have seen an important example of this in Archimedes' solution of the division of the sphere, while the cube duplication of Menaechmus is an older and more obvious instance.[26] Working out the details of this claim relating to the solid constructions would demand only the investigation of a number of additional cases, the type of inquiry characteristic of the treatises in the Euclidean and Apollonian tradition of analysis. Under this alternative view, it would be an intuition of the algebraic structure of the problems, as it were, which motivated the selection of terminology. Many of the cases taken up in Apollonius' *Plane Loci* seem quite remote from geometric interest, yet are easily appreciated as resulting through the urge to set out the variety of forms which give rise to solving lines and circles.[27] This view need hardly press us to suppose that the ancients attained, or even sought, a complete reduction of the field of problems to such algebraic forms. But it provides a sense in which an ostensibly arbitrary classification of problems, one whose rationale evaded the later commentators, might have originally been based on a real insight into the nature of these questions.[28]

Terms like Euclid's "locus with respect to surfaces" and "locus with respect to three and four lines" and Eratosthenes' "locus with respect to means," for instance, are purely *descriptive* in their intent. In particular, they distinguish certain classes of problems from others by virtue of the conditions which define the loci, rather than by considerations of the methods used in their construction. The terms "planar locus" and "solid locus" thus differ from the others in that they do seek a distinction on the basis of the construction method. Even so, their introduction need not at first have been intended for other than descriptive purposes. With Pappus, however, the distinction obtains a *normative* aspect:

> The following sort of thing somehow appears to be no small error to geometers: whenever a planar problem is found by someone via conic or linear [lines], and on the whole whenever [some problem] is solved from a class other than its own, for instance, the problem on the parabola in the fifth book of Apollonius' *Conics* and the solid *neusis* toward a circle assumed by Archimedes in the book *On the Spiral*; for by using no solid one is able to find the theorem proved by him.[29]

This rather timid pronouncement is the only ancient statement I know of which articulates the formal requirement to seek planar constructions in preference to others. Yet it is almost invariably presented in modern accounts as the principal objective of problem solving throughout the ancient tradition.[30] Is there evidence that the ancients ever actually subscribed to this rule? Pappus' remark indicates at once that neither Archimedes nor Apollonius felt bound by it, for their own constructions are singled out for criticism. One can safely assume that the suitable alternative constructions could have been produced, had either geometer admitted the need.[31] The remark relative to Archimedes is especially confusing, first in that it appears to misconstrue what a "solid" construction is, and second in that the alternative construction later given by Pappus employs conics, and thus is itself of the solid kind.[32] Surely Pappus meant to object to the appeal to the *neusis*, on the grounds that when its replacement by a solid construction is possible, as here, that would be preferable. But then, this would not illustrate Pappus' chief point about maintaining the appropriate boundaries between the planar, solid, and linear classes. Either Pappus is following an addled source, or else he is betraying a certain confusion on his own part. Pappus' responsibility for this passage is suggested by the fact that it is an insertion into his general account of the classification of the problems, but only in one of its appearances (text *B* cited above), not the other (*A*).

A review of the solutions devised for problems reveals little if any concern over this formal issue. Hippocrates constructs his third lunule by means of a *neusis*, with not the slightest indication that a planar method (which can be given here) is to be supplied.[33] In the 4th century geometers attacked the problem of the cube duplication with a host of different techniques, including the intersection of solids, the construction of special curves, and the use of mechanical motions.[34] As befits an early stage in the technical development of the field, the ambition is to find solutions by whatever means one can, to conceive of new methods and explore their potential freely, but hardly to restrict artificially the domains within which one or another method is deemed valid. To be sure, Euclid's *Elements* investigates only one segment of the field of problems, those constructible by planar methods. But that neither exhausted the planar field, as we become fully aware through the investigations of Apollonius, nor did it deter Euclid from pursuing extensive studies of solid problems and the theory of conics.[35] The successful discovery of a solid construction for the cube duplication did not discourage geometers after Menaechmus from the search for alternative solutions of the linear kind. When Nicomedes castigates Eratosthenes over his method of cube duplication, it is not for reasons of formal correctness, but rather for practical feasibility.[36] These 3rd–century geometers share an interest in finding new mechanical procedures for solving geometric problems, and this seems to distinguish them from the geometers of the preceding century; but the formal ideal voiced by Pappus does not concern them.

The class of *neusis* constructions undergoes a notable transition during the 3rd century. In the work of Archimedes they are used in the study of spirals, as auxiliary to the formal proofs of the properties of the tangents to these curves. Given that Archimedes is often sensitive to matters of formal procedure, as one sees in his delicate management of the introduction of his mechanical method in his *Quadrature of the Parabola* and the *Method*, this surely attests to his acceptance of *neuses* as formally proper. Elsewhere he solves the problems of the inscription of the heptagon and the trisection of the angle via *neuses*, and geometers in the generation after him follow his example in devising *neuses* for the cube duplication and defining curves like the conchoids via *neusis* conditions.[37] But with Apollonius a new element enters. Although he himself is credited with an alternative construction of the cube duplication via *neusis*, he is also said to have used a method via conics for solving this problem. His *On Neuses* takes up only cases of the construction which may be effected by planar methods. Among his other analytic works, a division between planar and solid methods is indeed in evidence. The writings *On Planar Loci* and *On Tangencies* deal only with planar constructions, while others, like the *Section of an Area*, the *Section of a Ratio*, and the *Determinate Loci*, seem to have their principal role within the study of problems on the construction of conics.[38] It is thus clear that around Apollonius' time the field of geometry had matured to the effect that it had become important to regularize the formal methods of problem solving. The investigation of planar and solid problems as distinct classes would carry the expectation of discovering in detail the relative ranges of these methods, as would the effort to reduce other kinds of constructions, like the *neuses*, to one or the other class.

In this way the relative ordering of the different kinds of problems served to guide the research efforts of Apollonius and his contemporaries.[39] Other methods, such as the generation of special curves via motions, would continue to be of interest, since problems like the circle quadrature and the arbitrary division of the angle remain outside the scope of both the planar and the solid methods. But even with this, it is not clear that Pappus' rule of method had attained formal recognition, for we may note that Pappus himself does not abide by it. For instance, having prefaced his section on methods of cube duplication with the statement about the classification of problems (A), he goes on to present four different solutions: two are mechanical (the methods of Eratosthenes and of

Hero) and two are linear (those of Nicomedes and of Pappus himself).[40] Yet Pappus quite conspicuously informs us that the problem is of the solid type.[41] Apparently, neither he nor Hero, who also states the solid nature of this problem, senses that in presenting solutions other than solid ones he has committed a formal error.

There is a much deeper difficulty, however: in the context of the Greek geometry the formal rule proposed by Pappus and the classification scheme corresponding to it are unworkable. If one can produce a planar construction for a problem, that does indeed establish the planar character of the problem. But the discovery of a solid construction for a problem does not yet suffice for securing the solid classification of this problem; one must go on to demonstrate that no planar construction is possible. Although comparable impossibility proofs were an essential feature of the ancient theory of irrational lines, we have no grounds for supposing that the Greek geometers ever did or could set forth the kind of proofs required for the classification of problems. The forms ultimately worked out in the 19th century depend on algebraic techniques of a sort alien to the Greek geometric approach.[42] Despite this, Pappus and Hero seem perfectly comfortable in asserting the solid nature of the problems of cube duplication and angle trisection and the linear nature of the problem of the generalized angle division, oblivious, it would appear, even to the realization that such a claim required a proof. Indeed, Pappus prefaces Book III of the *Collection* with scornful remarks on the attempt by a colleague to solve the cube duplication via planar methods.[43] To be sure, the method is unsuccessful, being at best a procedure for the arbitrarily close approximation of the solution via a recursive planar construction. But Pappus speaks almost as if the very notion of trying to find such a solution for the "solid" problem is absurd *per se*. Like the mathematical authorities of a much later age who effectively ruled against the possibility of the circle quadrature a century before the transcendentality of the problem was actually demonstrated,[44] he appears to have accepted the long record of failed attempts as the basis for affirming the solid character of the cube duplication:

> Admitting the problem to be solid, they [e.g., Eratosthenes, Philo, and Hero] effected its construction only by means of instruments—in agreement with Apollonius of Perga who also effected its analysis by means of the sections of the cone, while others [did so] by means of the solid loci of Aristaeus, but no one by means of properly so-called planar [methods].[45]

However convenient the classification might be, we would surely have appreciated from him some recognition that the formal placement of a solid problem like this is an open question subject to proof.

The situation with respect to loci is subtly different. A "planar locus" is by definition a straight line or a circle; a "solid locus" is one of the conic sections. If, then, the determination of a problem of locus results in one of the former, the locus is planar; if one of the latter, it is solid. There can be no ambiguity here, so that the classification of loci and problems of locus into planar, solid, and linear categories is perfectly feasible. As indicated above, the terminology is already well attested with Apollonius and well suits the aim of systematizing the study of constructions in the analytic works by him and his contemporaries. What remains difficult to explain, however, is the rationale of the terms "planar" and "solid" in the instance of loci, unless we are to accept the commentators' very weak proposals, as considered above. It seems to me that an account of the following sort is possible. Well before Apollonius, one already knew that loci of the former sort, namely lines and circles, had their special domain of application toward the solution of problems in which a line segment must be found to satisfy a specified second-order relation with respect to given lines. Similarly, loci which are conics were known to be applicable toward the solution of problems in which the line segment must satisfy a third-order relation. This aspect of the applicability of these two types of loci could justify naming them "planar" and "solid," respectively, even if one did not have a clear proof that all problems expressible via such second- and third-order relations can be solved via loci of the former and latter types, respectively. The naming would surely be descriptive, rather than normative, and would be convenient for focusing research into the relative domains of problems solvable via the different methods of construction.

It is not difficult to see how one might go on to extrapolate this viable scheme for problems of locus to the whole field of geometric problems, in this way obtaining the classification used by Hero and Pappus. Although this move brings with it the logical difficulties just mentioned, the commentators seem unaware of this fact. It is thus possible that one of them bears responsibility for introducing the defective conception of the classification scheme, rather than its having been a fixture within the earlier research tradition. Since neither Proclus nor Eutocius follows this scheme, it would appear not to have received a conspicuous statement in the works of the earlier geometers or in the later compilations by Geminus and Menelaus. Far from being the "standard" ancient view, then, Pappus' tripartite division of problems and the criterion for the choice of construction techniques which accompanies it emerge as a minority opinion held by a few of the late commentators through their misconception of the nature of the 3rd-century enterprise of problem solving.

PROBLEMS, THEOREMS AND THE METHOD OF ANALYSIS

The distinction between problems and theorems is maintained meticulously by the authors of the principal treatises of classical geometry, Euclid, Archimedes, and Apollonius, as well as in the miscellaneous materials preserved by the later commentators, Hero, Pappus, and Eutocius. As a matter of *form*, a problem is cast as an infinitive expression seeking the construction of a geometric term in a specified relation to other given terms (e.g., "to construct the triangle whose sides are three given line segments").[46] By contrast, a theorem is typically set in the form of a conditional asserting a property of a specified geometric configuration (e.g., "if in any triangle two lines are drawn from the endpoints of its base and meet within it, then the lines will have a combined length less than that of the two remaining sides").[47] A point not missed by the ancient commentators is that the theorem refers to a general class of entities (e.g., "*any* triangle"), whereas the problem usually results in the production of a unique figure. Indeed, it may happen that the problem cannot be solved unless certain auxiliary conditions are met. One thus typically supplies a "diorism" asserting those conditions (in the above case, "where the three lines are such that no two have a combined length less than the third"). This might be motivated through a separate theorem (as in Euclid, "in any triangle the combined length of any two of its sides is greater than its third side")[48] or alternatively, usually in the context of an analysis of the problem, through an allusion to a condition for a certain step of the construction (e.g., that two given circles must have a point of intersection not lying on the line joining their centers). When the problem has many solutions, it may be expressed as a theorem of locus (e.g., "the points whose distances from two given points have a given ratio lie on a given circle"), or as a locus problem (e.g., "to find the locus of points such that...").[49] Such loci provide an important instrument for the solution of problems. For instance, the problem of constructing the triangle of three given sides might alternatively be found via the intersection of a circle (the locus of points whose distances from two given points are in a given ratio) and an ellipse (the locus of points whose distances from the same two given points have a given sum). Of course, since this particular problem permits of solution within the context of elementary geometry, one could not here wish to appeal to the more advanced methods of conics, as this alternative construction does.[50]

These illustrations indicate that from the purely formal viewpoint the distinction between problems and theorems is largely artificial. One can easily recast any problem as a theorem, merely by incorporating into the protasis of the theorem all the details of the construction of the problem. Conversely, a theorem might be rephrased as the demand to produce a construction or to discover some property of a given figure. In most instances, there would appear to be a natural preference for one form over the other. In the case of propositions of locus, however, the distinction is not entirely clear. Like problems, they require a construction; but like theorems, they aim to prove a property of what is constructed, namely, its identification as a given line, circle, conic, or other figure. Pappus and Proclus thus reserve a third category, "porisms," in separation from those of problems and theorems,[51] and I suspect that this is nothing other than the class of loci. The commentators are neither clear nor convincing in explaining how porisms differ from problems, but they do assure us that Euclid's *Porisms* was a compilation of theorems of locus. I would thus suppose the "porism" was Euclid's term for what later geometers designated as "locus," and that, not perceiving this, the later commentators attempted their own explanation on etymological grounds.

A formal treatise in geometry might thus be organized as a body of theorems presented in the order of their deductive dependence. As Proclus observes, Euclid strikes a roughly even balance between theorems and problems in his first book.[52]

In this way, the problems often serve as justification for the introduction of auxiliary terms in later propositions. For instance, the theorem that the angles of any triangle equal two right angles (I, 32) requires the introduction of a line parallel to a side of a given triangle and passing through one of its vertices; this can be effected through the construction set out in the preceding problem (I, 31). Commentators, both ancient and modern, have thus found it natural to assign to the Euclidean problems the special role of securing the proofs of existence of the geometrical entities investigated in the theorems.[53] We shall consider this issue further below. But one should note that this view offers, at best, an incomplete picture of the diverse role of problems within the ancient geometry. Some problems are included not for their applications in later theorems, but for their intrinsic interest. Book IV and most of Book XIII, for instance, are devoted to the exposition of problems: the inscription of regular polygons in the former, that of the regular polyhedra in the latter. Here, the subordination is reversed, in that any theorems enter as auxiliary lemmas toward the constructions.[54] Pappus' treatment of the angle trisection suggests a provenance in a deductively ordered structure of solved problems: the trisection assumes a *neusis* whose solution as a solid problem is known from a prior lemma; the latter, in turn, assumes the construction of a hyperbola of given asymptotes, solved in another lemma; this refers back to the construction of the conic sections answering to given parameters, as solved by Apollonius in Book I of the *Conics*.[55] Clearly, the structure of these problems is determined by the deductive requirements of the problems themselves, not by the demands arising from their use in other theorems. Indeed, Apollonius remarks that theorems like those expounded in *Conics* III have their principal value within proofs dealing with locus problems.[56] It is noteworthy that problems are few in number in the first four books of the *Conics*. As we have seen, the appearance of the hyperbola construction in II, 4 is due to an interpolation by or after Eutocius; the sections of problems at the end of Book I and closing Book II cannot easily be viewed as contributing to the deductive needs of the theorems in other parts of the *Conics*. Their removal would in no way affect the logical structure or soundness of the main body of theorems.[57] The same may be observed of Euclid's Book I. The validity of its theorems does not depend on the presence of the problems.[58] Were one to segregate the problems, however, the proofs of their constructions would require appeal to the theorems in the rest of the book. One may see that the three postulates of construction—to draw the line through two given points; to extend a given line indefinitely; and to draw the circle of given center and radius—are stated in the form of *problems* whose construction may be assumed as irreducible, and so may initiate the ordered sequence of problems. By contrast, the two remaining postulates—the equality of all right angles and the nonparallelism of two lines when the interior angles formed with them by a transversal line sum to less than two right angles—are stated in the form of *theorems* and are introduced as premises in theorems in Book I.[59]

Further examples of this kind could be supplied in profusion. They reveal that problem solving was the essential part of the geometric enterprise marked off by the works of Euclid, Apollonius, and those in their tradition, and that for them the compilation of bodies of theorems was an effort ancillary to this activity. It is important to realize this as one turns to the views proposed by ancient thinkers on the nature of problems and theorems and their relative significance for geometry. Pappus and Proclus draw from a centuries-long tradition of discussion of the definitions of these and associated terms, of which the following, from Pappus, is a representative sample:

> Those wishing to discriminate more precisely the things sought in geometry... think fit to call a "problem" that in whose case it is proposed (*proballein*) to make or construct something, but a "theorem" that in which, on the supposition of certain things, one investigates (*theōrein*) what is implied by them and is contingent on them in every way. Among the ancients, some call [them] all problems, others [call them all] theorems.[60]

On this latter division of opinion, Pappus tells us no more; but Proclus reports the positions of half a dozen writers widely distributed over the span of antiquity. As one might anticipate, those emphasizing the role of theorems draw their inspiration from the ideas of Plato: since the objects of geometry are eternal, unchanging verities, it is improper to speak of one's bringing such things into being, constructing them, and so on, the way one typically does in a geometric problem.[61] Rather, their study is the contemplation of what they truly are, and thus only the form of the theorem is appropriate to this discipline. Proclus ascribes this view to Plato's nephew Speusippus, as well as to a mathematician named Amphinomus.[62] Much the same view is assigned to Plato himself by Plutarch, on the authority of the Platonic essay by Eratosthenes[63]; even if the incident of Plato's chiding Eudoxus for his approach to solving the cube duplication is surely apocryphal, this conception is indeed faithful to Plato's own thought on the nature of geometry. As director of the later Academy and exponent of Neoplatonism, Proclus naturally finds his sympathies allied with the same view. In agreement with Geminus, he makes it the aim of geometry to move through problems to theorems, as it were, rising from sensibles to ideals.[64]

By contrast, writers closer to the activity of geometers emphasize the contrary view. According to Proclus, the "mathematicians in the following of Menaechmus" classed all the propositions as "problems" (*problēmata*)[65]:

> while the problem (*probolē*) is of two sorts. Sometimes the thing sought is produced (*porizein*); but sometimes by taking what has been delimited we see either what it is, or what kind of thing it is, or what property it has, or what relations it has to something else.

Here we perceive the same distinction which results in the separation of locus problems (and "porisms") from the body of problems of construction, although Menaechmus hardly subscribes to Euclid's terms. Nevertheless, it is an important insight that the methods of finding constructions might be applied toward the discovery of properties of given configurations. Proclus does not provide the rationale behind Menaechmus' view, although it would appear to be epistemological, in contrast with the ontological position of the Platonists. That is, the "problem" is the superior vehicle in the heuristic effort of gaining knowledge about geometric figures. Proclus attempts a compromise by admitting that we must engage in an activity which employs a vocabulary of construction, but that geometry deals only with the "intelligible matter" of figures, rather than with their "sensible matter," as one does in mechanics.[66] The actual being of abstract geometrical entities, however, is subject neither to generation nor to change of any sort. Proclus, through his preoccupation with the Platonic doctrines on the nature of being, seems to have missed the key issue of heuristic method.

Another mathematical writer, Carpus of Antioch "the mechanician," defends the priority of problems on pedagogical grounds. Elsewhere he is known to us for an account of the mechanical work of Archimedes and for having solved the circle quadrature via "the curve called simply 'of double motion'."[67] To Carpus, problems have an obvious advantage over theorems in their being simple to state. Proclus cites, at some length, certain views drawn from his work "on the subject matter of astronomy":

> on the distinction of these, he says that the problematic genus is prior to that of the theorems. For the subjects about which properties are sought are found through the problems. Moreover, the statement (*protasis*) of the problem is simple and without further need of any specialized understanding.... But that of the theorem is effortful and demanding of great precision and expert discrimination, in order that it appear neither more nor less than the truth.[68]

Carpus illustrates this point in the particular context of the problems and theorems which open Euclid's Book I, and so has in mind the pragmatic question of how best to present geometric materials to learners. Proclus concedes the greater effectiveness of the problematic form in this case, as also in the case of the application of geometry to the mechanical arts. But he musters support from Geminus for the view of the greater perfection of theorems, on the grounds, as noted above, that these better capture the essence of abstract geometric truth. Proclus thus seems not to sense the significance of another statement by Carpus, that the usual methods for problems are well suited for "the more obscure cases." From the viewpoint of geometric research, this is surely the striking advantage of the problematic form: that one may, with relative ease, articulate a problem whose solution is not known and in this way define a direction for gaining new knowledge in geometry. By contrast, the statement of a theorem must already incorporate the full description of what has been discovered. The theorem, then, is the appropriate mode when one is concerned with the formal organization of known results, rather than with the discovery of what is not yet known. In their respective positions on the evaluation of the problematic and theoretic forms, Proclus and Carpus thus signal the contrasting objectives appropriate to their differing manners of involvement in geometric studies.

Proclus calls attention to another distinction between problems and theorems[69]:

> According to Zenodotus[70]... the theorem is distinguished from the problem in that the theorem seeks what is the property predicated of its subject matter, while the problem [seeks] what is to exist [in order that] something be. Whence also, ac-

cording to Posidonius,[71] the one is defined as a proposition according to which one seeks whether it is or is not [possible], the other[72] a proposition in which one seeks what it is or what sort of thing it is ... For it is different either to seek simply and indefinitely whether it is [possible to construct a line] at right angles from this point to this line, or to investigate [*theōrein*] what is [the nature of] the line at right angles.

Despite the highly elliptical terms, it is clear that both views assign to problems an existential function. In the former case, they effect the configuration such that a specified condition may be said to obtain. In the latter, they determine whether or not a construction satisfying the specified condition is possible. They thus accord quite comfortably with an account in the manner of modern logic, which would transcribe theorems via universal quantifiers, but problems via existential quantifiers. Indeed, Proclus subscribes to this view of problems in his own reconstruction of Euclid's dialectical purposes in first setting out three problems of construction (I, 1-3) before taking up the first theorem on congruence of triangles (I, 4):

> Without a prior construction of the triangles and a production of their genesis, how could he presume to teach about their essential properties and of the equality of their angles and their sides? ... For let someone, happening along before the making of those things, say that if two triangles have a property like that, they shall also have this one in every case. Then, would it not be easy for everyone to respond to him: For do we at all know whether a triangle is capable of being constructed?...To cut off these difficulties in advance, then, the author of the *Elements* has also handed down the construction of the triangles...and to these he quite reasonably attaches the theorem by which is proved [the property of congruent triangles asserted here].[73]

However nicely this account suits the needs raised by Proclus' Platonist ontology, it hardly presents a convincing portrait of Euclid's intent. If he had indeed wished to affirm the existence of triangles, he need only have taken three arbitrary points and then introduced the three line segments which connect them two by two in accordance with his first postulate. The assumption of arbitrary points or lines as "given" is not founded on any explicit postulate, but is in fact an element in practically all of Euclid's problems. For instance, in I, 1, the line segment which is to serve as the base of the constructed equilateral triangle is such a given term. What is especially inappropriate about this construction as an alleged proof of the existence of triangles is that it brings about only one particular kind of triangle and might thus risk misleading the learner into conceiving of this special case, when in the context of general theorems on triangles, such as I, 4, he ought to conceive a case better representative of the general class. But as a simple yet interesting first exercise in the mobilization of postulates and other general principles for working out a formal geometric proposition, the problem Euclid has chosen to open Book I is beyond reproach.[74]

It is of course an important insight on the part of Proclus and the commentators on whom he depends to perceive this nuance in the logical conception of problems, and we would certainly not wish to deny that Euclid and other geometers might themselves have been aware of it. But as we have seen in the many examples referred to earlier, this view cannot cover the full range of contexts served by the introduction of problems in the ancient geometry. Indeed, this existential aspect is strongly marked only in the case of "diorisms," which articulate the limits on the solvability of problems. Note, however, that the condition of solvability for the problem of constructing a triangle of given sides (I, 22) is framed by Euclid as a *theorem* (I, 20): that any two sides of a triangle have a combined length greater than the third side; not as a *problem*.[75] Euclid does not actually show how this result secures the existence of the point of intersection of the two circles by which the construction of the triangle is effected in the subsequent problem. It has thus been left to Proclus and other commentators to fill this gap.[76] In view of Euclid's omission, one becomes doubly wary of the suppositions that he appreciated the need for existence proofs in cases like this, and that he inserted the problems to serve this need.[77]

Throughout the present study, we have seen the heuristic power which the method of analysis afforded the ancients in their search for solutions to geometric problems. The most detailed account of this method surviving from antiquity is given by Pappus as an introduction to his survey of the analytic corpus in *Collection* VII. To this we may add two very brief statements on the nature of this method, the one in a scholium prefacing some alternative proofs to the opening theorems of Euclid's Book XIII, the other from Hero's commentary to Euclid's Book II as extant in the Arabic translation by al-Nairīzī.[78] A comparison of the three versions reveals some important aspects of Pappus' account which seem not to have been considered in the profuse discussion of it by modern scholars.[79]

Pappus opens with a statement summarizing the nature and objectives of the analytic corpus (*topos analyomenos*):[80]

> The so-called 'analytic [corpus]' ... is a certain special body prepared next after the making of the common elements for those wishing to acquire a power in lines [i.e., lines and curves] conducive to the finding of problems proposed to them, and it has been compiled as useful for this purpose alone. It has been written by three men: Euclid, author of the *Elements*, and Apollonius of Perga, and Aristaeus the Elder, and it adopts the method according to analysis and synthesis.

He now sets out to characterize this special method for problem-solving:

> (*Pappus*) Now analysis is a way from what is sought, as if admitted, through the things that follow in order, up to something admitted in the synthesis.[81]

> (*Scholium*) Now analysis is a taking of what is sought, as if admitted, through the things that follow, up to something admitted as true.[82]

> (*Hero*) Now analysis is, when any question has been proposed, that we first set it in the order of the thing sought, which has been found [!], and we will then reduce (it to something) whose proof has already preceded. Then therefore it is manifest, we say, that the thing sought has now been found according to analysis.[83]

The statements by Pappus and the scholiast are obviously in virtual literal agree-

Appraisal of the Analytic Field in Antiquity   355

ment. An actual dependence of the scholiast on Pappus is entirely possible; but the conformity in sense (if not in specific wording) with Hero reveals that at least this much of the account has a firm precedent in sources older than Pappus. From here, Pappus amplifies thus:

> (*Pappus*) For in the analysis, having hypothesized what is sought as if already in effect, we examine that from which this results and in turn the antecedent of that, until proceeding in this backward manner we come down opposite some one of the things already known or having the order of first-principle. And such a method we call 'analysis,' as being like a backward resolution.

Through our experience of the ancient uses of analysis, we should have expected it to be described as a searching through the geometrical consequences deduced from what is sought, on the hypothesis that the latter is already in effect. This is the natural reading of the opening lines, even if the term "things that follow" (*ta akoloutha*) need not bear the stronger sense of logical deduction in a context like this one.[84] Pappus, however, seems to view the procedure in quite a different manner, as a finding of successive antecedents, so that the result initially hypothesized ends up being a deduction from the derived sequence. But before pursuing this further, let us note that Pappus' remarks have the appearance of being a gloss on a statement on the analytic method received from a prior source. A similar pattern emerges in the next section of his account. The two alternative accounts provide the anticipated analogues to their statements of analysis:

> (*Scholium*) But synthesis is a taking of the thing admitted through the things that follow up to something admitted as true.[85]

> (*Hero*) But synthesis is, that we begin from a thing known, then we will compose, until the thing sought is found. Thus, the thing sought will then be manifest according to synthesis.

Pappus omits any such corresponding line, but instead provides the complement to his own amplified statement on analysis:

> (*Pappus*) But in the synthesis, in reverse order, having posited as already in effect what had been obtained last in the analysis, and having ordered in the natural manner as consequents what there [were ordered as] antecedents,[86] and having composed them to each other, we at last arrive at the construction of what is sought. And this we call 'synthesis.'

For Pappus, then, the difference between the two methods is merely one of the order of exposition. Despite his earlier insistence on the singular significance of analysis for the study of problems, he now accords with Hero and the scholiast in considering its application in the instance of theorems. To establish a theorem analytically, then, Pappus would hypothesize as true the desired conclusion A and then reason, "A will follow, if B holds; and B, if C; etc." Arriving at something known to be true (e.g., a previous theorem or a postulate), he can then affirm the truth of A. Now, Hero and the scholiast go on actually to work out demonstrations in the double mode for a series of theorems.[87] In the former instance, this results in alternative proofs for eight of the first 10 theorems in

Euclid's Book II; in the latter, for the five theorems which open his Book XIII. The theorems are all quite straightforward, and both writers treat them in the same manner. Far from adopting the mode suggested by Pappus, both frame the analysis as a sequence of deductions leading from the hypothesis of the result to be proved and terminating in something known to be true. In effect, their analyses are nothing other than regular proofs of the converse of the proposed theorems. Furthermore, one may note that both writers share an interest in securing their theorems without recourse to the geometric diagrams, this being possible here in that only the formal manipulation of identities is at issue. Such affinities between these two treatments must surely indicate mutual dependence, either through common authorship or through the scholiast's referring to Hero for his model. In the light of our earlier observation of the agreement between the scholium and the opening lines of Pappus' account, this would suggest that it is Pappus who has borrowed from a source (for instance, Hero himself or a writer following him) to establish a familiar context for the elucidation of this method. The remarks which he attaches to it, however, would reflect his own interpretation of the ancient geometry, but have no particular standing within the older tradition of mathematical commentary.

A fundamental difficulty in Pappus' view of the order of deductive sequence in the analysis is that, if he were correct, there ought to be no need for the synthesis; for the analysis of itself would constitute a demonstration of the theorem. It happens that a treatment adopting precisely the form implied by Pappus appears among his lemmas to Apollonius' *Determinate Section*.[88] In the context of a specified figure, it is to be proved that a stated relation A obtains; the writer passes through a series of steps of the form "B is the same as C" and "if D obtains, it will be necessary alternately to seek whether E obtains." He ultimately hits upon a relation known to obtain. In his paraphrase of this lemma, Heath speaks of it as "a case of *theoretical analysis* followed by *synthesis*," and after the analysis he adds the words, "The synthesis is obvious." To be sure, this tag line appears now and again among the analyses in the *Collection*.[89] But in the present case it is purely Heath's addition; for it is not used here by Pappus, and there is nothing which overtly indicates that this proof is intended as an analysis or that a synthesis ought to be supplied. Thus, if the writer of the lemma is following the procedure prescribed by Pappus, replacing each derived relation with another equivalent to it, he treats the synthesis not merely as "obvious," but as superfluous.[90]

One can see how Pappus might construe a proof technique of this sort to be a form of analysis. Reflecting on passages in the philosophical literature where the analytic manner is pictured as a search for antecedent premises or conditions, he might then adopt this as his own view, as he sets out to amplify the sketchy remarks made by earlier mathematical commentators. The following passage from Aristotle's discussion of the nature of moral deliberation (*bouleusis*) is especially interesting in this connection[91]:

> The statesman does not deliberate over good order, nor does anyone else over the end [to be achieved]. But *positing* the *end*, they *investigate* the questions of how

and through what things it shall be. If it appears to result from several things, they *investigate* through what it will result in the easiest and best manner; but if it seems to result from a single thing, they consider how it shall result from that, and what that in turn results from, until they *arrive* at the first cause which is last in the order of discovery. For the deliberator seems to *seek* and to *analyze* in the said manner, just as in the case of a diagram [i.e., geometric figure or proposition] ..., and that what is last in the *analysis* is first in the genesis. Now if they *happen upon* something *impossible*, they leave off; for instance, if it requires money, but this cannot be *provided*. But if it appears *possible*, they try to do it.

Forms of the terms indicated in italics appear in Pappus' account of analysis as well.[92] It fixes the method as a search for antecedent causes in a way that examples drawn from the mathematical literature never would. Even Aristotle's consideration of the two alternative outcomes of the analysis figures prominently in Pappus' subsequent remarks. We are thus readily led to suppose that Pappus drew the basis of his account of analysis from the philosophical literature. Of course, Pappus does not cite Aristotle here.[93] But nowhere in the *Collection* does he cite any philosopher other than Plato, and that only once for a philosophical view.[94] The several references to the "five Platonic solids," that is, the regular polyhedra (in Pappus' Book V), can hardly be viewed as citations of Plato. One can discern several Aristotelian doctrines in Pappus' *Commentary on Euclid's Book X*, even though he does not expressly cite Aristotle there either.[95] It seems to me most probable that Pappus could pick up such general views through the medium of commentators like Geminus and others, conversant with a syncretizing form of Platonism.[96] In such a case, Pappus himself might not be fully aware of the ultimate provenance of some of his views.

One thus approaches the concluding section of Pappus' account with suspicions that it, too, despite his access to the whole ancient corpus of analysis, may present not a distillation of that ancient tradition, but rather a rephrasing of standard philosophical views. He is concerned here with distinguishing between the analysis of theorems and that of problems:[97]

> The class of analysis is of two sorts: the one which is called 'theoretic' seeks what is true, while the other which is called 'problematic' is productive (*poristikon*) of what has been proposed. In the case of the theoretic kind, having hypothesized what is sought as being and as true, and then having gone forward through the things which follow in order, as true and as being in accordance with hypothesis, up to something admitted, then if that thing admitted is true, so also will the thing sought be true, and the proof will be converse to the analysis; but if we come upon a thing admitted as false, so also will the thing sought be false.
> In the case of the problematic kind, having hypothesized the thing proposed as known, and then having gone forward through the things which follow in order as true up to something admitted, then if that admitted thing is possible and producible, what those in mathematics call 'given' (*dothen*), so also will the thing proposed be possible, and in turn the proof will be converse to the analysis; but if we happen upon something admitted as impossible, so also will the problem be impossible.[98]

And a 'diorism' is a preliminary distinction of when and how and in how many ways the problem also will be possible.
So much then concerning analysis and synthesis.

This amounts to hardly more than an elaboration of the preceding remarks in the obvious way. Pappus remains uncertain about the order of implication in the analyses, and in the strict sense he would require all steps to be simply convertible. Otherwise, he could not maintain both that the truth of what is derived implies the truth of the hypothesis and also that its falsity implies the falsity of the hypothesis. As we have seen, the effort of logically converting the steps in the analysis of a problem to obtain its synthesis is the basis for discovering the conditions under which the solution is actually possible, and so leads to the articulation of the *diorism*. But what Pappus has to say about this critical element of the method is so extraneous to his account that the modern editors typically bracket this line as a later interpolation.[99] Similarly, the extremely subtle role of the "givens" in the analysis, as we have observed in Chapter 4, is barely hinted at in Pappus' feeble aside. Applying insights derived from the modern logical investigation of heuristics, Hintikka and Remes have perceived nuances in the logical status of auxiliary constructions in the analysis of figures.[100] From Pappus' account one may gather, however, that the commentator was at best but subliminally aware of such difficulties.

What Pappus emphasizes is the distinction between the theoretic and problematic forms of analysis. But examined more closely, one begins to see that his notion of an analysis of theorems is gratuitous. To the extent that he is viewing the expository form of the method, rather than its heuristic role, he seems not to sense that such analyses are entirely redundant[101]; for consider any one of the cases worked out by Hero or the scholiast to *Elements* XIII, for instance, the analysis and synthesis of the theorem XIII, 1, to the effect that if a line is divided into segments x and y in extreme and mean ratio (i.e., $y : x = x : x + y$), then the square of $\frac{1}{2}(x + y) + x$ is five times the square of $\frac{1}{2}(x + y)$. In the analysis one adopts the hypothesis that $(z + x)^2 = 5z^2$ and then deduces through the condition that x and y are in extreme and mean ratio that $z = \frac{1}{2}(x + y)$; in the synthesis one starts from $z = \frac{1}{2}(x + y)$ and deduces that $(z + x)^2 = 5z^2$. Clearly, the presentation of the analysis here is completely artificial. In the case of the more complicated geometric configuration given by Pappus in a lemma toward the construction of a case of Apollonius' three-circle problem, the steps leading to the desired proof might be less obvious, so that the artificiality of the analysis is less evident.[102] But what the commentators lose sight of is that such theorems arise within the context of other theorems or problems which not only reveal what the claim made by the theorem ought to be—that is not merely a matter of conjecture—but also will display the elements of the desired proof. It is no wonder, then, that analyses of theorems are so rarely found in the mathematical literature. These are either artificial and unnecessary (as in those given by Hero) or serve for finding proofs to theorems whose stated result is somehow known in isolation from the context which gives rise to them. Note, furthermore, that in Pappus' alternative outcome of the analysis of the theorem, where the derived result turns out to be false, the formal demonstration may be produced merely by negating the hypothesis to obtain a new assertion and then submitting the steps of the analysis, as already worked out, in proof of this. The term "reduction to the impossible" designating this form of proof was already familiar at the time of Aristotle, and doubtless even before then.[103] By its very name, it invites comparison to the method of "reduction" used, for instance, by Hippocrates in the examination of the cube duplication, and discernible as a precursor of the method of analysis. The connection was not missed by the later commentators.[104] What this reveals is that for a large body of theorems, the analysis of itself yields a proof, so that no synthesis will be called for. Once again, the artificiality of Pappus' theoretic type of analysis and synthesis is apparent.

It is not difficult to explain how Pappus came to advance this notion of theoretic analysis.[105] If we again consider the theorem of the scholiast, the result it establishes is needed in XIII, 11 for showing that if a regular pentagon is inscribed in a circle of rational diameter, its side will be a "lesser" irrational line.[106] The construction here gives rise to the division of a given line into extreme and mean ratio and the consideration of the ratio of $(z + x)^2$ to z^2, for $z = \frac{1}{2}(x + y)$. One might well adopt the format of an analysis to determine what that ratio is, namely 5 : 1. But in doing so, one would be analyzing not a theorem, but a *problem*.[107] We recall that Menaechmus had indicated two types of problems: those that lead to the construction of a figure having certain specified properties, and those that lead to the determination of certain properties obtaining for a specified figure. In the latter case, the result would naturally be stated as a theorem, of which the proof would be identical to the synthesis of what was earlier proposed as a problem. But having made this determination, one would have neither expository nor heuristic reasons for fabricating an analysis in addition to the synthesis of this derived theorem.

In this regard, special interest attaches to an observation made by Carpus in his advocacy of the superiority of problems over theorems[108]:

> And in the case of problems there is a certain single way, that found in common through the analysis, according to which by going forward we can achieve success. For in this way one hunts after the more obscure of the problems. But in the case of theorems the handling is hard to get hold of, as up to our times, he says, no one has been able to offer a common method for the discovery of these, so that by virtue also of their facility the class of problems would be simpler.

Although Carpus does not insist that analysis is a method appropriate to problems to the exclusion of theorems, that is the natural inference to be drawn from his remark. He is known to antedate Pappus, who cites him, and we would surely set him before Hero as well. For the effort by Hero and Pappus to accommodate the method of analysis to theorems, indeed to suggest in subtle ways that its use for theorems might even be prior to that for problems, would surely have pressed Carpus not to deny the *existence* of a proposal for a general method of investigating theorems, but rather to question the alleged *utility* of analysis in these cases. Viewed thus in the light of Carpus' position, Pappus' account of analysis gains a dialectical dimension. By insisting on the relevance of this method to the study of theorems, he effectively neutralizes one of the more telling arguments on the side of the problems in the continuing debate over the relative importance of problems and theorems.[109] But he does this without convincing support from the mathematical literature, where the instances of theoretic analysis, rare on any account, all seem to be due to later commentators, like Pappus himself, as they supply minor lemmas to the great treatises of the older geometric tradition.

Blinded by the glare of Plato's ontology, the later commentators seek to impose on the classical geometry an interpretation it does not easily sustain. Since problems seek to construct geometrical entities, that is, to bring into being things which are forever existing, or seek to cut, augment, diminish, or otherwise modify things which are in reality immutable, they must assume a position subordinate to theorems. For only the latter are philosophically correct in form, proving what is true of the ideal entities of geometry. Problems are allowed a role at the tentative stage when knowledge is incomplete, or for the convenience of learners. Or they might render the service of effecting proofs of the existence of terms needed in the formal demonstrations of theorems. Even the method of analysis, the method *par excellence* of problem solving, is transferred to theorems through the artificial ploy of "theoretic analysis," while in their quest for formal proofs the commentators regularly choose to delete the analyses from older treatments cast in accordance with the double method.[110] In all these ways the later tradition betrays its insensitivity to the outlook and objectives of the older enterprise of geometry. To be sure, there are rare occasions when the commentators somehow grasp the discrepancy. In his discussion of Euclid's *Porisms*, for instance, Pappus observes[111]:

> ...the ancients best knew the difference between these three kinds, as is clear from the definitions; for they said that a "theorem" is what is put forward for the proof of the thing proposed, a "problem" is what is projected for its construction, and a "porism" is what is put forward for its production (*porismos*). But this definition has been changed by the more recent [writers] who can't produce (*porizein*) everything, but use these elements and prove only this, that what is sought exists,[112] but don't produce this.

This statement must discourage any supposition that the ancients conceived of problems merely as existence lemmas for the proofs of theorems. In so distancing himself from those contemporaries who supposed this very thing, Pappus for once displays a feeling for the heuristic motives of the older generation of geometers. For Apollonius and the other authors of the analytic corpus, the discovery of solutions to problems was the heart of their geometric activity, and by their own account, the compilation of theorems was subordinate to this. But for the later generations, not sharing this heuristic motive, the power of the analytic method became merely superfluous—impressive, but no longer valuable.

" ... AND MANY AND THE GREATEST SOUGHT, BUT DID NOT FIND."

The search for solutions to the three problems of cube duplication, angle trisection, and circle quadrature spanned the entire period of the classical geometry from Hippocrates to Apollonius and retained great interest for commentators throughout the later period of antiquity. Did the ancients view this effort as having succeeded in attaining its goal?

In the cases of the cube duplication and the angle trisection, the commentators seem pleased that the solutions put forward answer to the nature of the problems. Hero and Pappus designate them as "solid" problems, in view of the successful application of conic sections toward their solution. Of course, this need not of itself rule out the possibility of finding a more elementary "planar" method. But Pappus' almost abusive treatment of a certain colleague "pretending to be a great geometer" for his alleged solution of the cube duplication by planar means indicates that, to him, the book on this question would appear to have been closed.[113] By analogy, we might expect that the several known constructions of the circle quadrature would leave him content to class the problem as "linear" and so to cease any search for a more elementary method. Indeed, he does just this in the case of the arbitrary division of the angle, classifying it as "linear" without a word on the possibility or advisability of looking for solutions other than the ones via the quadratrix and the spiral which he presents.[114] But he never actually makes this move for the circle quadrature, so that he may well have wished to keep the question open. At least one of the ancient solutions, the fallacious attempt by Bryson to equate the circle with a rectilinear figure somehow intermediate between rectilinear figures (squares?) drawn around and within the circle, suggests the aim to find a solution via the familiar elementary methods.[115] Is this indeed the geometers' objective in their research on this problem? The theme of the circle quadrature surfaces on several occasions in the discussions of the Aristotelian commentators, thanks to a few passing references to the problem in Aristotle's *Categories* (Ch. 7), *Posterior Analytics* (I, Chs. 9 and 12), and *Physics* (I, Ch. 2). Despite the nontechnical character of these sources, the commentators often avail themselves of authorities from the technical literature as well as from the historical and philosophical writings on geometry. Thus, one might expect that they would reflect whatever conclusions had been drawn concerning the status of the circle quadrature among the ancient geometers. In this indirect way, then, they are potential witnesses to the progress of these studies.

Philoponus, as we have noted, makes the important distinction between the *existence* of the square equal to the circle and its actual construction; geometers *assume* the former, he maintains, as the basis of their investigation of the latter.[116] A similar observation is made by Eutocius in connection with Archimedes' theorem on the area of the circle. In proving that the circle equals the right triangle whose legs equal, respectively, the radius and the circumference of the circle, has he not assumed that one can take a line equal to the circumference, a thing which has been proved neither by him nor by anyone else?

> But one must immediately observe that Archimedes writes nothing outside of the things which are appropriate. For it is somehow clear to everyone that the circumference of the circle is some magnitude, I believe, and this is among those extended in one [sc. dimension], while the straight line is of the same kind. Even if it seemed not yet possible to produce a straight line equal to the circumference of the circle, nevertheless, the fact that there exists some straight line by nature equal to it is deemed by no one to be a matter for investigation.[117]

This is explicit admission of an appeal to intuitions of continuous magnitude, like several we have already cited, where the actual construction of the entity required is sharply separated from the issue of its existence.[118] Speaking as a mathematical commentator, Eutocius is clearly of the view that no satisfactory construction of the solution to the circle quadrature has yet been achieved. Such is also the view of the philosophical commentator Ammonius, disciple of Proclus and mentor of Eutocius and Philoponus. In his remarks on Aristotle's distinction between potential and actual knowledge (*Categories*, Ch. 7), he observes:

> The geometers, on constructing the square equal to the given rectilinear figure, sought whether it was possible to find a square equal to the given circle. And many and the greatest [of them] sought, but did not find it. Only the divine Archimedes found an extremely good approximation, but the exact construction has not been found to this day. And this is perhaps impossible; for on these grounds even he [sc. Aristotle] has said "if indeed it is knowable (*epistēton*)."—And perhaps for this reason he has made the straight line not dissimilar to the circular arc, if it is knowable or if it is not.—Then he says that [if indeed] the quadrature of the circle is knowable, although no manner of its knowledge yet exists, then also from this what is knowable is prior to knowledge.[119]

The remark is reproduced with further elaborations by Philoponus and by Simplicius.[120] From this it would appear that the commentators leave the issue open: the search for a construction is valid, that is, it is in principle possible, even if no successful result has yet been achieved. In maintaining the latter, Ammonius might seem merely to be ignorant of the more advanced constructing efforts, for he mentions only Archimedes' derivation of the estimate (3 $\frac{1}{7}$) in the *Dimension of the Circle*. But the same view, as we have seen, is held by Eutocius, who can hardly be charged with such ignorance. Simplicius too is of this opinion, while at the same time knows through a report derived from Iamblichus of the constructions via special curves proposed by Archimedes, Nicomedes, Apollonius, and Carpus.[121]

The same distinction between the known and the knowable appears in the commentary on Euclid's *Data* by Marinus, a disciple of Proclus, active late in the 5th century A.D. In specifying the variety of senses in which terms may be spoken of as "given," Marinus introduces the class of "produceables" (*porima*), namely, those entities whose construction can in fact be presented.[122] Contrary to these are the "nonproduceables" (*apora*), of which the circle quadrature is an example. But Marinus goes on to distinguish two sorts of entities in this latter class: those not yet constructed, but capable of construction (these are termed *poriston idiōs*, "produceable in the special sense," as contrasted with those in the former class, termed *kyriōs porimon*, "produced in the strict sense"); and the *apora*, opposed to the *porima*, whose constructibility (*zētēsis*) has not yet been decided (*adiakritos*). Alternatively, he terms the circle quadrature "unknown" (*agnōston*) yet "ordered" (or "fixed," as it were of constant value—*tetagmenon*). However pedantic this straining over terminological niceties might seem, the distinction Marinus is here striving to make between what is unknown and what is unknowable is an important one. Despite the long history of inconclusive efforts on the circle quadrature, the ancient mathematical tradition remained unwilling to pronounce on the impossibility in principle of finding a solution.

Closely related to this is the question of whether the diameter and the circumference of the circle are commensurable with each other. The development of geometric and arithmetic techniques for deciding on the rationality or irrationality of given terms, such as the incommensurability of the side and diameter of the square, was an important factor underlying the rigorizing movement within the pre-Euclidean geometry.[123] Nevertheless, Greek geometers might sometimes assert a conclusion of this kind without the firm support of a demonstration. Consider Hero's treatment of the problem of cutting off a segment equal to one–third a given circle (*Metrica* III, 18).[124] He explains that he will give only a convenient approximation to the solution, "since it is clear that the problem is not rational." That his solution is not exact is plain; for by finding two chords drawn from a point on the circumference, he shows that the area lying between the chords is exactly one-third of the circle, so that a small additional segment must be neglected in viewing this to be a solution to the stated problem. But on what grounds Hero can conclude the irrationality of the problem we learn nothing more. In the matter of the irrationality of π, comparable statements are made by medieval and Renaissance writers like Maimonides, Maurolico, Stifel, and Gregory, all long before the first actual proof of this result was obtained by Lambert in the 18th century.[125] Might these conjectures have been based on ancient precedents? Another remark from Eutocius' commentary on the *Dimension of the Circle* is suggestive. In defending Archimedes against the charge raised by some critics that his values for the ratio were only rough approximations, as compared with the closer values worked out by Apollonius and Ptolemy, Eutocius writes:

> But if anyone felt strongly about effecting this in minute detail, he could have used what is said in the *Mathematical Syntaxis* of Claudius Ptolemy..., and I myself would have done this, save that I had in mind, as I have said many times, that it is not possible exactly to find the straight line equal to the circumference of the circle by the things said here.[126]

Given that Eutocius earlier observes, quite correctly, that the square root of a nonsquare number "cannot be found exactly,"[127] his reader might well be led to infer that Eutocius likewise maintained the irrationality of the circumference and diameter of the circle, even if he does not flatly say that. Eutocius himself probably intends something weaker: that by qualifying his statement with the phrase "by the things said here," he means only to claim that *the specific procedure* set out by Archimedes cannot yield an exact value, since among other things it requires a series of root approximations. Doubtless, within the long tradition of practical computation the constant recourse to approximations for the measurement of circles and similar figures would occasion surmises concerning their irrationality. But the ancient geometers did indeed make the essential distinction between experience and proof in this instance, as is revealed in this comment by Eutocius' contemporary, Simplicius:

> The reason why one still investigates the quadrature of the circle and the question as to whether there is a line equal to the circumference, despite their having remained entirely unsolved up to now, is the fact that no one has found out that these are impossible either, in contrast with the incommensurability of the diameter and the side (of the square).[128]

In view of this firm statement recognizing the need for a demonstration, not yet at hand, of the impossibility of solving these problems, we may infer that Simplicius maintained the same caution relative to the question of their irrationality. On these matters the ancient tradition thus admitted that the goal of their search remained elusive.

What we never learn from the ancient writers on the circle quadrature, however, from the time of Aristotle and Eudemus in the 4th century B.C. right through to the time of the commentators in the 5th and 6th centuries A.D., is what the precise constraints on a construction would be for any solution of this problem to be judged acceptable. But certain responses to the constructions which were proposed offer us some insights. Speaking of the methods used by Archimedes and others, Simplicius remarks that "it would appear that they all made the construction of the theorem mechanical (*organikē*)".[129] In the same context, but now responding to a claim made by Porphyry (that "most learned" neo-Platonist commentator of the 3rd century A.D.), he remarks that "perhaps a certain mechanical discovery of the theorem had been made, but not a demonstrative (*apodeiktikē*) one".[130] Simplicius appears to maintain that a merely mechanical construction, corresponding to the use of curves of the "linear" class, does not satisfy the conditions for a geometric demonstration. In this, he conforms to an attitude already expressed by Pappus in his reservations over the "rather mechanical manner of generation" of the quadratrix, for which he can find a precedent in the criticisms made earlier by Sporus.[131] A similar view arises among the Arabic geometers, who distinguish between the "mobile" and the "fixed" sorts of geometry, doubtless by way of transferring the ancients' distinction between "mechanical" and "demonstrative," and they too manifest a certain preference for the latter.[132] We have already seen how Plutarch, a thinker with strong Platonist leanings, cast Plato himself in the role of criticizing the supposedly mechanical methods of Eudoxus and Menaechmus for solving the cube duplication, and he even presumes to impute to Archimedes an antimechanical outlook on the field of geometry.[133] But the later commentators are hardly unanimous in adopting this biased attitude. Geminus and Carpus acknowledge the great importance of mechanical pursuits in the mathematical work of Archimedes, even if his writings stress the precision of geometric theory.[134] In offering a mechanical method of his own for solving the cube duplication, Pappus shares in the spirit of the geometers of the 3rd century B.C., like Eratosthenes, Nicomedes and Diocles, whose search for practical solutions to such problems led them to explore mechanical approaches.[135] Both Hero and Pappus wrote extensively on the application of geometric techniques in mechanics. It is such a context which occasions Hero's presentation of the *neusis* for the cube duplication, for instance, and one readily detects the practical interest attached to his accounts of the measurement and section of figures in the *Metrica*.[136] When Pappus writes that "those problems in mechanics called 'organic' arise through being separated from the geometric domain," his meaning might not be entirely clear.[137] But it becomes so when we consider the item for which this serves as preface: the solution of the problem of gauging the diameter of a mutilated cylindrical column; for this leads, through a series of lemmas on ellipses, to the determination of the ellipse passing through five given points.[138] This is but one of many instances where the ancient geometers found in mechanical situations a source of interesting problems for geometric research.

When commentators like Simplicius voice their dissatisfaction with such mechanical approaches to the circle quadrature, they still do not clearly say what the alternative "demonstrative" or "geometric" construction ought to consist of. Presumably, their hope was for an elementary planar solution, or even a solid one. But if efforts along these lines actually were launched, evidence of them has not survived. To the extent that they insist on some such method, they also reveal an insensitivity to the origins of geometric constructions and means of construction; for how are the "planar" techniques of circles and straight lines any other than mechanical, that is, derived by way of abstraction from the practical use of compasses and rulers?[139] Similarly, the conics are no more or less mechanical than the curve formed by Archytas through the sectioning of solids. It is only through an arbitrary process of *fiat* by postulate that Euclid transforms these mechanical means into the starting points of a geometric theory. But for the conics and the field of solid geometry, even this process remains incomplete; for neither Euclid nor Apollonius eliminates the appeal to mechanical conceptions, specifically the rotations of plane figures about fixed axes, for the generation of cones and other solid figures, nor does either ever specify what the process of sectioning solids like these involves, whether by postulate or by theorem.[140] That the process is not merely conceptual is evident from the discussions of solid figures by Hero and Pappus, mentioned just above.

The situation relative to the constructing technique via *neusis* is the same. In some instances, as in the cube duplications of Philo, Hero, and Pappus, an actual ruler (*kanonion*) is conceived as being manipulated about a pivot until the desired configuration is achieved.[141] But in other instances, the operation is expressed in abstract terms: to draw a line such that, while inclining (*neuousa*) toward a given point, it marks off segments having a specified property. Such is the mode adopted by Hippocrates (in the account from Eudemus) for the construction of the third lunule, by Archimedes in the lemmas to the theorems on spirals, and by Sporus in his cube duplication (an alternative equivalent to the manner followed by Diocles and Pappus).[142] Such also is the manner of the cube duplication of Apollonius, as reported by Philoponus; it is nothing other than the abstract equivalent of the ruler construction given by Philo. Where the condition of the inclining line enters, one reads, "but this is assumed as a postulate (*aitēma*) without proof (*anapodeikton*)."[143] Now, this is surely a strategy one might have adopted: to posit a few of the generalized configurations of *neuses* as formal postulates and from this begin to elaborate a sequence of problems and theorems. But the ancients never seem to have taken this approach. For Apollonius and his colleagues, *neuses* were treated as problems to be effected in some instances, as for the cube duplication here or for the angle trisection, by means of conic sections, while in other instances by planar means.[144] Similarly, the motion-generated "linear" curves, like the spirals, conchoids and quadratrices, each have their own "primitive property" (*archaikon symptōma*); for instance, for the plane spiral it is that the rays drawn to the curve from the origin have the same ratio as the angles they make with the initial ray. All further properties and applications of these curves are effected as deductions from their *symptōmata*. If these curves are naturally conceived through mechanical motions, they need not have been treated as such in the formal exposition of their properties. That this formalization was not done, with the result that these curves came to occupy a subordinate position in the hierarchy of construction techniques, is thus largely accidental, the fault of their relative unfamiliarity, their lateness of invention, their more advanced character, or their specialized range of applicability, by comparison with the constructions presented by Euclid and Apollonius.

The central issue still eludes us: the sheer diversity of solutions proposed for each of the three special problems would indicate that the ancient geometers were engaged in a search never consummated to their satisfaction. What was the object of that search? The response, that they sought planar solutions for these problems, attempts to provide a simple motive behind the ancient geometry. It is a view which the commentators in late antiquity appear to maintain, as do many—indeed, most—modern scholars who write on this subject. But it is surely a brutal oversimplification; for in trying to probe the motives of researchers, we face an issue which has psychological aspects as well as formal mathematical ones, and this is likely to require a more complex account. Part of the answer, I believe, lies in recognizing a certain symbiosis between problems and the methods for solving them. The activity of studying and solving (or even not solving) problems gives rise to new techniques; the latter in their turn may gain a life of their own, and through further elaboration give rise to new problems and new techniques. For instance, Archimedes' *neusis* for the angle trisection could suggest to Nicomedes the mechanical mode for generating the conchoids; but then, in addition to the initiating context of the problem, the curves draw attention to their asymptotic behavior and other properties and to the possibility of defining related curves, like the circle-based conchoids, and examining the applications of these. The later commentators do not sense this flexible interaction between problems and methods, this participation in a heuristic process, but instead attempt to describe it in terms of a specific product, the goal of research. But their inability actually to articulate this goal, to state precisely what form of construction was being sought within the earlier geometric tradition, reveals that no such specific goal operated as the motivating element for those geometers.

EPILOGUE

The Greek historical writer Herodotus tells the story of how Pharaoh Psammetichus set out to determine which of the races of man was the oldest.[145] He had two newborn babes carried off to be reared in isolation from society by a herdsman who was to report back as soon as they started uttering recognizable words. In the meantime he was under orders not to speak in their presence, but only to see to their physical needs, such as bringing in the goats every now and again to provide them milk. After about two years the herdsman noticed that the children would say *bekos* whenever he came to look in on them. He duly informed the king who found through further investigation that *bekos* was the Phrygian word for bread. For this reason, Herodotus says, the Egyptians from that day onward acknowledged the Phrygians as an older race than their own.

However much one might hope to discern in this account a prototype of the experimental method in the social sciences, it is clear that Herodotus is joking. The children were not speaking Phrygian, they were speaking goat. The ancient commentators on the analytic tradition of geometry slipped in much the same way. For the Aristotelian commentators, as also for the Neoplatonist Proclus, philosophical questions were naturally the principal interest. Geminus, significant for his encyclopedic survey of mathematics, was an adherent of the Stoic philosophy. Through him and his like, other writers such as Hero and Pappus, while engaged primarily in the presentation of mathematical texts, might thus import philosophical elements into their more general observations on the nature of mathematics. In this way, the later tradition of commentators applied the language and concepts of philosophy in their effort to describe, interpret, and criticize the works of the older tradition of geometers.

We have seen the fruits of their inquiry. Notions of the ideal nature of geometric entities lead them to emphasize the priority of theorems over problems for the exposition of geometric truths. The method of analysis, an effective instrument for discovering the solutions of problems, is purported to serve as well in the investigation of theorems, a claim little encouraged through a consideration of the actual literature of geometry. Others, interested in giving an account of the logical structure of geometry, assign to problems the role of justifying the assumptions on the existence of entities made in theorems. Stimulating as an insight into the logical account of geometry, this view captures neither the diversity and power of the problematic form, nor the methods actually adopted by geometers for addressing issues of existence when they arose in their theorems. What the early geometers introduced as a useful division of loci becomes, among the commentators, a normative division of problems in general, imposing in an unworkable form the recommendation to seek planar methods of solution. At the same time, the rich field of mechanical methods is assigned to the periphery of the domain of geometry. In all these ways, the philosophical preoccupations of the commentators render them insensitive to the nature and motives of the older tradition of research.

Much the same applies to more recent scholarship. Examining the ancient geometry under the conception that its objectives were primarily formal, for instance, or that it subscribed to a constructivist metamathematical program or that it otherwise reflected awareness of concerns characteristic of later movements in mathematics and mathematical philosophy might yield an account more sophisticated than that proposed by the commentators from late antiquity, but is not for this more likely to be faithful to the earlier research. Indeed, the modern accounts typically attempt to build on the start provided by the commentators, assigning greater weight to a single passage by Pappus on the nature of the method of analysis or to a single short comment on the relative ordering of the techniques of construction than to the whole corpus on the analysis of problems. Once it is recognized that such statements by the commentators are founded not on the inspection of and meditation upon geometrical materials, but on paraphrases of stock positions on the related general questions discussed in the philosophical literature, one will know better how to approach the geometrical writers on the one hand and the commentators on the other in order to gain the appropriate insights into the issues which happen to be of interest. To be sure, the geometers were involved in a philosophically interesting effort, just as Herodotus' children were uttering sounds interpretable as Phrygian words. But the ones were not engaging in philosophy, just as the others were not speaking in Phrygian. Inferences on such a basis as to the relative significance of philosophical considerations within the ancient geometry will be no more valid than the Egyptians' deductions on the relative antiquity of their race.

The present survey of the ancient geometry has called attention to the prominence of the activity of investigating problems of construction. Perceiving that the problems of circle quadrature and of cube duplication did not lend themselves to solutions comparable to those more elementary problems, like the rectangle quadrature or the interpolation of a single mean proportional, effectible via familiar techniques like the use of circles and straight lines,

Hippocrates advanced the study of both, the former through his quadrature of the lunules, the latter by reducing it to the finding of two mean proportionals. Subsequent efforts by Eudoxus and his disciples developed new geometric techniques in the context of these and related problems: the generation of special curves, for instance; the implementation of the theory of the application of areas, later serving as a pillar of the theory of conic sections; the limiting techniques appropriate for effecting the measurement of curvilinear figures; and the versatile method of analysis, to become the most characteristic feature of the later problem-solving effort.

In the 3rd century B.C. this activity undergoes a split, maintained with remarkably little overlap. One branch, centering on the circle quadrature and related inquiries, with the concomitant emphasis on the use of the Eudoxean limiting methods, is dominated by the discoveries of Archimedes, who not only applied these methods successfully to new figures, like the conoids of revolution, but also effected technical refinements on the methods themselves. His findings by means of *neusis* constructions and his studies of the spiral, a model instance for the class of curves generated via mechanical motions, inspired several of the geometers in the following generation in their search for alternative constructions of the three "classical" problems.

But a second branch of the problem-solving activity during this period developed in a manner little influenced by the work of Archimedes. Spurred by the studies of locus and the applications of the theory of conics initiated by the work of Euclid and his contemporaries, geometers extended the domain of problems effected via the method of analysis. By the time of Apollonius, about a century later, the wealth of results permitted one to gain a sense of the structure of the field: which forms of conditions give rise to planar loci (circles and lines), which others to solid loci (conic sections), and accordingly, which kinds of problems are amenable to planar or solid solving methods. The surviving evidence, even in its highly imperfect condition, makes clear that these geometers aimed at exhaustion in their researches into specific classes of problems. For instance, Apollonius in one treatise secured the construction of circles answering to conditions of tangency over the whole range of possible orientations and devoted other works toward effecting the same for the conics. He explored in detail the manner of effecting forms of *neusis* constructions via the planar methods, while he and other geometers pursued the investigation of solid problems like the solution of the three- and four-line locus.

Throughout this two-century span of research, the preeminent aim is toward finding solutions to problems and discovering new properties of geometric figures. The geometers themselves stress this heuristic goal: Archimedes describes with evident pride his "mechanical method" for finding the measures of surfaces and solids and centers of gravity; Apollonius presents his compendium on the conics as an instrument for investigating locus problems; even the commentator Pappus acknowledges that the corpus of analysis is organized wholly for abetting the solution of problems. This ideal, even without such explicit testimonies, is manifest throughout the geometrical literature despite its awesome commitment to the rigors of formal exactness in proof. The latter mesmerized the later commentators into such a concern over formal questions that they lost sight of this heuristic objective, and it was hardly before the 17th century that geometers, impatient with the older formal style, revived this goal and set out to extend the ancient findings through their new algebraic, projective, and infinitesimal methods. Without access to techniques such as these, the ancient geometers after Apollonius could not have hoped to extend significantly the massive body of established results. It was this body, but-neither the mind nor the spirit, of the older tradition which the commentators studied and transmitted, but only imperfectly understood.

NOTES TO CHAPTER 8

[1] Heath, *History of Greek Mathematics*, II, p. 358.

[2] Eutocius, in Apollonius, *Opera*, ed. Heiberg, II, p. 176. He expressly cites the alternative proofs and other items in his "copies" throughout the commentary.

[3] Pappus, *Collection* (VII), II, p. 682; cf. Chapter 6.

[4] For a survey of this field, see my *Evolution of the Euclidean Elements*.

[5] Proclus, *In Euclidem*, p. 74; Pappus, *Commentary on Euclid's Book X*, Pt. I, 1, and II, 1. For an account of the reconstruction of Apollonius' theory of irrationals proposed by F. Woepcke, see the edition of Pappus' *Commentary* by G. Junge and W. Thomson, pp. 26-29, 64, 119; and T. Heath, *Euclid's Elements*, III, pp. 255-259. Pappus constructs two cases of irrational lines of the Euclidean type in *Collection* (IV), I, pp. 178-186. In the extant corpus of Archimedean writings, incommensurable magnitudes enter only once, at *Plane Equilibria* I, 7; I show how this relates to the techniques of a pre-Euclidean proportion theory in "Archimedes and the Pre-Euclidean Proportion Theory."

[6] See Chapters 5 and 7.

[7] One may consult Heath's *History* for brief accounts of these figures. The nontrivial significance of Eutocius' editorial work on the transmission of the *Conics* is not generally recognized; this is a major point in my "Hyperbola-Construction in *Conics* II."

[8] Text *A: Collection* (III), I, pp. 54-56; *B*: (IV), I, pp. 270-272.

[9] Here Pappus inserts a few bibliographical notes on prior studies of curves.

[10] Here text *B* holds a passage asserting a criterion on the proper choice of construction methods; this is translated and discussed below (see note 29).

[11] It is difficult to say precisely what the close conformity of these two passages implies in the matter of their provenance. However, the opening of *A*, "The ancients say ...," must indicate that Pappus is following a source for the content, if not the exact wording, of his remarks here. (Note this, despite the "We say ..." in *B* which might, if viewed in isolation, suggest Pappus' responsibility for the view expressed.) As shown below, his "planar" and "solid" classes do indeed have precedents in the older literature, and Pappus might well have been able to draw his account from Geminus or Menelaus, both of whom wrote extensively on geometric problems and curved lines. I would thus suppose that where *A* and *B* are in agreement, Pappus is transcribing from a source.

Now, Pappus applies this passage in two different, not fully consonant, ways in the two contexts. In *A* he goes on to remark that the difficulty lay in *drawing* the conic sections as plane curves, and so leads into a presentation of mechanical methods for finding the two mean proportionals; in *B* it is the ancients' early *unfamiliarity* with the properties of the conics which he says frustrated their efforts to trisect the angle, whence he proceeds to show three methods of solution via conics. Presumably, then, Pappus has left his source as he attempts to indicate the relevance of the opening remarks for the technical materials which follow. Thus, the view he expresses of that relevance is his own, and we cannot hope to use it as a gauge of the attitude in the older geometric tradition.

[12] *Collection* (VII), II, pp. 660–662.

[13] As cited by Pappus, *ibid.*, (III), I, p. 62. This is confirmed in the Arabic version of Hero's *Mechanica* I, 11.

[14] *Conics*, Book I, Preface; Apollonius, *Opera*, ed. Heiberg, I, p. 4.

[15] *Ibid.*, II, p. 178. The three examples of locus Eutocius cites at this point are all of the planar type, one being the two-point locus discussed in Chapter 4.

[16] *Ibid.*, p. 184.

[17] *Collection* (VII), II, pp. 652, 662, 672. See the discussion of Zeuthen's proposed reconstruction in Chapter 6.

[18] *In Euclidem*, pp. 394 f. Proclus uses as an illustration of a planar locus that all equal parallelograms having the same base will have their upper edge lying along the same line parallel to the base; as a solid locus he cites the fact that equal parallelograms drawn in a given angle will have the opposite vertex lying along a given hyperbola.

[19] *Ibid.*, p. 395. The fact a philosopher like the 3rd-century Stoic Chrysippus should find such a useful metaphor in the geometers' locus, while earlier philosophers do not mention it, suggests to me that this notion of locus was relatively recent in the 3rd century B.C.

[20] *Collection* (VII), II, pp. 660–662. Here Pappus cites "the preface of Apollonius' *Elements*" which may refer to the general treatise on the principles of mathematics attributed to him by other writers; cf. Heath, *History*, II, p. 193.

[21] The terms might be translated as "restrained," "passing through," and "turning up," respectively.

[22] See note 11 above.

[23] Note that the latter criticism would not apply against the less restrictive sense of "solid" locus maintained by Eutocius and Proclus.

[24] *Collection* (VII), II, p. 672. Hultsch brackets this line, as he does many others throughout his edition of the *Collection*. To the extent that his motive is to impose a certain consistency upon the whole work, however, his effort may be in vain; for discrepancies will be inevitable, given Pappus' dependence on a wide variety of sources in the compilation of these geometric materials.

[25] Cf. *Conics* I, Preface (note 14) and Apollonius' writing *On Planar Loci*. The term *topos* ("locus") is not used in its geometric sense by Archimedes, nor does it appear with the locus propositions of the Aristotelian *Meteorologica* III, 3 and 5. It appears in the title of one of Euclid's works (*Surface Loci*) cited by Pappus (cf. *Collection* (VII), II, p. 636), but is notably absent from the title of Eratosthenes' work cited at the same place (i.e., *On Means* rather than *On Loci with Respect to Means*). This is risky ground

to base an inference, however, since such cited titles need not be original to the works. Proclus reports of the geometer Hermotimus of Colophon, shortly before the time of Euclid, that in addition to discovering many of the things in the *Elements*, he "compiled certain things on the loci" (*In Euclidem*, p. 67). This might give Hermotimus a role in the solution of problems like that in the *Meteorologica* III, 5; but again, we cannot infer from Proclus' report that the term "locus" was already then used to designate such problems. It is of interest to note the series of theorems from Euclid's *Optics*, Props. 37, 38, 44–47, of which the following is representative: "There is a place (*topos*) where, when the eye is fixed while the object seen is moved, the latter will always appear equal (in size)" (*ibid.*, 37; note the differences in the Theonine versions, Props. 41–47). These theorems all deal with the specification of *topoi* where certain optical phenomena obtain. But at the same time they yield a geometrical figure corresponding to each *topos*. It seems to me possible that contexts of this sort in optics, where *topos* retains its basic sense of "physical place," served as background for the early study of geometrical locus. This would be but one of several ways in which optical studies contributed to new developments within the geometric field in antiquity (cf. the discussion of the early theory of conics in Chapter 4; and Diocles' work on burning mirrors in Chapter 6).

[26] On these efforts by Menaechmus and Archimedes, see Chapters 3 and 5.

[27] See the résumé by Heath, *History*, II, pp. 185–189.

[28] This view is elaborated below.

[29] *Collection* (IV), I, pp. 270–272; cf. note 10 above.

[30] See, for instance, Heath, *History*, I, pp. 218 f, who is markedly more circumspect in this matter than most writers. The assumption of the quest for planar constructions is carefully scrutinized and convincingly discredited by A. D. Steele, "Ueber die Rolle von Zirkel und Lineal," *Quellen und Studien*, 1936, 3, pp. 289–369.

[31] On the problem from Apollonius cited here, see Chapter 7.

[32] See Chapter 5(ii) for the solid construction of Archimedes' *neusis*.

[33] See Chapter 2. Steele provides a very thorough account of this construction, arguing that the *neusis* cannot have been intended to be effected in any way other than as a *neusis*; "Zirkel und Lineal," pp. 319–322.

[34] See Chapter 3.

[35] See Chapter 4.

[36] See Chapter 6.

[37] See Chapters 5 and 6. Note that Archimedes' solution to the circle quadrature via drawing tangents to the spiral is not unrelated to the procedure of *neusis*.

[38] See Chapter 7.

[39] Such classifications can define the project of specifying the relative domains of problems solvable by the various constructing means. An impression of the wide range of results to be obtained in this area can be derived from the articles in Enriques' survey (cf. note 42).

[40] Pappus, like Hero, actually employs a *neusis*; but his method is equivalent to Diocles' use of an auxiliary curve (see Chapter 6).

[41] See the statement at the end of text *A* cited above, and his reproduction of Hero's statement to the same effect (see note 13 above).

[42] See, for instance, A. Conti in Enriques, *Fragen der Elementargeometrie*, II,

art. 7, pp. 193–195. The irreducibility of the third-order relations for the cube duplication and the angle trisection was first shown by L. Wantzel (1837); see Liouville's *Journal*, 2, pp. 366–372.

[43] *Collection*, I, p. 30. His lengthy and unsympathetic account of the proposed method (*ibid.*, pp. 32–48) should begin to raise doubts among those who would assume in Pappus himself a high level of geometric expertise. See the brief summary by Heath, *History*, I, pp. 268–270.

[44] The *Académie française* ruled in 1775 against the further consideration of contributions relating to the circle quadrature (cf. P. Beckmann, *A History of Pi*, 2nd ed., 1971, p. 173); but the first successful proof of the transcendentality of π came only with Lindemann's publication in 1882, extending an earlier result by Hermite (1873) (cf. art. 8 in Enriques, *op. cit.*). Similarly, Descartes pronounced that "it will be an error to strive in vain to wish to construct a problem by a class of lines simpler than its own nature permits" (*Géométrie*, ed. Smith, p. 157) precisely two centuries before the first specific results on the reducibility of the associated polynomials were presented (see note 42).

[45] *Collection* (III), I, p. 56.

[46] Cf. *Elements* I, 22. The implied governing construction might be "it is proposed..." (*proballetai*, or *proteinetai*), as sometimes supplied in discussions by Pappus, Proclus, and others, or alternatively "it is possible ... "(*dunaton esti*). Of course, the actual construction establishes far more than the mere possibility of the construction.

[47] Cf. *Elements*, I, 21.

[48] *Ibid.*, I, 20.

[49] Cf. the locus problem in the *Meteorologica* III, 5 (Chapter 4 above).

[50] For an example of just this sort from Pappus, see Chapter 4, note 16.

[51] Pappus, *Collection* (VII), II, pp. 648 ff; Proclus, *In Euclidem*, pp. 212, 302. Proclus recognizes an alternative sense of "porism," namely "corollary" (*ibid.*, pp. 301–303). The term derives from *porizein*, "to provide" or "to devise." An appearance of a form of the verb in the *Meteorologica* (III, 5, 376 a 15) may carry the technical sense, although another appearance in a mathematical passage in the *Ethics* (1112 b 26; discussed below) seems not to.

[52] *In Euclidem*, p. 81.

[53] This is a view proposed, among others, by Proclus and by Zeuthen. See note 77.

[54] Note the propositions of the form "there are places (*topoi*) such that..." in Euclid's *Optics* (see note 25 above). These are undeniably existential, both in sense and in form. But they are framed as *theorems* and serve as the basis for a *problem* (Prop. 48): "to find the places where the equal magnitude appears half as big, etc." Similarly, the materials in the pseudo-Euclidean *Catoptrica* culminate in a *problem* (Prop. 29): "it is possible to construct a mirror such that many faces are seen in it, some bigger, some smaller, some nearer, some further away, some reversed, some not." This serves merely to bring together the various results already established on plane, convex, and concave mirrors; it is hardly the equivalent of an existence proof.

[55] Pappus, *Collection* (IV), I, pp. 272 ff. See Chapter 7 and my "Hyperbola–Construction in *Conics* II."

[56] *Conics*, Preface to Book I (ed. Heiberg, I, p. 4): "the third book (contains) many remarkable theorems useful for the syntheses and diorisms of solid loci." Similarly, the contents of Book IV are said to be "useful for the analyses of diorisms," while Book V also is "needed for obtaining a knowledge of the analysis and determination of problems, as well as for their synthesis" (cf. Heath, *Apollonius*, p. lxxiv). The same attitude is maintained by Pappus when he describes the *topos analyomenos* as "a certain special corpus ... conducive to the finding of problems ... and compiled as useful for this purpose alone" (*Collection*, VII, Preface, ed. Hultsch, II, p. 634).

[57] See Chapter 7.

[58] Archimedes does without any such existence proof in the area theorem of *Dimension of the Circle*, Prop. 1; the omission was criticized by some, but defended by Eutocius (Archimedes, *Opera*, III, p. 230; see discussion below). Further examples relating to the alleged existential function of problems appear in chapter 3.

[59] Note that it was the theorematic form of Postulates 4 and 5 which encouraged ancient geometers like Apollonius and Ptolemy, respectively, to seek proofs for them; cf. Proclus, *In Euclidem*, pp. 188 ff, 191 ff, 365 ff. Proclus asserts flatly that the "postulate" on parallels *is* a theorem (*ibid.*, p. 191).

[60] *Collection* (III), I, p. 30; cf. (VII), II, pp. 648 ff. Note that the notion of displacing *all* problems by theorems or *vice versa* receives no support from the technical literature, where both forms appear and serve clearly marked purposes. Comparable distinctions in the dialectical roles of problems and theorems (e.g., the distinction between *problēmata* and *protaseis* in debating contexts, *Topics* I, 4, 10, 11) or in the different priority of terms in the order of knowing and the order of proving (e.g. *Posterior Analytics* I, 2) may suggest a loose connection between these discussions on geometry and the Aristotelian epistemology.

[61] *Republic* 527; cf. 510d and *Euthydemus* 290c.

[62] *In Euclidem*, pp. 77 f. On Amphinomus, see Chapter 3.

[63] See the discussion of the *Platonicus* fragment, Chapter 2.

[64] *In Euclidem*, p. 243.

[65] *Ibid.*, p. 78.

[66] See Aristotle's *Metaphysics* Z.10, 1036 a 9; and the discussion by Heath, *Mathematics in Aristotle*, pp. 213–216, 224–226.

[67] Proclus, *In Euclidem*, pp. 241 f; Pappus, *Collection* (VIII), III, p. 1026; Simplicius, *In Physica*, ed. Diels, p. 60. The identity of this curve is not known; Tannery conjectures it was the cycloid, but I have argued that it may be the cylindrical helix (see my "Archimedes and the Spirals," pp. 73 f).

[68] *In Euclidem*, p. 242.

[69] *Ibid.*, pp. 80 f.

[70] Here Zenodotus is described as "in the succession of Oenopides" and "among the disciples of Andron." While Oenopides would be the 5th-century geometer-astronomer, this helps little in placing Zenodotus. Given the later reference to Posidonius in this same passage, an identification with a Stoic writer of this name from the 2nd century B.C. might be indicated (cf. *Pauly Wissowa*, "Zenodotus (4) and (6)"), but this is entirely conjectural. So also is the degree to which such late writers reflect the actual ideas of the earlier writers with whom they are associated. K. von Fritz has observed that Zen-

odotus' observation about problems (i.e., that they seek "that on the assumption of which something is," *tinos ontos ti esti*) has to do with the *analysis* of problems ("Oenopides" in *Pauly Wissowa*, reprinted in his *Schriften zur griechischen Logik*, Stuttgart, 1978, 2, pp. 151 f). Note, however, that the implied conception of analysis accords with the view of Pappus, following the Aristotelian manner, that is, as a search for suitable *antecedents*, rather than as an investigation of consequences (see discussion below). Of course, one can hardly hope to ascribe to Oenopides a role in the introduction of the analytic method on the basis of this evidence.

[71] This is surely Posidonius, the distinguished Stoic teacher of the late 2nd century B.C.

[72] The manuscripts here insert *problēma* which Morrow understandably rejects as an interpolation, since it violates the sense of the passage (cf. his *Proclus A Commentary...*, p. 66n).

[73] *In Euclidem*, pp. 233 f.

[74] A view minimizing the formal intent underlying Euclid's geometry is argued by A. Seidenberg in "Did Euclid's Elements, Book I, Develop Geometry Axiomatically?" *Archive for History of Exact Sciences*, 14, 1975, pp. 263–295.

[75] Oddly, Proclus can here cite some Epicureans who criticize proving this result at all, given its perfectly obvious character (*ibid.*, p. 322).

[76] See Heath, *Euclid's Elements*, I, pp. 293 f. Note that if the analysis of this problem had been given first, the necessity of this condition and its role within the synthesis would have been clear. It would so appear that Euclid's formal difficulties in cases like this follow from his decision to present only the syntheses of the problems.

[77] I believe that the *locus classicus* for the thesis that the ancient geometric constructions were meant to serve as proofs of existence is Zeuthen's "Die geometrische Construction als 'Existenzbeweis' in der antiken Geometrie," *Mathematische Annalen*, 47, 1896, pp. 222–228. His case rests primarily on Euclidean materials, although he draws from more advanced work as well. But when he supposes, for instance, that the Greeks had to seek a construction of the conics (e.g., as sections of a cone) because they would not admit mechanical or pointwise methods, he commits a plain error (cf. the discussions of Menaechmus and Diocles in Chapters 3 and 6). It is thus hard to understand how the thesis has gained such credence. To be sure, one can find telling instances where problems enter to justify the introduction of terms necessary for subsequent theorems. An instance not cited by Zeuthen is *Elements* XII, 16, 17: two problems assumed in steps in the proof of Prop. 18 (that spheres are as the cubes of their diameters). But we have seen other instances where existence is established via *theorems*, subordinate to problems (cf. note 54 above), as well as many cases where the problems of construction have standing in their own right, without reference to associated theorems (cf. note 56 and Chapter 3). We can only conclude that the ancients did not distinguish proofs of existence from other geometric propositions in the way modern mathematicians do; that apart from diorisms (which are by their nature existential in force) no special format was reserved for questions of existence as they arose, these being handled now as theorems, now as problems, but often only as tacit assumptions. For a nicely balanced view on the relation of ancient geometric constructions and existence proofs, see I. Mueller, *Philosophy of Mathematics...*, pp. 15 f, 27–29.

[78] For the scholia, see Euclid, *Opera*, ed. Heiberg, IV, pp. 364 ff. The commentary by al-Nairīzī (Annaritius) survives both in Arabic and in its Latin translation by Gerard

of Cremona. I have consulted the latter in the edition by M. Curtze, in Euclid, *Opera*, ed. Heiberg, IX; for references to the former, see F. Sezgin, *Geschichte des arabischen Schrifttums*, V, pp. 283-285.

[79] See the contributions cited by J. Hintikka and U. Remes, *The Method of Analysis*, 1974, Ch. ii. The authors have compiled an extensive survey of ancient comments on analysis in their Ch. viii.

[80] *Collection* (VII), II, p. 634.

[81] *Ibid.*

[82] Euclid, *Opera*, IV, p. 364.

[83] I follow the Latin of Gerard in *ibid.*, IX, p. 89. He seems here to have made a valiant attempt to render a difficult passage. The Arabic (Leiden, Or. 399, f. 25v, lin. 10-12) can be given thus:

> As for analysis (*taḥlīl*), it is when any question has been proposed to us, we say (that) we take it down (*nunziluhā*) (to) the degree (*manzila*) of the thing sought (*maṭlūb*), that it is found (*innahu mawjūd*), then we break it up (*nafudduhu*) to something whose proof has already preceded, so that since it is manifest to us, we say that what was sought (*maṭlūb*) has now been found (*wujida*) through analysis.

Hintikka and Remes cite a translation made for them by T. Harviainen from the Arabic: "we suppose the thing sought as being (*mawjūd*), then we resolve it to something already proved." (*Method of Analysis*, p. 93) They thus wish to read *innahu mawjūd* as a version of Greek *hōs on*. But in context, *mawjūd* must surely here mean "found," as the verbal form *wujida* clearly does later in the passage. The underlying Greek would thus be *hōs zētoumenon*, that is, "as if found." The Arabic translator has lost the nuance of *hōs* (present in the parallel Greek passages cited: *hōs homologoumenon*). Consequently, Gerard renders as *que est inventa* what, in light of the presumed Greek, would more accurately be *quasi inventa*. As for terminology, "analysis" becomes *taḥlīl* ("*dissolutio*"), "synthesis" is *tarkīb* ("*compositio*").

[84] This looseness of the senses of *akolouthein* is emphasized by Hintikka and Remes, *op. cit.*, pp. 13 ff; they opt for the rendering "concomitants."

[85] Note that the scholiast speaks of the thing admitted as being "true," and so would seem to have in mind a theorematic context of analysis, like that described later by Pappus. Hero's term "known" could apply equally well to the analysis of problems and that of theorems. The phrase "up to something admitted as true" here only poorly suits the context, and is likely to be a scribal error due to the appearance of the same phrase in the preceding sentence (see note 82). Some manuscripts here have an alternative phrase, "up to the ending up or taking of what is sought" (cf. Heiberg, *Euclidis Opera*, IV, p. 366n).

[86] It seems difficult to assign to "antecedents" and "consequents" here their usual logical senses and yet to remain compatible with Pappus' notion of the order of analysis given before; for there the terms found were logical antecedents, even though they happened to be discovered later in the analysis. Of course, in the usual presentation of an analysis, terms are derived as logical consequences, so that a reversal of the logical order does indeed have to be secured in producing the synthesis.

[87] In the *Metrica*, Hero presents many examples of "analyses" of an interesting variant sort. In these, the objective is to derive a computational rule or procedure by

means of what is called an "analysis" and then in the following "synthesis" to work out a solution in particular numbers via the derived rule. In *Metrica* III, 4–14, 20–22, these are all problems in the division of figures, and one can appreciate the appropriateness of the analysis/synthesis distinction. But in *Metrica* I, 8–25 and II, 6–9, 13, the rules all involve expressions for the areas and volumes of figures. These are all stated as *problems* (e.g., "to find the area of the regular heptagon" in I, 20), but they have no analytic hypothesis of the sort which dominates the analysis of a problem. On the other hand, these analyses do start from the whole figure and work through to a relation of a computable sort pertaining to the givens (e.g., the given length of 10 for the side of the heptagon), while the syntheses begin with the givens and calculate with these in accordance with the derived relation to produce what is sought. Doubtless, this format was adapted from the analysis of problems in the more advanced geometry. Hero's version may be based on a prior tradition following this same format; he depends, for instance, on values drawn from "the books on chords" (a reference to Hipparchus? cf. *Opera*, ed. Schöne, p. 58 n), but his heading to the section of analyses and syntheses is ambiguous (*ibid.*, p. 16). At any rate, Pappus seems to know of this format; for he invites the reader to work out numerically a value for the cube duplication "in accordance with the analysis" through consultation of Ptolemy's table of chords (*Collection*, I, p. 48).

[88] *Collection* (VII), II, pp. 708–710; cf. Heath, *History*, II, pp. 406 f.

[89] See, for instance, analyses of the problems on the inscription of the regular polyhedra in *Collection* III, Prop. 54 ff. The same appears in Diocles' analysis of Archimedes' solid problem (cf. Toomer, *Diocles*, p. 86). In some instances, the synthesis might merely be omitted without comment, as in Pappus' treatment of the analyses of the "solid *neusis*" used by Archimedes (*Collection* (IV), I, Prop. 42 f). It is interesting, however, that in most such cases the commentators Eutocius and Pappus will supply the missing synthesis, or at least fill in a sketch of it.

[90] It is of course possible that Pappus himself is responsible for this lemma. It would then merely be a reflection of his own view of the nature of analysis, rather than evidence from the more general technical field.

[91] *Nicomachean Ethics* III.3, 1112 b 15–27.

[92] The passage is well noted by Hintikka and Remes (*op. cit.*, pp. 85 f) who draw attention to the remarkable conformity in terminology between it and the account of analysis by Pappus. But they seem reluctant to admit that Pappus might have referred to it as a model for his own statement.

[93] This observation is made by Hintikka and Remes, *op. cit.*, p. 86.

[94] *Collection* (III), I, p. 86 (i.e., that the geometric mean is a cause of harmony in all things; this would appear to relate to *Timaeus* 31–32, as noted by Hultsch, *ibid.*, p. 87n).

[95] These are noted by Thomson in his introduction to the *Commentary of Pappus...*, pp. 51–57.

[96] An extreme example of the mathematical commentary conceived within the framework of Platonist doctrines is that by Proclus on Euclid's Book I; see the introductory remarks by Morrow, *op. cit.*, pp. xxiv–xliii.

[97] *Collection* (VII), II, pp. 634–636.

[98] Note how Pappus' view here of the two possible outcomes of the analysis recalls Aristotle's observation to the same effect in the *Ethics* passage cited above.

[99] Cf. Hultsch in *Collection*, II, p. 636n. This line is deleted without comment in the versions of Pappus' account given by Heath (*History*, II, p. 401) and by I. Thomas (*Greek Mathematical Works*, II, p. 598).

[100] *Op. cit.*, Chs. iv–vi.

[101] Hintikka and Remes (*ibid.*, Ch. iv) insist that the theoretic form of analysis be taken quite seriously and attack in sharp terms any view that the heuristic processes underlying problem solving and theorem proving are somehow distinct (*ibid.*, p. 47, criticizing M. Mahoney, "Another Look at Greek Geometrical Analysis," *Archive for History of Exact Sciences*, 1968–69, 5, pp. 319–348, especially p. 328). But surely two different things are at issue here. Let us of course accept the authors' findings on the logical description of the problems and theorems, to the effect that the description is the same in either case. Nevertheless, the analysis of a theorem in the sense they require must begin with the hypothesis that the result claimed in the theorem is true. How is that result arrived at? In the pedagogical context, the teacher might present a claimed theorem to the student and ask him to prove (or disprove) it; in this case, the student might well employ an analytic procedure to assist him in working out a proof. Otherwise, the analytic hypothesis must be a matter of mere conjecture. One can well doubt whether the ancients approached the finding of proofs of theorems *as theorems* in such a context-free manner. Furthermore, in cases like Archimedes' theorem on the area of the circle (that it equals one-half the radius times the circumference), the proof results from hypothesizing that the claimed result is *false*. It is hard to see the value, from the historical point of view, of assimilating the techniques used for investigating such theorems with the analytic method used for problems, even if this proves to be rewarding from the logical point of view.

[102] *Collection* (IV), I, p. 186. This example is discussed in detail in Hintikka and Remes, *op. cit.*, Ch. iii. The lemma which closes *Collection* VII (II, pp. 1016–20) is also an attempt toward the analysis of the theorem, although the text has been severely garbled. Other than these, I know of no extant analyses of theorems, save for those of the trivial sort mentioned earlier (see note 78 and note 87).

[103] Cf. *Prior Analytics* I, 23, 41 a 21 ff.

[104] Cf. Proclus, *In Euclidem*, pp. 212 f.

[105] Pappus is not alone in proposing the theoretic form of analysis; there are comparable statements from Proclus (cf. *In Euclidem*, p. 42; if indeed the "analytic faculty," or *analytikē dynamis*, mentioned there refers to the geometric analysis) and from his disciple Marinus (*In Euclidis Data*). The latter writes, for instance, that

> the knowledge of the *data* is most necessary for the purposes of the so-called analytic corpus (*topos analyomenos*)...analysis is the finding of proof, it is compiled for us for finding the proof of the similar things, and acquiring the analytic faculty is greater than having many proofs of things in part. (Menge, *Euclidis Opera*, VI, pp. 252-254)

In citing these passages, Hultsch notes their resemblance to Pappus' account of analysis, the verbal reminiscences hardly assignable to mere coincidence (*Collection*, III, pp. 1275 f). Hultsch infers, not the later writers' use of Pappus, but their common dependence on a much older account of analysis. Since Euclid, Apollonius, and Aristaeus were the principal authors of this corpus, he reasons, one of them must have been responsible for this alleged account, and doubtless Euclid himself was the one, since he was the oldest among them. Apparently, one must suppose that geometers could not apply the method of analysis without also preparing and handing down a form of philosophical statement on its nature; and apparently, a later commentator like Geminus was incapable of composing such a statement without an explicit model in hand deriving ultimately from Euclid or his disciples.

[106] See my account of this theorem in *Evolution of the Euclidean Elements*, Ch. VIII/IV, especially pp. 279 ff. Proposition XIII, 1 is also used in the proof of XIII, 6; but the latter has been convincingly argued to be an interpolation (cf. Heath, *Euclid's Elements*, III, p. 451).

[107] Even with this modification, the analysis in this case would be unnecessary, since the claimed result is an obvious feature of the construction of this section of the line in *Elements*, II, 11. That is, x is produced by taking the line $x + z$ equal to the hypotenuse of a right triangle whose legs are z and $2z$ ($= x + y$); see Chapter 3. One is thus led to suspect that XIII, 1, like XIII, 6, is an interpolation—whence perhaps the entire series of Lemmas 1–6 may be a later addition to the body of constructions and theorems of that book.

[108] Proclus, *In Euclidem*, p. 242. See note 67 for other references to Carpus.

[109] Proclus, *ibid*. For further references to this debate, see notes 60ff.

[110] One may cite Eutocius' removal of the analysis in his version of the hyperbola construction (see Chapter 7, note 33) and Pappus' stated preference for the synthetic mode by virtue of its greater conciseness in comparison with the analytic mode (*Collection*, I, pp. 410–412). One notes that the angle trisections presented by Thābit ibn Qurra and Aḥmad ibn Mūsā (see my "Transmission of Geometry from Greek into Arabic") eliminate the analyses Pappus presents (cf. *Collection*, I, pp. 270 ff). Other examples of the commentators' preference for syntheses are cited in note 89 above.

[111] *Collection* (VII), II, pp. 650–652.

[112] An alternative translation: "only that this is what is sought," seems possible, but does not conform to the sense of the passage.

[113] *Collection* (III), I, p. 30.

[114] *Ibid.*, (IV), I, p. 284.

[115] For references to the accounts by ps.-Alexander and by Alexander as reported by Philoponus, see Chapter 3.

[116] Philoponus, *In Analytica Posteriora*, ed. M. Wallies (*CAG* 13), p. 112; see Chapter 3.

[117] Commentary on *Dimension of the Circle*, Prop. 1, in Archimedes, *Opera*, III, p. 230.

[118] See Chapter 3.

[119] *In Categorias*, ed. A. Busse (*CAG* 4, Pt. 4, 1895), p. 75.

[120] Philoponus, *op. cit.*, p. 120; Simplicius, *In Categorias*, ed. K. Kalbfleisch (*CAG* 8, 1907), p. 192.

[121] Cf. Simplicius, *loc. cit.* and also *In Physica*, ed. Diels, p. 60.

[122] Commentary on the *Data* in Euclid, *Opera* VI, ed. Menge, p. 240.

[123] See the survey in my *Evolution of the Euclidean Elements*, e.g., Ch. IX.

[124] *Opera* III, ed. Schöne, p. 172; for an account, see Heath, *History*, II, pp. 339 f. Hero's construction here seems related to another problem in the sectioning of a circle in Euclid's *Division of Figures*, Prop. 29 (cf. Heath, *ibid.*, I, p. 429). Note that in both

cases, although the problem seeks a part which is *one-third* the circle, the method is clearly a general one.

[125] This conjecture is made by Maimonides in his comment to *Mishnah Eruvin* I, 5 (for which reference I thank Daniel Lehmann, Department of Mathematics, University of Southern California). Maurolico bases his own conjecture on the isoperimetric property of the circle; cf. M. Clagett, *Archimedes in the Middle Ages*, III, pp. 779n, 782n. For references to Stifel and Gregory, see Hobson, *op. cit.*, p. 31. Note that Descartes seems to be of the same view when he observes that the motions which yield the spiral, the quadratrix, and other such curves, that is, a circular motion synchronized with a linear motion, do not have a ratio which is exactly measurable (*Géométrie*, II, ed. Smith, p. 45). He may well have in mind the similar remark made by Sporus, as reported by Pappus (*Collection*, I, p. 254), that the synchronization requires that this ratio be known, yet the object of the use of these curves is toward finding that ratio. On Lambert's proof of the irrationality of π (1761/68), see Hobson, *op. cit.*, pp. 43 ff; and Enriques, *op. cit.*, art. 8.

[126] Eutocius, in Archimedes, *Opera*, III, p. 260.

[127] *Ibid.*, p. 232.

[128] *In Physica* VII.4, ed. Diels, pp. 1082 f. Simplicius also observes here that the circle quadrature "has not yet been found; but even if it might seem at this time to have been found, it is only through certain disputed hypotheses." One seems to hear an echo of this view in a remark by the Banū Mūsā: that one can compute an approximation to the value of the ratio of the circumference and diameter as closely as desired via the given (Archimedean) method, while "no one has to this day found anything beyond what has appeared to us" (*Verba filiorum*, Prop. 6; Clagett, *Archimedes*, I, p. 264).

[129] *In Physica*, p. 60.

[130] *In Categorias*, p. 192.

[131] *Collection* (IV), I, pp. 252 ff; cf. note 125 above.

[132] Cf. the survey of angle trisections by al-Sijzī (cited in Woepcke, *Omar*, p. 120), where the "mobile" manner of the "ancient" (i.e., Archimedean) construction is criticized. Most forms of the Arabic terms, *thabt* ("fixed") and *bayyin* ("manifest"), are indistinguishable in unpointed text. In the account of Nicomedes' cube duplication (see chapter 6) by Abū Ja'far Muḥ. b. al-Ḥusain it is the ancient "method of instruments" which is replaced by a "geometric" procedure, the latter being the same as Pappus' use of conics for effecting the *neusis* for the angle trisection; cf. Sezgin, *op. cit.*, p. 306.

[133] Cf. *Vita Marcelli*, xiv and the references cited in Chapter 2.

[134] See the citation by Pappus in *Collection* (VIII), III, p. 1026.

[135] See Chapter 6.

[136] On the cube duplication, see the references in Chapter 5, note 115. Note his inclusion of practical methods for the measurement of figures in *Metrica* I, 39 and II, 20; his remarks on the practical interest of measures of the torus and of the sectioned cylinder (II, 13 and 16); his devotion of the entire writing *On the Dioptra* to the practical use of this angle-sighting device in surveying.

[137] *Collection* (VIII), III, pp. 1072-74. Hultsch's translation of this as *problemata...sine demonstratione geometrica solvuntur* (that is, "problems solved without geometric demonstration"; *ibid.*, p. 1075) cannot be right.

[138] See the remarks on this problem in Chapter 7.

[139] This is a point made with considerable force by Descartes in the opening section of Book II of his *Géométrie*.

[140] See, for instance, Euclid's Book XI, Def. 18 and Apollonius' Book I, Def. 1. In their comments on this part of Apollonius, neither Eutocius nor Pappus indicates knowledge of any ancient attempt to formalize these conceptions.

[141] On these *neuses*, see note 38 in Chapter 7.

[142] See, respectively, Chapters 2(ii); and 5(ii); and 6 (note 106).

[143] For the text, see Philoponus, *In Ana. Post.*, p. 105n; reproduced by Heiberg in Apollonius, *Opera*, II, p. 106.

[144] See Chapter 7.

[145] *Histories* II, 2.

Bibliography of Works Consulted

I. Ancient works

Alexander of Aphrodisias, *In Aristotelis Meteorologica Commentaria*, ed. M. Hayduck, *CAG* III, pt. 2, Berlin, 1899

ps.-Alexander of Aphrodisias, *In Aristotelis Sophisticos Elenchos Commentaria*, ed. M. Wallies, *CAG* II, pt. 3, Berlin, 1898

Ammonius, *In Aristotelis Categorias Commentaria*, ed. A. Busse, *CAG* IV, pt. 4, 1895

Anthemius of Tralles, *Peri Paradoxôn Mêchanêmatôn* (fragment): see *Mathematici Graeci Minores* (pp. 78-87); Huxley

Apollonius, *Opera*, ed. J. L. Heiberg, Leipzig: B. G. Teubner, 2 vol., 1891-93 (repr. Stuttgart, 1974)

Apollonius: see also Balsam, 1861; Halley, 1710; Heath, 1896; Ver Eecke, 1924

Archimedes, *Opera*, ed. J. L. Heiberg, Leipzig: B. G. Teubner, 2. ed., 3 vol., 1910-15 (repr. Stuttgart, 1972)

Archimedes, *On the Inscription of the Regular Heptagon*: see Hogendijk

Archimedes: see also Heath, 1897; Dijksterhuis, 1956; Stamatis, 1970-74; Ver Eecke, 1921

Aristarchus, *On the Sizes and Distances of Sun and Moon*: see Heath, *Aristarchus of Samos*

Aristophanes, *Comoediae*, ed. F. W. Hall and W. M. Geldart, 2. ed., 2 vol., Oxford: Clarendon Press, 1906-07

Aristotle, *Opera*, ed. I. Bekker, et al., 5 vol., Berlin: Academia litterarum regia borussica ap. G. Reimerum, 1831-36, 1870 (repr. Berlin, 1960-61)

Aristotle: see Ross; Lee

Banû Mûsâ, *Verba filiorum*: see Clagett, *Archimedes in the Middle Ages*, I; al-Ṭûsî; Suter

al-Bîrûnî: see Suter; Schoy

Bobbio Mathematical Fragment: see *Mathematici Graeci Minores* (pp. 87-92); Huxley

Bibliography 383

CAG = *Commentaria in Aristotelem Graeca*, Berlin: Academia litterarum regia borussica ap. G. Reimerum: see Alexander of Aphrodisias; Ammonius; Joannes Philoponus; Olympiodorus; Simplicius; Themistius

Diocles, *On Burning Mirrors*: see Toomer

Dioscorides, *Materia Medica*, facsimile of the 6th cent. A.D. ms., *Cod. Vindobon. med. gr. 1*, Österreichische Nationalbibliothek, Vienna, 5 pts., with commentary volume, ed. H. Gerstinger, Graz: Akademische Druck- und Verlaganstalt, 1966-70

Euclid, *Opera*, ed. J. L. Heiberg and H. Menge, Leipzig, 9 vol. (I-IV: *Elementa*, 1883-85; V: *Scholia*, 1888; VI: *Data*, 1896; VII: *Optica, Catoptrica*, 1895; VIII: *Phaenomena, Scripta Musica*, 1916; IX: *Anaritii Commentaria*, 1899)

Euclid, *Elementa post I. L. Heiberg*, ed. E. Stamatis, 5 vol., Stuttgart: B. G. Teubner, 1969-77

Eutocius, *In Apollonium Commentaria*: see Apollonius, vol. II

Eutocius, *In Archimedem Commentaria*: see Archimedes, vol. III

Hero of Alexandria, *Opera*, Leipzig: B. G. Teubner, 5 vol. (I: *Pneumatica*, ed. W. Schmidt; II: *Mechanica*, ed. L. Nix; *Catoptrica*, ed. W. Schmidt; III: *Metrica*, ed. H. Schöne, 1903; IV: *Definitiones, Geometrica*, ed. J. L. Heiberg, 1912; V: *Stereometrica*, ed. J. L. Heiberg, 1914)

Herodotus, *Historiae*, ed. C. Hude, 3. ed., 2 vol., Oxford: Clarendon Press, 1927

Iamblichus, *In Nicomachi Introductionem Arithmeticam*, ed. H. Pistelli, Leipzig: B. G. Teubner, 1894

Joannes Philoponus, *In Aristotelis Analytica Posteriora Commentaria*, ed. M. Wallies, *CAG* XIII, pt. 3, 1909

Marinus, *In Euclidis Data*: see Euclid, vol. VI

Mathematici Graeci Minores, ed. J. L. Heiberg, (Kongelige Danske Videnskabernes Selskab, Historisk-filosofiske Meddelelser, 13.3), Copenhagen: A. F. Høst, 1927

al-Nairîzî, *Commentary on Euclid's Elements*, ms. Leiden, *Or. 399*, 1 (cf. edition by J. L. Heiberg and R. O. Besthorn et al., 1893-1932)

Nicomachus of Gerasa, *Introductio Arithmetica*, ed. R. Hoche, Leipzig: B. G. Teubner, 1866

Olympiodorus, *In Aristotelis Meteora Commentaria*, ed. W. Stüve, *CAG*, XII, pt. 2, Berlin, 1900

Pappus, *Collectionis quae supersunt*, ed. F. Hultsch, 3 vol., Berlin: Weidmann, 1876-78 (repr. Amsterdam: Adolf M. Hakkert, 1965)

Pappus, *The Commentary of Pappus on Book X of Euclid's Elements*, ed. G. Junge and W. Thomson, Cambridge, Mass.: Harvard University Press, 1930 (repr. New York/London: Johnson Reprint Co., 1968)

Pappus, *Commentary on Books V and VI of Ptolemy*: see Rome, vol. I

Pappus: see also Ver Eecke, 1933

Philoponus: see Joannes Philoponus

Plato, *Opera*, ed. J. Burnet, Oxford: Clarendon Press, 5 vol., 1899-1906

Plutarch, *Moralia*, ed. W. R. Paton, I. Wegehaupt et al., Leipzig: B. G. Teubner, 7 vol., 1925-67

Plutarch, *Vita Marcelli*, in *Vitae Parallelae*, II, pt. 2, ed. K. Ziegler, 2. ed., Leipzig: B. G. Teubner, 1968

Plutarch: see de Lacy and Einarson; Cherniss; Perrin

Presocratics: see Diels and Kranz

Proclus, *In Primum Euclidis Elementorum Librum Commentarii*, ed. G. Friedlein, Leipzig: B. G. Teubner, 1873

Proclus, *In Platonis Timaeum*, ed. E. Diehl, 3 vol., Leipzig: B. G. Teubner, 1903-06

Ptolemy, *Syntaxis Mathematica (Almagest)*, ed. J. L. Heiberg, 2 pts., Leipzig: B. G. Teubner, 1898-1903 (= Ptolemy, *Opera*, vol. I)

al-Sijzî, *On the Construction of the Heptagon*: see Hogendijk

Simplicius, *In Aristotelis Categorias*, ed. K. Kalbfleisch, *CAG* VIII, Berlin, 1907

Simplicius, *In Aristotelis de Caelo*, ed. J. L. Heiberg, *CAG* VII, Berlin, 1894
Simplicius, *In Aristotelis Physica*, ed. H. Diels, 2 vol., *CAG* IX-X, Berlin, 1882, 1895
Themistius, *In Aristotelis Physica*, ed. H. Schenkl, *CAG* V, pt. 2, Berlin, 1900
Theon of Alexandria, *Commentary on Books I-IV of Ptolemy*: see Rome, vol. II-III
Theon of Alexandria, Recension of Euclid's *Optics*: see Euclid, vol. VII
Theon of Smyrna, *Expositio rerum mathematicarum ad legendum Platonem utilium*, ed. E. Hiller, Leipzig: B. G. Teubner, 1878
Theophrastus, *Enquiry into Plants*, ed. A. Hort, (Loeb Classical Library), London: William Heinemann/New York: G. P. Putnam's Sons, 1916
al-Ṭūsī, Naṣīr al-Dīn, *Rasā'il*, 2 vol., Hyderabad: Osmania University, 1939-40
'Umar al-Khayyāmī: see Woepcke
Vitruvius, *De Architectura*, ed. V. Rose, 2. ed., Leipzig: B. G. Teubner, 1899

II. Modern works: Books

Allman, G. J., *Greek Geometry from Thales to Euclid*, Dublin/London: Hodges, Figgis for the Dublin University Press, 1889 (repr. New York: Arno Press, 1976)
Archibald, R. C., *The Cardioide and some of its Related Curves*, (diss.: Kaiser-Wilhelms-Universität Strassburg), Strassburg i. E.: Josef Singer, 1900
Balsam, H., *Des Apollonius von Perga sieben Bücher über Kegelschnitte*, Berlin: Georg Reimer, 1861 (German translation, with notes, of Apollonius' *Conics* from the edition of Halley)
Baron, M., *Origins of the Infinitesimal Calculus*, Oxford: Pergamon Press, 1969
Becker, O., *Das mathematische Denken der Antike*, 2. ed., Göttingen: Vandenhoek & Ruprecht, 1966 (1. ed., 1957)
Becker, O., *Mathematische Existenz*, Halle a. d. Salle, 1927 (Sonderdruck aus *Jahrbuch für Philosophie und phänomenologische Forschung*, 8)
Becker, O., *Zur Geschichte der griechischen Mathematik*, Darmstadt: Wissenschaftliche Buchgesellschaft, 1965 (includes reprints of contributions by H. Hankel, H. G. Zeuthen, J. E. Hofmann, A. D. Steele, et al.)
Beckmann, P., *A History of Pi*, 2. ed., New York: St. Martin's Press, 1971
Bieberbach, L., *Theorie der geometrischen Konstruktionen*, Basel: Birkhäuser, 1952
Bluck, R. S., *Plato's Meno*, Cambridge: Cambridge University Press, 1961
Bourbaki, N. (pseudonym), *Éléments d'histoire des mathématiques*, Paris: Hermann, 1969
Boyer, C., *The History of the Calculus and its Conceptual Development*, New York: Dover, 1959 (repr. of 1949 ed.)
Boyer, C., *The Rainbow from Myth to Mathematics*, New York: T. Yoseloff, 1959
Breidenbach, W., *Das delische Problem*, Stuttgart: B. G. Teubner, 1953
Brown, P., *The World of Late Antiquity*, London: Thames & Hudson, 1971
Bruins, E. M., *Codex Constantinopolitanus*, 3 vol., Leiden: E. J. Brill, 1964 (facsimile with text, English translation and commentary)
Burkert, W., *Lore and Science in Ancient Pythagoreanism*, Cambridge, Mass.: Harvard University Press, 1972 (trans. of *Weisheit und Wissenschaft*, Nuremberg: Hans Carl, 1962)
Burnet, J., *Early Greek Philosophy*, 4. ed., London: Macmillan, 1930 (repr., New York: The World Publ. Co., 1957; 1. ed., London, 1892)
Buschor, E., *Griechische Vasen*, new ed., Munich: R. Piper & Co., 1969
Cavalieri, B., *Geometria indivisibilibus continuorum nova quadam ratione promota*, Bologna, 1635; 1653 (trans. as *Geometria degli Indivisibili*, L. Lombardi-Radice, Turin, 1966)
Chasles, M., *Aperçu historique des méthodes en géométrie*, Paris: Gauthier-Villars et fils, 3. ed., 1889 (2. ed., 1875; 1. ed., 1837)

Chasles, M., *Les Trois livres des Porismes d'Euclide*, Paris: Mallet-Bachelier, 1860
Cherniss, H., ed., *Plutarch's Moralia* (1033a-1086b), vol. 13, pt. 2, (Loeb Classical Library), Cambridge, Mass.: Harvard University Press/London: William Heinemann, 1976
Clagett, M., *Archimedes in the Middle Ages*, I: *The Arabo-Latin Tradition*, Madison, Wisc.: The University of Wisconsin Press, 1964; II: *The Translations from the Greek by William of Moerbeke*, Philadelphia: American Philosophical Society, 1976; III: *The Fate of the Medieval Archimedes*, Philadelphia, 1978; IV: *A Supplement on the Medieval Latin Traditions of Conic Sections*, Philadelphia, 1980
Clagett, M., *The Science of Mechanics in the Middle Ages*, Madison, Wisc.: The University of Wisconsin Press, 1959
Coolidge, J. L., *A History of the Conic Sections and Quadric Surfaces*, Oxford: Oxford University Press, 1945 (repr., New York: Dover, 1968)
Coxeter, H. M. S., *Projective Geometry*, New York: Blaisdell, 1964
Cremona, L., *Elements of Projective Geometry*, Oxford: At the Clarendon Press, 1893 (2. English ed.; 1. Italian ed., 1876)
Desargues, G.: see Taton
Descartes, R., *La Géométrie*, Paris, 1637 (facsimile with English translation by D. E. Smith and M. L. Latham, 1925; repr., New York: Dover, 1954)
Diels, H. and W. Kranz, *Die Fragmente der Vorsokratiker*, 3 vol., Dublin/Zurich: Weidmann, 1966 (repr. of the 6. ed., Berlin: Weidmann, 1951-52)
Dijksterhuis, E. J., *Archimedes*, Copenhagen: Munksgaard, (Acta historica scientiarum naturalium et medicinalium, 12), 1956 (also New York: The Humanities Press, 1957; trans. from the Dutch edition, Groningen: P. Noordhoff, comprising essays of 1938-44)
Dilke, O. A. W., *The Roman Land Surveyors: An Introduction to the Agrimensores*, Newton Abbot (Devon): David & Charles, 1971
Drachmann, A., *Ktesibios, Philon and Heron*, Copenhagen: Munksgaard, (Acta historica scientiarum naturalium et medicinalium, 4), 1948
Drachmann, A., *The Mechanical Technology of Greek and Roman Antiquity*, Copenhagen: Munksgaard, (Acta historica scientiarum naturalium et medicinalium, 17), 1963
Dreyer, J. L. E., *A History of Astronomy from Thales to Kepler*, 2. ed., New York: Dover, 1953 (rev. repr. of the 1906 ed.)
DSB = *Dictionary of Scientific Biography*: see Gillispie
Enriques, F., *Fragen der Elementargeometrie*, 2. Germ. ed., 2 vol., Leipzig/Berlin: B. G. Teubner, 1923 (trans. of *Questioni riguardanti la geometria elementare*, 2. ed., Bologna, 1912-14)
Finley, M. I., *Ancient Sicily*, New York: Viking Press, 1968
Fishback, W. T., *Projective and Euclidean Geometry*, 2 ed., New York: John Wiley and Sons, 1969
Fitzgerald, A., *The Letters of Synesius of Cyrene*, London: H. Milford for Oxford University Press, 1926
Freeman, K., *The Pre-Socratic Philosophers: A Companion to Diels, Fragmente der Vorsokratiker*, Oxford: B. Blackwell, 1946 (2. ed., Cambridge, Mass.: Harvard University Press, 1952)
Gillings, R., *Mathematics in the Time of the Pharaohs*, Cambridge, Mass.: The MIT Press, 1972
Gillispie, C. C., ed., *Dictionary of Scientific Biography*, 15 vol., New York: Charles Scribner's Sons, 1970-78
Guthrie, W. K. C., *A History of Greek Philosophy*, Cambridge: Cambridge University Press, I: *The Earlier Presocratics and the Pythagoreans*, 1962; II: *The Presocratic Tradition from Parmenides to Democritus*, 1965; III: *The Fifth-Century Enlightenment*, 1969

Halley, E., *Apollonii Pergaei Conicorum libri octo*..., Oxford: Theatrum Sheldonianum, 1710

Hankel, H., *Beiträge zur Geschichte der Mathematik im Alterthum und Mittelalter*, Leipzig: B. G. Teubner, 1874 (2. ed., ed. J. E. Hofmann, Hildesheim: G. Olm, 1965; chapter: "Die Mathematiker im 5. Jahrhundert," reprinted in O. Becker, 1965, pp. 1-17)

Heath, T. L., *Apollonius of Perga: Treatise on Conic Sections, edited in modern notation*..., Cambridge: Cambridge University Press, 1896 (based on Heiberg's edition of 1891-93; repr. Cambridge: W. Heffer & Sons, 1961)

Heath, T. L., *Aristarchus of Samos: The Ancient Copernicus*, Oxford: At the Clarendon Press, 1913

Heath, T. L., *A History of Greek Mathematics*, 2 vol., Oxford: At the Clarendon Press, 1921

Heath, T. L., *Mathematics in Aristotle*, Oxford: At the Clarendon Press, 1949

Heath, T. L., *The Thirteen Books of Euclid's Elements*, 2. ed., 3 vol., Cambridge: Cambridge University Press, 1926 (based on the edition of Heiberg, 1883-85)

Heath, T. L., *The Works of Archimedes...with a Supplement [,] The Method of Archimedes*..., New York: Dover, n.d. (repr. of the Cambridge University Press edition of 1897 with the supplement of 1912; based on Heiberg's first edition of Archimedes, 1880-81, and his first edition of the *Method*, 1907)

Heiberg, J. L., *Geschichte der Mathematik und Naturwissenschaften im Altertum*, Munich: C. H. Beck, 1925

Heiberg, J. L., *Philologische Studien zu griechischen Mathematikern*, Leipzig, 1880 (extracts from the *Jahrbücher für classische Philologie*, Supplementband 11, pp. 355-398; 12, pp. 377-402; 13, pp. 543-577)

Heiberg, J. L., *Quaestiones Archimedeae*, Copenhagen: Rudolphus Kleinius, 1879

Hintikka, J. and U. Remes, *The Method of Analysis: Its Geometrical Origin and its General Significance*, Dordrecht, Neth.: D. Reidel, 1974

Hobson, E. W. et al., *Squaring the Circle and Other Monographs*, New York: Chelsea Publ. Co., 1953 (repr. of essays by E. W. Hobson, H. P. Hudson, et al.)

Huxley, G., *Anthemius of Tralles: A Study in Later Greek Geometry*, (Greek, Roman and Byzantine Studies, monograph no. 1), Cambridge, Mass., 1959

Ito, S., ed., *The Medieval Latin Translation of the Data of Euclid*, Tokyo: University of Tokyo Press, and Boston/Basel/Stuttgart: Birkhäuser, 1980

Iushkevich, A. P.: see Youschkevitch

Karamazou, S., *The Amasis Painter*, Oxford: At the Clarendon Press, 1956

Kline, M., *Mathematical Thought from Ancient to Modern Times*, Oxford/New York: Oxford University Press, 1972

Knorr, W. R., *Ancient Sources of the Medieval Tradition of Mechanics*, (Monograph no. 6 of the *Annali* of the Istituto e Museo di Storia della Scienza), Florence, 1982

Knorr, W. R., *The Evolution of the Euclidean Elements*, Dordrecht, Neth.: D. Reidel, 1975

Kretzmann, N., ed., *Infinity and Continuity in Ancient and Medieval Thought*, Ithaca, N.Y.: Cornell University Press, 1982

de Lacey, P. H. and B. Einarson, ed., *Plutarch's Moralia*, vol. 7, (Loeb Classical Library), Cambridge, Mass.: Harvard University Press/London: William Heinemann, 1959

Lee, H. D. P., *Aristotle: Meteorologica*, (Loeb Classical Library), Cambridge, Mass.: Harvard University Press/London: William Heinemann, 1952

Loria, G., *Spezielle algebraische und transcendente ebene Kurven: Theorie und Geschichte*, 2. Germ. ed., Leipzig/Berlin: B. G. Teubner, 1910-11 (1. Ital. ed., 1902; 1. Germ. ed., 1902; 2. Ital. ed., 1909)

Loria, G., *Le scienze esatte nell' antica Grecia*, Milan: Ulrico Hoepli, 1914

Mahoney, M. S., *The Mathematical Career of Pierre de Fermat*, Princeton, N.J.: Princeton University Press, 1973
Marsden, E. W., *Greek and Roman Artillery: Technical Treatises*, Oxford: At the Clarendon Press, 1971
Mau, J., *Zum Problem des Infinitesimalen bei den antiken Atomisten*, Berlin: Akademie Verlag, 1954
Meserve, B. E., *Fundamental Concepts of Geometry*, Reading, Mass.: Addison-Wesley, 1955
Mogenet, J., *l'Introduction à l'Almageste*, Brussels: Académie royale de Belgique, (*Mémoires*, Classe des lettres et des sciences morales et politiques, 51, fasc. 2), 1956
Montucla, E., *Histoire des recherches sur la quadrature du cercle*, Paris: Ch. Ant. Jombert, 1754 (2. ed., Paris: Bachelier père et fils, 1831)
Morrow, G., *Proclus: A Commentary on the First Book of Euclid's Elements*, Princeton, N.J.: Princeton University Press, 1970
Mueller, I., *Philosophy of Mathematics and Deductive Structure in Euclid's Elements*, Cambridge, Mass.: The MIT Press, 1981
Neugebauer, O., *The Exact Sciences in Antiquity*, 2. ed., Providence, R.I.: Brown University Press, 1957
Neugebauer, O., *A History of Ancient Mathematical Astronomy*, 3 vol., Berlin/Heidelberg/New York: Springer-Verlag, 1975
Neugebauer, O. and A. Sachs, *Mathematical Cuneiform Texts*, New Haven, Conn.: American Oriental Society and American Schools of Oriental Research, 1945
Neugebauer, O., *Mathematische Keilschrift-Texte*, 3 vol. (*Quellen und Studien zur Geschichte der Mathematik, Astronomie und Physik*, Abt. A: 3), Berlin: Springer-Verlag, 1935-37
Neugebauer, O., *Vorlesungen über Geschichte der antiken mathematischen Wissenschaften*, Berlin: J. Springer, 1934
Niebel, E., *Untersuchungen über die Bedeutung der geometrischen Konstruktionen in der Antike*, (*Kant-Studien*, Ergänzungsheft 76), Cologne, 1959
Orlandos, A., *Les Matériaux de construction et la technique architecturale des anciens grecs*, (École française d'Athènes: *Travaux et Mémoires*, 16, pt. 2), Athens, 1968
Oxford Classical Dictionary, 2. ed., ed. N. G. L. Hammond and H. H. Scullard, Oxford: Clarendon Press, 1970
Parker, R. A., *Demotic Mathematical Papyri*, Providence, R.I.: Brown University Press, 1972
Perrin, B., *Plutarch's Lives*, vol. 5 (incl. the *Life of Marcellus*), (Loeb Classical Library), London: William Heinemann/New York: Macmillan, 1917
PW = *Paulys Real-Encyclopädie*: see Wissowa
Reimer, N. T., *Historia problematis de cubi duplicatione*, Göttingen: Joann. Christian. Dieterich, 1798
de Roberval, G., *Observations sur la composition des mouvemens et sur le moyen de trouver les touchantes des lignes courbes* (1668), repr. in the *Mémoires de l'Académie royale des sciences*, Paris, 6, 1730
Rome, A., ed., *Commentaires de Pappus et de Théon d'Alexandrie sur l'Almageste*, 3 vol., (*Studi e Testi*, vol. 54 and 72, 106), Rome and Vatican City: Bibliotheca Apostolica Vaticana, 1931 and 1936/43
Ross, W. D., ed., *Aristotle: de Anima*, Oxford: Clarendon Press, 1961
Ross, W. D., ed., *Aristotle: Physics*, Oxford: Clarendon Press, 1936
Rudio, F., *Der Bericht des Simplicius über die Quadraturen des Antiphon und des Hippocrates*, Leipzig: B. G. Teubner, 1907
Schneider, C. K., *Illustriertes Handbuch der Laubholzkunde*, Jena: Gustav Fischer, 1912
Schneider, I., *Archimedes: Ingenieur, Naturwissenschaftler und Mathematiker*, Darmstadt: Wissenschaftliche Buchgesellschaft, 1979

Schoy, C., *Die trigonometrischen Lehren des...al-Bîrûnî*, Hanover: Orient-Buchhandlung Heinz Lafaire K.-G., 1927

Sezgin, F., *Geschichte des arabischen Schrifttums*, V: *Mathematik*, Leiden: E. J. Brill, 1974

Shorey, P., *Plato: The Republic*, 2 vol., (Loeb Classical Library), Cambridge, Mass.: Harvard University Press/London: William Heinemann, 1935

Solmsen, F., *Aristotle's System of the Physical World*, Ithaca, N.Y.: Cornell University Press, 1960

Stamatis, E., *Archimêdous Hapanta*, 3 vol. (in 4), Athens: Technikou Epimelêtêriou tês Hellados, 1970-74

Struik, D. J., *Analytic and Projective Geometry*, Reading, Mass.: Addison-Wesley, 1953

Szabó, Á., *Die Anfänge der griechischen Mathematik*, Budapest: Akadémiai Kiadó and Munich/Vienna: R. Oldenbourg, 1969 (Engl. trans. by A. M. Ungar as *The Beginnings of Greek Mathematics*, Dordrecht, Neth.: D. Reidel, 1978)

Tannery, P., *La Géométrie grecque, comment son histoire nous est parvenue et ce que nous en savons*, Paris: Gauthier-Villars, 1887

Tannery, P., *Mémoires scientifiques*, ed. J. L. Heiberg and H. G. Zeuthen, Paris: Gauthier-Villars/Toulouse: Édouard Privat, 17 vol., 1912-50

Taton, R., *l'Oeuvre mathématique de G. Desargues*, Paris: Presses Universitaires de France, 1951

Thomas, I., *Greek Mathematical Works*, 2 vol., (Loeb Classical Library), Cambridge, Mass.: Harvard University Press/London: William Heinemann, 1957 (repr. of edition of 1939-41)

Thomson, W.: see Pappus, *The Commentary...on Book X*

Toomer, G. J., ed., *Diocles: On Burning Mirrors*, Berlin/Heidelberg/New York: Springer-Verlag, 1976

Ver Eecke, P., *Les Coniques d'Apollonius de Perge*, Bruges: Desclée, De Brouwer et Cie, 1924

Ver Eecke, P., *Pappus: La Collection mathématique*, 2 vol., Paris/Bruges: Desclée, De Brouwer et Cie, 1933

Ver Eecke, P., *Les Oeuvres complètes d'Archimède*, Paris/Brussels: Desclée, De Brouwer et Cie, 1921 (2. ed., 2 vol., Paris, 1960)

Vogel, K., *Vorgriechische Mathematik*, 2 vol., Hanover: H. Schroedel, 1958-59

van der Waerden, B. L., *Science Awakening*, Groningen, Neth.: P. Noordhoff, 1954 (repr., New York: John Wiley & Sons, 1963; 1. Dutch ed., 1951)

Waschkiess, H. J., *Von Eudoxos zu Aristoteles*, Amsterdam: B. R. Grüner, 1977

Whittaker, E. T. and G. Robinson, *Calculus of Observations*, 4. ed., London: Blackie & Son, 1944 (repr. New York: Dover, 1967)

Wieleitner, H., *Spezielle Ebene Kurven*, Leipzig: G. J. Göschen, 1908

Wissowa, G. et al., ed., *Paulys Real-Encyclopädie der classischen Altertumswissenschaft*, new ed., 1. ser., 24 vol. in 47 pts., Stuttgart: J. B. Metzler, 1894-1963; 2. ser., 10 vol. in 19 pts., 1914-72; Suppl. I-XIV, 1903-74

Woepcke, F., *l'Algèbre d'Omar Alkhayyâmî*, Paris: Benjamin Duprat, 1851

Young, J. W. A., ed., *Monographs on Topics of Modern Mathematics*, New York: Longmans, Green and Co., 1911 (repr. New York: Dover, 1955)

Youschkevitch, A. P., *Les Mathématiques arabes (VIIIe-XVe siècles)*, Paris: Librairie philosophique J. Vrin, 1976 (trans. from the Germ. ed. of 1964, with additional references and notes)

Zeuthen, H. G., *Geschichte der Mathematik im Altertum und Mittelalter*, Copenhagen: A. F. Høst, 1896 (chs. 4 and 14 reprinted in Becker, 1965, pp. 18-44)

Zeuthen, H. G., *Die Lehre von den Kegelschnitten im Altertum*, Copenhagen: R. v. Fischer-Benzon, 1886 (repr. Hildesheim: G. Olm, 1966)

Zwikker, C., *Advanced Plane Geometry*, Amsterdam: North-Holland, 1950 (repr. New York: Dover, 1963)

III. Modern works: Articles

Barnes, J., "Aristotle, Menaechmus, and Circular Proof," *Classical Quarterly*, 1976, 26 (N.S.), pp. 278-292.
Becker, O., "Eudoxos-Studien I: Eine voreudoxische Proportionenlehre und ihre Spuren bei Aristoteles und Euklid," *Quellen und Studien*, 1933, 2:B, pp. 311-333
Becker, O., "Eudoxos-Studien II: Warum haben die Griechen die Existenz der vierten Proportionale angenommen?" *Quellen und Studien*, 1933, 2:B, pp. 369-387
Becker, O., "Eudoxos-Studien III: Spuren eines Stetigkeitsaxioms in der Art des Dedekind'schen zur Zeit des Eudoxos," *Quellen und Studien*, 1936, 3:B, pp. 236-244
Becker, O., "Eudoxos-Studien IV: Das Prinzip des ausgeschlossenen Dritten in der griechischen Mathematik," *Quellen und Studien*, 1936, 3:B, pp. 370-388
Becker, O., "Eudoxos-Studien V: Die eudoxische Lehre von den Ideen und den Farben," *Quellen und Studien*, 1936, 3:B, pp. 389-410
Björnbo, A. A., "Hippias (13) aus Elis," in *PW*, VIII, pt. 2, 1913, coll. 1706-11
Björnbo, A. A., "Hippokrates (14) aus Chios," in *PW*, VIII, pt. 2, 1913, coll., 1780-1801
Böker, R., "*Neusis*," in *PW*, suppl. IX, 1962, coll. 415-461
Böker, R., "Winkel- und Kreisteilung," in *PW*, ser. II, 9, 1961, coll. 127-150
Böker, R., "Würfelverdoppelung," in *PW*, ser. II, 9, 1961, coll. 1193-1223
Brown, M., "Pappus, Plato and the Harmonic Mean," *Phronesis*, 1975, 20, pp. 173-184
Bulmer-Thomas, I., "Oenopides of Chios," in *DSB*, X, 1974, pp. 179-182
Bulmer-Thomas, I., "Hippocrates of Chios," in *DSB*, VI, 1972, pp. 410-418
Busard, H. L. L., "Der Traktat de Isoperimetris," *Mediaeval Studies*, 1980, 42, pp. 61-88
Clausen, T., "Vier neue mondförmige Flächen, deren Inhalt quadrierbar ist," *Journal für die reine und angewandte Mathematik (Crelle)*, 1840, 21, pp. 375-376
Dicks, D. R., "Eratosthenes," in *DSB*, IV, 1971, pp. 388-393
Dickson, L. E., "Constructions with Ruler and Compasses; Regular Polygons," in Young, 1911, pp. 351-386
Drachmann, A., "Fragments from Archimedes in Heron's Mechanics," *Centaurus*, 1963, 8, pp. 91-146
Engels, H., "Quadrature of the Circle in Ancient Egypt," *Historia Mathematica*, 1977, 4, pp. 137-140
Fischler, R., "A Remark on Euclid II, 11," *Historia Mathematica*, 1979, 6, pp. 418-422
Fowler, D., "Archimedes' Cattle Problem and the Pocket Calculating Machine," University of Warwick, Mathematics Institute, preprint, 1981
Fowler, D., "Ratio in Early Greek Mathematics," *Bulletin of the American Mathematical Society*, 1979, 1 (n.s.), pp. 807-846
Freudenthal, H., "Y avait-il une crise des fondements des mathématiques dans l'antiquité?" *Bulletin de la Société mathématique de Belgique*, 1966, 18, pp. 43-55
von Fritz, K., "Oenopides aus Chios," in *PW*, 17, pt. 2, 1937, coll. 2258-2272 (repr. in *Schriften zur griechischen Logik*, Stuttgart, 1978, 2, pp. 149-161)
Goldstein, B. R. and A. C. Bowen, "A New View of Early Greek Astronomy," *Isis*, 1983, 74, pp. 330-340.
Hasse, H. and H. Scholz, "Die Grundlagenkrisis der griechischen Mathematik," *Kant-Studien*, 1928, 33, pp. 1-27
Heiberg, J. L., "Die Kenntnisse des Archimedes über die Kegelschnitte," *Zeitschrift für Mathematik und Physik*, hist.-litt. Abth., 25, 1880, pp. 41-67
Heiberg, J. L., "Ueber Eutokios," in Heiberg, *Philologische Studien*, I, pp. 357-384
Heiberg, J. L., "Zum Fragmentum mathematicum Bobiense," *Zeitschrift für Mathematik und Physik*, hist.-litt. Abth., 1883, 28, pp. 121-129

Hobson, E. W., "Squaring the Circle," in Hobson et al., 1953
Hofmann, J. E., "Über die Annäherung von Quadratwurzeln bei Archimedes und Heron," *Jahresbericht der Deutschen Mathematiker-Vereinigung*, 1934, 43, pp. 187-210 (repr. in Becker, 1965, pp. 100-124)
Hogendijk, J. P., "Greek and Arabic Constructions of the Regular Heptagon," University of Utrecht Mathematics Department, preprint no. 236, April 1982 (revised version in *Archive for History of Exact Sciences*, 1984, 30, pp. 197-330)
Hudson, H. P., "Ruler and Compass," in Hobson et al., 1953
Huxley, G., "Ôkytokion," *Greek, Roman and Byzantine Studies*, 1967, 8, pp. 203-204
Knaack, G., "Erastosthenes (4) aus Kyrene," in *PW*, 6, pt. 1, 1909, coll. 358-389
Knorr, W. R., "Archimedes' Lost Treatise on the Centers of Gravity of Solids," *Mathematical Intelligencer*, 1978, 1, pp. 102-109
Knorr, W. R., "Archimedes' *Neusis*-Constructions in *Spiral Lines*," *Centaurus*, 1978, 22, pp. 77-98
Knorr, W. R., "Archimedes and the *Elements*: Proposal for a Revised Chronological Ordering of the Archimedean Corpus," *Archive for History of Exact Sciences*, 1978, 19, pp. 211-290
Knorr, W. R., "Archimedes and the Measurement of the Circle: A New Interpretation," *Archive for History of Exact Sciences*, 1976, 15, pp. 115-140
Knorr, W. R., "Archimedes and the Pre-Euclidean Proportion Theory," *Archives internationales d'histoire des sciences*, 1978, 28, pp. 183-244
Knorr, W. R., "Archimedes and the Spirals: The Heuristic Background," *Historia Mathematica*, 1978, 5, pp. 43-75
Knorr, W. R., "La Croix des mathématiciens: The Euclidean Theory of Irrational Lines," *Bulletin of the American Mathematical Society*, 1983, 9(N.S.), pp. 41-69
Knorr, W. R., "The Geometry of Burning-Mirrors in Antiquity," *Isis*, 1983, 74, pp. 53-73
Knorr, W. R., "The Hyperbola-Construction in the *Conics*, Book II: Variations on a Theorem of Apollonius," *Centaurus*, 1982, 25, pp. 253-291
Knorr, W. R., "Infinity and Continuity: The Interaction of Mathematics and Philosophy in Antiquity," in N. Kretzmann, ed., *Infinity and Continuity*, 1982, pp. 112-145
Knorr, W. R., "Observations on the Early History of the Conics," *Centaurus*, 1982, 26, pp. 1-24
Knorr, W. R., "On the Early History of Axiomatics: The Interaction of Mathematics and Philosophy in Greek Antiquity," in J. Hintikka et al., ed., *Pisa Conference Proceedings*, Dordrecht, 1981, I, pp. 145-186
Knorr, W. R., "Techniques of Fractions in Ancient Egypt and Greece," *Historia Mathematica*, 1982, 9, pp. 133-171
Landau, E., "Über quadrierbare Kreisbogenzweiecke," *Sitzungsberichte, Berliner Mathematische Gesellschaft*, 1902-03, pp. 1-6 (supplement to *Archiv für Mathematik und Physik*, 4)
Loria, G., "Sopra una relazione che passa fra due antiche soluzioni del problema di Delo," *Bibliotheca Mathematica*, 1910/11, 11, ser. 3, pp. 97-99
Luria, S., "Die Infinitesimaltheorie der antiken Atomisten," *Quellen und Studien*, 1933, 2:B, pp. 106-185
Mahoney, M. S., "Another Look at Greek Mathematical Analysis," *Archive for History of Exact Sciences*, 1968/69, 5, pp. 319-348
Mueller, I., "Aristotle and the Quadrature of the Circle," in N. Kretzmann, ed., *Infinity and Continuity*, 1982, pp. 146-164
Nelson, H. L., Remarks on Archimedes' 'Cattle Problem', as reported in *Scientific American*, June 1981, p. 84
Neuenschwander, E., "Die ersten vier Bücher der Elemente Euklids," *Archive for History of Exact Sciences*, 1973, 9, pp. 325-380

Neuenschwander, E., "Die stereometrischen Bücher der Elemente Euklids," *Archive for History of Exact Sciences*, 1974/75, 14, pp. 91-125
Neugebauer, O., "Apollonius-Studien," *Quellen und Studien*, 1933, 2:B, pp. 215-254
Neugebauer, O., "The Equivalence of Eccentric and Epicyclic Motion according to Apollonius," *Scripta Mathematica*, 1959, 24, pp. 5-21
Neugebauer, O., "On the Astronomical Origin of the Theory of Conic Sections," *Proceedings of the American Philosophical Society*, 1948, 92, pp. 136-138
Philippson, R., "Philonides (5)" in *PW*, 20, pt. 1, 1941, coll. 63-73
Riddell, R. C., "Eudoxan Mathematics and the Eudoxan Spheres," *Archive for History of Exact Sciences*, 1979, 20, pp. 1-19
Rudio, F., "Der Bericht des Simplicius über die Quadraturen des Antiphon und des Hippokrates," *Bibliotheca Mathematica*, 1902/03, 3, ser. 3, pp. 7-62
Schoy, C., "Graeco-Arabische Studien," *Isis*, 1926, 8, pp. 21-40
Seidenberg, A., "Did Euclid's *Elements*, Book I, Develop Geometry Axiomatically?" *Archive for History of Exact Sciences*, 1975, 14, pp. 263-295
Seidenberg, A., "Origin of Mathematics," *Archive for History of Exact Sciences*, 1976, 18, pp. 301-342
Seidenberg, A., "The Ritual Origin of Geometry," *Archive for History of Exact Sciences*, 1963, 1, pp. 488-527
Seidenberg, A., "Some Remarks on Nicomedes' Duplication," *Archive for History of Exact Sciences*, 1966, 3, pp. 97-101
Simon, M., "Lunulae Hippocratis," *Archiv für Mathematik und Physik*, 1904/05, 8, p. 269
Smith, D. E., "History and Transcendence of Pi," in Young, 1911, pp. 387-416
Soedel, W. and V. Foley, "Ancient Catapults," *Scientific American*, March 1979, 240, pp. 150-160
Steele, A. D., "Über die Rolle von Zirkel und Lineal in der griechischen Mathematik," *Quellen und Studien*, 1936, 3:B, pp. 287-369 (Part II, pp. 313-369, repr. in Becker, 1965, pp. 146-202)
Stolz, O., "Zur Geometrie der Alten, insbesondere über ein Axiom des Archimedes," *Mathematische Annalen*, 1883, 22, pp. 504-519
Suter, H., "Das Buch der Auffindung der Sehnen im Kreise von...el-Bîrûnî," *Bibliotheca Mathematica*, 1910/11, 11, ser. 3, pp. 11-78
Suter, H., "Über die Geometrie der Söhne des Mûsâ ben Schâkir," *Bibliotheca Mathematica*, 1902, 3, ser. 3, pp. 259-272
Tannery, P., "Eutocius et ses contemporains," *Bulletin des sciences mathématiques et astronomiques*, 1884, 19 (=8, ser.2), pp. 315-329 (in *Mémoires scientifiques*, II, 1912, no. 36, pp. 118-136)
Tannery, P., "Le fragment d'Eudème sur la quadrature des lunules," *Mémoires de la Société des sciences physiques et naturelles de Bordeaux*, 1883, 2. ser., V, pp. 217-237 (in *Mémoires scientifiques*, I, 1912, no. 25, pp. 339-370)
Tannery, P., "Pour l'histoire des lignes et surfaces courbes dans l'antiquité," *Bulletin des sciences mathématiques et astronomiques*, 1883, 18 (=7, ser. 2), pp. 278-291; 1884, 19 (=8, ser. 2), pp. 19-30, 101-112 (in *Mémoires scientifiques*, II, 1912, no. 30, pp. 1-47)
Tannery, P., "Sur les solutions du problème de Delos par Archytas et Eudoxe: Divination d'une solution perdue," *Mémoires de la Soc. sci. phys. nat.*, Bordeaux, 1878, 2. ser., II, pp. 277-283 (in *Mémoires scientifiques*, I, 1912, no. 5, pp. 53-61)
Thaer, C., "Die Würfelverdoppelung des Apollonios," *Deutsche Mathematik*, 1940, 5, pp. 241-243.
Toomer, G. J., "Apollonius," in *DSB*, I, 1970, pp. 179-193
Toomer, G. J., "The Chord Table of Hipparchus and the Early History of Greek Trigonometry," *Centaurus*, 18, 1973, pp. 6-28

Toomer, G. J., "The Mathematician Zenodorus," *Greek, Roman and Byzantine Studies*, 1972, 13, pp. 177-192

Tóth, I., "Das Parallelenproblem im Corpus Aristotelicum," *Archive for History of Exact Sciences*, 1967, 3, pp. 249-422

Tropfke, J., "Die Siebeneckhandlung des Archimedes," *Osiris*, 1936, 1, pp. 636-651

Tschakaloff, L., "Beitrag zum Problem der quadrierbaren Kreisbogenzweiecke," *Mathematische Zeitschrift*, 1929, 30, pp. 552-559

Tschebotaröw, N., "Über quadrierbare Kreisbogenzweiecke. I," *Mathematische Zeitschrift*, 1935, 39, pp. 161-175

Unguru, S., "History of Ancient Mathematics: Some Reflections on the State of the Art," *Isis*, 1979, 70, pp. 555-565

Vogel, K., "Aristaeus," in *DSB*, I, 1970, pp. 245-246

van der Waerden, B. L., "Pre-Babylonian Mathematics. I-II," *Archive for History of Exact Sciences*, 1980, 23, pp. 1-46

van der Waerden, B. L., "Pythagoreer-1.D. Pythagoreische Wissenschaft," in *PW*, 24, 1963, coll. 277-300

van der Waerden, B. L., "Zenon und die Grundlagenkrise der griechischen Mathematik," *Mathematische Annalen*, 1940/41, 117, pp. 141-161

Wantzel, L., "Recherches sur les moyens de reconnaître si un Problème de Géométrie peut se résoudre avec la règle et le compas," *Journal de mathématiques pures et appliquées (Liouville)*, 1837, 2, pp. 366-372

Wieleitner, H. and J. E. Hofmann, "Zur Geschichte der quadrierbaren Kreismonde," *Wissenschaftliche Beiläge*, Realgymnasium München, 1933/34

von Wilamowitz-Moellendorf, U., "Ein Weihgeschenk des Eratosthenes," *Göttinger Nachrichten*, Phil.-hist. Kl., 1894, pp. 15-35 (repr. in *Kleine Schriften*, Berlin/Amsterdam, 1971, II, pp. 48-70)

Zeuthen, "Die geometrische Construction als 'Existenzbeweis' in der antiken Geometrie," *Mathematische Annalen*, 1896, 47, pp. 222-228

Indices

The indices are separated according to names (A), subjects (B) and passages (C). In each case, references are to the pages where the specified items are discussed. An entry bearing the suffix "n" followed by a numeral refers to the note with that number appearing on the indicated page. The terms "*v.*" (for *vide*) and "*v. also*" denote cross-references. When the letter "*A*," "*B*" or "*C*," is attached, the entry of the cross-reference appears in the respective *Index A, B* or *C*.

Index A: Names

Entries are to individuals, traditions of learning (e.g., "Egyptian mathematics") and institutions (e.g., "Academy"). Works no longer extant are listed in subheadings to authors in this index (e.g., Eratosthenes, *On Means*), but entries for extant works appear in *Index C*. Certain authors (in particular, T. L. Heath, B. L. van der Waerden, and H. G. Zeuthen) are cited frequently in the notes, but only substantive discussions are indexed.

Abu 'l-Jūd 181-182, 204n97
Abū Bakr 308, 332n40
Abū Jaʿfar 380n132
Académie française 373n44
Academy: *v.* Plato, and associates in Academy
Aḥmad ibn Mūsā 331n28, 379n110
al-ʿAlā ibn Sahl 182-183
Albert of Saxony 200n38
Alexander of Aphrodisias 7, 28-31, 34-35, 38-39, 45n53, 76, 79, 95n67-68, 106,
141n18, 379n115
[Alexander] 95n68
Alhazen (ibn al-Haytham) 46n66, 183-184, 205n104, 285n99, 286n101
Allman, G. J. 46n68, 97n94
Amasis painter 249, 250-251, 253-254
Ammonius 40, 48n93, 76, 95n67, 362
Amphinomus 71, 74-75, 94n61, 95n63, 351, 374n62
Amthor, A. 294

Anaxagoras 26, 29, 45n62
Anon., *Book of Chords* 205n112
Anon. isoperimetric writer 290n166
Anon. scholiast to Euclid, *El.* XIII 354-355, 358-359, 375n78, 376n85
Anthemius 63, 237-239, 285n93, n98-99, n101, 321, 327, 336n80, 338n110
Antiochus IV (king) of Syria 275
Antiphon 27-29, 45n53, 81, 87, 95n69
Apollonius 43n20, 283n61, ch. 7 (*passim*)
analytic works of 213, 293, 313, 328-329, 340, 344, 346, 354, 360, 369, 378n105
on angle trisection 305, 324
and Archimedes 294-297, 313, 363
and Aristaeus 321-329
on astronomy 308-311, 332n47
on asymptotes to conics 224
and commentators *v.* Pappus, and Apollonius; Eutocius, and Apollonius
on conic sections 114, 136, 335n63, 364
conic theory, of: predecessors 111, 113, 143n44, 297-298
conic theory, of: terminology 62, 187, 304, 324, 326-328, 337n87
and contemporaries 218, 234, 282n35, 302, 329
on cube duplication 188, 190-191, 219, 282n40, 305-308, 332n39, n43, 347, 366
dating of 274-276, 291n176-177
and Euclid 116, 138, 313-321
and Eudoxean geometry 195, 340
on foci of conics 145n72, 205n98, 284n80
formal style of 213, 335n73
on hyperbola construction 304, 324
on irrationals 370n5
on loci 343, 348
on locus (two-point) 103-108, 139n9
on *neusis* 144n67, 218, 346, 366
and Nicomedes 303-312
on normals to conics 173, 315-320
on normals to parabola 176, 282n42, 345
on postulates of Euclid 374n59
and predecessors 125, 210, 322, 328-329, 334n54
"problem of parabola" of 319-321, 336n75, 345; *v. also:* on normals to parabola
on problems 328, 350-351, 374n56
projective theorems of 337n93; *v. also (B)* projective geometry
on regular solids 291n174, 323, 338n98
on spiral 166, 283n61, 296; *v. also: (B) cochlias*; spiral, cylindrical
works of (lost):
Determinate Section 322, 327, 337n92, 356
Elements 371n20
Okytokion 157, 159, 294-295, 330n5
Neuses 47n86, 218, 281n34, 298, 300-302, 329, 346
Planar Loci 103, 105, 108, 139n9, 302, 329, 342-344, 346, 371n25
Section of an Area 321
Tangencies 90n19, 161, 302, 322

Arabic mathematics 8, 285n101, 341, 364, 376n83
Archibald, R. C. 287n124
Archimedes 20, ch. 5 (*passim*), 369
on analysis 170-174, 196
on angle trisection 185-187, 346, 380n132
and Apollonius 293-297, 313, 363
on arcs and sectors 82, 98n104, 200n49, 283n63
on area of circle (*Dim. Circ.*, 1) 83, 197n12, 228, 233, 272, 361-364, 374n58
on arithmetical computations 155-159, 194, 207n134, 294-295
biography of 151, 197n1-2, 43n30
on burning mirrors 238
"cattle problem" of 294-295, 330n6
chronology of writings of 153, 202n65, n68
on circle quadrature: approximation 155-159, 329
on circle quadrature: via spiral 297, 372n37
on circular segments 201n63
on *cochlias* 166, 296
on conics 62, 91n33, 111, 113-114, 121, 142n42, 202n80, 297-298, 315, 328, 335n59, n62, n65, 337n87
on conoids 136
on continuous magnitude ("Archimedes' axiom") 78, 198n13
convergence method of 84, 96n74; *v. also:* on limits
on curvilinear measure (axioms) 155, 198n14, n16, 283n64
on division of sphere (*Sph. Cyl.* I, 4) 170-174, 176, 202n71, 234, 239, 263, 274, 276, 304, 344
and Eratosthenes 191, 197n2, 209, 211, 274, 295
and Euclid 139n1, 151-152, 197n2
and Eudoxean geometry 194-196; *v. also:* on limits
on Eudoxus 26, 78
grave of 42n17
and Hero 98n109, 157-159, 195, 197n12, 205n112, 271
and Hero's cube duplication 188-191, 196, 225, 282n40, 305
and Hero's rule for cube roots 194
and Hero's rule for triangles 290n160
on inscription of heptagon 67, 178-181
on isoperimetric figures 273-274
and later geometers (influence on) 85, 209-210, 219, 274-277, 292n179-180
on limits 79, 96n76-77, n80, 197n11, 224
on limits: early method 152-155, 194-195, 198n17
on measurement of cosmos 20, 211; *v. also: (C)* Archimedes, *Sand Reckoner*
"mechanical method" of 152, 195, 197n7, 273, 369
on mechanics 3-4, 195, 264-267, 346, 352, 365
on mechanics: writings 205n112, 207n139

on *neusis* 178-187, 258, 263, 274
on *neusis*: for angle trisection 185-187, 221-222, 303, 329
on *neusis*: for spirals 38, 176-178, 282n42, 326, 336n75, 345-346, 366
on *neusis*: its status as a solving method 184-185, 187, 196
on parabolic segments 202n64
On Plinthides of 157, 295
practical efforts of 151, 190
on practical geometry 188, 195, 206n114
on proportion theory 80, 370n5
on semiregular solids 272, 290n169
on solid problem: *v.* on division of sphere
on sphere measurement 272
on spiral 85, 129, 143n49, 161-169, 245, 274, 308
on square roots 156, 194, 199n21, 207n134
summation lemma of (*Con.* 1) 266, 289n145
works of 152
and Zenodorus 234
Archytas 3, 12n7, 48n90, 50, 216
on cube duplication 4-5, 17, 19-22, 50-54, 57, 59, 86, 187, 212
v. also: (*B*) curve, Archytas' cylindric section
Arias, P. E. and M. Hirmer 249
Aristaeus 338n95, 378n105
and Apollonius 321-329
on conics 111, 113, 138, 217, 327-328
"the elder" 323-324, 335n62, 354
on regular solids 91n44, 291n174, 323, 327, 338n98
on solid loci 332n39, 347
Solid Loci of 213, 323-329, 338n108, 342
Aristarchus 55, 205n106a, 211, 308-311
Aristotelian commentators 7, 361, 367
Aristotle,
on analysis 95n64, 356-357, 375n70, 377n98
on causes in geometry 75, 94n62
on circle as least path 290n170
on circle quadrature 26, 28-31, 34-36, 38, 76, 81
on circle rectification 83-85, 98n106
on continuity 99n121, 102
on converses 75
on isoperimetric figures 273
on locus problem (*Meteor.* III 5) 102-108
and mathematics 83, 86, 98n108, 107-108, 374n60
on reduction 29-30, 359
terminology of 108, 141n28
his theory of science 23, 43n35, 50, 374n60
on vision 107
on zoology 261
Artemis (goddess) 330n5

Babylonian: *v.* Mesopotamian
Balsam, H. 146n87, 335n60

Baltzer, R. 203n84
Banū Mūsā 90n26, 187, 290n159, 380n128
Barnes, J. 95n65
de Beaugrand, J. 146
Becker, O. 46n68, 88, 88n5-6, 96n82, 206n123-124, n126, 283n52, 284n70
Beckmann, P. 373n44
Berlin painter 249, 255
al-Bīrūnī, Abu 'l-Raiḥān 140n16, 199n34, 203n90, 205n106, 206n113, 290n160
Björnbo, A. A. 46n67, 47n77, 81, 87, 97n94
Bluck, R. S. 92n55, n57
Bobbio (monastery) mathematical fragment 236, 239, 244, 284n86, 285n99
Böker, R. 5, 46n75, 282n48, 331n20
Bowen, A. C. 333n52a
Boyer, C. 141n24
[Bradwardine] 200n38
Brieskorn, E. and H. Knörrer 289n149, n158
Brown, M. 278
Bruins, E. M. 199n27, 206n123, n126
Bryson 27, 76-78, 87, 95n67, n69, 361; *v. also*: (*B*) continuity, and Bryson's principle
Burkert, W. 48n91
Burnet, J. 45n60
Busard, H. L. L. 290n166
Buschor, E. 249, 288n134

Callippus 90n20
Cantor, M. 81, 97n93
Carpus of Antioch 14n23, 352, 359-360, 365, 379n108
Cassini, G. D. 289n158
Castillon, G. F. 263, 287n124
Cavalieri, B. 266, 267, 288n141-142
Chasles, M. 118, 137, 144n63, n66-67, 146n88
Chinese mathematics 157, 199n30
Chrysippus 343, 371n19
Cicero 42n17
Clausen, T. 37
Cleomedes 277n3
Commandinus, F. 331n30
Conon 85, 111, 129, 162, 169, 233-234, 275, 282n41, 313, 329
Coolidge, J.L. 90n31, 146n88, 285n98, 333n50
Cremona, L. 144n63

Darboux, G. 289n158
Demetrius (king) of Syria 275
Democritus 27, 41n3, 87
Desargues, G. 116, 137, 146n89, 149n108, 322, 337n90, n92-93; *v. also*: (*B*) Desargues configuration
Descartes, R. 14n23, 89n10, 287n114, 373n44, 380n125, 381n139
Dicks, D. R. 277n2
Diels, H. 46n68

Dijksterhuis, E. J. 47n87, 144n58, notes to ch. 5 (*passim*), 204n94, 292n180, 330n4, 335n63
Dinostratus 50, 80, 83-86, 98n111, 129, 233
Diocles 43n20, 64-65, 209, 233-246, 336n80
  on Archimedes' solid problem 174-177, 181, 202n71, 234, 274, 304, 377n89
  on burning mirrors 285n99, n101
  on conics 113, 315, 335n65
  on cube duplication 190, 219, 234, 240-244, 282n40, 286n106, 365-366, 372n40
  on curve for cube duplication 242-247, 270, 274, 284n79, 286n111
  dating of 275-276
  mechanical method of; v. pointwise construction
  and Menaechmus 64-65, 240
  on normals 143n44
  notation of 284n81
  on parabola: focus 234-235, 284n80
  on parabola: focus-directrix construction 128-129, 137, 147n94, 236-237, 240
  on pointwise constructions 44n46, 63, 91n36, 284n88
  on predecessors 291n179, 292n180, 327
Dionysodorus 209-210, 263-267, 340
  dating of 275-276
  on torus 264-267, 270, 274
  on Archimedes' solid problem 171, 174-176, 183, 202n71, 263, 274
Dionysos (god) 249-254, 262
Dioscorides 262, 288n132
Dositheus 20, 169, 233-234, 239, 275-276
Drachmann, A. G. 201n58, 205n112

Egyptian mathematics 16, 25, 193, 195, 202n63, 207n129
Eleatics 16
Engels, H. 44n44
Enriques, F. 13n13, n18, 14n25, 372n39
Epicureans 198n15, 375n75
Eratosthenes 169, 201-218
  and Archimedes 191, 197n2, 209, 211, 274, 295
  on astronomy 277n4
  on cube duplication 17-23, 42n6, 211-213, 288n138, 346, 365
  on cube duplication (as source) 4-5, n26, 39, 43n32, 50, 60, 190-191, 226
  on Delian problem 17-23, 88; v. also: *Platonicus*
  life of 277n2
  on loci relative to means 134, 147n100, n103, 213-218, 276, 279n27, 281n35, 343-344; v. also: (B) loci, relative to means
  On Means 213
  measurement of earth 206n112, 277n3, n5
  on Menaechmus 62-63, 86
  and Nicomedes 206n121, 219
  as the "new Plato" 90n22, 278n13

Platonicus 3, 17-18, 20-22, 57, 59, 61, 213, 278n13, 351, 374n63
Euclid ch. 4 (*passim*)
  analytic works of 108-120, 213, 340, 344, 354, 369, 378n105
  and precursors 39, 86, 92n50, 101-102, 198n17
  and Apollonius 116, 138, 313-321
  and Archimedes 139n1, 151-152, 197n2
  critics of 138
  biography of 138n1
  circle theorem of (*El.* XII, 2) 78-79, 96n78
  on conic problems 120-127, 276-277
  on conics 62, 202n80, 298, 327-329, 335n62, 337n93, 346
  disciples of 291n177, 294, 313, 321, 329
  formal style of 70-71, 213, 375n76
  on inscription of pentagon 92n52, 181
  on locus (three- and four-line) 116, 322-323, 327, 337n87
  mechanical writing of 197n4
  on optical loci 372n25, 373n54
  projective theorems in 116, 148n105; v. *also*: works of, *Porisms*
  role of problems in 349-351, 353-354
  on sphericity of cosmos 143n48
  works of (lost):
    Conics 102, 111
    Porisms 102, 109, 116-120, 127, 132-133, 136-138, 141n16, 144n66-67, 349
    Surface Loci 127-128, 132, 136-138, 144n67, 236, 324, 342
Eudemus of Pergamon 275
Eudemus of Rhodes 15-16, 21, 26, 29, 31-32, 34-35, 38-39, 45n53, 46n68, 48n91, 50, 66, 108, 278n25
Eudoxus 3, 12n7, 369
  and analysis 67, 86
  and Aristotle 50, 83
  astronomy of 54-55, 89n15, 90n20, 308, 333n52a
  on circle theorem (cf. *El.* XII 2) 96n80
  on cube duplication 4, 17, 19-22, 52-57, 86, 212, 242-244, 333n51, 351, 364
  on cube duplication: and ps.-Platonic method 59-61, 190
  on hippopede 129, 270, 310-311
  on limits 26, 29, 32, 78-80, 82, 86, 96n76-77, 99n121, 101, 152, 154, 198n13, n17, 210, 224, 228, 274, 276, 283n52, 293, 340
  on means 278n16
  and philosophy 7, 11
  and Plato 2, 50
  on proportion theory 39, 79-80, 86-87, 101
Euergetes: v. Ptolemy III
Eutocius 10, 339-341
  and anon. *Introduction to Ptolemy* 290n166
  and Anthemius 238
  and Apollonius 139n9, 171, 313, 338n108, 381n140

on Archimedes' circle theorem 361-364
on Archimedes' solid problem 171, 174, 176, 183
and Archytas 51
as a commentator 341, 370n7
on conics 113
on cube duplication 4-6, 188, 209, 308, 332n43
and Diocles 233, 246
and Eratosthenes 17-23, 212
and Eudoxus 52-53, 60
formal manner of 181, 377n89, 379n110
and Hero 226
on hyperbola construction 304, 331n33, 335n69, 350
on loci 342-343, 371n15, n23
on locus (three-line) 144n59
on locus (two-point) 104
and Menaechmus 61, 66-67
and Nicomedes 219, 223, 226, 283n54
and "Plato" 57, 59
pointwise constructions of 63, 91n36, 284n88, 285n91
terminology of 90n28, 171, 187
Exekias (in the manner of) 249, 254

de Fermat, P. 287n114
Fischler, R. 92n49
Foley, V. 206n120
Fontana, G. 284n69
Fowler, D. 199n21-22, 330n6
von Fritz, K. 374n70

Gelon (king) 20
Geminus 5-6, 143n48, n53, 296, 341, 343, 348, 351-352, 357, 365, 367, 370n11, 379n105
Gerard of Cremona 90n26, 205n107, 375n78, 376n83
Gillings, R. 44n44
Goldstein, B. R. 333n52a
Gregory, J. 363, 380n125
Guldin, P. 264-266, 288n135, n142
Guthrie, W. K. C. 45n53, n61, 81

Halley, E. 314, 335n60, 336n82
Hankel, H. 87, 97n93
ibn al-Haytham: *v.* Alhazen
Heath, T. L. *passim* in the notes (esp. ch. 4, 5, 7); 5, 11, 81, 88n3, 92n55, 126, 139n2, n9, 140n15, 144n67, 145n76, 153, 197n1, 270, 285n99, 291n177, 300, 320, 335n63, 335n70, n72, 336n75-76, n81, 337n92-93, 338n100, 356, 372n30
Heiberg, J. L. 8, 17, 81, 91n33, 142n42, 153, 199n27, 239, 283n54, 285n99, 291n173, n177
Heijboer, A. 92n55
Heraclides 294-302, 313, 329, 334n55
Heraclitus, geometer 144n67, 298-300
Heraclitus, Presocratic philosopher 331n20
Heraclius 297-298

Hermes (god) 249, 255
Hermite, C. 373n44
Hermotimus 142n30, 372n25
Hero 142n28, 277n3, 339-341
on analysis 354-356, 358, 376n85, n87
and Archimedes 83, 98n109, 157-159, 188, 195, 197n12, 205n112, 271
on area of segments 201n63, 202n64
on circle division 363, 379n124
on classification of problems 347-348, 361
on cube duplication 188-191, 274, 282n40, 286n107, 305-308, 336n77, 347, 365, 372n40
on cube duplication: and Nicomedes 225-226
on dioptra 42n6, 380n136
on figures of revolution 265
on indivisibles 289n147
on mechanical geometry 365
metrical writings of 199n25
on mixtilinear figures 287n119
on rule for catapults 206n120
on rule for cube roots 191-194
on rule for square roots 192-194, 207n127, n130, n134
on rule for triangles 187, 271, 290n159
on torus (Dionysodorus) 264-266, 288n135
Herodotus 367-368
Hintikka, J. and U. Remes 142n36, 358, 376n79, n84, n92-93, 378n101-102
Hipparchus 157, 199n31, 308, 310, 332n49, 377n87
Hippasus 88
Hippias of Elis 80-81, 83-84, 87, 97n93-94, n97, 283n61
Hippias, geometer 80-81, 85, 227, 230-231, 268-270
Hippocrates of Chios ch. 2 (*passim*)
circle theorem of 28, 76, 78, 86
on cube duplication 17, 22-24, 50, 359, 369
and *Elements* 41, 102
false quadrature by 26, 31, 34-36, 87
fragment on lunules: *v.* (C) Simplicius, *In Phys.* (1.61-68)
on lunules 24, 26, 29-40, 46n66, 68, 102, 160, 287n119, 369
neusis of 37, 40, 187, 345, 366, 372n33
Hippodamos 292n179
Hofmann, J. E. 207n134
Hogendijk, J. P. 203n90, 204n96, 205n99, n103
Hoppe, E. 199n27
Hort, A. 287n130
Hultsch, F. 282n48, 287n120, n121, 290n166, 291n177, 331n30, 336n75, 338n100, 371n24, 378n105, 380n137
Husserl, E. 88
Huxley, G. 285n93, n99, 330n5
Huygens, C. 201n63, 271
Hypatia 19, 42n18
Hypsicles 291n174, 323-324, 327, 338n97, n98

Iamblichus 80-81, 201n60, 330n12, 362
Isidore of Miletus 42n6, 338n97
Ito, S. 142n33

Joannes de Muris 46n66

Kepler, J. 88n6, 267, 288n135, 289n158
al-Kindī, Yaʿqūb 285n99, 286n101
Kleophrades painter 249, 252-253
Knorr, W. R. 12n3, 46n68, 91n43, 96n77, n82, 98n110, 139n2, 141n21, ch. 5 notes (*passim*), 285n99, 286n101, 289n143, 292n180, 331n33, 338n95, n105, n110, 370n5, n7, 374n67
Ktesibius 206n112
Kuhn, T. S. 207

de La Hire, P. 333n50
Lambert, J. H. 363, 380n125
Landau, E. 37
Lee, H. D. P. 139n9, 141n26
Lehmann, D. 380n125
Leodamas 50, 86
Leon 50, 74, 86, 142n37
Lindemann, F. 373n44
Linnaeus, C. 262
Livy, 197n1, 206n122
Loria, G. 5, 247, 262, 284n71, 286n107, 287n111, n118, 289n158, 290n162
Luria, S. 99n120
Lysippides painter 249, 253

Mahoney, M. S. 14n23, 142n36, 378n101
Maimonides 363, 380n125
Marinus 362-363, 378n105
Mau, J. 99n120
Maurolico, F. 363, 380n125
Mamercus 81
Menaechmus 12n7, 50
  on conics 90n30-31, 144n54
  on cube duplication 4-5, 17, 19-22, 50, 61-67, 72-73, 75, 86, 102, 176, 189-190, 202n76, 212, 234, 240, 304, 306, 344, 346, 364
  on cube duplication: curves 62-66, 94n61, 111, 284n88
  on problems 75-76, 351, 359
Mendell, H. 44n46, 141n19
Menelaus 5, 59, 341, 348, 370n11
Menge, H. 142n33
Mesopotamian (Babylonian) astronomy 42n5, 310, 332n49
Mesopotamian mathematics 16, 25, 67, 92n48, 102, 139n5, 158, 194-195, 202n63, 207n133
Meton 26
Mogenet, J. 290n166
Morrow, G. R. 375n72
Mueller, I. 14n21, 96n75-76, 99n121, 139n2, 375n77

al-Nairīzī (Anaritius) 354, 375n78
Nelson, H. L. 330n6
Neuenschwander, E. 139n2
Neugebauer, O. 91n42-43, 111, 139n1, 207n133, 235, 332n47, n49, 334n53
Newton, I. 286n111
Nicomedes 209-210, 219-233, 282n41, 283n61
  and Apollonius 303-312
  and Eratosthenes 206n121, 219
  on angle trisection 187, 221, 329
  on conchoids 143n49, 245, 247, 258, 263, 268-270, 274, 367
  on cube duplication 42n6, 188, 190-191, 225-226, 274, 305, 308, 346-347, 365, 380n132
  his lemmas on conchoids 223-225, 283n52, n54
  and *neuses* 221-222, 225-226, 310-311
  on quadratrix 80-81, 84, 226-233, 274, 296-297
Nicoteles 282n41, 329, 335n63

Oenopides of Chios 15-16, 41, 41n5, 374n70
Olympiodorus 106, 141n18
Omar Khayyām 174, 202n76

Paeonius 19
Pappus,
  on analysis 196, 213, 354-360, 375n70, 376n85-86, 377n87, n90, n92, n98, 378n105; v. also: (*B*) *topos analyomenos*
  on angle trisection 85-86, 128, 146n89, 187, 221, 303-304, 324, 327, 331n26, 335n71, 350, 371n11, 380n132
  on anonymous cube duplication 207n128, 347, 373n43
  and Apollonius 176, 300-302, 304, 313, 315, 319-320, 324, 330n5, 335n69, 336n75, n83, 338n108, 381n140
  on *arbelos* 160
  and Archimedes 83, 176-178, 195, 197n12, 336n75
  and Aristaeus 111, 323
  on chord lemma 140n16
  on circle quadrature via spiral 178, 203n89
  on classification of problems 176-178, 217, 341-348, 361
  as commentator 11, 14n27, 339, 371n24
  on conchoids 222, 282n48
  on conics 113
  on cube duplication 242, 286n106, 347, 365-366, 372n40
  and Eratosthenes 19
  on focus-directrix 136
  formal manner of 181, 377n89, 379n110
  and Guldin's theorem 264-266
  on harmonic mean 214
  on irrationals 370n5
  on isoperimetric figures 272-274, 290n166, 291n171, n173
  his lemma on surface locus 146n93

his lemmas to Euclid's *Porisms* 116-119, 124, 132-138, 144n65, 145n68, n72, 215, 278n27, 279n30, 281n35, 322, 337n91, n93
on loci 342-345
on locus (three-line) 288n138
on mechanical geometry 365
on *neusis* relative to two lines 218
and Nicomedes 223
on porisms 360
on problems 94n61, 350, 370n11, 374n56
and philosophy 357
his projective theorem of hexagon 117-119
on quadratrix 80, 84-85, 226, 228, 230, 364
on regular solids 67, 91n44
on sectors 200n49; *v. also*: Archimedes, on arcs and sectors
on segment theorem 26
on spiral *neusis* 176-178, 203n87, 377n89
on spirals 162, 166, 169, 195, 200n47, 266
on surface locus 129
and Theaetetus 278n25
Parker, R. A. 202n63
Parmenides 87
Parmenion 305
Pascal, B. 116, 118, 137, 145n68, 149n100, 337n91, n93
Pascal, E. 222, 258, 263
Perseus 81, 263, 267-272, 276
Philo of Byzantium 141n28, 206n112, n120
on cube duplication 305-308, 332n43, 347, 365-366
Philonides 274-276, 291n175, n179
Philoponus, Joannes 188, 361-362
on Bryson 76-77, 95n67-69
Plato,
and associates in Academy 2-3, 17, 20-24, 49-50, 86
chronology of works of 73
on hypotheses 71
on Hippias 81
and mathematics 67, 88
philosophy of 5, 7, 21-22, 57, 59, 74, 76, 95n70, 213, 351-353, 360, 364-365, 377n94
[Plato], cube duplication of 42n6, 57-61, 63, 66, 86, 190, 212, 240, 242-244, 286n107
Plutarch 2-5, 12n1, 17-18, 21-22, 59, 66, 197n1, 213
on Archimedes 3-4, 197n1, 206n122
on Plato 2-5, 351, 364
Polybius 197n1-2, 206n122
Porphyry 364
Posidonius 206n112, 353, 374n70-71
Proclus,
and Amphinomus 95n63
on analysis 67, 74, 378n105
on application of areas 66
on archaic terminology for conics 143n53
on Archimedes' circle theorem 197n2, n12
and Bryson 76-77, 95n67, n69

on cissoid 246-247, 261, 263
on conchoid 282n48
on converses 75
on early geometry 15-16, 25
on Euclid's *Elements* 86
and Hippias 80-81
and Hippocrates 39
on horn angle 48n93, 95n71
on isoperimetric figures 290n166
on loci 343, 348, 371n18, n23
on mixed curves 287n119
on parallel lines 374n59
and Plato 49
on problems 76, 351-354
on Pythagorean theorem 102
sources of 191
on spiric sections 289n148
on surface locus 129
Protagoras 27
Ptolemy 143n48, 157-158, 199n34, 210, 291n171, n173, 310, 363, 374n59, 377n87
Ptolemy III Euergetes (king) 17-18, 275, 291n177
Pythagoras 59, 66, 81
Pythagoreans 16, 47n88, 49, 66, 87-88, 213; *v. also*: Pythagorean theorem; Pythagorean triplets
Pythion 233, 275, 292n179

al-Qūhī, Abū Sahl 42n6, 205n104

Remes: *v.* Hintikka
Riddell, R. 89n12, n18, n19, 54-57, 333n51
de Roberval, G. 201n55, 246-247, 258, 286n110, 287n114
Rudio, F. 45n68

Schiaparelli, G. 90n20
Schneider, C. K. 248, 287n129
Schneider, I. ch. 5 notes (*passim*)
Schoy, C. 203n90, 205n103-104
Seidenberg, A. 12n3, 43n32, 283n56, 375n74
Sezgin, F. 200n42, 204n96, 376n78
al-Shannī, Abū ʿAbdallāh 183
al-Sijzī, Abū Saʿīd Aḥmad 181-183, 204n96, 205n104, n106, 338n109, 380n132
Simon, M. 46n66
Simplicius 26, 29, 31-32, 35-36, 38, 40, 45n53, 46n66, n68, 95n67, 273, 290n170, 291n171, 362-365, 380n128
Simson, R. 104-105, 140n13, n15, 144n62, n66
de Sluse, R. 271
Soedel, W. 206n120
Solmsen, F. 139n9, 141n20, n23
Speusippus 75, 351
Sporus 82, 203n89, 230, 364, 380n125
on cube duplication 242, 286n106, 366
Steele, A. D. 13n17, 47n84, n86, n89, 207n128, 372n30, n33

Steiner, J. 126, 146n88
Stifel, M. 363, 380n125
Stoics 367
Strabo 210, 277n3
Suidas 90n22
Suter, H. 205n107, 206n113
Synesius 19
Szabó, Á. 12n3, 41n4

Tannery, P. 5, 8, 46n67-68, 47n77, 53, 57, 81, 87, 89n10, 97n94, 147n93, 199n27, n30, 200n51, 203n83, 247, 262, 267, 278n17, 287n117, 289n151, 332n48, 374n67
Taton, R. 337n90
Thābit ibn Qurra 159, 178, 181, 183, 187, 282n51, 331n30, 379n110
Thaer, C. 332n46
Thales 48n91
Theaetetus 50, 86, 101, 215, 278n25, 340
Themistius 7, 45n53
Theodorus 49
Theon of Alexandria 42n18, 158, 197n12, 289n147, 339
  on isoperimetric figures 272-274, 290n166, 291n171, n173
Theon of Smyrna 3, 17-18, 21, 213
Theophrastus 261-262
Thomas, I. 44n46
Thomson, W. 204n96, 377n95
Toomer, G. J. 43n20, 139n1, 143n42, 144n58, 147n94, 157, 199n31, 202n71, n80, 233-236, 239, 246, 276, 284n74, n75, n79, n85, n88, 285n99, 286n103, 287n114, 291n175, n177, n179, 336n81
Tóth, I. 139n7
Tropfke, J. 178, 203n90, n93
Tschakaloff, L. 37
Tschebotaröw, N. 37
al-Ṭūsī, Naṣīr al-Dīn 90n26, 205n107, 307, 317, 332n46, 333n50

Tycho Brahe 310
Tzetzes, Joannes 197n1-2

Unguru, S. 204n94

Ver Eecke, P. 264, 331n25, 338n100
Vieta, F. 47n89, 233
Villapaudo 88n6
Vitruvius 45n62, 206n114, 305
Viviani, V. 88n6
Vogel, K. 338n98

van der Waerden, B. L. 12n3, 43n32, 88n5, 89n7, 90n24, 91n42, 92n48, 143n54, 335n63
Wallenius, M. J. 37
Wantzel, L. 373n42
Waschkiess, H. J. 88n3, 99n121, 141n21
Wegner, U. 37
Weierstrass, K. 77
Werner, J. 143n50
Whiteside, D. T. 199n21
von Wilamowitz, U. 12n10, 14n26, 18-21
Woepcke, F. 202n76, 205n104, n106

Youschkevitch, A. P. 202n76

Zeno of Elea 11, 87
Zenodorus 263
  dating of 233-234, 275-276, 291n175, n179
  on isoperimetric figures 198n14, n18, 290n169, 272-274
Zenodotus 352, 374n70
Zeuthen, H. G. 13n13, 47n87, 80, 90n30-31, 97n89, 116, 118-119, 123-127, 134, 137, 143n49, n54, 144n55, n66, 145n68, n76, n78, 146n87, n88, 147n103, 203n94, 213-218, 278n15, 279n28, 281n30, 322, 329n1, 334n53, 335n63, 336n75, n86, 337n87, n91-92, 373n53, 375n77

Index B: Subjects

aitēma (postulate) 336
akolouthein (follow) 355, 376n84
Alexandria 43n20, 138n1, 210, 291n177
algebraic relations for curves: v. *symptōma*
analyein (analyze) 75
analysis, 9, 23, 354-360, 369
  ancient accounts of 354-360
  of antecedents 375n70, 376n86
  in Archimedes 151, 170-174
  and convertibility 95n65, 142n36
  in Euclid 109, 142n35, 151
  in 4th century (B.C.) 66-76, 86
  and problem solving 329n1, 313-315, 354, 359, 374n56
  in Pappus: v. *topos analyomenos*
  and reduction 23, 49

  in 17th century 14n23
  and synthesis contrasted 9, 51-52, 377n89, 379n110
  of theorems 376n85, 378n101-102, n105, 356-360, 367-368
  analytic corpus v. *topos analyomenos*
angle division:
  via spiral 98n112
  via quadratrix 84, 226-227
angle trisection,
  Arabic 205n106-107, 221, 327, 380n132
  of Archimedes 185-187, 346, 380n132
  via *neusis*: v. *neusis*, for angle trisection
  of Nicomedes 187, 221, 329
  in Pappus 128, 303-304; v. *also*: (A) Pappus, on angle trisection

practical procedures 41, 85, 107
  via quadratrix 84, 97n94, 226, 233
  as solid construction 85
  via surface locus 128
apagōgē (reduction) 23
apodeiktikē (demonstrative) 364
apora (nonproduceables) 362-363
Apollonian circles 103; *v. also* (A) Apollonius, on locus (two-point)
application of areas 37, 63, 66-69, 71-75, 86, 91n41, 102, 142n38, 203n94, 344, 369
applied geometry: *v.* practical geometry
arbelos 159-160, 287n119
Archimedean axioms: *v.* (A) Archimedes, on continuous magnitude; on curvilinear measure
astronomy 16, 41n5, 54, 20, 210, 308-311, 340
asymptote,
  ancient definition of 283n53
  of conchoids 221, 223-224
  of hyperbola 207n136

bisection (convergence) 79, 83, 96n79, 198n13
botany 261
burning mirrors 63, 141n22, 233-239, 263, 275-276, 285n99, n101, 292n180, 321, 327, 336n80, 372n25

cardioid 258, 263, 310, 333n51-52
Cassel cup 249, 256-257
catapults 206n120
center of gravity 152, 195, 206n112, 263-267, 274, 289n146, 340, 369
circle,
  area of: *v.* (A) Archimedes, on area of circle
  division of area 205n112, 363, 379n124
  isoperimetric property 198n14, 272-274
circle quadrature,
  approximate: *v.* π
  by Archimedes 153-159, 297, 329, 372n37
  conditions of solution of 364-365
  in 5th century (B.C.) 25-39
  in 4th century (B.C.) 76-86
  via quadratrix 82, 226-233
  60- and 360-division of 277n5
  via special curves 362
  via spirals 167-170, 178, 196
circular segments,
  area of 159, 168, 188, 200n39, 201n63
  isoperimetric 290n169
cissoid:
  ancient curve 246-247, 258-263, 287n117, 310
  modern curve 246-247
cochlias:
  Archimedean screw 166, 201n58
  cylindrical spiral 166, 282n49, 283n61, 295-297
"cochlioid, sister of" 283n61, 296; *v. also*: spiral, cylindrical

cochloids 222, 282n48-49
compass and straightedge 15-16, 37, 40, 365; *v. also*: planar constructions
compound ratios 265-266, 288n138, 326
conchoids 187, 219-226, 245, 258-261, 270, 283n53, 287n121, n127, 346, 367
  circle-based 222-223, 258-263, 287n127, 310-312, 333n52, 367
  and cissoids 247, 263
  naming of 282n48-49
conics,
  archaic construction of 113-115, 143n52
  conjugate diameters of 313, 327
  construction problems of 126-127, 145n83, 321-323, 337n87, n91
  construction via focus and directrix 128-132, 136, 143n44, 144n67, 147n94, 236-237, 240, 324-326, 337n87
  diameter of (definition) 114
  early terminology of 143n53, 171, 202n80, 275
  early theory of 61-62, 111-116, 120, 143n44, 315, 319, 326, 335n62-63, 337n87, 338n103, 372n25
  focus of 321
  latus rectum of (definition) 114, 144n58
  as means loci 133-134, 147n103, 216, 279n30
  normals to: *v.* normal, to conics
  polar of: *v.* polar, of conics
  as projective locus 145n68, 146n87
  as second-order curves 65
  similar 313, 335n59
  as solid sections 113, 143n54, 148n106, 270
  as solid solving method 170, 182
  tangents of: *v.* tangent, to conics
  as three- and four-line locus: *v.* locus, three- and four-line
conoids 159
continued fractions 156-158, 199n22, n24
continuity 29, 78, 99n121
  and Bryson's principle 76-77, 80, 83, 95n70
convergence: *v.* bisection; (A) Archimedes, on limits; Eudoxus, on limits
Copernican hypothesis 310
cube duplication 364-366
  of Apollonius 305-308
  Arabic 308
  of Archytas 50-52
  of Diocles 240-242
  of Eratosthenes 211-213
  of Eudoxus 52-57, 59-61
  of Hero 188-191, 305-307
  of Menaechmus, 61-66
  via *neusis*: *v. neusis*, for cube duplication
  of Nicomedes 225-226
  of Pappus 242
  of Philo 305-307
  of Plato (attributed) 57-61
  as solid construction 341-342, 347
  of Sporus 242

and two mean proportionals 3, 23-24, 40, 62
  v. also: Delian problem; (A) Apollonius; Archytas; Diocles; Eratosthenes; Eudoxus; Eutocius; Hero; Menaechmus; Nicomedes; Pappus; Philo; [Plato]; Sporus—on cube duplication
cube roots 191-194
cubic relations 174
curvature, center of 318
curve,
  Archytas' cylindric section 51, 129, 170
  from cylindric section (Tannery) 53
  "of double motion" (Carpus) 352
  v. also: cissoid; conchoid; conics; hippopede; quadratrix; spiral; spiric section
curves,
  classification of 258, 341-348; *v. also:* problems, classification of
  from solid sections 112, 143n48, 341-345, 365; *v. also:* solid locus
  mechanical construction of 59, 82, 86, 112, 165-166, 201n55, 210, 227, 230, 246, 308, 310, 333n50, 345-346, 366-367, 369, 375n77, 380n125
  "mixed" 247, 261, 287n127, 289n156
  planar: *v.* planar locus
  linear: *v.* linear constructions
  pointwise construction of 44n46, 54, 63-64, 66, 73, 91n35-36, 143n51, 236-239, 284n88, 285n91, 327, 375n77
  for problem solving 270
cycloid 374n67

dapanan (exhaust) 28
data (givens) 109-111, 357-358, 378n105
Delian problem 2-3, 17, 20-22, 24, 39, 49, 57, 88
Desargues configuration 116, 136, 148n105
dimensionality 271, 288n138, 290n159
dioptra 42n6, 380n136
diorismos (determination) 110; *v.* diorisms
diorisms 73-74, 76, 94n58, 142n37, 172-173, 335n73, 349, 354, 358, 374n56
dothen (given): *v.* data
dynamis (power) 26, 38

earth, measurement of 206n112, 210, 277n5
eccentric model of planetary motion 308-311, 332n47
elleipein (fall short) 66, 72
ellipse 62, 111
  area of 159
  via five-point construction 365
  as locus of constant sum 270, 285n98
  mechanical generation of 333n50
  as oblique circle 90n32, 145n67, 327
  as solid section 143n48
  v. also: conics
empirical procedures: *v.* practical geometry

epharmozein (conform) 27
epicyclic model of planetary motion 308-311, 332n47
epicycloid 89n18-19, 287n117, 310-312, 333n52
epigram 267, 295, 330n7
Euclidean constructions: *v.* planar constructions
evolute 318
"exhaustion, method of": *v.* (A) Eudoxus, on limits
existence, assumptions of 361-362; *v. also:* nonconstructive assumptions
existence of fourth proportional 80, 96n76, 283n52
existence proof, and constructions 74, 77, 143n49, 373n46; *v. also:* problems, as existence proof
extreme and mean ratio 67-70, 91n48, 182, 358-359

formal style in geometry 86, 181
foundations, modern study of 99n122
foundations crises 87-88, 207n140

generalization in mathematics 173-174
geography 210
"geometric algebra" 102, 203n94, 331n20
givens: *v. data*
gnōmōn (sundial) 16

harmonic division (definition) 132
hedera helix (ivy) 248, 262
helices (spirals) 258, 296
heptagon, problem of inscription 178-185, 203n90, 204n94, 346
hippopede 54, 57, 129, 247, 260-261, 267, 270, 289n150, 310-311
homoeomeric property of curves 296
horn angle 48n93, 77, 96n71
hyperballein (exceed) 66
hyperbola,
  asymptote property of 90n31
  construction via asymptotes of 144n67, 187, 304, 324, 331n33, 335n69, n71, 350
  for cube duplication 61-62
  intercepts of 306, 315, 332n44, 335n71
  for *Meno* problem 73
  tangent to 94n59, 173, 335n71
  as two-branched curve 116, 122, 125, 315, 335n63
  v. also: conics
hypercycloid 333n51
hypocycloid 287n117, 333n50
hypothesis: *v.* (A) Plato, on hypotheses
hyption (supine figure) 116

impossibility, proof of 347, 364
incommensurables 363-364
indivisibles 87, 288n135, n141, 266-267, 289n147
infinitesimals 99n120; *v. also:* indivisibles

inscription of triangle (in Plato's *Meno*) 71-73
instruments of construction 16, 20, 26, 40-41, 42n6, 44n46, 187; *v. also*: ruler, flexible
intuitionism 88
involution 322, 337n92
irrationals 102, 278n25, 340, 347, 359, 363, 370n5
   as means 215
isoperimetric figures 198n14, n18, 234, 263, 272-274, 290n166, n169, 291n173-174, 380n125
isoperimetric writing (anonymous) 290n166, 291n171
isorrhopia (equilibrium) 288n143
ivy 246-257, 261-263, 288n134

kanōn ("rule" of conchoids) 219-221
kanonion (ruler) 365; v. *neusis*, mechanical
kissos (ivy) 261
koilogōnion (concave polygon) 290n166
konchē (shell) 263
kōnotomein (conic sectioning) 62
koskinon (sieve) 211

lemniscate 89n13
limaçons (circle-based conchoids) 222, 258, 263, 311
limits: *v. (A)* Eudoxus, on limits; Archimedes, on limits
linear constructions 341-343, 346-347, 361, 364, 366
locus,
   classification of 342-345, 347-348, 368
   of constant product 270-271, 289n158
   of constant sum 270
   early references to 138, 371n19, n25
   ephectic, diexodic, anaphoric 343
   five-point (ellipse) 365
   of harmonic division of secants 134-136
   planar: *v.* planar locus
   of point-line conic 324-326
   as problem or theorem 349
   relative to means 147n103, 278n15, 279n30, 281n35, 343-344
   solid *v.* solid locus
   three- and four-line 116, 120-127, 145n76, n79, 146n85, 213, 271, 288n138, 322-323, 337n91, 344, 369
   two-point 102-108, 139n9, 300-301, 329, 373n49
lunules 26, 30-40, 46n66, 102
lyein (solve) 341

mawjūd (found) 376n83
maximum property,
   of equilateral triangle 92n58
   of sphere and circle: *v.* isoperimetric figures
mean proportionals 212, 277n9, 288n138
means 133, 213-217, 278n16
   harmonic 214-215, 278n27
   loci relative to: *v.* locus, relative to means

"perfect proportion of" 214-215
mechanical constructions 3-5, 17, 22, 42n6, 57, 59-60, 90n26, 211-213, 278n13, 347, 364-368, 371n11, 380n132;
   v. also: instruments of construction
mechanics (equilibrium) 288n143; *v. (A)* Archimedes, on mechanics
mēniskos (lunule) 34
mesolabe 17, 19, 211
mixed angles 96n72
mobile vs. fixed geometry (Arabic) 205n106, 364, 380n132
moment 265-266, 288n143
Moscow Papyrus 44n45, 207n129

neuousa (inclining) 34, 366; v.: *neusis*
neusis (inclination) 34, 46n75, 210, 372n37
   and alternative constructions 345-346
   for angle trisection 282n51, 311, 327, 350, 380n132
   of Archimedes 178-187; *v. also: (A)* Archimedes, on *neusis*
   and cissoids 246
   and conchoids 221, 225-226, 233, 258
   via conics 303-308
   for cube duplication 188-191, 226, 242, 306-308, 311, 332n43, 346, 365, 372n40
   for heptagon inscription 178-180, 184-185, 204n94
   of Hippocrates 37, 40, 187, 345, 366, 372n33
   as mechanical 41, 365-366
   as nonconstructive assumption 97n88
   for pentagon inscription 47n88, 67-68
   planar construction of 47n86, 302, 311
   relative to conics 217-218
   relative to rhombus 298-302, 329
   relative to square 298-300, 329, 331n20
   as solid problem 196
   for spirals 165, 176-178, 221, 326, 336n75, 345-346, 369
   third-order 47n89
Newton-Raphson rule 193, 207n131
nonconstructive assumptions 80, 96n76, 97n88, 282n52; *v. also*: convergence, and Bryson's principle; existence of fourth proportional
normal,
   to conics 173, 304, 313, 315-320, 328, 332n44, 335n73, 336n75
   as minimal line 315, 319, 335n64-65
   to parabola 143n44, 176, 282n42, 315-320, 345
   to quadratrix 230
   to spiral 200n51

ophiuride 59, 86, 242-244, 286n107
optics 284n81, 372n25, 373n54
   and conics 111, 137
organikē (mechanical) 364
ovals of Cassini 289n158

paraballein (apply) 66
parabola,
  as burning mirror 234-237, 321
  for cube duplication 61-62, 64
  focus of 236-237, 240, 244, 284n80, 285n99
  normals to 203n82, 315-320
  parameter of 114
  quadrature of 168, 200n63
  semicubic 318
  subnormal of 143n44, 235
  tangent to 173, 234, 279n27, 314
  v. also: conics
parallel postulate 374n59
parespasmenos (drawn in) 111; *v. (C)* Euclid, *Optics* (P 36)
parhyption (hypersupine figure) 116
patrons: *v. (A)* Gelon; Paeonius; Ptolemy III
pedal curve, of parabola 244
Pellian equation 294
pentagon, inscription of 47n88, 67-71, 181
"perfect proportion" of means 214-215
philosophy and mathematics 3, 5-8, 14n21, 28, 50, 74, 87, 357, 367, 374n60
π,
  approximation of 25, 155-159, 198n20, 199n30, n34, 201n63, 231-233, 294-295, 362, 380n128
  irrationality of 363-364, 380n125
  transcendentality of 373n44
planar constructions 37, 182, 298, 302, 319-322, 328, 332n39, 341-347, 361, 365-366, 368, 372n30
planar locus 329, 347-348, 369, 371n15, n18
planetary motion 308-311, 333n52a
Platonic philosophy: *v. (A)* Plato, philosophy of
plectoids 129, 342
Plimpton 322 (cuneiform tablet): *v.* Pythagorean triplets
polar, of conics 135, 147n103, 213, 215, 279n29
pole,
  of circle-based conchoid 310, 333n52
  of cissoid 246
  of conchoid 258-260, 219-221
polychōrētoteros (more spacious) 291n171
polygōnoteros (more polygonal) 291n173
polyhedra 272
  regular solids 67, 91n44, 101, 290n169, 291n174, 323-324, 327, 338n98, 357, 377n89
porima (produceables) 362-363
porismos (porism) 142n31, 349, 360, 373n51
poristikon (productive) 357
porizein (produce; provide) 103, 109, 351, 360, 373n51
postulates (in *Elements*) 350, 374n59
practical geometry 25, 41, 44n45, 158, 188, 190-191, 209, 219, 239, 365, 380n136
prime numbers 37, 211
proballein (propose) 351, 373n46

problēma, problēmata 351, 374n60
problems 9, 348-349
  classification of 6, 74, 94n61, 176-178, 217-218, 258, 319-323, 326-327, 341-348, 361, 368, 370n11, 372n39
  criteria of their solution 176, 361-367
  as existence proof 74, 77, 80, 143n49, 353-354, 360, 368, 373n46, n54, 374n58, 375n77; *v.* existence proof
  irreducibility of 373n42, n44
  relation of planar and solid 176-178
  and theorems 75, 348-354, 374n60
probolē (problem) 351
projective geometry 116, 118-119, 126, 136-137, 144n63, 145n71, 279n27, 322, 337n90-91, n93
proportion theory 45n59, 80, 102, 370n5
protasis (statement) 352, 374n60
Pythagorean theorem 102
Pythagorean triplets 139n5

quadratrix 80-86, 143n49, 196, 203n89, 270, 274, 287n127, 330n12, 364, 380n125
  for angle trisection 84, 97n94, 226, 233
  for circle quadrature 226-233
  and cissoid 258
  definition of 82
  for division of angles 226
  generalized 231
  its implicit Archimedean assumptions 98n104
  naming of 84-85, 219
  and Nicomedes 80-81, 84, 226-233, 274, 296-297
  via solid section 129
  and spirals 85, 147n97, 201n51, 226-227, 230, 233, 342
  as surface locus 283n59
  tangents to 227, 230

rainbow 102, 106-108
ratio, in Hippocrates 45n59
rectification of circle 83
reduction 23-24, 29-30, 36, 44n40, 71, 359
reduction to impossible 359
regular solids: *v.* polyhedra
Renaissance (16th-century) mathematics 157, 198n20, 199n30
Rhind Papyrus 25
rhopē 288n143: *v.* moment
ritual origins of geometry 3
"rope stretchers" 16
ruler, flexible 44n46, 48n95, 63, 85, 91n36, 284n88

salinon 160, 287n119
scholium, on analysis: *v. (A)* Anon. scholiast to Euclid, *El.* XIII
solid constructions 319-321, 332n39, 338n108, 341-344, 346-347, 361; *v. also*: angle trisection; cube duplication; *neusis*, for spirals

solid locus 323-324, 327, 347-348, 369, 371n18, n23, 374n56
solid projection 147n97, n103, 278n23, 279n30, 281n35, 283n59, n61
solids of revolution 264-266; *v. also*: torus
sphere,
  division of 170-176; *v. also*: (A) Archimedes, on division of sphere
  isoperimetric property of 272-273, 290n169
spiral, 245, 270, 287n121, n127, 342, 380n125
  for circle quadrature 167-170, 178, 196
  conical 167, 226
  cylindrical 166-168, 226, 282n49, 283n61, 296, 374n67
  for division of angle 98n112
  and linear curves 287n120-121
  plane (Archimedean) 161-168
  and quadratrix 85, 147n97, 201n51, 226-227, 230, 233, 342
  spherical 162
  tangents to 163-167, 200n51
spiric sections 247, 267-272, 287n121
  as locus of constant product 270-271
  as locus toward hyperbola 271
square roots 156, 192-194, 199n21, 207n127, n134
straight line, as shortest distance 154, 198n14
sundial 16, 305
surface locus 129, 136-137, 166-167, 230, 236, 283n59, 342-344
Syene 277n4
symptōma (defining property of curve) 112-113, 143n49, 228, 268, 271, 285n91, 315, 366
Syracuse 152, 238

taḥlīl (analysis) 376n83
tangent,
  ancient definition of 200n51, 201n53, 237
  to conchoids 287n125
  to conics 207n136, 314-315
  to cylindrical spiral 296-297
  to hyperbola 94n59, 173, 335n71
  to parabola 173, 234, 279n27, 314
  to spirals 372n37
  tangent inequality 272-273, 290n164
tarkīb (synthesis) 376n83
terminology,
  archaic (pre-Euclidean) 38-39, 47n90, 62, 71, 108, 141n27, 284n81
  of conics (Apollonian) 62, 304
  of conics (pre-Apollonian) 113-114, 143n53, 171, 202n80, 275
tetragōnizousa 81, 233, 283n61; *v. also*: quadratrix
theōrein (investigate) 351, 353
thyreos (shield; ellipse) 112, 143n48
topos:
  locus 343, 371n25
  (physical) place 372n25, 373n54
topos analyomenos (analytic corpus) 109, 119, 217, 293, 313, 322, 340, 354, 374n56, 378n105
torus 51-53, 57, 263-272, 288n135, 344, 380n136
  sections of 287n121, 289n149-150; *v. also*: spiric sections
trochoids 333n51
truncated solids 207n129
two mean proportionals 3, 23-24, 40, 62

vase painting 249-257, 262, 288n134

zētēsis (constructibility) 363

Index C: Passages

The index lists discussions of specific passages as well as more general accounts of extant works by ancient authors. For discussions of lost works, one should consult *Index A*. Passages are specified in accordance with the cited editions. Numerals in parentheses usually denote page numbers, unless preceded by "c." (for "chapter") or "P" (for "proposition"). For major works of Apollonius, Archimedes and Euclid, entries consisting of a Roman numeral followed by an Arabic numeral denote book and proposition (e.g., "I 33" denotes the 33rd proposition of Book I); the same format for works by Aristotle, Hero and Ptolemy, however, denotes book and chapter. In the instances of multi-volume editions, as of works by Eutocius, Pappus and a few others, citations are to volume and page in a decimal format (e.g., "2.54-56" denotes pages 54-56 of volume 2). Citing Pappus' *Collection* is complicated by the presence of at least four independent subdivisions of its contents. I have preferred to cite the volume and page in Hultsch's edition (in the decimal format), but attach indications of chapter or proposition number where these alternatives appear in my discussions.

Alexander, *In Arist. Meteor.* (Hayduck, 162-172): 141n18
[Alexander], *In Arist. Soph. Elen.* (Wallies, 90): 95n68

Alhazen, *Inscr. of Heptagon* (Schoy; Woepcke): 205n104
Ammonius, *In Arist. Cat.* (Busse, 75): 379n119

Index   405

Anaxagoras (*DK* 59 A38): 45n61
Anon., *Isoper. Fig.* (in *Pappi Coll.*, ed.
    Hultsch, 3.1138-65): 272, 290n166
Anthemius, *Paradox. Mech.* (Heiberg, 78-87):
    91n38, 237-239, 285n93, n95-n98,
    336n80
Apollonius,
  Conics (Heiberg): 10, 138, 171, 176, 213,
    238, 288n138, 293, 297-298, 313-314,
    328-329, 334n53, 335n59-60, 350
  Book I: 62
  I pref.: 143n43, 144n59, 298, 329n1,
    330n17-18, 336n73, n84, 338n113,
    371n14, n25, 374n56
  I Def. 1: 381n140
  I 11-14: 149n106, 314
  I 15-16: 327
  I 20-21: 285n91, 315
  I 21: 124, 324
  I 33: 132, 207n136, 284n83
  I 34: 132, 207n136, 216
  I 33-36: 314
  I 36: 124
  I 52: 336n75
  I 52-60: 148n106, 314
  I 54-55: 304
  Book II: 62
  II pref.: 275-276
  II 1-3: 315
  II 4: 144n67, 304, 335n69, 350
  II 8-10: 94n59
  II 9-11: 315
  II 12: 94n59, 331n29
  II 13: 224
  II 14: 207n136, 224
  II 15-23: 335n61
  II 31-43: 335n61
  II 44-48: 314
  II 49: 314
  II 49-53: 314
  Book III: 218, 328-329
  III 16: 120-122, 126
  III 16-23: 125, 322
  III 17: 120, 123, 327
  III 18-19: 122
  III 27: 124
  III 37: 147n101, 215-216, 279n30
  III 37-40: 135
  III 41: 127
  III 41-43: 321
  III 45: 145n72
  III 45-52: 284n80
  III 48: 285n98, 321
  III 51: 321
  III 52: 285n98, 321
  III 53-56: 125-126, 145n76, n79, 146n85
  Book IV: 322, 329
  IV pref.: 282n41, 298, 329n1, 335n63,
    336n73
  IV 24-55: 319
  IV 25: 145n83

  Book V: 143n44, 304
  V pref.: 330n17, 335n64
  V 8-10: 315
  V 13-14: 315
  V 27: 284n82
  V 27-29: 315
  V 51: 203n82, 317-319, 336n73, n75
  V 52: 308, 317-319, 336n73
  V 55-63: 319
  V 62: 315
  Book VI: 335n59
  VI pref.: 330n17
  VI 11: 335n59
  Book VIII: 335n60
  Okytokion frag. (ed. Heiberg, 2.124-132):
    157, 159, 294-295, 330n5
  Section of a Ratio: 321
Archimedes, *Opera* (Heiberg):
  Cat. Prob.: 191, 211, 294-295
  Conoids: 152, 159, 161, 168-169, 205n112,
    266, 313
  (pref.): 43n19, 335n59
  (P 1): 29, 266
  (P 3): 121
  (P 5-6): 200n41
  (P 21-32): 200n41
  Dim. Circ.: 78, 152-159, 168, 194, 197n9,
    n11-12, 198n14, n17, 202n68, 205n112,
    228, 294-295
  (P 1): 44n43, 83, 98n102, n104, 272-273,
    283n64, 374n58, 379n117
  (P 3): 295, 362
  DC (Latin versions): 200n38
  Floating Bodies (II 5): 335n64; (II 10):
    335n59
  Inscr. of Heptagon (Hogendijk; Schoy):
    152, 178-181, 203n90, 204n95-96
  Lemmas: 152, 159-161, 178, 185-187,
    200n42
  Method: 29, 152, 191, 202n64, 205n112,
    266-267, 274, 346; (pref.): 169, 202n66,
    211; (P 2): 202n68
  Plane Equilibria: 152; (I 7): 370n5; (II 3,
    7): 335n59
  Quadrature of Parabola: 45n52, 152, 169,
    346
  (pref.): 43n19, 168, 201n61, 202n65
  (P 1-3): 142n42
  (P 2): 335n62
  (P 6-17): 202n64
  (P 18-24): 168, 202n64
  Sand Reck.: 20, 43n21-23, 295, 330n8; (I
    9): 197n5; (I 21): 290n164; (II 2): 273,
    291n173
  Sph. Cyl.: 5, 84, 170, 205n112
  (*Book I*): 45n52, 152, 272
  (I pref.): 43n19
  (I post.): 198n13-14, 154-155, 283n64
  (I 3): 230
  (I 5): 153
  (I 5-6): 96n74

(I 14): 159
(I 33): 200n40, n50
(I 42-44): 163, 200n40, n49, 202n67
(*Book II*): 152, 169, 313
(II pref.): 43n19
(II 1): 188, 308
(II 3): 170
(II 4): 170, 177, 181, 183-184, 196, 202n69, 204n94
(II 9): 273
Spir. 99n113, 139n1, 152, 161-162, 167, 177-178, 183, 185, 204n94, 326
(pref.): 43n19, 169, 200n47, 202n65, 330n9
(P 2): 143n49
(P 5): 38
(P 5-9): 176, 221
(P 12): 143n49
(P 13): 165
(P 18-20): 163-165, 176
(P 21-24): 200n46
Archytas, fr. 2 (*DK* 47 B2): 48n90, 278n26
Aristarchus, *Sizes...of Sun and Moon* (Heath, P 5-7, 10): 290n164
Aristophanes,
  Birds (lin. 1001-05): 26, 42n6, 44n46
  Clouds (lin. 178): 44n46
Aristotle, *Opera* (Bekker):
  de Anima (II 2, 413a16-20): 24, 44n40; (II 7): 107
  de Caelo: 107
  (I 2, 283a6): 142n40
  (I 12): 141n28
  (II 4): 141n28
  (II 4, 287a23-30): 198n14, 273, 290n170
  (III 2): 141n28
  Cat. (c. 7): 361-362
  Ind. Lin. (970a8): 142n39
  Metaphys. (B 2, 997b32): 45n55; (Z 10, 1036a9): 374n66; (L 8): 98n108, 107; (M 3):43n35
  Meteor.: 11
  (III 3): 141n17, 371n25
  (III 5): 102-108, 110, 138, 139n9, 140n10, n15, 329, 371n25, 373n49
  (III 5, 376a4-6): 139n8
  (III 5, 376a15): 373n51
  (IV): 141n26
  Nic. Eth. (III 3, 1112b11-27): 95n64, 373n51, 377n91
  Phys. 29, 107
  (I 2, 185a16): 44n48, 47n79, 361
  (II 2, 194a30): 14n19
  (IV 8): 141n28
  (VI): 141n28
  (VII 4, 248a19-b7): 83, 98n107
  Pol. (IV 1, 1288b28): 111
  Post. An.
  (I 2, 71b8-32): 95n62
  (I 2, 72a25-b4): 374n60
  (I 4-5): 75
  (I 7): 43n35

(I 9): 361
(I 12, 78a7-14): 75, 95n64, 142n36, 361
(I 22, 84a6ff): 95n64
(I 24, 85b23-28): 95n62
(II 13, 79a15-16): 198n14
Pr. An.: 108, 141n28
(I 9): 95n66
(I 23, 41a21ff): 378n103
(I 24): 284n81
(II 2-4): 95n64
(II 25): 45n63
Soph. Ref. (c. 11, 171b15): 44n48, 95n66
Top. (I 4, 10, 11): 374n60
[Aristotle], *Mech.* (pref., 847b25): 95n70; (c. 20, 854a12-15): 48n90
[Aristotle], *Prob.* (XV 7; XVI 6): 62, 143n48

Banū Mūsā, *Verba filiorum* (Clagett; Suter) (P 16): 90n27; (P 17): 90n26; (P 18): 187, 205n107
al-Bīrūnī,
  Chords (Suter): 140n16, 206n113, 290n160
  Trigonometry (Schoy): 199n34
Bobbio math. fr. (Heiberg, 87-92): 284n86, 285n99, 286n109

Cicero, *Tusc. Disp.* V 64ff: 42n17

Diocles, *Burn. Mir.* (Toomer): 233-234
  (P 1-3): 141n22
  (P 1): 143n44, 234-235, 239-240, 335n62, n65
  (P 2): 285n99
  (P 4): 91n36, 147n94, 236-237, 284n87-88
  (P 5): 147n94, 236-237
  (P 6): 284n85
  (P 7): 202n71
  (P 8): 202n71, 377n89
  (P 9): 284n85
  (P 10): 91n39, 240, 286n104
  (P 11-16): 240
  (P 12): 91n36
Dioscorides, *Mat. Med.*: 262, 288n132

Euclid, *Opera* (Heiberg-Menge)
  Data (Menge): 102, 109-110, 138, 142n33, n35, n38, 213; (P 28): 142n38; (P 58): 142n38; (P 59): 37, 301; (P 93): 142n34
  Division of Figures (Archibald): 205n112, 380n124
  Elem. (Heiberg): 7-8, 69, 101-102, 109-110, 138, 139n1-2, n4, 142n38-39, 151, 195, 293, 323
  (*Postulates*): 374n59
  Book I: 350
  I 1-3: 353
  I 4: 353
  I 5: 297
  I 12: 15
  I 20: 154, 354, 373n48
  I 21: 373n47

I 22: 142n39, 354, 373n46
I 23: 15
I 26: 48n91
I 30: 142n38
I 31: 350
I 32: 350
I 43: 180
I 44: 66
I 45: 25
I 47: 46n69, 102
Book II: 354, 356
II 5-6: 66
II 6: 37
II 11: 67-69, 91n48, 92n49, 379n107
II 12: 46n69, 289n153
II 13: 46n69
II 14: 24, 63
Book III: 39
III 16: 48n93, 201n53
III 20: 186
III 31: 186
III 32: 69
III 35: 52, 121
III 36: 121
III 37: 69
Book IV: 350
IV 10-11: 67-71, 181
Book V: 39, 69, 79-80
V 12: 45n58
Book VI: 69
VI Def. 3: 68
VI 3: 104, 139n9, 155
VI 11: 98n101
VI 13: 24, 63, 214
VI 17: 24
VI 25: 66
VI 26: 142n38
VI 28: 66
VI 29: 37, 66
VI 30: 67-69
Book X: 271, 278n25, 340
X 1: 79, 96n79
XI Def. 18: 381n140
XI 22-23: 336n73
XI 33 por: 44n38
Book XII: 39, 76, 78-80, 96n80, 198n17, 265, 340
XII 2: 26, 28, 78, 97n86, 154-155
XII 16: 97n86, 375n77
XII 17: 375n77
XII 18: 154, 375n77
Book XIII: 350
XIII 1: 358, 379n106-107
XIII 1-5: 354, 356, 379n107
XIII 1-5 scholia (ed. Heiberg IV, 364ff): 375n78, n82, n85
XIII 6: 379n106-107
XIII 11: 359
XIII 16-17: 67
Book XIV: 338n97; v. Hypsicles
Book XV: 338n97

Optics (Heiberg): 96n72, 145n67, 284n81; (P 8): 290n164; (P 36): 111, 137, 327; (P 37-38): 372n25, 373n54; (P 44-47): 372n35; (P48): 373n54
Phaenomena (Menge) 62, 143n48
[Euclid], Catoptrics (Heiberg); 96n72, 284n81; (P 29): 373n54; (P 30): 141n22
Eudemus: v. Hippocrates fr.; Simplicius, In Phys. (1.61-68)
Eutocius,
Comm. on Apol. (Heiberg): 10, 338n108, 340
(2.168): 285n94, 330n16
(2.168-170): 143n52
(2.170-171): 291n77
(2.178): 371n15
(2.180-184): 139n9
(2.184): 371n16
(2.186): 144n59
(2.230-234): 91n37, 285n91
(2.290, 312, 354): 285n94
Comm. on Arch. Sph. Cyl. (Heiberg): 5, 17
(3.54-106): 13n14
(3.56): 89n8
(3.56-58): 42n6, 90n21
(3.58): 90n28
(3.58-60): 205n110
(3.60-62): 331n38
(3.62-64): 90n28
(3.64-66): 90n28, 331n35
(3.66-70): 91n36, 284n73
(3.70): 90n28
(3.70-78): 286n106
(3.78-84): 12n11, 90n29
(3.84): 42n6, 43n26
(3.84-88): 12n11, 88n5
(3.88): 12n5, 43n33, n36
(3.88-90): 43n24, n36
(3.88-96): 42n9, n11, 277n7
(3.90): 12n9, 42n6, n14, n16, 277n9
(3.92-94): 42n6, 90n28, 278n11
(3.94-96): 277n8
(3.96): 42n14, n16, 43n27, 88n4, 91n34, 277n9
(3.98): 206n121, 282n39
(3.98-100): 42n6, 90n28
(3.98-106): 282n36
(3.130-176): 171-176, 183
(3.130): 202n77
(3.130-132): 202n70
(3.132-152): 202n71
(3.152-160): 202n71
(3.160-174): 202n71, 284n73
(3.176): 331n33, 370n2, 379n110
Comm. on Arch. Dim. Circ. (Heiberg)
(3.228-260): 98n109
(3.228): 330n2
(3.230): 199n35, 200n38, 374n58, 379n117
(3.232): 380n127
(3.258): 199n26, 330n3
(3.260): 380n126

Hero,
  Belopoeica (Marsden): 188, 190, 206n112, n120
  Catoptrics (Schmidt): 96n72, 284n81
  Comm. on Eucl.: v. al-Nairīzī
  Definitions (Heiberg) (c.7, 4.20): 296, 330n14; (c. 30, 4.34): 287n119; (c. 38, 4.38): 287n119; (c. 97, 4.62): 288n140
  Dioptra (Schöne): 42n6, 206n112, 380n136; (c. 30): 290n159
  Geometr. (Heiberg, IV): 98n109, 199n25
  Mech. (Nix): 188, 190, 205n112, 207n139; (I 10): 206n119; (I 11): 206n119, 371n13; (I 24): 206n112
  Metr. (Schöne): 188, 195, 199n25, 205n112, 365, 376n87
  I 8: 192, 194, 207n127, 290n159
  I 22: 205n112
  I 24: 205n112
  I 26: 98n109, 199n26, 206n114, 295
  I 27-32: 201n63
  I 32: 202n64
  I 34: 200n41
  I 36: 296, 201n59
  I 37: 98n109, 200n39, 201n59, 296
  I 39: 44n45, 380n136
  II pref.: 289n147
  II 6: 207n129
  II 9: 207n129
  II 13: 205n112, 264
  II 20: 44n45, 206n114, 380n136
  III 13: 380n136
  III 16: 380n136
  III 18: 205n112, 363, 379n124
  III 20: 191-192, 206n123
  Stereometr. (Heiberg, V): 98n109, 199n25
Herodotus, *Hist.* II 2: 381n145
Hippocrates fr.: 44n41, n49; v. Simplicius, *In Phys.* (1.61-68)
Hypsicles: [Euclid], *Elem.* XIV (pref.): 338n97

Iamblichus, *Comm. on Nic.* (Pistelli, 111): 278n20
Ioannes Philoponus: v. Philoponus

Maimonides, *Mishnah Eruvin* I 5: 380n125
Marinus, *Comm. on Eucl. Data* (Menge): 362; (240): 379n122; (252-254): 378n105
Math. Gr. Min. (Heiberg): v. Anthemius; Bobbio math. fr.

al-Nairīzī (Anaritius), *Comm. on Euclid* (ed. Curtze, IX, 89): 376n78, n83; (Arabic, ms. Leiden, Or. 399/1, f. 25v): 376n83
Nicomachus, *Intr. Arith.* (Hoche) (II 25): 278n20; (II 26-28): 278n16; (II 29): 278n20

Olympiodorus, *In Meteor.* (Stüve, 243-259): 141n18

Pappus,
  Coll. (Hultsch): 8, 83, 195-196, 339, 356-357
  Book II: 294
  (1.2-28): v. Apollonius, *Okytokion*
  Book III: 5, 278n16
  (1.30): 347, 373n43, 374n60, 379n113
  (1.32-48): 207n128, 373n43
  (1.48): 377n87
  (1.54-68): 13n14
  (1.54-56): 282n49, 287n120, 370n8, n11, 372n41
  (1.56): 282n37, 332n39, 373n45
  (1.56-58): 42n13, 277n8
  (1.58-62): 282n36
  (1.60): 282n49
  (1.62): 371n13
  (1.62-64): 141n28, 205n110
  (1.64-68): 286n106
  (1.68-70): 278n19
  (1.80-82): 278n21
  (1.86): 377n94
  (1.132-162): 91n44
  (1.142-162; P54-58): 377n89
  (1.150-162; c. 51-52): 291n174
  Book IV: 5, 160
  (1.178-186): 370n5
  (1.186): 378n102
  (1.208-232): 200n45
  (1.234): 143n49, 147n96, 200n47
  (1.234-242; c. 21-25): 99n113, 207n137
  (1.242-250): 282n36
  (1.244): 143n49, 282n43, n49
  (1.246-248): 282n49
  (1.250-252): 97n90, 98n111
  (1.250-258): 293n58
  (1.250-262): 98n103
  (1.252): 143n49, 282n38
  (1.252-254): 283n65, 380n125
  (1.252-256): 98n105, 380n131
  (1.256-258): 98n103, 283n62
  (1.258): 98n109
  (1.258-262; P28): 147n97, 283n59, 330n12
  (1.262-264; c. 34, P 29): 147n97, 201n56, 330n12
  (1.264-268; c. 35): 200n48, 207n137
  (1.270): 282n49, 287n121
  (1.270-284): 13n14
  (1.270-272): 99n114, 203n81, 282n42, 336n75, 372n29, 370n8, n10-11, 379n110
  (1.272): 282n51
  (1.272-276; c. 36-40): 146n91, 281n33, 303
  (1.272-280; P 31-33): 187, 205n108, 331n26, 373n55
  (1.276-280): 331n31
  (1.282-284): 146n91, 338n100
  (1.284): 146n91, 379n114
  (1.284-288): 98n112
  (1.298-302; c. 52-54, P 42-44): 203n84, 338n105, 377n89
  (1.302): 203n88
  Book V: 272, 290n169, 357

(1.308-334): 290n167
(1.312-316): 98n109
(1.336-340; P 12): 198n18, 200n49
(1.340-342): 45n50
(1.362-410; c. 20-43): 207n137
(1.410-412): 379n110
(1.434-438; c. 55): 291n174
(1.444-446; c. 57): 291n174
Book VI:
(2.542): 94n61
(2.560-562): 205n106
(2.588-592; P 53): 143n47, 145n67, n72, 338n112
(2.592-594; P 54): 338n112
Book VII: 142n32, 144n67, 196, 213, 293, 313, 323, 338n108, 340, 354
(2.634): 329n1, 374n56, 376n80-81, 377n97
(2.636): 278n14, 336n73, 371n25, 377n97
(2.640-642): 336n83
(2.642-644): 337n88
(2.648-652): 144n61, 145n73, 373n51, 374n60,
(2.650-652): 379n111
(2.652): 278n15, 371n17
(2.652-654): 149n109
(2.660-662): 371n12, n20
(2.662): 278n15, 371n17
(2.666): 139n9
(2.672): 142n41, 278n15, 281n31, 338n96, n104, 371n17, n24
(2.672-674): 143n52, 144n59
(2.674): 144n57
(2.676): 142n41
(2.676-678): 149n109, 334n56
(2.678): 291n177
(2.680-682): 207n138, 288n137
(2.682): 288n139, 370n3
(2.684-704): 336n83
(2.704-770): 337n88
(2.708-710): 377n88
(2.730): 140n14
(2.778-780): 331n22
(2.780-782): 330n19
(2.782-784; P 72): 144n67, 330n19
(2.866-874): 337n89
(2.866-918): 116-120, 124-125, 132-136, 138, 144n65, n67
(2.874-876): 278n27
(2.882; lemma 11): 145n70
(2.884): 144n65
(2.884-886): 145n69
(2.894; lemma 19): 147n102
(2.904; lemma 28, P 154): 145n72, 215, 278n22
(2.904-906): 140n16
(2.906-908; P 157): 145n72
(2.918-1004): 321
(2.952-960): 335n66
(2.954-958; P204, lem. to Con V): 144n67, 335n69
(2.1004): 146n93

(2.1004-1014; P 236-238): 128, 137, 144n67, 146n92, 338n102
(2.1016-1020):378n102
Book VIII: 207n139
(3.1026): 374n67
(3.1026): 380n134
(3.1070): 90n28
(3.1072-1074): 380n137
(3.1076-84; P 13-14): 322
(3.1078-1080): 337n87
(3.1106): 98n109
Comm. on Eucl. Book X (Junge, Thomson): 357, 370n5
Comm. on Ptolemy (Rome, 254-260): 200n39; (254-260): 98n109, 198n18
Lemmas to Apol. Conics: v. Coll. VII (2.918-1004)
Lemmas to Eucl. Porisms: v. Coll. VII (2.866-918)
Lemmas to Eucl. Surface Loci: v. Coll. VII (2.1004-1014)
Philo, Belopoeica (Marsden, 110): 141n28, 305, 331n38
Philoponus, In An. Post. (Wallies, 104-105): 331n36; (105n): 381n143; (111-114): 95n67, n69; (112): 95n70-71, 96n73, 379n116; (120): 379n120
Plato,
  Euthydemus (290c): 374n61
  Hipp. Min.: 81; (367c-368e): 97n97
  Meno (86e-87b): 69-74, 110, 142n37
  Philebus (23c-27c): 95n70; (51c, 56c, 62b): 42n6
  Protag. (315c, 318e): 97n97
  Rep. 5, 22, 43n34, (510c-e, 525-527): 4, 90n23, 374n61
  Tim. (31-32): 377n94
Plutarch,
  Life of Marcellus (14.5-6): 12n7, 43n30, 278n13, 380n133
  Moralia
  (386e): 42n8, 43n29
  (579a-d; Socr. Dem.): 2, 12n2, 42n8, 43n29
  (607f): 45n61
  (718e-f): 12n7, 43n30, 278n13
  (720a): 91n40
  (1094b): 91n40
Proclus,
  Comm. on Eucl. Book I (Friedlein): 7, 15, 377n96
  (42): 378n105
  (65-67): 97n93
  (65): 45n62, 97n95
  (66): 48n92, 74, 88n1, 99n116, 142n37
  (67): 91n45, 142n30, 372n25
  (74): 370n5
  (77-78): 75, 374n62, 65
  (80-81): 373n52, 374n69, n72
  (105): 201n58, 330n13
  (106): 333n50
  (111): 143n48, n53, 287n115, n126

(112): 89n16, 289n148, n150, n156
(113): 287n121
(119): 289n148-149
(122): 48n93
(126): 143n48, 287n115, n128
(127): 48n93, 89n16, 287n119, 289n150
(128): 287n115
(134): 48n93
(152): 287n115
(163): 287n119
(164): 287n115
(177): 287n115, n126
(187): 287n115
(188ff): 374n59
(191ff): 374n59
(202): 75, 94n62, 336n73
(212): 373n51, 378n104
(213): 12n5, 44n39
(220-221): 94n60
(233-234): 95n71, 375n73
(241-242): 374n67-68, 379n108-109
(243): 374n64
(251): 296
(254): 75
(272): 97n91, 282n37-38, 283n61, n68, 287n127
(283): 41n1
(301-303): 373n51
(322): 198n15, 375n75
(333): 41n1
(352): 48n91
(356): 97n91, 98n98, 143n49, 283n61, n68, 289n148, n152
(365ff): 374n59
(394-395): 371n18-19
(419): 91n40
(422-423): 44n43, 98n109
(426): 139n5
Comm. on Pl. Tim. (Diehl, 2.173-174): 278n20
Ptolemy, *Syntaxis* (Heiberg), 323, (*Book I*): 158, 272, (I 3): 143n48, 290n163, (I 10): 290n164, (VI 7): 199n32, (XII 1): 308

al-Qūhī, Abū Sahl, *Inscr. of Heptagon* (Woepcke): 205n104

al-Sijzī, Abū Saʿīd Aḥmad, *On Angle Trisection* (Woepcke): 205n106, 380n132
Inscr. of Heptagon (Hogendijk): 205n99, n101; (Schoy): 205n104
Simplicius,
In de Caelo (Heiberg, 411-414): 198n14, 291n171
In Cat. (Kalbfleisch, 192): 379n120-121, 380n130
In Phys. (Diels)
(1.53-69): 95n67
(1.54-55): 45n53-55
(1.56-57): 45n65
(1.57-58): 46n67
(1.59-60): 48n93
(1.60): 46n67, 47n90, 97n90, 201n60, 282n38, 283n61, 330n12, 374n67, 380n129
(1.60-69): 44n49
(1.61): 45n59
(1.64): 46n74
(1.66): 46n76
(1.67): 47n77
(1.68): 47n78
(1.69): 46n67, 47n79
(2.1082-83, ad VII 4): 380n128

Themistius, *In Phys.* (Schenkl, 3-4): 45n53
Theon of Alexandria,
rec. of Eucl. *Opt.* (Heiberg): 96n72, 284n81, (P41-47): 372n25
Comm. on Ptol. (Rome): 272; (355-379): 290n163; (358): 290n164; (359-364): 98n109, 290n165; (398-399): 289n147; (492-495):199n34
Theon of Smyrna, *Math. Expo.* (Hiller): 3, 21, (2): 12n6, 42n8, 43n28, 278n12
Theophrastus, *On Plants* (Hort, I 10): 288n131, (III 10): 288n131, (III 18): 261, 288n131
al-Ṭūsī, Naṣīr al-Dīn, schol. to *Con.* V 52: 307, 317

Vitruvius, *Arch.* (IX pref.): 206n114, (IX 8.1): 331n37